Handbook of Spintronic Semiconductors

Handbook of Spintronic Semiconductors

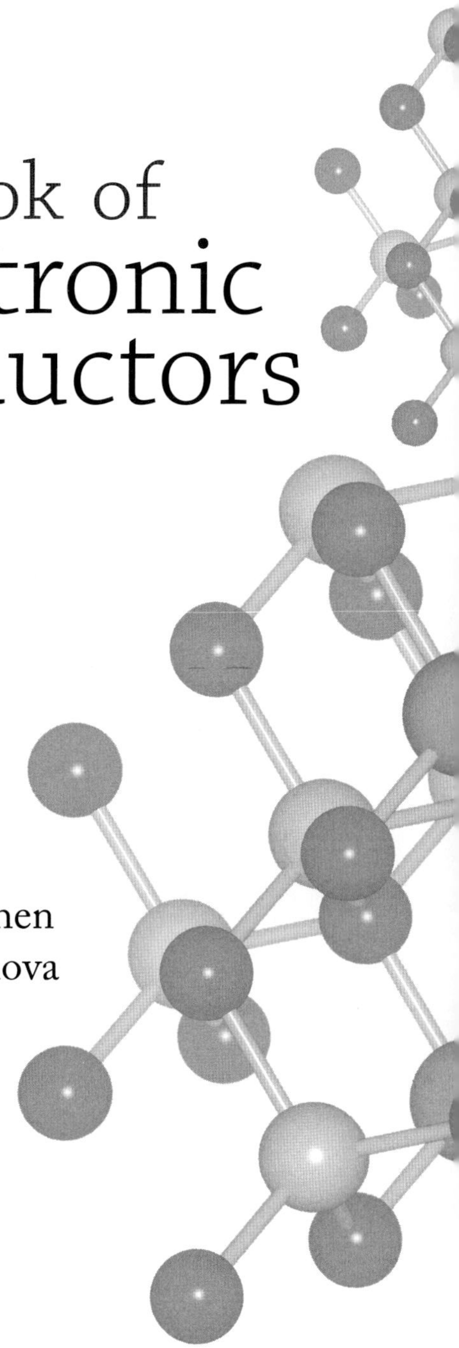

edited by
Weimin M Chen
Irina A Buyanova

Published by

Pan Stanford Publishing Pte. Ltd.
Penthouse Level, Suntec Tower 3
8 Temasek Boulevard
Singapore 038988

Email: editorial@panstanford.com
Web: www.panstanford.com

British Library Cataloguing-in-Publication Data
A catalogue record for this book is available from the British Library.

HANDBOOK OF SPINTRONIC SEMICONDUCTORS

ISBN-13 978-981-4267-36-6
ISBN-10 981-4267-36-8

Printed in Singapore.

Preface

With the limits of microelectronic miniaturization in sight there has been an enormous drive to innovate radically new technologies. As an addition or alternative to electron charge, the storage and transport of electron spin in "spintronics" can not only improve the performance of and add new functionalities to existing devices but also could revolutionize electronics, leading to new spin-enabled devices such as magnetic RAM, spin transistors, spin optoelectronic devices, and spin quantum computers. Spintronics opens the door to a new generation of very high speed, very low power devices for computation and data transmission, to the integration of processing and storage capabilities thus far carried out separately, and to the merging of electronics, photonics and magnetics into single technologies for multifunctional devices.

The success of spintronics relies on the ability to create, control, maintain, manipulate and detect spin orientation and coherence over practical time and length scales. An essential element of a spintronic device, and perhaps the most challenging, is a spin source through which a desired spin orientation and coherence can be generated. Recent demonstration of ferromagnetism in dilute magnetic semiconductors (DMS) and spin injection from DMS has shown the promise of DMS as a spin source that is compatible with existing semiconductor technology. Driven by the potential of future applications, there have been intensive research efforts worldwide during recent years in the development of spintronic semiconductors, in understanding their physical properties, and in exploring their spin-enabling functionality for device applications. Significant progress has been made both theoretically and experimentally, while many open questions still remain.

This book provides an in-depth review of the rapidly developing field of spintronic semiconductors by a group of leading experts in the field. It covers a broad range of topics, including growth and basic physical properties of dilute magnetic semiconductors based on II–VI, III–V and IV semiconductors, recent developments in theory and experimental techniques, and potential device applications. The aim is to provide postgraduate students, researchers and engineers a comprehensive overview of our present knowledge and future perspectives of spintronic semiconductors.

We would like to thank all of the contributors for sharing their experience and expertise with interested readers and for conveying, through their excellent chapters, their passions for this exciting area of research. We would also like to express our gratitude to Pan Stanford Publishing for our effective collaboration.

Weimin M. Chen and Irina A. Buyanova

Contents

Chapter One

Computational Materials Design in Semiconductor Nano-spintronics

H. Katayama-Yoshida[*,†,‡], K. Sato[*,†], M. Toyoda[†], V. A. Dinh[†], T. Fukushima[†] and H. Kizaki[*,†]

[*]*Graduate School of Engineering Science, Osaka University, 1-3 Machikaneyama, Toyonaka, Osaka 560-8531, Japan*
[‡]*hiroshi@mp.es.osaka-u.ac.jp*
[†]*The Institute of Scientific and Industrial Research, Osaka University, 8-1 Mihogaoka, Ibaraki, Osaka 567-0047, Japan*

We start by discussing the need for semiconductor nano-spintronics based on the present status and future limitations in Si-CMOS technology. We consider how we can design and realize a semiconductor nano-spintronics based on the combination of top-down (nano-lithography) and bottom-up (self-organization) nanotechnologies. In order to go beyond the limitations in Si-CMOS technologies, we discuss a possible new methodology in semiconductor nano-spintronics to realize a design-based new class of nano-spintronics with high Curie temperature ($T_C > 1000$ K): high-density [Tbit/(inch)^2] nano-spintronics devices caused by spinodal nano-decomposition, high-speed-switching (THz) spintronics devices, and energy-saving (non-volatile) spin-based devices. We also propose a colossal magnetic response by electric field or photonic excitation as well as colossal-thermal management (spincaloritronics) on the nanoscale. In order to realize a high Curie temperature (T_C) in diluted magnetic semiconductors (DMS) ($T_C > 1000$ K), we discuss the electronic structure and ferromagnetic mechanism in III–V and II–VI-based DMS, spinodal nano-decomposition, and blocking phenomena in the semiconductor nanomagnet based on the local density approximation (LDA). We also discuss the spinodal nano-decomposition in inhomogeneous DMS and a new fabrication method of 100 Tbits/inch^2 for semiconductor nano-magnets using two-dimensional and three-dimensional spinodal nano-decomposition in the self-organization, and for a colossal magnetic response controlled by electric field or photonic excitation. We look at ZnO and GaN-based DMS by self-interaction corrected LDA (SIC-LDA), which is very important for highly correlated systems, comparing the calculated electronic structure with photoemission spectroscopy. The final sections are devoted to $CuAlO_2$ and TiO_2-based DMS by SIC-LDA, and non-transition metal-doped DMS by SIC-LDA. We conclude by summarizing computational nano-materials design.

Handbook of Spintronic Semiconductors by W M Chen & I A Buyanova

www.panstanford.com
978-981-4267-36-6

1.1 INTRODUCTION

We begin by discussing why we need semiconductor nano-spintronics based on the present status and future limitations in Si-CMOS technology according to Gordon Moore's law. We show how we can design and realize a semiconductor nano-spintronics based on the combination of top-down (nano-lithography) and bottom-up (self-organization) nanotechnologies. In order to go beyond the limitations in Si-CMOS technologies, we propose a possible new methodology in semiconductor nano-spintronics to realize a design-based new class of nano-spintronics with high Curie temperature ($T_C > 1000\,\mathrm{K}$) or high-blocking (B) temperature (T_B): high-density (Tbit/(inch)^2), high-speed-switching (THz), and energy-saving (non-volatile) spintronics devices. We also propose a colossal magnetic response by electric field or photonic excitation, as well as colossal-thermal management (spincaloritronics) nano-spintronics devices.

1.1.1 Gordon Moore's Law and Beyond Si-CMOS

As predicted by Moore's law, proposed in 1965 by Gordon Moore, and the International Technology Roadmap for Semiconductors, the progress of silicon CMOS technology (Si-CMOS) utilizing lithography has increased exponentially (the number of transistors that can be placed on an integrated circuit doubles every 18 months). The density of transistors has already reached $10^9 \sim 10^{10}/(\mathrm{inch})^2$ with switching speeds up to 10 GHz. With the increasing density of transistors, most of the essential problems for the future Si-CMOS concern nanoscale thermal-energy management and energy management. Hence, we are now approaching the "death" of Si-CMOS technology in 2018 due to the physical limitations in the SiO_2-based Si-CMOS gates. In order to allow Moore's law to continue in the face of these limitations, we need to create a new class of energy-saving, ultra-high-density, and ultra-fast-switching devices based on the today's semiconductor nanotechnology (see Fig. 1.1).

1.1.2 Semiconductor Nano-spintronics

An electron is an elementary particle, which carries the charge (e^-) and spin ($S = 1/2$) degrees of freedom. For the realization of ultra-fast switching (THz), high-integration (more than Tbits/inch^2), and energy-saving (non-volatile) semiconductor nano-spintronics (see Fig. 1.2) that goes beyond Si-CMOS technology,[1] we need to (i) realize a high Curie temperature (T_C) in diluted magnetic semiconductors (DMS) ($T_C > 1000\,\mathrm{K}$), (ii) develop a new fabrication method of 100 Tbits/inch^2 of semiconductor nano-magnets by a self-organization, and (iii) develop a colossal magnetic response controlled by the electric field or photonic excitations.

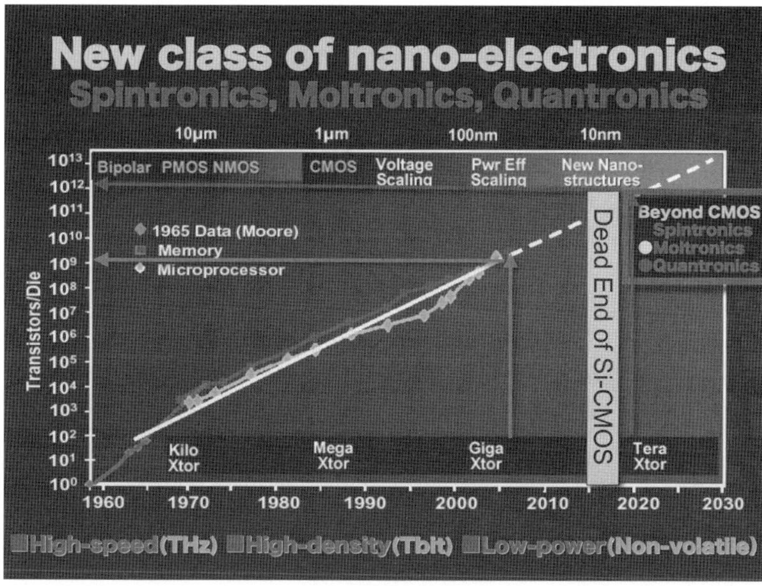

Figure 1.1. Gordon Moore's law for microprocessor power and memory proposed in 1965. We are currently achieving Gbit/(inch)^2 densities and GHz switching speed and we should reach densities and up to Tbit/(inch)^2 and THz switching speed by, 2020 based on Moore's law. The physical limitations in Si-CMOS technology (Si-CMOS barrier) suggest that Si-CMOS will reach a dead-end in 2018. In order to go beyond these limitations, we require a new class of electronics, such as *spintronics* (spin-based electronics), *moltronics* (molecular electronics), and *quantronics* (quantum electronics such as quantum computation and quantum information technology).

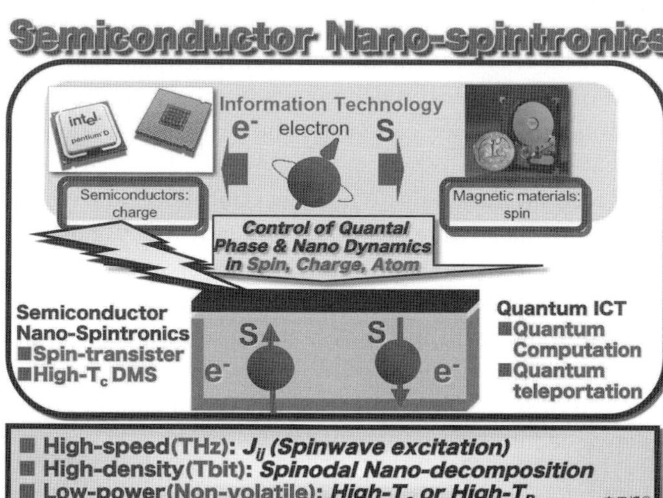

Figure 1.2. Semiconductor nano-spintronics is based on the control of spin (S) and charge (e^-) degrees of freedom by applying the gate voltage at the same time. Today's semiconductor devices (Si-CMOS) utilize charge-based electronics, and, magnetic memory is based on the control of magnetization at the nanoscale.

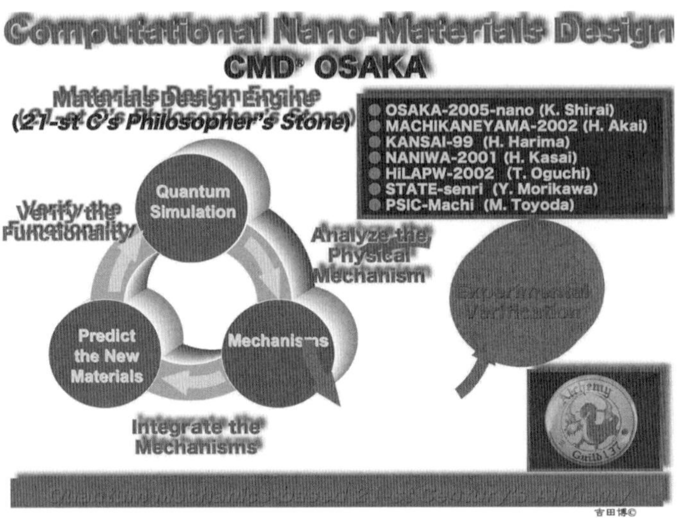

Figure 1.3. Computational nano-materials design system (CMD®). Using quantum simulation, we can analyze a physical mechanism, predict a new material by integrating the physical mechanisms, and verify the functionality. By itinerating the materials design and experimental verification, we can design a new functional material.

1.1.3 Computational Nano-materials Design and the CMD® System

We have developed a computational nano-materials design system for semiconductor nano-spintronics called CMD® System at Osaka University (see Fig. 1.3). Using a quantum simulation, we can analyze the physical mechanisms of ferromagnetism in the DMS. Then, we can design new functional nano-materials based on the integration of the physical mechanisms obtained through the analysis by quantum simulation. Finally, we can verify the proposed functionality by quantum simulation.

If the newly-designed materials do not meet our demands, we can analyze the reasons and propose an alternative candidate as a new functional material through the analysis. Using this circulation of design, realization and analysis in the CMD® system, we can ultimately design the desired new functional materials and have an experimental group fabricate and realize the design. If the experimentally fabricated materials do not satisfy our demands, we can analyze the real reasons for which. Then, we can design new, more appropriate functional materials by using the CMD® system (see Fig. 1.3).

1.1.4 Ab initio Calculation of Electronic Structure and Magnetic Mechanisms in Dilute Magnetic Semiconductors (DMS)

Wide band-gap semiconductors such as II–VI (ZnTe, ZnS, ZnSe and ZnO) and III–V (GaAs, GaP, GaSb and GaN) compound-based DMS systems are disordered systems, where Zn or Ga atoms are randomly replaced by transition-metal (TM) impurities. We treat the substitutional disorder by using the Korringa–Kohn–Rostoker coherent-potential approximation (KKR-CPA) method[2,3] with the local density approximation (LDA) and the self-interaction corrected LDA (SIC-LDA)[4–6] to go beyond the LDA. We use the KKR-CPA package MACHIKANEYAMA-2002 developed by Akai (http://sham.phys.sci.osaka-u.ac.jp/kkr/). In the KKR-CPA, an effective medium, which describes the configuration average of the disordered system, is calculated self-consistently within the single-site approximation proposed by Shiba[7] in 1971 at Osaka University. The LDA is well known to be sometimes insufficient for correlated electron systems such as ZnO, where the LDA band-gap is $\sim$1 eV in contrast with the experimentally observed band-gap of 3.3 eV.[4] In order to go beyond the LDA, we have developed a new method to take into account the self-interaction correction to the LDA (SIC-LDA), where the band gap of ZnO is 2.8 eV.[4–6]

Based on calculations using KKR-CPA-LDA[2,3] and KKR-CPA-SIC-LDA,[4–6] we propose a unified physical picture of the magnetism in the II–VI and III–V-based DMS, where Zener's double-exchange mechanism is dominant in the wide-band gap semiconductors,[3,7–12] On the other hand, Zener's *p–d* exchange mechanism is dominant in GaAs and GaSb-based DMS.[13,14] The super-exchange interaction mechanism[15,16] (Kanamori and Goodenough Rule) competes with the ferromagnetically dominant Zener's double-exchange or Zener's *p–d* exchange mechanism. In a homogeneous system, we calculate almost exactly the Curie temperature (T_C) by using a Monte Carlo simulation,[17,18] combined with the magnetic force theorem, and obtain good agreement with experiment [T_C and X-ray photoemission spectroscopy (XPS)].

1.1.5 Inhomogeneous DMS Caused by Spinodal Nano-decomposition

Since the solubility of 3D-transition metal impurities in III–V and II–VI compound semiconductors is low in the thermal-equilibrium condition, we use the thermal non-equilibrium crystal-growth condition such as low temperature MBE or the sputtering method for the fabrication of DMS. In this case, we can expect spinodal nano-decomposition or clustering. Sato and Katayama-Yoshida proposed spinodal nano-decomposition while controlling the three-dimensional and two-dimensional crystal growth conditions (see Figs. 1.4–1.7).[19–24]

1.1.6 3D Dairiseki Phase and 1D Konbu Phase as a Quantum Dot and Nano Wire

In an inhomogeneous system, we propose a three-dimensional *Dairiseki phase*[19,21–24] and a one-dimensional *Konbu phase*[20–24] caused by spinodal nano-decomposition (see Figs. 1.4–1.7). Spinodal nano-decomposition is responsible for the high-T_C [or high blocking (B) temperature (T_B)][25] phases in real

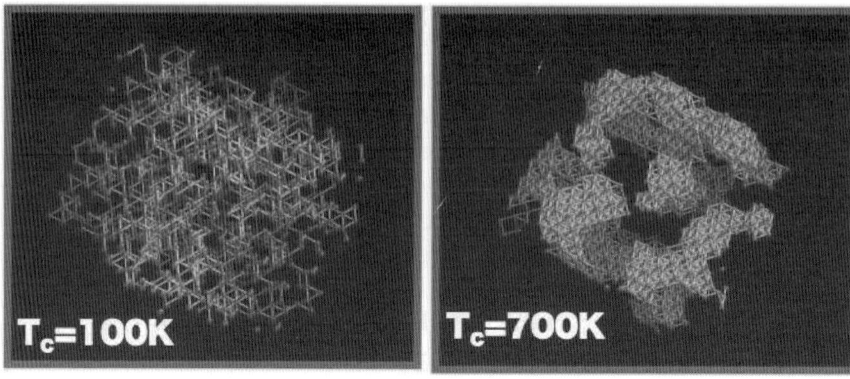

(a) (b)

Figure 1.4. Three-dimensional (3D) spinodal nano-decomposition above the ferromagnetic percolation limit (Cr $>$ 20%).[26] (a) Random configuration of (Ga, Cr)N, Cr 30% (T_C = 100 K), and (b) three-dimensional spinodal nano-decomposition (Dairiseki Phase) of (Ga, Cr)N, Cr 30% (T_C = 700 K) caused by the recovery of percolation path due to spinodal nano-decomposition.

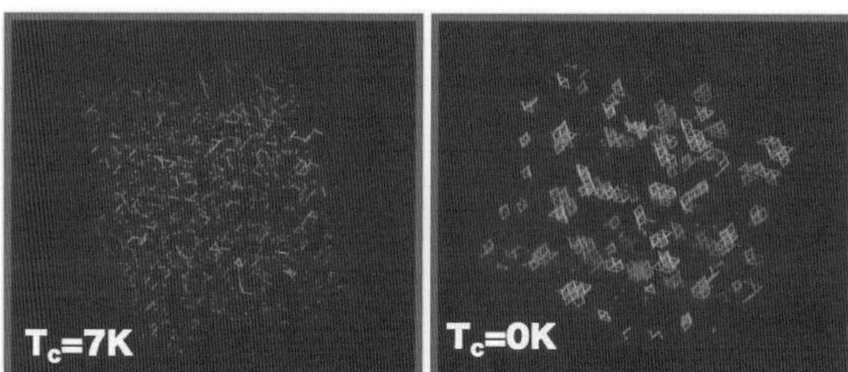

(a) (b)

Figure 1.5. An example of 3D spinodal nano-decomposition below the percolation limit ($<$ 20%). (a) Random configuration of (Ga, Cr)N, Cr 5% (T_C = 7 K), and (b) 3D spinodal nano-decomposition (Dairiseki phase) of (Ga, Cr)N, Cr 5% with super-paramagnetic phase caused by the lack of the ferromagnetic percolation path due to spinodal nano-decomposition in the case of low concentration (Cr 5%).

DMS. We design a position control method that uses seeding (top-down nanotechnology) and a shape control method that uses the vapor pressure in self-organization (bottom-up nanotechnology). We show the self-organized fabrication method for nano-magnets with 100 T-bit/inch^2 densities by using thermal non-equilibrium crystal growth methods such as molecular-beam epitaxy (MBE), metal-organic chemical-vapor deposition (MOCVD), or metal-organic vapor-phase epitaxy (MOVPE).[20–24]

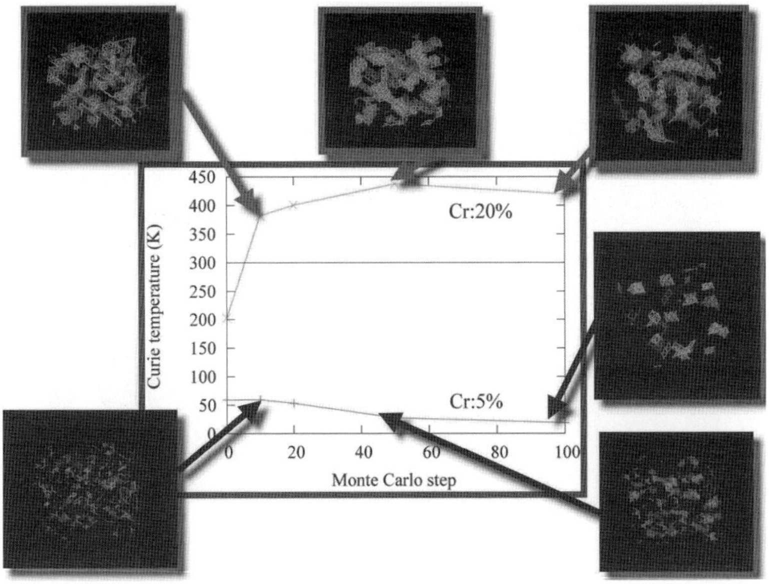

Figure 1.6. T_C versus spinodal nano-decomposition as a function of Monte Carlo step (spinodal nano-decomposition) for high concentration (Cr 20%) and low concentration (Cr 5%) in (Zn, Cr)Te.

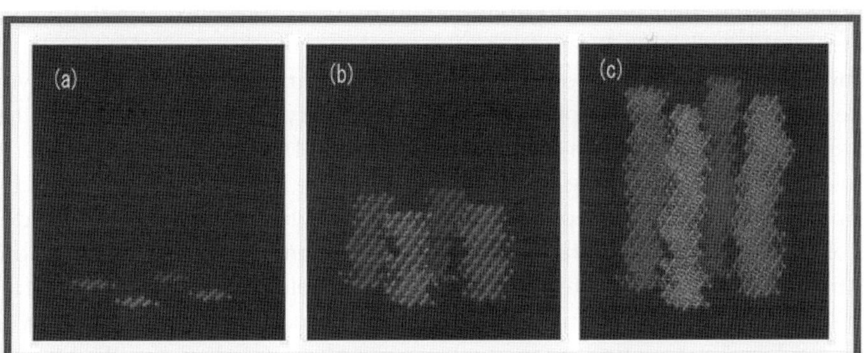

Figure 1.7. (a) Periodic and nanoscale seeding using periodic 13 Cr atoms by nanoscale lithography in (Zn, Cr)Te. (b) (c) Konbu phase caused by two-dimensional (2D) spinodal nano-decomposition in the layer-by-layer (2D) crystal growth conditions (Monte Carlo simulation) in (Zn, Cr)Te with Cr 5%. T_C is zero due to the super-paramagnetism. However, blocking (T_B) temperature is very high due to the shape-anisotropy and magneto-crystal anisotropy in the nano-wire magnet.

1.1.7 Colossal Magnetic Response by Electric Field and Photonic Excitation

For the realization of semiconductor nano-spintronics, we should colossalize the magnetic response in the nano-magnet by applying the gate voltage (electric field)

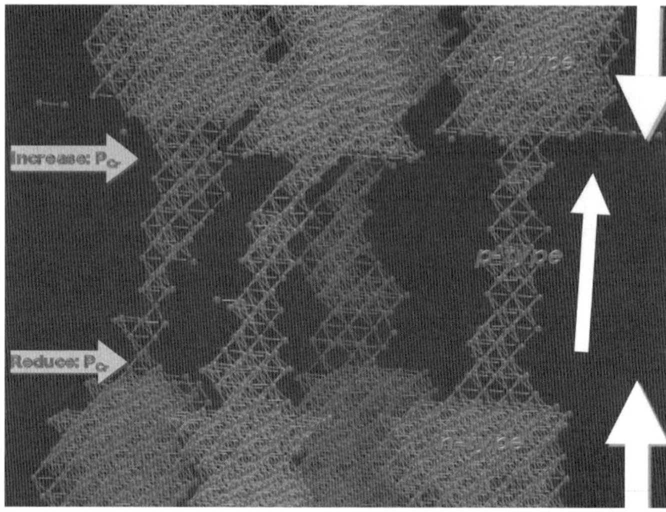

Figure 1.8. Shape control of nano-magnets by controlling the vapor pressure of Cr in the two-dimensional spinodal nano-decomposition of (Zn, Cr)Te. By making an *n*-, *p*-, and *n*-type junction (*n-p-n junction*), we can control the direction of the magnetic domain by using spin-current injection without charge current. Only the Cr atom is plotted in the ZnTe matrix.

or photonic excitation (electron-hole-pair formation) (see Figs. 1.8–1.10). To do so, we should design and realize the inhomogeneous quantum dot or quantum wire by using self-organization, such as spinodal nano-decomposition. If we can realize the spatial variation of the band-gap by varying the valence-band maximum (VBM) and conduction-band minimum (CBM) on the nanoscale, we have the possibility of colossalizing the magnetic response in the inhomogeneous DMS. By applying the negative or positive gate voltage, we can concentrate the doped holes or electrons in the quantum nano-dot or quantum nano-wire. Using photonic excitation, we can concentrate the excited holes or electrons in spatially different regions of the nanoscale quantum dot or quantum nano-wire. Therefore, we can expect a colossal magnetic response where one photon orders a million spins, or one electron (or hole) induces a magnetic phase transition.

1.1.8 Spincaloritronics

In addition to the conventional Peltier effect, we propose a new class of thermoelectric-cooling mechanism based on adiabatic spin-entropy expansion in a quasi-one-dimensional nano-superstructure by injecting the spin current from the ferromagnetic metal to the paramagnetic one.[30] The spin-entropy expansion cooling mechanism dominates to enhance the thermoelectric cooling-power dramatically in current perpendicular to plane (CPP) giant-magneto-resistance (CPP-GMR) in the ferromagnetic-metal/nonmagnetic-metal nano-interface. Based upon the spin-entropy expansion mechanism, we design new thermoelectric-cooling

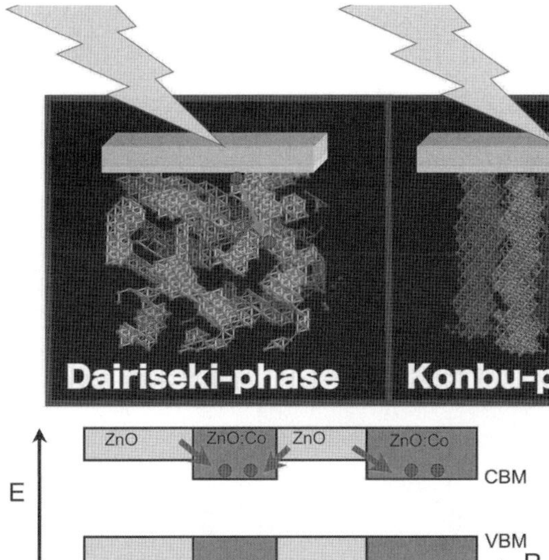

Figure 1.9. Colossal magnetic response caused by the self-organized *3D Dairiseki phase* in (Zn, Co)O, and *1D Konbu phase* caused by two-dimensional (2D) spinodal nano-decomposition in (Zn, Co)O. Applying a positive gate voltage, we can create n-type doping through the gate. Due to the spatial (R) distribution of the local band-gap caused by the spinodal nano-decomposition, the doped electrons can be accumulated near the Co high-concentration region; therefore, we can expect a colossal magnetic response and a phase transitions by electron doping through the gate.

nano-superstructures using the newly-designed half-Heusler ferromagnets NiMnSi ($T_C = 1050\,\mathrm{K}$) and quasi-one-dimensional Konbu-phase (Zn, Cr)Te with very high blocking temperature ($> 1000\,\mathrm{K}$) by spinodal nano-decomposition (see Fig. 1.11).

In order to consider the current direction of the R–I shifts, we take into account that the shift is caused by adiabatic spin-entropy expansion in addition to the conventional Peltier effect in *Konbu phase* or CPP nano-superstructure interface. When the spin current flows through a CPP nano-superstructure interface, Joule heating (RI^2), adiabatic spin-entropy expansion cooling ($T\Delta\sigma I$), and conventional Peltier cooling (ΠI) occurs at the same time. The spin-entropy ($\Delta\sigma$) expansion cooling term becomes dominant in the adiabatic spin injection in the extreme and thermal non-equilibrium conditions with ultra-high spin current injection. We can extend the theory for the change in resistance ΔR based on the charge and spin current by taking into account the conventional Peltier effect and newly-introduced adiabatic spin-entropy expansion cooling by assuming the constant heat capacity of the CPP elements. The thermoelectric-cooling power per unit area of the CPP element (10^5–$10^6\,\mathrm{W/cm^2}$) is colossally larger than that of conventional thermoelectric materials ($\sim 5\,\mathrm{W/cm^2}$).

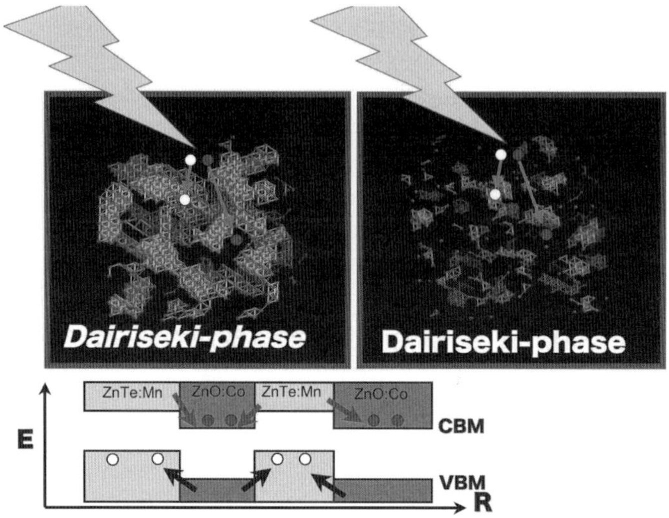

Figure 1.10. Colossal magnetic response caused by the self-organized *3D Dairiseki phase* in (Zn, Co)O: (a) Co 30%, and (b) Co 5%, caused by three-dimensional (3D) spinodal nano-decomposition in (Zn, Co)O. Using a photo-excitation, we can dope electrons and holes, which are separated in the spatial (R) nanoscale region with different band-gaps. Due to the spatial (R) distribution of the local band-gap caused by the spinodal nano-decomposition, the doped electrons can be accumulated near the Co high-concentration region; therefore, we can expect a colossal magnetic response and phase transitions by photo-excitations.

1.1.9 Organization of This Chapter

First, in Sec. 1.1, we have discussed the role and realization of semiconductor nano-spintronics considering the progress and limitations of Si-CMOS (*H. Katayama-Yoshida*). In order to realize a high Curie temperature (T_C) in diluted magnetic semiconductors (DMS) ($T_C > 1000\,K$), we discuss the electronic structure and ferromagnetic mechanism in III–V and II–VI-based DMS, spinodal nano-decomposition, and blocking phenomena in the nano-magnet by the local density approximation (LDA) in Sec. 1.2, (*K. Sato, T. Fukushima and H. Katayama-Yoshida*). Section 1.3 is devoted to the development of a new fabrication method of 100 Tbits/inch^2 semiconductor nano-magnets by self-organization, and to a colossal magnetic response controlled by the electric field or photonic excitations. Section 1.4 is concerned with ZnO and GaN-based DMS by self-interaction corrected LDA (SIC-LDA), comparing the calculated electronic with photoemission spectroscopy (*M. Toyoda*). Section 1.5 is devoted to TiO_2-based DMS by SIC-LDA (*H. Kizaki*). Section 1.6 is on non-transition metal-doped DMS by SIC-LDA[26–28] (*V. A. Dinh*). In the following section, we discuss the realistic ferromagnetic mechanism in DMS compared with Zener's *p–d* exchange mechanism.[31–34] We also discuss the co-doping method[35,36] to control and increase the solubility of 3*d*-TM impurities in semiconductors.

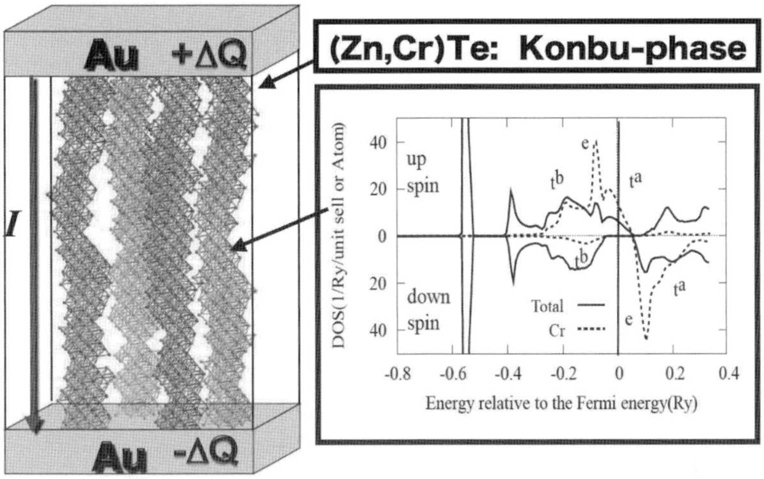

Figure 1.11. Konbu phase sandwiched between Au contacts designed by the self-organized two-dimensional (2D) spinodal nano-decomposition in the layer-by-layer crystal growth simulation by the Monte Carlo method based upon the *ab initio* calculation of chemical pair interactions between Cr–Cr, Zn–Cr and Zn–Zn pairs. Averaged Cr concentration is 20% in (Zn, Cr)Te. Growth-position control of the Konbu phase is made by the seeding of the periodic 25 Cr atoms on the Au contact and substrate. The pink atom corresponds to Cr. The regions between the nano superstructures of columns are pure ZnTe. The inserted density of states (DOS) of the Cr high concentration region caused by 2D spinodal nano-decomposition indicates the half-metallicity (100% spin-polarized ferromagnets). Applying a high current of charge and spin from the upper to the lower-side between the Au contacts, the upper Au contact is heated up and the lower Au contact is cooled down.

1.2 III–V AND II–VI COMPOUND SEMICONDUCTOR BASED DMS BY LDA

In this section, the electronic structure and magnetism of typical III–V and II–VI DMS systems are discussed. For the first half of this section, we focus on homogeneous DMS systems in which magnetic impurities are distributed randomly, i.e., without any correlation. For such systems, the Korringa–Kohn–Rostoker coherent potential approximation (KKR-CPA) method can be conveniently applied to describe the configuration average of the electronic structure.[37] For the present calculations we use the KKR-CPA package MACHIKANEYAMA-2002 developed by Akai.[38] By using this method, the electronic structure and the stability of the ferromagnetic state of the DMS systems are calculated.[39,40] Moreover, by using the magnetic force theorem and by mapping the system on a classical Heisenberg model, the exchange interactions between magnetic impurities in DMS are estimated.[41,42] Once the exchange interactions are obtained, exact Curie temperatures of DMS systems can be estimated by performing the Monte Carlo simulation.[43,44] It will be shown that high-T_C (above 300 K) is very difficult to realize for the normal concentration range ($1 \sim 10\%$) due to the magnetic percolation problem.

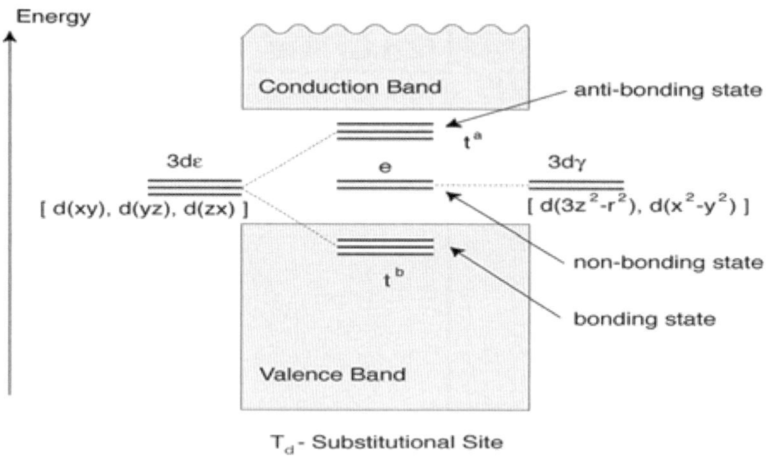

Figure 1.12. Schematic picture of the electronic states of $3d$-TM in semiconductors at the tetrahedral substitutional site.

In order to propose a materials design for high-T_C DMS, in the second half of this section we discuss the effect of inhomogeneity of DMS systems. In particular, it will be shown that the DMS systems have a tendency towards spinodal decomposition.[45–48] We will explain how to simulate the spinodal decomposition and its effect on the ferromagnetism of DMS. In addition, the co-doping technique is proposed to control spinodal decomposition in DMS.[49]

1.2.1 Electronic Structure and Chemical Trends

Before discussing the ferromagnetism of DMS, the electronic structure of a 3d transition metal (3d-TM) impurity in semiconductors is explained.[39] In general, 3d-TM atoms occupy the cation site which is the tetrahedral substitutional site. Due to tetrahedral symmetry, the 3d states of the impurity split into the three-fold degenerate $d\varepsilon$ states (xy, yz, zx) and the two-fold degenerate $d\gamma$ states $(x^2 - y^2, 3z^2 - r^2)$. Because of the compatibility of the symmetry, $d\varepsilon$ states hybridize strongly with p states of ligands which make host valence bands, resulting in the bonding states, t^b, in the valence bands and the anti-bonding states, t^a, in the band-gap. On the other hand, the hybridization between the $d\gamma$ states and the host valence states are very weak, leading to the non-bonding states, e, in the gap. This situation is schematically shown in Fig. 1.12. For finite concentrations, due to the anti-bonding nature of the t^a states, they make rather broad (i.e., delocalized) impurity bands in the gap. On the other hand, the impurity bands made from e states are relatively narrow (i.e., localized), reflecting their non-bonding character. This ordering of the impurity bands and their localization degree are very important in understanding the ferromagnetism of DMS and its chemical trend.

Figures 1.13 and 1.14 show calculated total density of states per unit cell and partial density of $3d$ states per TM atom at TM sites. We show (Ga, Mn)N and (Ga,

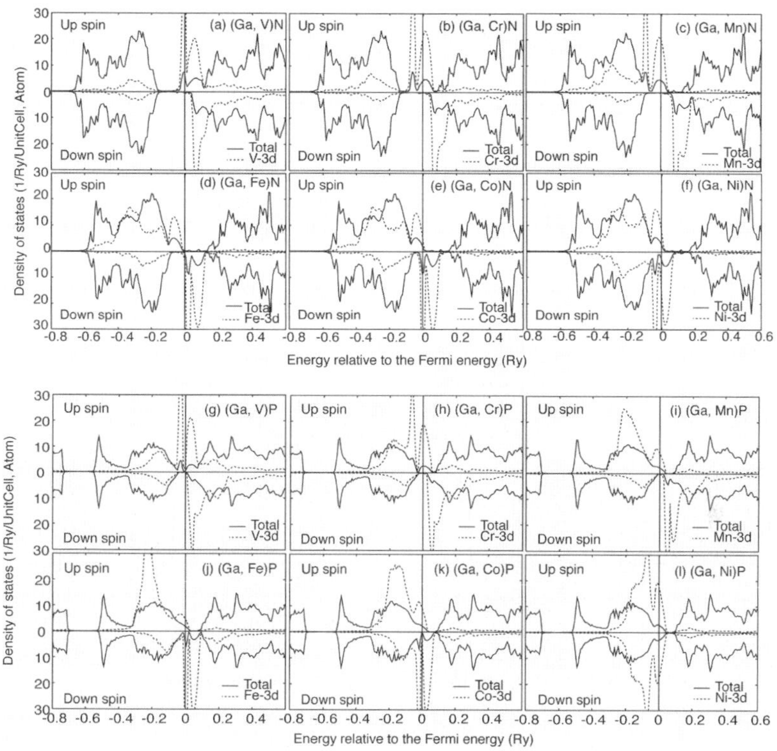

Figure 1.13. Calculated density of states of GaN-based DMS [upper panels, (a)–(f)] and GaP-based DMS [lower panels, (g)–(l)]. As 3d-TM impurities, we consider V to Ni. Solid lines show total DOS and dotted lines show partial DOS of 3d components at TM sites. A ferromagnetic configuration is assumed and the TM concentration is 5% in all cases.

TM)P as typical III–V DMS in Fig. 1.13 and (Zn, TM)O and (Zn, TM)S as typical II–VI DMS in Fig. 1.14. The concentration of TM impurities is 5% and the electronic structure is calculated assuming a ferromagnetic configuration in each case. We can understand the DOS reasonably well based on the explanation given above. For example, in (Ga, Cr)N we can clearly distinguish rather broad anti-bonding t^a states and very sharp non-bonding e states in the gap. The Fermi level is located in the t^a impurity band and two-thirds of the peak is occupied. The bonding states are completely buried but still distinguished in the valence bands. Due to the large exchange splitting, the t^a and e impurity bands for down (minority) spin state is located above the Fermi level.

The occupation of the impurity bands depends on the number of d electrons which each 3d-TM impurity introduces. For example, in (Ga, V)N the Fermi level (E_F) is located in the valley between the t^a and the e states, and in (Ga, Fe)N all of the up (majority) spin states of the impurity bands are occupied and all of the down spin states are left unoccupied. In (Ga, Co)N and (Ga, Ni)N, down spin e states are occupied approximately by one and two electrons, respectively; however, the very small but finite amplitude of up spin states is found at the E_F, and thus the systems

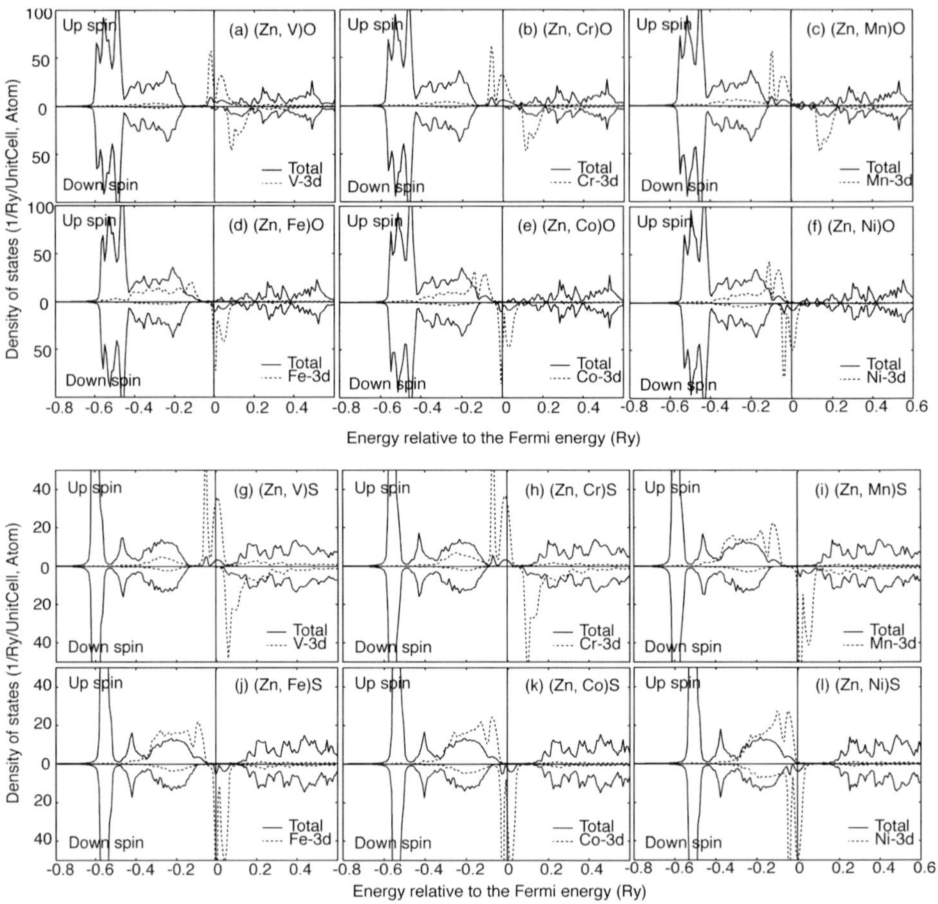

Figure 1.14. Calculated density of states of ZnO-based DMS [upper panels, (a)–(f)] and ZnS-based DMS [lower panels, (g)–(l)]. As 3*d*-TM impurities, we consider V to Ni. Solid lines show total DOS and dotted lines show partial DOS of 3*d* components at TM sites. A ferromagnetic configuration is assumed and the TM concentration is 5% in all cases.

are not completely half-metallic. (Ga, V)N, (Ga, Cr)N and (Ga, Mn)N are calculated to be half-metallic as shown in Figs. 1.13(a), 1.13(b) and 1.13(c), and (Ga, Mn)Fe shows insulating DOS.

In Figs. 1.13(g)–1.13(l), the calculated DOS of GaP-based DMS systems are shown. The main difference to the GaN cases is the position of the 3*d* states of TM impurities relative to the host valence bands. Compared to the N-2*p* states, P-3*p* states are located higher in energy; thus, in GaP-based DMS systems 3*d* states of TM impurities have larger amplitude in the host valence bands than in the GaN-based DMS systems. In other words, the relative weight of 3*d* states in the bonding t^b states becomes larger in GaP-based DMS than in GaN-based DMS. For example, in (Ga, Mn)P the main peak of Mn-3*d* states is in the valence band and at E_F we can find only a partially occupied small peak which is already merged to the valence band. The structure of the impurity bands is clearer in (Ga, V)P and (Ga, Cr)P than in (Ga, Mn)P, but the position of the impurity bands is much nearer to the valence band compared to the (Ga, V)N and (Ga, Cr)N cases. (Ga, Fe)P and (Ga, Co)P are not half metallic and (Ga, Ni)P is non-magnetic, according to the present calculations as shown in Figs. 1.13(j), 1.13(k) and 1.13(l). The (Ga, TM)As and (Ga, TM)Sb DMS systems show rather similar DOS to (Ga, TM)P shown in Figs. 1.13(g)–1.13(l) and the electronic structure of (Ga, TM)N systems is exceptional. This is due to the large binging energy of $2p$-electrons in N.

The calculated DOS for II–VI DMS systems are shown in Fig. 1.14. As a typical example we calculate ZnO-based DMS and ZnS-based DMS. The overall structure of the calculated DOS is similar to that of III–V DMS systems, i.e., we can clearly distinguish anti-bonding broad t^a states and non-bonding narrow e states. However, the occupation of the impurity bands is different. For example, in GaN-based DMS the impurity bands for up spin states are filled up for the Fe impurity, but in ZnO- and ZnS-based DMS this situation happens for the Mn impurity case. This is due to the different charge state of the cation in II–VI DMS. In II–VI DMS, TM impurities have a 2+ charge state nominally, and therefore they have one more electron to occupy the impurity bands compared to TM impurities in III–V DMS, where the nominal charge state of the TM impurities is 3+. As a result, 1/3 and 2/3 of the t^a states are occupied in (Zn, V)S and (Zn, Cr)S, respectively, and the down spin e states are occupied by one and two electrons in the case of (Zn, Fe)S

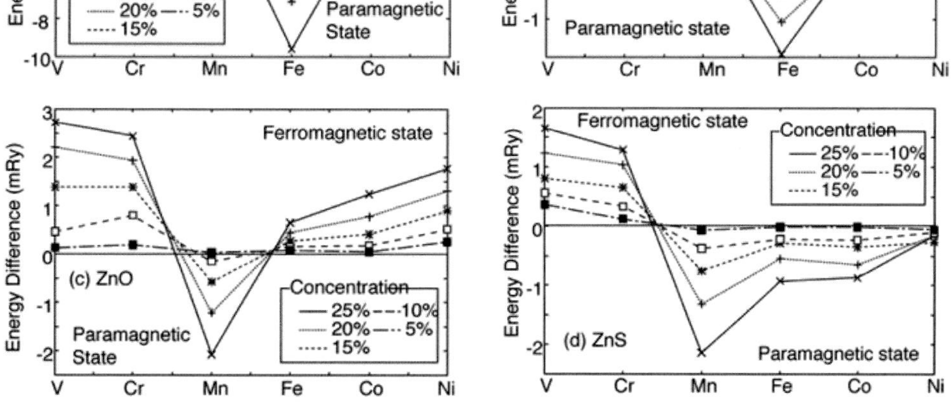

Figure 1.15. Chemical trends of the stability of the ferromagnetic state. The energy difference between the ferromagnetic state and the DLM (paramagnetic) state is shown for (a) GaN-, (b) GaP-, (c) ZnO- and (d) ZnS-based DMS.

and (Zn, Co)S, respectively. In the Ni-doped case, down spin t^a-states are partially occupied. The same is also true for ZnO-based DMS, except for V- and Cr-doped cases where the t^a-states are less occupied compared to V- and Cr-doped ZnS due to the significant underestimation of the band-gap energy.

1.2.2 Ferromagnetic Mechanism in the DMS

In order to understand the origin of the ferromagnetism, the chemical trend of the stability of the ferromagnetic state in the III–V and II–VI DMS is investigated.[39] For this purpose, the concept of the disordered local moment (DLM) state is conveniently used to study the magnetic state of magnetic materials.[37] By using the CPA, the ferromagnetic $(\mathrm{Ga}_{1-x}, \mathrm{Mn}_x^{\mathrm{up}})$N is calculated by introducing one impurity component, Mn, with up spin, at the Ga site with concentration x. Similarly, by introducing two impurity components, $\mathrm{Mn}^{\mathrm{up}}$ with an up spin state and $\mathrm{Mn}^{\mathrm{down}}$ with a down spin state, the DLM state can be described as $(\mathrm{Ga}_{1-x}, \mathrm{Mn}_{x/2}^{\mathrm{up}}, \mathrm{Mn}_{x/2}^{\mathrm{down}})$N. The DLM corresponds to the configuration average of the magnetic system with randomly oriented local magnetic moments. It is well known that the DLM state describes the paramagnetic state or the spin-glass state.[37] Since we calculate total energies of the ferromagnetic state and the DLM state within the same framework, the stability of the ferromagnetic state is estimated simply by comparing the total energies.

Figure 1.15 shows the calculated total energy difference, ΔE, between the ferromagnetic state and the DLM state. ΔE is calculated as $\Delta E = TE(\mathrm{DLM}) - TE(\mathrm{FM})$, where TE(DLM) and TE(FM) are the total energy of the DLM state and the ferromagnetic state, respectively. As typical cases, the results for GaN-, GaP-, ZnO- and ZnS-DMS are shown in the figure. As a whole, they show very similar chemical trends in terms of the stability of the ferromagnetism, i.e., for the first half of $3d$-TM series the ferromagnetism is favored, while for the latter half the ferromagnetism is not stable, except for ZnO-DMS where Fe-, Co- and Ni-doped ones show ferromagnetism. The curves shown in Figs. 1.15(a) and 1.15(b) are shifted to the larger atomic number side compared to those in Figs. 1.15(c) and 1.15(d). For example, in the case of the III–V DMS, the minimum, where the ferromagnetic state is most unstable, appears in the Fe-doped cases; on the other hand in case of the II–VI DMS, the minimum appears in the Mn-doped cases.

In general, there are two important mechanisms which govern the magnetic interactions between magnetic ions in DMS. One is the ferromagnetic double-exchange interaction and the other is the anti-ferromagnetic super-exchange interaction. Suppose that there are two neighboring magnetic impurities at sites i and j, and they have partially occupied impurity bands. If their magnetic moments are parallel to each other, the d electrons at one site i can hop to the other site j. As a result, the system can lower the kinetic energy. The hopping of the electron between anti-parallel magnetic moments costs energy as large as exchange splitting and is almost prohibited. Thus, only the ferromagnetic configuration gains kinetic energy, and this mechanism is called the double-exchange mechanism.[39,50–52] This mechanism is particularly effective when half of the impurity band for one spin state is occupied, and does not work for d^0, d^5 and d^{10} high-spin configurations.

When the magnetic moments are anti-parallel, there is another mechanism that stabilizes this configuration. As is easily recognized, there is a hybridization between the majority (minority) spin states at site i and the minority (majority) spin states at site j, because they have the same spin state. As a result, the majority spin states at each site lower the energy to stabilize this configuration. For the d^5 high spin configuration, the anti-ferromagnetic super-exchange is most effective, but for the d^0 and d^{10} configurations this mechanism does not work. In the language of band theory, the energy gain due to the double exchange comes from the band broadening of the partially occupied impurity bands. On the other hand, the energy gain by the super-exchange originates from the shift of the center of gravity of the occupied impurity bands.

The calculated chemical trend is reasonably well explained by considering the competition between above mentioned magnetic interactions.[51,52] As explained above, the super-exchange interaction is particularly effective for the d^5 configuration, but for this electron configuration the double exchange is suppressed. As a result, the DLM state is expected to be stabilized for this configuration. The d^5 electron configuration is expected for Mn impurities in II–VI semiconductors and Fe impurities in III–V semiconductors, and for these compounds the DLM state is most favored, as shown in Fig. 1.15. Deviating from the d^5 configuration, the ferromagnetic state becomes stable due to the contribution of the double exchange. Though one might think the stabilization of the ferromagnetism should show a symmetrical curve around the d^5 configuration, the calculated curves are not, except in the case of ZnO-based DMS. This asymmetry reflects the different localization nature of the impurity bands. As already explained, in the tetrahedral symmetry the impurity bands show the characteristic structure and appear in the order of up spin e states, up spin t^a states, down spin e states, and down spin t^a states from the lower energy side. Therefore, for the d^3 and d^4 configurations the t^a states are partially occupied and for the d^6 configuration the e-states are partially occupied. Remembering that the t^a-states have a delocalized nature and the e states have a localized nature, it is reasonable that the t^a states contribute to the ferromagnetism much more than the e states. For the d^2 and d^7 configurations, E_F falls in the valley between the t^a peak and the e peak, and therefore the double exchange is not effective. However, the ferromagnetic super exchange works for these configurations and the ferromagnetism is stable in V-doped III–V DMS.[52] In Co-doped II–VI, the ferromagnetic super exchange does not compete with the anti-ferromagnetic super exchange, resulting in the paramagnetic ground state.

For a more detailed discussion, we calculate the concentration dependence on Curie temperature for Mn-doped III–V DMS by using the mean field approximaion (MFA).[40] It is well known that the MFA sometimes overestimates T_C and does not give reliable predictions. Therefore, we use the MFA only to analyze the underlying ferromagnetic mechanism and a hybrid method, which uses

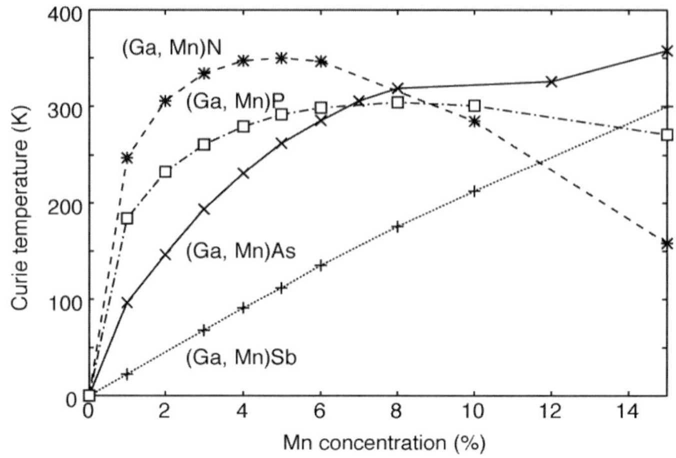

Figure 1.16. Calculated Curie temperature of (Ga, Mn)N, (Ga, Mn)P, (Ga, Mn)As and (Ga, Mn)Sb by using the mean field approximation (MFA). T_C is calculated as a function of Mn concentration.

first-principles electronic structure calculations and the Monte Carlo method, will be developed later in this section to calculate T_C accurately.

By using the MFA, the Curie temperature is estimated from the total energy difference ΔE as $k_B T_C = 2\Delta E/3x$.[40,51] By using this equation, Curie temperatures in the MFA, T_C^{MFA}, are calculated for (Ga, Mn)N, (Ga, Mn)P, (Ga, Mn)As and (Ga, Mn)Sb as a function of Mn concentration. The results are summarized in Fig. 1.16. In the figure, we can find a very interesting chemical trend in the concentration dependence. In (Ga, Mn)N, T_C shows a sharp increase for low concentrations and reduces for higher concentrations, reading a maximum at around 5%. In (Ga, Mn)P, after the increase for low concentrations, T_C saturates at about 300 K. Compared to (Ga, Mn)N and (Ga, Mn)P, the increase of T_C in (Ga, Mn)As is much more moderate. For these three compounds, for low concentrations the concentration dependence of T_C is approximately proportional to the square root of Mn concentration. On the other hand, in (Ga, Mn)Sb T_C increases almost linearly with Mn concentration. This contrast in the concentration dependence indicates that the dominant exchange mechanism is different in (Ga, Mn)N and (Ga, Mn)Sb. Moreover, the gradual transition in the concentration dependence of T_C from (Ga, Mn)N to (Ga, Mn)Sb suggests the systematic transition in the mechanism governing the ferromagnetism in this series of compounds.

The transition in the mechanism of the ferromagnetism can be found in the calculated electronic structure of Mn-doped III–V DMS systems. Figure 1.17 shows the total DOS and partial DOS of Mn-3*d* states in these DMS systems. As shown in Fig. 1.17(a), in (Ga, Mn)N the impurity *e* and t^a bands appear clearly in the band-gap and the 2/3 of the t^a bands is occupied as explained already. Going from N to Sb, the Mn-3*d* states shift their amplitude to lower energies. As a result, in (Ga, Mn)N the t^a bands mainly consist of Mn-3*d* states, but in (Ga, Mn)Sb the

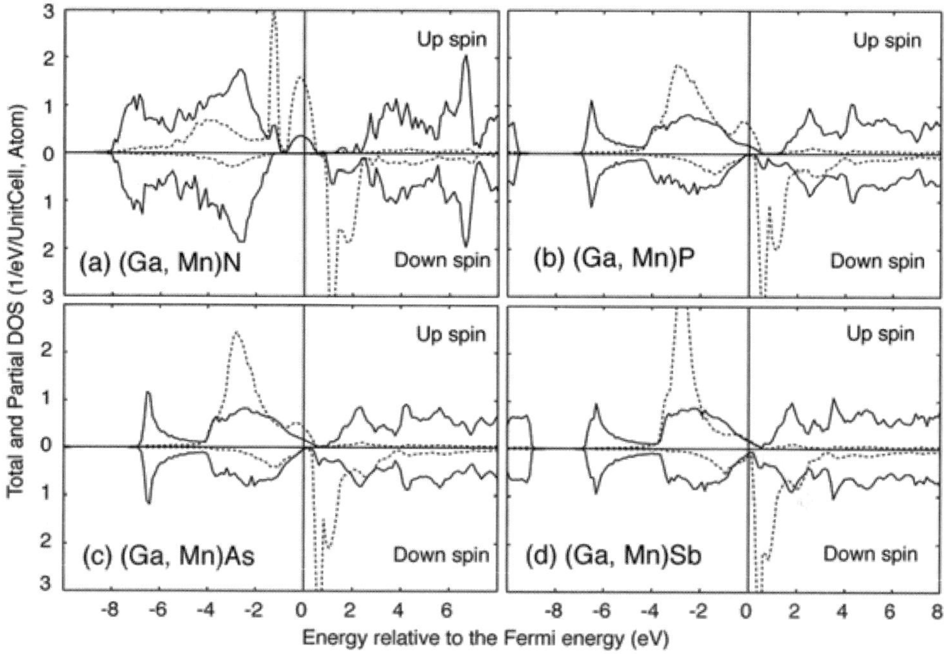

Figure 1.17. Total DOS (solid curves) and partial DOS of 3*d* states at Mn (dashed curves) in (a) (Ga, Mn)N, (b) (Ga, Mn)P, (c) (Ga, Mn)As and (d) (Ga, Mn)Sb. Mn concentration is 5% and the ferromagnetic configuration is assumed.

t^b bands have a strong Mn-3*d* character. Since the electron around E_F has the *d*-character in (Ga, Mn)N as shown in Fig. 1.17(a), the double exchange reasonably explains the ferromagnetism in this compound. As already pointed out, the energy gain in the ferromagnetic configuration is due to the band broadening of the partially occupied impurity bands. It is known that within the tight-binding picture the width of the impurity band is proportional to the square root of the impurity concentration, and this is why the calculated T_C is proportional to the square root of Mn concentration in (Ga, Mn)N for low concentrations.[51]

In (Ga, Mn)Sb, the main amplitude of Mn-3*d* appears as the t^b bands in the valence bands; thus the nominal electronic configuration of Mn is $Mn^{2+}(d^5)$. As a result, one hole per Mn is introduced in the host valence bands. Due to the hybridization between the exchange split Mn-3*d* states and the host valence bands, the valence bands are polarized anti-parallel to the Mn moments. This polarization of the host valence bands produces an effective magnetic field that aligns the Mn magnetic moments and stabilizes the ferromagnetic state. This is called the *p–d* exchange mechanism, another ferromagnetic mechanism playing a role in the magnetism of DMS.[51–55] In this case, the stability of the ferromagnetic state is proportional to the polarization of the valence bands, which is proportional to the Mn concentration. This is why the (Ga, Mn)Sb shows a linear concentration dependence in T_C. The ferromagnetic mechanisms in the DMS are summarized schematically in Fig. 1.18.

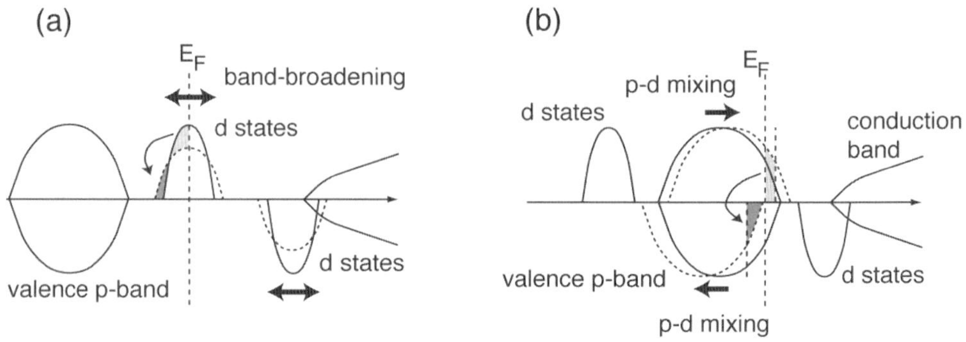

Figure 1.18. (a) Schematic diagram of the electronic structure of the double-exchange dominated DMS systems. The partially occupied *d*-level is located in the band-gap. The full lines show the impurity bands for a low concentration and the dashed lines for a high concentration. By transferring the states in the gray area, the system gains the band energy. (b) Schematic diagram of the electronic structure of the *p*–*d* exchange dominated DMS systems. The dashed lines show the valence bands after the hybridization with the magnetic impurities.

Another important factor in the magnetism of DMS is the anti-ferromagnetic super exchange. Since this mechanism is due to the hybridization between the occupied majority states and the unoccupied minority states, as explained in the previous sub-section, the strength of this interaction should be proportional to the concentration of the magnetic impurities and inversely proportional to the energy difference between the majority and the minority states, which is the exchange splitting. Therefore, the anti-ferromagnetic contribution is normally larger for higher concentrations and for systems with smaller exchange splitting. This is why we find strong suppression of the ferromagnetism in (Ga, Mn)N for higher concentrations but the suppression is not significant in (Ga, Mn)As and (Ga, Mn)Sb.

1.2.3 Curie Temperature (T_C) of Homogeneous Systems

In general, the mean field approximation is justified for systems with infinite configuration numbers and with infinite ranges of the interactions. This means that it might be dangerous to apply the approximation for systems with dilute magnetic impurities and with short-ranged exchange interactions. In this sub-section, the validity of the MFA is assessed and an accurate method for calculating T_C of the DMS systems is proposed.

One of the methods that goes beyond the MFA in simulating the phase transition is to perform Monte Carlo simulations on an effective model. For this purpose, we employ a classical Heisenberg model to describe the DMS systems because the local magnetic moment of the magnetic impurities is large and well defined. A prescription to calculate the exchange interactions for the projected Heisenberg model is proposed in Ref. 41 by using the magnetic force theorem and applied to the

DMS systems in Refs. 42–45. Figure 1.19 shows the effective exchange interactions calculated using the method. In the figure, the distance dependence of the interactions is plotted for several DMS systems. It is found that the range of interactions depends on which mechanism, the double exchange or the *p–d* exchange, is dominant. In typical double-exchange systems such as (Ga, Mn)N, (Ga, Cr)N and (Zn, Cr)Te, the interactions are short-range interactions and strong only for the first few neighbors. Due to these strong interactions for near neighbors, the calculated T_C by using the mean field approximation becomes very high already for low concentrations. However, it will be shown that these high-T_Cs are mostly due to artifacts from the unjustified use of the mean field approximation

For these systems with short-range interactions, the magnetic percolation problem is important for accurate T_C predictions. It is known that the percolation threshold for the FCC lattice is 20%,[56] and therefore if the concentration of magnetic impurities is below 20%, the ferromagnetic phase transition cannot occur for systems with only nearest-neighbor interactions. In real systems, the magnetic percolation threshold should be lower than the percolation threshold owing to the longer-range interactions, but the reduction of T_C can be significant. In order to see this effect quantitatively, in Fig. 1.20 Curie temperatures are estimated with the inclusion of the magnetic percolation effects by using Monte Carlo simulations. In these Monte Carlo simulations, we set up large super-cells ($6 \times 6 \times 6$, $10 \times 10 \times 10$ and $14 \times 14 \times 14$ conventional FCC unit cells) and distributed magnetic impurities randomly in the super-cells. By calculating the temperature dependence of the magnetization and by using the cumulant crossing method,[57] T_C was estimated.[42,43] As shown in Fig. 1.20, the effect of magnetic percolation is actually significant, and the ferromagnetism is strongly suppressed. The reduction of the Curie temperature from the MFA values is very large, particularly for low concentrations, where magnetic percolation is difficult to achieve in double-exchange systems.

For *p–d* exchange systems such as (Ga, Mn)As and (Ga, Mn)Sb, due to their long-range nature in exchange interactions [Fig. 1.19(c) and 1.19(d)], the magnetic percolation is easier to realize compared to double-exchange systems. In fact, for large concentrations the reduction of T_C is more moderate than in, for example (Ga, Mn)N and (Ga, Cr)N. However, for low concentrations it is crucial to take into account the disorder effects precisely in order to predict T_C accurately.

In the comparison between experimental values and theoretical predictions, the Curie temperatures calculated by the Monte Carlo method agree well with the experimental observations particularly in (Ga, Mn)P, (Ga, Mn)As and (Zn, Cr)Te. In (Ga, Mn)N and (Ga, Cr)N, the experimental values deviate from one another, and some experiments reported very high T_C while the others did not. We believe that this sharp contradiction comes from the problem of sample preparation. As will be shown in the next sub-section, (Ga, Mn)N and (Ga, Cr)N are thermodynamically unstable and easily show phase separation.[46–49] As a result, depending on the details of the experimental conditions, the phase separation occurs uncontrollably and the ferromagnetism shows strong sample dependence. For

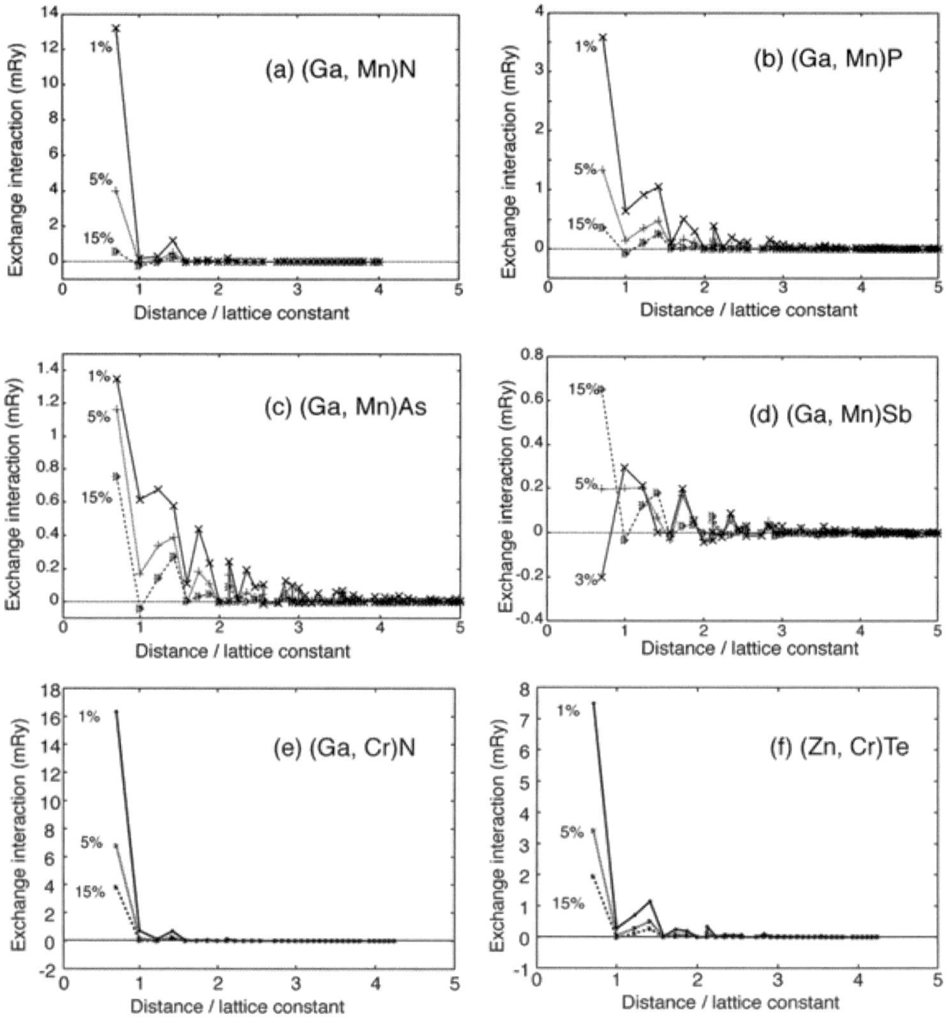

Figure 1.19. Effective exchange interactions as a function of distance in (a) (Ga, Mn)N, (b) (Ga, Mn)P, (c) (Ga, Mn)As, (d) (Ga, Mn)Sb, (e) (Ga, Cr)N and (f) (Zn, Cr)Te calculated by Liechtenstein's formula.

these cases, the low T_C values are to be compared to the present predictions with the assumption of homogeneous distribution of the magnetic impurities.

So far we have summarized our study on the homogeneous DMS systems and have shown that, in general, high-T_C is very difficult to realize for homogeneous DMS systems within a realistic concentration range (around 10%). In the next subsections, we will discuss how the inhomogeneous distribution affects the results obtained here and look into whether the magnetic properties of the DMS systems can be tuned by controlling the inhomogeneity of the impurity distribution.

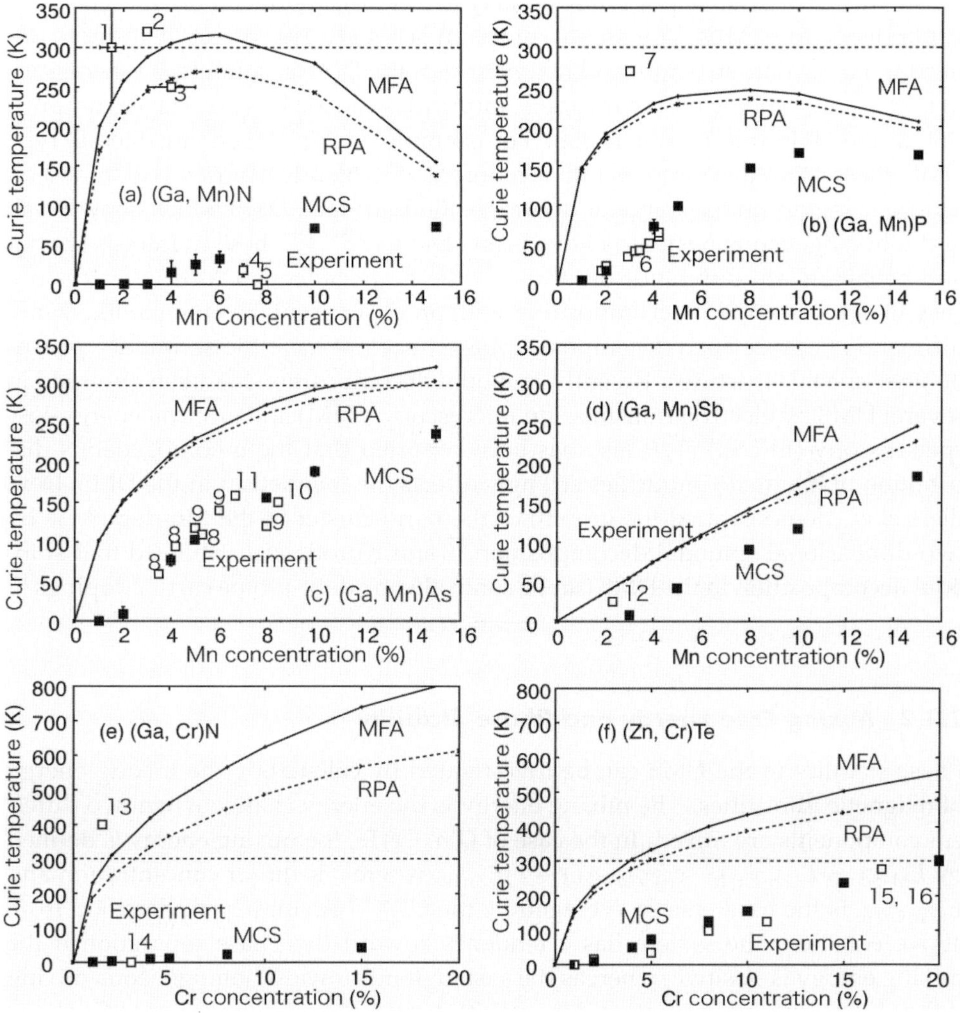

Figure 1.20. Calculated Curie temperature of (a) (Ga, Mn)N, (b) (Ga, Mn)P, (c) (Ga, Mn)As, (d) (Ga, Mn)Sb, (e) (Ga, Cr)N and (f) (Zn, Cr)Te by using the mean field approximation (MFA), the random phase approximation (RPA) and Monte Carlo simulations (MCS). Experimental values are also plotted.

1.3 SPINODAL NANO-DECOMPOSITION IN DILUTED MAGNETIC SEMICONDUCTORS (DMS)

1.3.1 Inhomogeneous Distribution of Transition Metal Impurities in DMS

In the above discussion, it was assumed that the distribution of magnetic impurities in the DMS is homogeneous. However, since the DMS are fabricated by thermal non-equilibrium crystal growth methods, such as the molecular beam epitaxy

(MBE), metal-organic vapor phase epitaxy (MOVPE), and metal organic chemical vapor deposition (MOCVD), this assumption is not always correct. In order to understand electronic structure and magnetism of the DMS accurately, it is necessary to take the inhomogeneity of magnetic impurities caused by spinodal decomposition in the DMS into consideration. For example, in Mn-doped zinc-blende type DMS, the system decomposes into a coherent zinc-blende Mn-rich region and an Mn-poor region under spinodal decomposition. In fact, DMS which show weak and inhomogeneous hysteresis have been observed.[59–60] These hysteresis shapes and behaviors are different from those of the homogeneous DMSs.[61–65] It is not easy to detect spinodal decomposition with an experimental technique like X-ray diffraction because such decomposed phase does not have the secondary phases of other crystal structures. Recently, spinodal decomposition has been observed in several DMS by electron transmission microscopy (TEM) and electron energy-loss spectroscopy (EELS).[66,67] It also has been reported that the inhomogeneous distribution of magnetic impurities strongly affects the magnetism in the DMS. Devillers *et al.* demonstrated the growth of the nano-cluster in the Mn-doped Ge by two-dimensional spinodal decomposition,[67] and Kuroda *et al.* showed that spinodal decomposition in the DMS can be controlled by n- or p-type carrier dopings.[66]

1.3.2 Mixing Free Energy and Phase Stability

Phase stability in the DMS can be investigated by calculating the mixing energy of magnetic impurities. The mixing energy is the energy change when two different components are mixed. In the case of (Zn, Cr)Te, the mixing energy is defined by $E_{\mathrm{M}}(c) = E_{(\mathrm{Zn,Cr})\mathrm{Te}} - cE_{\mathrm{CrTe}} - (1-c)E_{\mathrm{ZnTe}}$, where c is the Cr concentration and $E_{(\mathrm{Zn,Cr})\mathrm{Te}}$ is the total energy averaged by the CPA.[68] As may be understood from above equation, the system has a tendency toward the phase separation if the mixing energy is positive, whereas the system tends toward homogeneous mixing if the mixing energy is negative. Figure 1.21 depicts the mixing energy of Cr atoms in (Zn, Cr)Te as a function of the Cr concentration. As shown in Fig. 1.21, the mixing energy of (Zn, Cr)Te shows strong convexity and always has positive values for concentration region the investigated. This result leads to the phase separation in the ground state at $T = 0$: (Zn, Cr)Te has a miscibility gap. Note that in this calculation experimental lattice constants are employed, and the lattice relaxation effect is not taken into account because of the CPA calculation. The kinetic effect of the lattice strongly affects the impurity configuration in the substituted alloy.

As the temperature is increased, the entropy effect becomes prominent. Therefore, in order to understand the phase stability of the DMS at finite temperature, it is necessary to calculate the mixing free energy, $F(c, T) = E_{\mathrm{M}}(c) - TS$, where T is the temperature and S is the mixing entropy. In the case of a substituted binary alloy, the mixing entropy is easily obtained as $S = -k_{\mathrm{B}}[(1-c)\ln(1-c) + c\ln c]$, where k_{B} is the Boltzmann constant. Figure 1.22 indicates the mixing free energy F of (Zn, Cr)Te as a function of the Cr concentration. As shown in Fig. 1.22, with

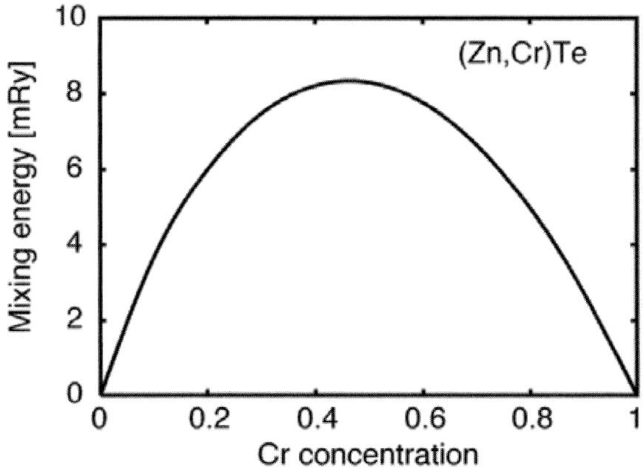

Figure 1.21. Mixing energy of Cr atoms in (Zn, Cr)Te as a function of Cr concentration.

increasing temperature, the mixing free energies show a gradual transition from convex to concave concentration dependence resulting in negative mixing free energy. At low temperatures, the mixing free energies have common tangents. Within these tangent points the DMS have energy gains with respect to the decomposition into the composition rich and poor phases. In particular, the region $\partial^2 F/\partial c^2 < 0$ is called the *spinodal region*, where the spinodal decomposition occurs spontaneously. On the other hand, the region $\partial^2 F/\partial c^2 > 0$ in the miscibility gap is called the *nucleation* or *binodal region*. Above the critical temperature (spinodal temperature), the two common tangent points disappear, such that the magnetic impurities in the DMS are distributed homogeneously.

The temperature (T) versus impurity concentration (c) phase diagram can be described from the calculated mixing free energy. At a temperature lower than the

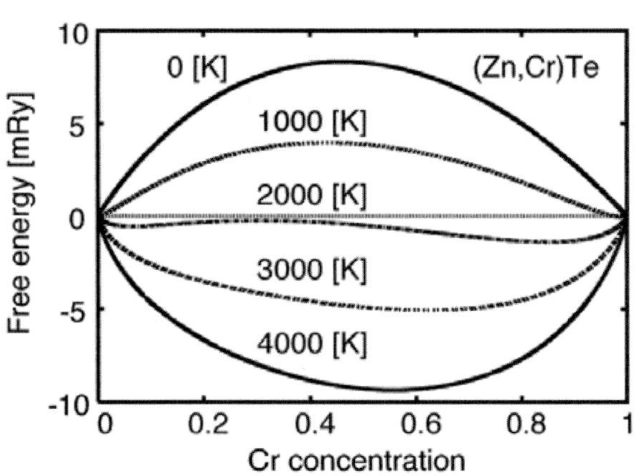

Figure 1.22. Mixing free energy of Cr atoms in (Zn, Cr)Te as a function of Cr atoms. The cases of 0, 1000, 2000, 3000 and 4000 K are shown in the figure.

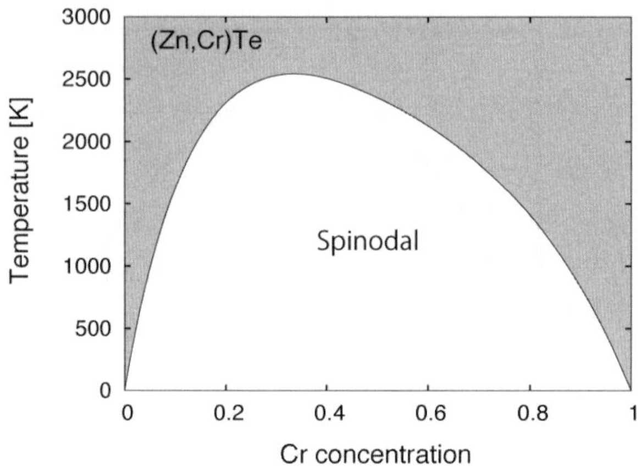

Figure 1.23. The phase diagram of (Zn, Cr)Te. The black line indicates the spinodal line.

spinodal temperature, the two spinodal points are determined from the second derivative of the mixing free energy with respect to the impurity concentration as zero ($\partial^2 F/\partial c^2 = 0$). Figure 1.23 shows the phase diagram of (Zn, Cr)Te. The black line in Fig. 1.23 signifies the phase boundary between the homogeneous and spinodal regions. Above the spinodal line, the system does not cause spinodal decomposition. On the other hand, below the spinodal line the system is unstable for an arbitrarily small fluctuation of the Cr concentration. As shown in Fig. 1.23, (Zn, Cr)Te tends to undergo spinodal decomposition at a commonly used crystal growth temperature in the experiment. For example, when (Zn, Cr)Te is grown at 500 K, Cr atoms in ZnTe are immiscible in the range of about $0.03 < c < 0.95$, such that in this range the system causes the phase separation into $(Ga_{0.97},Mn_{0.03})As$ and $(Ga_{0.05},Mn_{0.95})As$ by spinodal decomposition.

1.3.3 Effective Chemical-Pair Interactions in the DMS

As discussed above, in general, the DMS undergoes spinodal decomposition such that the distribution of magnetic impurities in the DMS is no longer homogeneous. In order to investigate the configuration of magnetic impurities in the DMS under spinodal decomposition and how the inhomogeneity of magnetic impurities affects the magnetism in the DMS, we have to calculate the effective pair interaction between two magnetic impurity sites i and j by using the generalized perturbation method within the KKR-CPA formalism.[69] According to this method, for a binary alloy $A_{1-x}B_x$, the effective pair interaction V_{ij} is defined as an energy difference between B–B and A–A pairs and two B–A pairs, i.e., $V_{ij} = V_{ij}^{BB} + V_{ij}^{AA} - 2V_{ij}^{AB}$, where V_{ij}^{AB} is the potential energy calculated by the KKR-CPA method when sites i and j are occupied by atoms A and B, respectively. As understood from the definition, negative V_{ij} means that the attractive interaction works between the same components at sites i and j, whereas positive V_{ij} leads to the repulsive interaction. Figure 1.24 shows the effective pair interaction in (Zn,

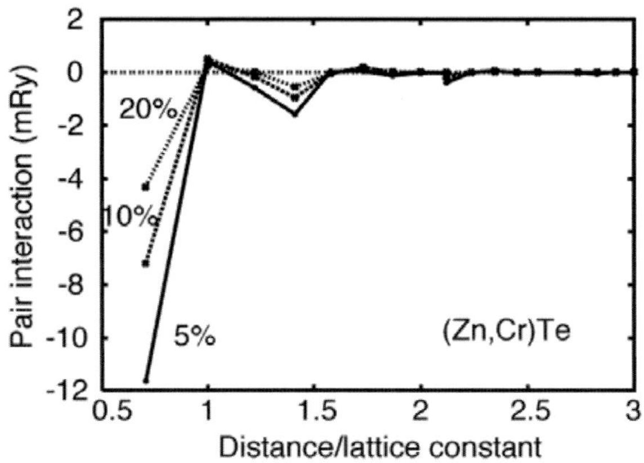

Figure 1.24. The effective pair interactions between Cr atoms in (Zn, Cr)Te. The horizontal axis is the distance between Cr atoms.

Cr)Te as a function of the distance between two Cr atoms. The cases of Cr 5%, 10% and 20% are depicted. As shown in Fig. 1.24, the effective pair interactions of (Zn, Cr)Te oscillate as well as the exchange coupling constant, and positive interactions firstly appear in the second nearest neighbor, but the main contribution to the phase state is a strong first nearest neighbor attractive interaction. Therefore, this result indicates that (Zn, Cr)Te undergoes phase separation. Such a tendency of the effective pair interaction is consistent with the calculated convex positive mixing energy in Fig. 1.21. Although not shown here, other DMS such as (Ga,Mn)N, (Ga,Mn)As and Co(Ti,Mn)Sb have negative effective pair interactions.

1.3.4 Simulation of the Spinodal Nano-decomposition

In this sub-section, based on the calculated effective pair interaction, we consider crystal growth simulation by the Monte Carlo method. The conserved order parameter Ising Hamiltonian is assumed here for describing the configuration of magnetic impurities in the DMS: $H = -\Sigma_{i,j} V_{ij}\sigma_i\sigma_j$, where V_{ij} is the effective pair interaction between sites i and j. σ_i is the occupation number at site i. In the non-equilibrium crystal growth technique used in the experiments, the system cannot reach thermal equilibrium. In order to simulate this situation, it is necessary to interrupt the Monte Carlo loop after certain steps. As a typical example, the (Zn, Cr)Te cases are shown; however, our simulation method is easily generalized to all of the DMSs.

1.3.5 Three-dimensional Spinodal Decomposition (Dairiseki Phase)

Firstly, simulations of the three-dimensional spinodal decomposition cases are performed. In the three-dimensional case, a large FCC super-cell is prepared, and the magnetic impurities in the DMS are randomly distributed as an initial

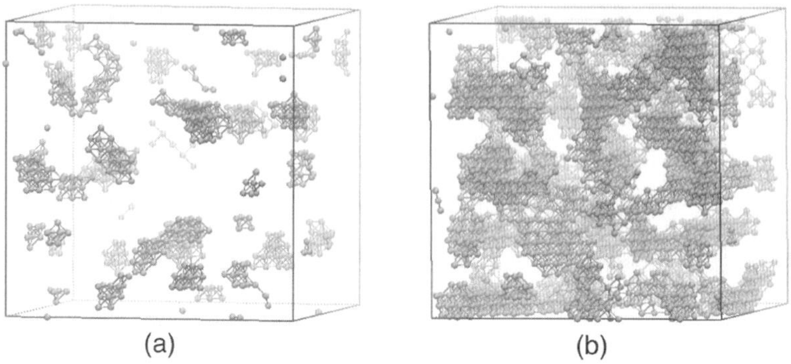

Figure 1.25. Snapshots of the simulations of three-dimensional spinodal decomposition (Dairiseki phase). The cases of Cr 5% (a) and 20% (b) are shown in the figure.

configuration. Starting from a random distribution of impurities, we choose one impurity site, then try to move the impurity to a site which is chosen from the unoccupied nearest-neighbor sites by obeying the Monte Carlo criterion. The trial site can be any one of the nearest-neighbor sites (e.g., one of the 12 sites in the FCC case) under three-dimensional spinodal decomposition. Figure 1.25 illustrates the simulation results of three-dimensional spinodal decomposition by the Monte Carlo method for (Zn, Cr)Te. Only Cr atoms in (Zn, Cr)Te after 100 Monte Carlo steps are shown in the figures. Nearest-neighbor Cr atoms are connected with bars. Note that Zn and Te atoms exist in the white region. In this simulation, the $14 \times 14 \times 14$ conventional FCC super-cells are taken as a simulation box. Simulations are performed at a scaled temperature of $k_B T/|V_{01}| = 0.5$, where k_B is the Boltzmann constant and V_{01} is the nearest-neighbor effective pair interaction. This corresponds to 919 K and 340 K for (a) $(Zn_{0.95},Cr_{0.05})Te$ and (b) $(Zn_{0.80},Cr_{0.20})Te$, respectively. The simulations are started from completely random configurations and the systems are quenched at the temperature $T = 0.5|V_{01}|/k_B$. As shown in Fig. 1.25(a), in the Cr 5% concentration case, at each Monte Carlo step the Cr atoms form more isolated clusters compared to the random configuration. On the other hand, the Cr atoms form large clusters which extend to the whole crystal for the Cr 20% case, as shown in Fig. 1.25(b). In this case, the large clusters consist of the zinc blende CrTe with half metallic ferromagnetism. It is easy to predict that these complicated configurations of Cr atoms affect the magnetism in the DMS profoundly. We call such spinodal phase *three-dimensional Dairiseki phase*, where Dairiseki means "marble" in Japanese. In fact, the three-dimensional Dairiseki phase has been experimentally observed in (Ga,Mn)N, (Ga,Cr)N, (Al,Cr)N, and (Zn, Cr)Te by using TEM, EDS and EELS analyses.[66,70–72]

The T_Cs of (Zn, Cr)Te under three-dimensional spinodal decomposition calculated by the RPA method[73,74] are shown in Fig. 1.26. The RPA method can take the magnetic percolation effect into consideration exactly, and provide a realistic estimation of T_C. The horizontal axis is the Monte Carlo step corresponding to the annealing time in the crystal growth experiment. Since the calculated results

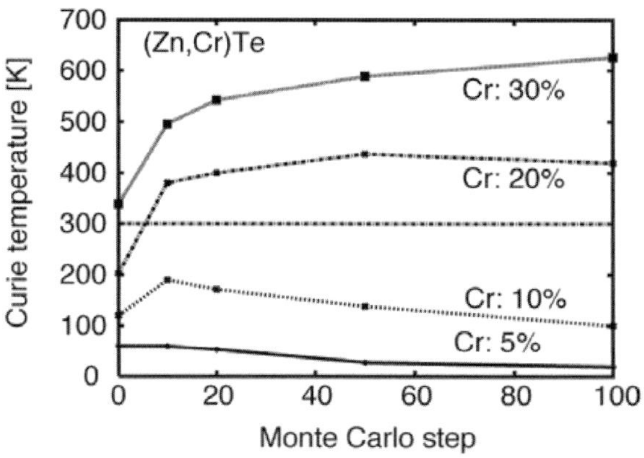

Figure 1.26. Curie temperatures of (Zn, Cr)Te under three-dimensional spinodal decomposition. The horizontal axis is the Monte Carlo step.

depend on the initial configuration of the Cr atoms in the simulation box, we use the 30 spinodal decomposition configurations to obtain a configuration average. As shown in Fig. 1.26, below the percolation threshold, the T_Cs of (Zn, Cr)Te decrease with increasing Monte Carlo step. Such a tendency can be understood from the fact that each spin inside the isolated clusters is aligned parallel, but magnetic interactions cannot work between isolated clusters and spread over the entire crystal, i.e., the superparamagnetism is realized in the low concentration case. On the other hand, in the high concentration region, the T_Cs rise with increasing Monte Carlo step. This is because the Cr atoms can establish magnetic percolating paths when forming large clusters in the high Cr concentration case; therefore, strong ferromagnetic interactions which originate from the first nearest-neighbor atoms can contribute to the high T_C. It is worthwhile comparing the present results with experiments. In experiment, even though the concentration of magnetic impurity is low, the high T_C is observed in DMS with the Diriseki phase. For instance, Kuroda *et al.* observed room temperature ferromagnetism in inhomogeneous n-type $(Zn_{0.95},Cr_{0.05})Te$.[66] In our simulation, the increase of T_C is only remarkable for high concentrations. Therefore, these simulations (three-dimensional spinodal decomposition) are not sufficient to explain the experimental results.

1.3.6 Two-dimensional Spinodal Decomposition (Konbu Phase)

Next, simulations of non-equilibrium layer-by-layer crystal growth with two-dimensional spinodal decomposition are introduced.[67] In these simulations, atomic diffusion is restricted to the surface, and only the nearest-neighbor sites on the same surface can be a candidate for a trial site in the layer-by-layer simulation [one of the four sites in the FCC (100) plane]. Moreover, ideal layer-by-layer growth is assumed, i.e., after the annealing process of the first layer, the resulting configuration of the first layer is fixed. Then, we deposit the second layer and start the annealing only for the second layer. Repeating this process for the required number of layers, we can obtain the spinodal decomposition phase under the layer-by-layer growth condition. Figure 1.27(a) shows the simulation result of layer-by-layer simulation with two-dimensional spinodal decomposition for (Zn, Cr)Te after 400 Monte Carlo steps. The impurity concentration is fixed at 5%, and the simulation box is the $16 \times 16 \times 16$ conventional FCC cell. Simulations are performed at a scaled temperature of $k_B T / |V_{01}| = 0.5$. As shown in Fig. 1.27(a), in contrast to the three-dimensional spinodal decomposition cases, characteristic quasi-one-dimensional structures appear when layer-by-layer growth is assumed. Due to the attractive interactions, magnetic impurities favor gathering together. Under the layer-by-layer condition, magnetic impurities cannot move out from the initial planes; thus, the shape of the cluster is spontaneously controlled, resulting in the quasi-one-dimensional shape. We call this quasi-one-dimensional phase "one-dimensional Konbu phase", where Konbu means seaweed in Japanese. By choosing temperature and growth speed, we can expect this kind of one-dimensional structure even though Cr concentration is low. The present simulations completely ignore migration barriers, and hence they do not give quantitative predictions, but they are describing some of the relevant physics qualitatively in DMS crystal growth. In fact, one-dimensional structures along the crystal growth direction with high-T_C have been observed experimentally in (Al, Cr)N (Ref. 70) and GeMn systems.[67]

The RPA calculation that predicts T_C for the quasi-one-dimensional Konbu phase is estimated to be about 20 K. Long chains of magnetic impurities exist; however, the magnetic network is only along the crystal growth direction and the magnetic correlations between Konbu phases are very weak. In the RPA calculations, it is assumed that the classical isotropic Heisenberg model can describe the magnetism of the DMS well. As is well known, the ferromagnetism is suppressed in the perfect one-dimensional isotropic Heisenberg model.[76] Therefore, T_C of the

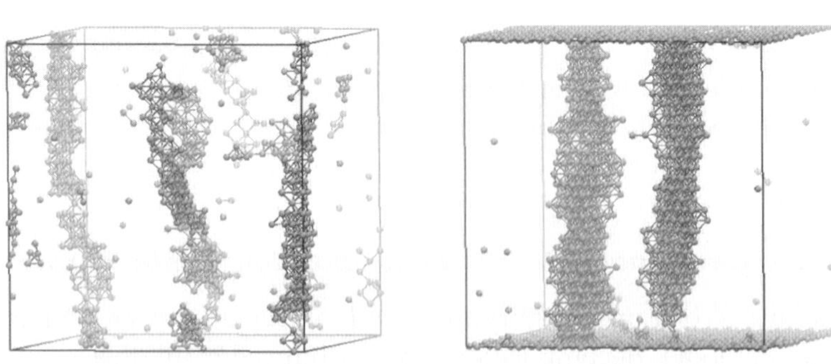

Figure 1.27. (a) Snapshots of the simulation of two-dimensional spinodal decomposition in Cr 5% doped ZnTe (Konbu phase). (b) Konbu phases are combined by the delta doping phase.

quasi-one-dimensional Konbu phase is low. If it is possible that due to some fluctuation the Konbu phases intersect each other somewhere in the crystal, the system is really ferromagnetic because the magnetic interaction can extend to the whole crystal. This idea leads us to a new materials design for high-T_C DMS by using delta doping method. In the delta doping, dopant atoms are confined to a single atomic plane within a semiconductor and are concentrated on one or two atomic layers in a crystal plane. The case of (Zn, Cr)Te with delta doping is illustrated in Fig. 1.27(b). In this simulation, the impurity concentration of the Konbu region is 5%. Artificially inserted one delta-doping layer which contains many magnetic impurities (surface density is 80%) links the quasi-one-dimensional Konbu phases together. The calculated T_C for this artificial structure is as high as 346 K.

1.3.7 Co-doping Method for Controlling Spinodal Decomposition

Spinodal decomposition, as explained in the previous chapters, can be a useful tool in forming nano-structures in DMS systems via self-organization. In the following sub-sections, by using this effect some applications of spinodal decomposition to fabricate functional materials, such as a DMS with high-blocking temperature, a prototype of a spintronics device and a spincaloritronics device, will be presented. However, the synthesis of homogeneous DMS with high T_C is still important and, for this purpose, we will discuss the co-doping method to control spinodal decomposition. In Sec. 1.2.3, it has been shown that the absence of percolation prevents ferromagnetism in homogeneous DMS systems for low concentrations. By controlling spinodal decomposition and suppressing the phase separation, we can dope magnetic impurities homogeneously in semiconductors up to high concentrations.

The doping problem also appears in the valence control of wide band-gap semiconductors. For example, it is well known that n-type doping is easy in ZnO, while p-type doping is extremely difficult. This is called unipolarity and prevents some of practical applications of wide band-gap semiconductors. Based on first-principles calculations, Yamamoto and Katayama-Yoshida have proposed the co-doping method to overcome this difficulty.[77] According to their calculations, by introducing compensating impurities in addition to the dopant impurities, it is possible to enhance the solubility of the dopant impurities, reduce activation energy and improve mobility. We develop this idea to overcome the solubility limit of magnetic impurities in DMS systems. Normally, magnetic impurities act as acceptors in DMS; we consider donor impurities as co-dopants.

In order to see the enhancement of the solubility by co-doping, we calculate the mixing energy of magnetic impurities given the existence of co-dopant impurities. In the case of (Ga, Mn)As, we have chosen the O donor as a co-dopant, and we calculate the mixing energy (ΔE) from the total energies (TE) of $(\mathrm{Ga}_{1-x}, \mathrm{Mn}_x)(\mathrm{As}_{1-y}\mathrm{O}_y)$, $\mathrm{Ga}(\mathrm{As}_{1-y}\mathrm{O}_y)$ and $\mathrm{Mn}(\mathrm{As}_{1-y}\mathrm{O}_y)$, as $\Delta E = TE[(\mathrm{Ga}_{1-x}, \mathrm{Mn}_x)(\mathrm{As}_{1-y}\mathrm{O}_y)] - TE[\mathrm{Ga}(\mathrm{As}_{1-y}\mathrm{O}_y)] - TE[\mathrm{Mn}(\mathrm{As}_{1-y}\mathrm{O}_y)]$. By definition, positive

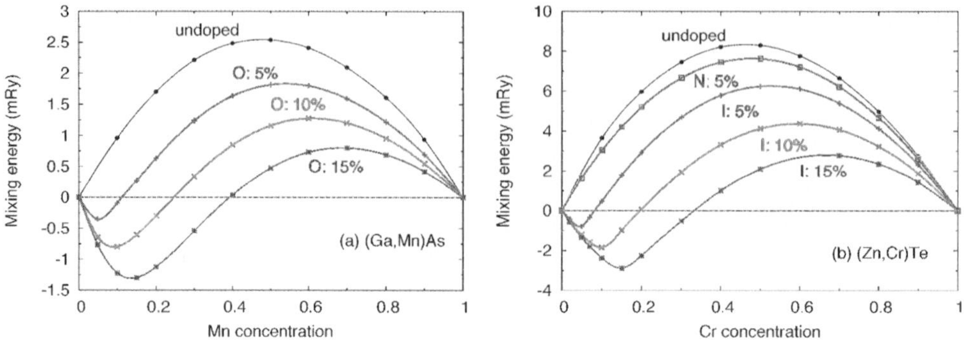

Figure 1.28. Calculated mixing energy of (a) (Ga, Mn)As and (b) (Zn, Cr)Te. As co-dopants, oxygen and iodine are chosen, respectively. For (Zn, Cr)Te, the N acceptor is also considered as a co-dopant.

ΔE indicates that the system has a tendency towards spinodal decomposition. The lattice relaxation is not taken into account, and we assume the lattice constant of the respective host materials.

Figure 1.28 shows the calculated mixing energy of (Ga, Mn)As and (Zn, Cr)Te with/without the co-doping treatment by substitutional O and substitutional I, respectively. For (Zn, Cr)Te we also calculated the co-doping effect by N acceptors. First of all, the mixing energy without co-doping shows strong convexity as a function of the concentration of magnetic impurities. This is consistent with experimental observations, namely, that DMS systems are easily decomposed and careful tuning of crystal growth conditions is necessary. The effect of co-doping is significant in both compounds. By introducing donor impurities the mixing energy is lowered and spinodal decomposition is suppressed. Thus, it is found that co-doping enhances the impurity solution and promotes a homogeneous distribution of magnetic impurities. Roughly speaking, the minimum of the mixing energy appears at around the concentration of the co-dopants. For the low concentrations, negative mixing energies are predicted and uniform mixing is expected under the thermal equilibrium condition. In contrast to this drastic change in the mixing energy due to the compensating co-dopants, acceptor co-doping by N into (Zn, Cr)Te has almost no effect, as shown in Fig. 1.28. This observation indicates that the energy lowering due to the co-doping with the compensating impurities originates from the bonding energy between the acceptor states (in the present cases, Mn and Cr) and the donor states (O and I, respectively).

The co-doping effect is also confirmed in the calculations of the pair interactions between magnetic impurities. Figure 1.29(a) shows the calculated pair interactions between Cr atoms in (Zn, Cr)Te. As shown in the figure, without co-doping, strong attractive interactions happen between Cr atoms and this is why the system shows spinodal decomposition. By introducing the co-dopant I, the interactions become repulsive, which means the Cr atoms favor homogeneous mixing. However, the N co-doping does not change the interactions so much, as shown in Fig. 1.29(c), and this is consistent with the calculation of the mixing energy shown in Fig. 1.28. In Figs. 1.29(d), (e) and (f), the Cr distributions are generated from the calculated pair interactions by using Monte Carlo simulation. It is clearly shown that the decomposition is strongly suppressed by co-doping. In principle, the same behavior is observed in the other DMS systems, such as (Ga, Mn)N and (Ga, Cr)N. Thus, we can expect that the co-doping method can be adopted generally to enhance the solubility limit in DMS systems and achieve high-T_C by overcoming the percolation threshold.[78]

1.3.8 Super-paramagnetic Blocking Phenomena

So far, we have seen that the DMS system is decomposed into two phases, the high concentration region and the low concentration region, due to spinodal decomposition. If the concentration of magnetic impurities is high enough, the high concentration region can spread all over the crystal resulting in high-T_C ferromagnetism, but for low concentrations the magnetic impurities form small clusters, whose crystal structure is coherent to the host material. The clusters are independent magnetically, and therefore, the whole system shows superparamagnetism. Thus, spinodal decomposition suppresses the ferromagnetism of DMS for low concentrations.

However, when the size of the clusters formed due to spinodal decomposition is large, the relaxation time of the magnetization can be quite long, leading to a hysteretic behavior in the magnetization process. This is called the *blocking phenomena* and originates from magnetic anisotropy. This effect is simulated by using the Monte Carlo method to discuss how clustering affects the hysteresis of the decomposed DMS systems via the superparamagnetic blocking phenomena.[48]

To simulate the magnetization process by using the Monte Carlo method, we employ the algorithm proposed by Dimitrov *et al.*[58] We have assumed uniaxial magnetic anisotropy. The hysteresis loops for a homogeneous and inhomogeneous (Konbu phase) (Zn, Cr)Te are simulated at the temperatures $T = 0.5T_C$, T_C and $2T_C$, where T_C (= 100 K) is the Curie temperature of the homogeneous (Zn, Cr)Te. The average concentration of Cr is 5% for both cases and we use a super-cell that is a $14 \times 14 \times 14$ conventional FCC unit cell. Notice that the Curie temperature of decomposed (Zn, Cr)Te is very low. The results are shown in Fig. 1.30. As shown in the figure, for the homogeneous (Zn, Cr)Te, which shows a square hysteresis loop below T_C, the hysteresis loop closes at T_C and the system shows a paramagnetic response above T_C.

On the other hand, the hysteresis phenomena of (Zn, Cr)Te is strengthened in the Konbu phase. As shown in Fig. 1.30(b), the width of the loop becomes wider in the Konbu phase compared to the homogeneous phase. Moreover, even at high temperature the ferromagnetic response still survives and the remanent magnetization is observed. This indicates that we can optimize the ferromagnetism by controlling spinodal decomposition. In the present simulation, there are ~550 Cr impurities in the super-cell and the size of the super-cell limits the

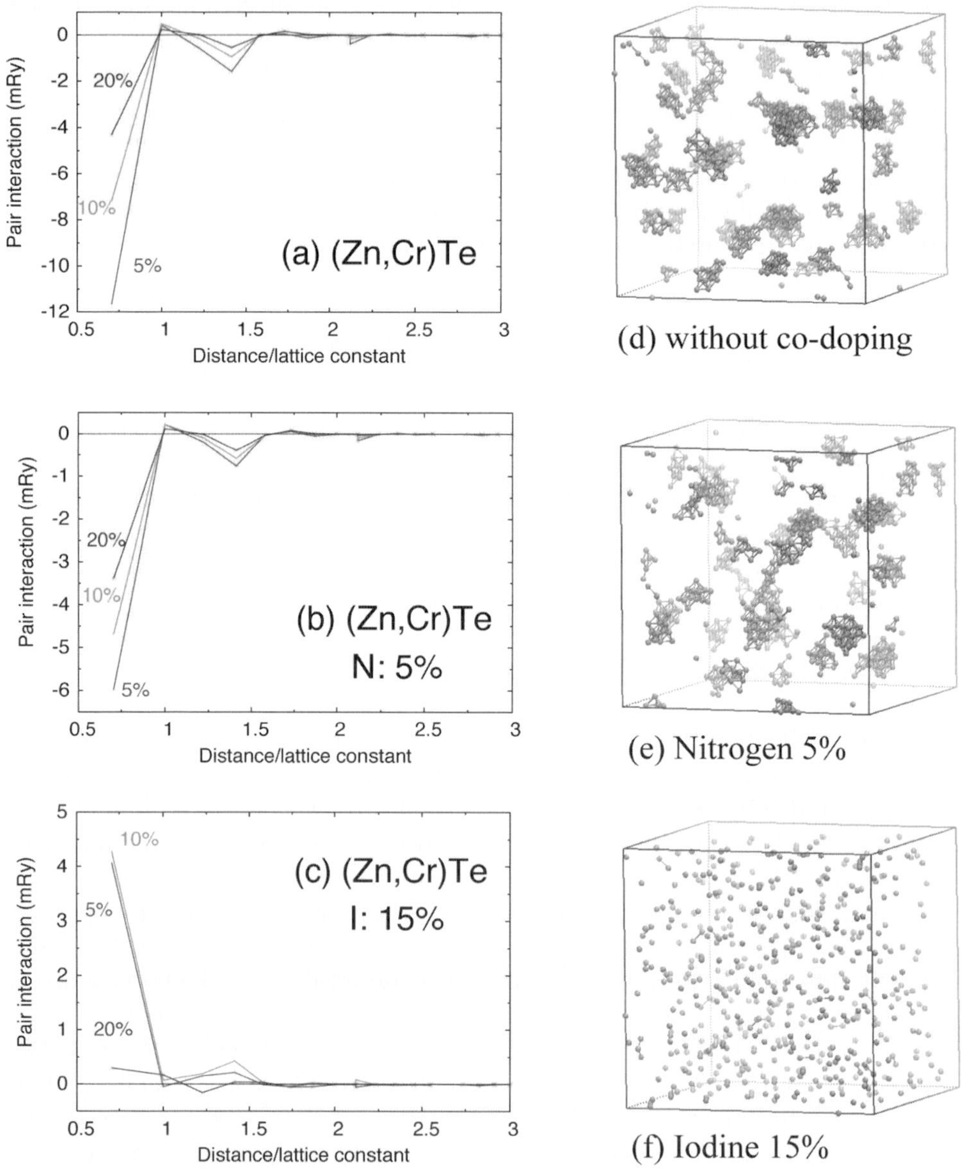

Figure 1.29. Calculated pair interactions between Cr atoms in (Zn, Cr)Te in the case of (a) without co-doping, (b) with N co-doping and (c) with I co-doping. (d), (e) and (f) show snapshots of the Cr distribution in (Zn, Cr)Te generated by using the interactions calculated in (a), (b) and (c), respectively. Green spheres indicate Cr positions in the host matrix.

maximum size of the clusters. It is known that the blocking temperature is proportional to the activation energy to flip the magnetization, and the activation energy is proportional to the volume of the cluster. Thus, we can expect rather high blocking temperature in real DMS systems if large clusters are formed by spinodal decomposition.

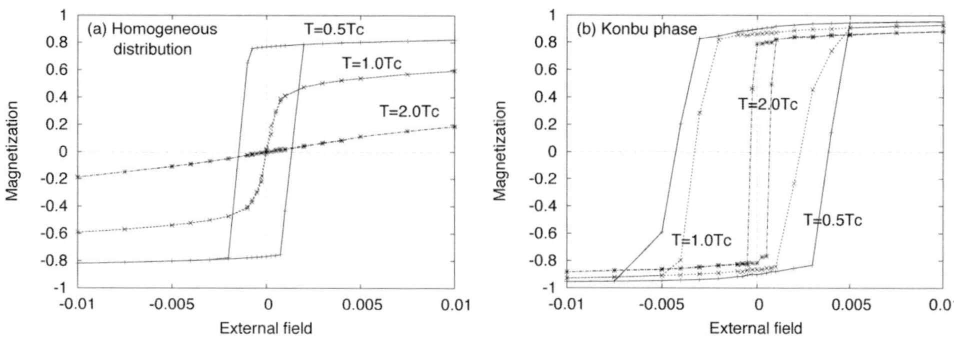

Figure 1.30. Monte Carlo simulations of the hysteresis loops for (a) homogeneous (Zn, Cr)Te and (b) (Zn, Cr)Te in the Konbu phase (inhomogeneous phase). For both cases, the average Cr concentration is 5%, and the simulations are performed at the temperatures $T = 0.5T_C$, T_C and $2T_C$, where T_C is the Curie temperature of the homogeneous (Zn, Cr)Te. Magnetization is normalized by the saturation magnetization, and the external field is normalized by the nearest-neighbor exchange interaction strength.

As already pointed out, normally the DMS systems have a strong tendency towards phase separation. Therefore, depending on the details of the experimental conditions the phase separation occurs uncontrollably and depending on the stage of the phase separation the system shows various magnetic properties (superparamagnetism, low-T_C, high-T_C, and so on). This consideration reasonably explains the reason for the experimental controversy about the ferromagnetism in wide band-gap semiconductor based DMS systems.

1.3.9 Applications of the Konbu Phase

As an application of the one-dimensional Konbu phase, terabit-density nanomagnets fabricated by bottom-up nanotechnology such as self-organized two-dimensional spinodal decomposition are introduced in this sub-section. As is well known, in practical use dynamic random access memory (DRAM) utilizes small capacitors which basically consist of two small metal plates separated by a thin insulator as a memory element. On the other hand, next-generation magnetic random access memory (MRAM) based on the TMR effect can realize a higher read and write speed than those of the DRAM. However, high integration of MRAM with the current technology is very difficult due to crucial problems such as the half-select, wiring, and low output voltage. Here, we show that Terabit-density nano-magnets fabricated by bottom-up nanotechnology like self-organized two-dimensional spinodal decomposition may possibly be used as a memory element and thereby solve such fundamental crucial problems. To do this, it is necessary to control the shape, growth position, and density of the quasi-one-dimensional Konbu phase in the DMS.

Firstly, we introduce a control simulation of the growth positions of the Konbu phase in the DMS by using periodic atomic patterns, which consist of magnetic atoms on the semiconductor substrate before the crystal growth ("nanoscale seeding"). After nanoscale seeding, layer-by-layer crystal growth simulation is performed by Monte Carlo simulation. Figure 1.31(a) shows the simulation result of growth position control. As shown in Fig. 1.31(a), Cr atoms in (Zn, Cr)Te form nanomagnets through nanoscale seeding via a 13-atom array of Cr atoms with an area of 1.44 nm^2. It can be clearly seen from Fig. 1.31(a) that the control of the growth positions of the nano-magnets by nanoscale seeding is almost perfect. The quasi-one-dimensional Konbu phase grows straight along the crystal growth direction when the vapor pressure or Cr concentration is constant during the layer-by-layer crystal growth.

The next requirement is to control the shape of the Konbu phase in the DMS. In this simulation, the shapes of the Konbu phase are controlled by changing the Monte Carlo step, Cr concentration, and growth temperature during the layer-by-layer crystal growth. These simulation parameters correspond to growth rate, vapor pressure of Cr atoms, and temperature in the experiment, respectively. Figure 1.31(b) shows the result of the shape control simulation of terabit density nanomagnets in (Zn, Cr)Te. Before the crystal growth, nanoscale seeding, which consists of a 25-Cr-atom array and covers 3.24 nm^2, is implemented to control the position of the nano-magnets. As shown in Fig. 1.31(b), the shape control of the quasi-one-dimensional Konbu phase in (Zn, Cr)Te is performed by reducing (lower black arrow) and increasing (upper black arrow) the vapor pressure of Cr atoms during the layer-by-layer crystal growth. The controls of the growth positions and shape of the terabit density nano-magnets in (Zn, Cr)Te appear to

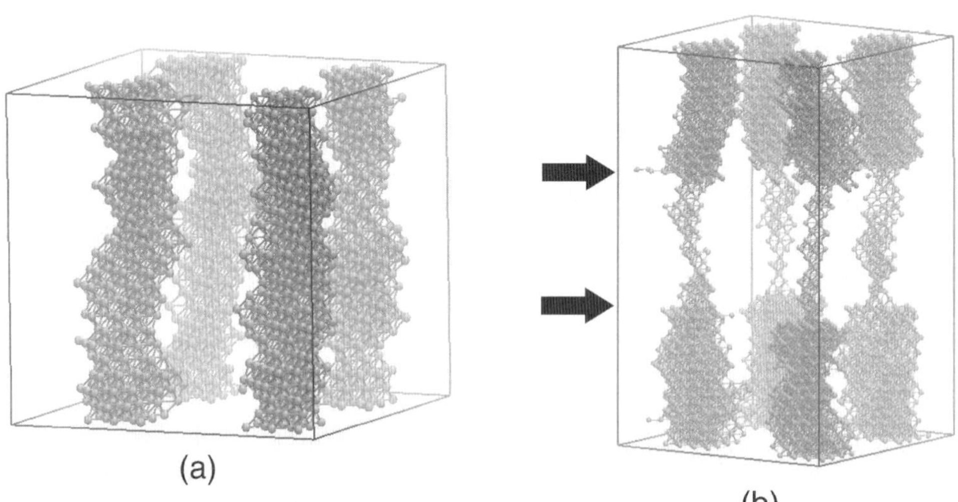

Figure 1.31. (a) Control of the growth positions of the Konbu phase in (Zn, Cr)Te by "Nanoscale Seeding" on the semiconductor substrate. (b) Shape control of the Konbu phase in (Zn, Cr)Te by changing the Monte Carlo step, Cr concentration, and scaled temperature during the layer by layer crystal growth.

work very well. Thus, the blocking temperature originating from the magnetic crystalline anisotropy or shape anisotropy in the terabit density nano-magnet can be controlled by changing the number of Cr atoms. Based on the simulation results presented in this sub-section, we suggest that terabit density nano-magnets along the crystal growth direction can be generated by using layer-by-layer crystal growth with two-dimensional spinodal decomposition, and the position and shape of nano-magnets can be controlled by using lithography on the semiconductor substrate and by changing the vapor pressure or concentration of the magnetic impurities during thermal non-equilibrium crystal growth, such as MBE, MOVPE, or MOCVD.

Finally, a prototype of the spintronics device based on the terabit density nano-magnets is introduced. In our prototype device, (Ga, Cr)N and (Ga, Al)N are alternately grown by using the layer-by-layer crystal growth technique, as shown in Fig. 1.32. Additionally, the shape of central (Ga, Cr)N is controlled by changing the vapor pressure and concentration of the Cr atoms. The (Ga, Al)N layers work as an insulating barrier. Under these conditions, we can the control spin directions of only central (Ga, Cr)N by applying the magnetic field or spin current because the coercive forces in the central (Ga, Cr)N region and the other (Ga, Cr)N regions are not the same value. Therefore, the TMR effect, which depends on the spin direction of the central (Ga, Cr)N, can be observed, and it can be expected that this device would work as a memory element of MRAM. The problem of wiring is solved by using self-organization. Furthermore, as mentioned above, since the

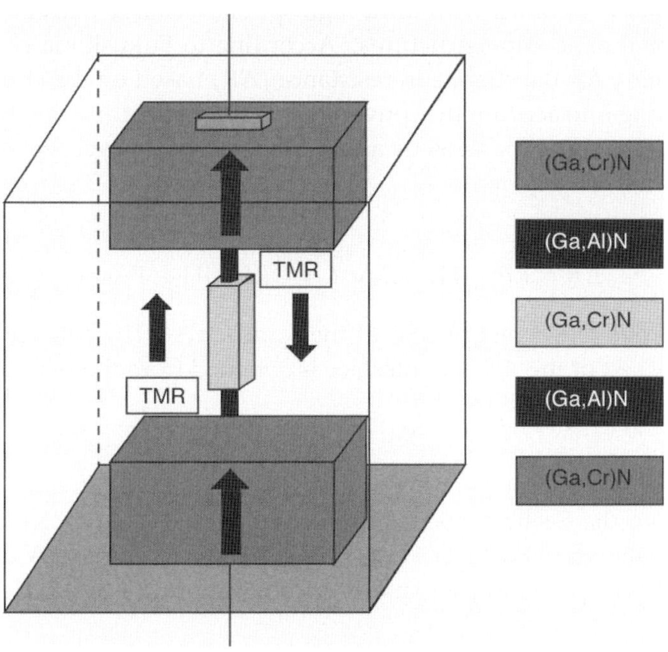

Figure 1.32. The prototype of the spintronics device based on the Konbu phase in the DMS.

area of this TMR element is quite small, it is possible to fabricate terabit density memory.

1.3.10 Spin Caloritronics Application in Colossal Spin-Entropy Expansion Cooling by the Konbu Phase

In addition to the conventional Peltier effect, we propose a new class of thermoelectric-cooling mechanism based on adiabatic spin-entropy expansion in a quasi-one-dimensional nano-superstructure (*Kobu phase*) by injecting the spin current from the ferromagnetic metal to the paramagnetic one.[79] The spin-entropy expansion mechanism dominates and enhances the thermoelectric-cooling power dramatically in CPP-GMR (*current perpendicular to plane giant magneto-resistance*) Co/Au nano-interface. Based upon the spin-entropy expansion mechanism, we can design new thermoelectric-cooling nano-superstructures using the newly-designed half-Heusler ferromagnet NiMnSi ($T_C = 1050\,\mathrm{K}$)[80] and the self-organized quasi-one-dimensional ***Konbu phase*** (Zn, Cr)Te with very high blocking temperature ($> 1000\,\mathrm{K}$) by spinodal nano-decomposition.

In order to consider the current direction of the *R–I* shifts, we take into account that the shift is caused by adiabatic spin-entropy expansion cooling in addition to the conventional Peltier effect in the *Konbu phase* or CPP nano-superstructure interface. When the spin current flows through a CPP nano-superstructure interface, Joule heating (RI^2), spin-entropy expansion cooling ($T\Delta\sigma I$), and conventional Peltier cooling (ΠI) occur at the same time. The spin-entropy expansion cooling term becomes dominant in the adiabatic spin injection in the extreme and thermal non-equilibrium conditions with ultra-high spin and charge current in the quasi-one-dimensional nano-superstructures. According to Fukushima *et al.*,[81] we can extend the theory for the change in resistance (ΔR) based on the charge and spin current by taking into account the conventional Peltiereffect and newly-introduced adiabatic spin-entropy expansion by assuming a constant heat capacity of the CPP elements. ΔR is described as

$$\Delta R \propto RI^2 - \Pi I - T\Delta\sigma I = I(RI - [\Pi + T\Delta\sigma]), \tag{1.1}$$

if the heating and the cooling occur at the same place. The conventional Peltier coefficient $\Pi_{\mathrm{Au/Co}}$ of the Au/Co interface is obtained from the relation

$$\Pi_{\mathrm{Au/Co}} = (S_{\mathrm{Au}} - S_{\mathrm{Co}})T = 9.8\,\mathrm{meV} \quad (\text{at } T = 300\,\mathrm{K}), \tag{1.2}$$

where S_{Au} and S_{Co} are the Seebeck coefficients of Co and Au, and T is the temperature. From the Seebeck coefficients of bulk Co ($-30.8\,\mu\mathrm{V/K}$) and bulk Au ($1.9\,\mu\mathrm{V/K}$), as shown in Eq. (1.2), $\Pi_{\mathrm{Au/Co}}$ is estimated to be 9.8 meV at 300 K.

The adiabatic spin-entropy expansion cooling term ($T\Delta\sigma$) is defined as follows:

$$T\Delta\sigma = Tk_{\mathrm{B}}\ln(2) = 17.8\,\mathrm{meV} \quad (\text{at } T = 300\,\mathrm{K}). \tag{1.3}$$

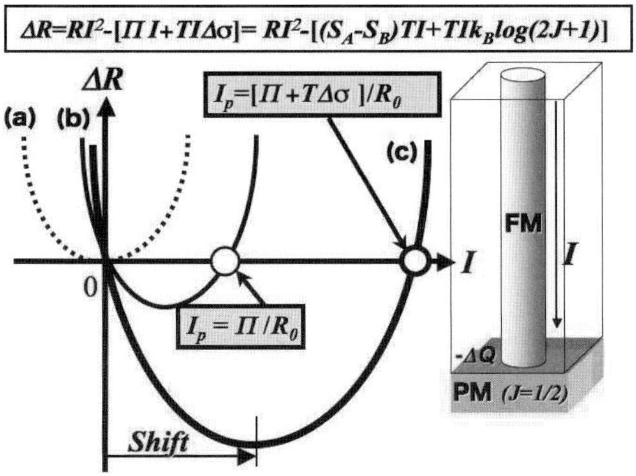

Figure 1.33. Typical R–I curves of a CPP-GMR element. Spin injection is from the ferromagnetic (FM) nano-wire to the paramagnetic (PM) metal contact with total spin J. (a) Joule heating (RI^2) term only (dotted line), (b) Joule heating (RI^2) and conventional Peltier effect (ΠI) terms (thin solid line), (c) Joule heating (RI^2), a conventional Peltier effect term (ΠI), and a spin-entropy expansion term ($T\Delta\sigma I$) (thick solid line). R is the zero-current resistance. The definition of the shift is shown in the figure.

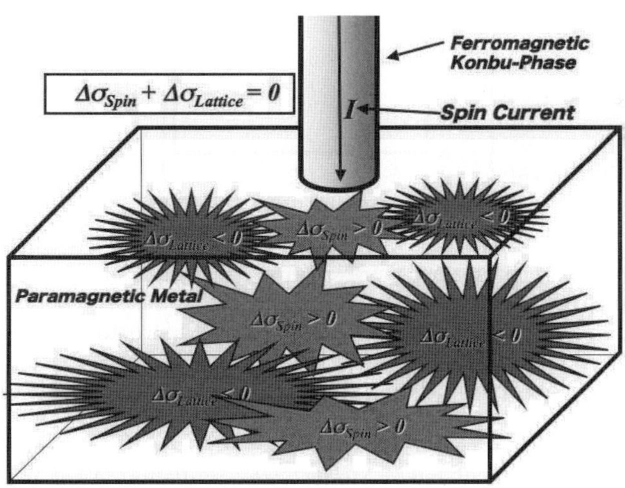

Figure 1.34. Schematic explanation of adiabatic spin-entropy expansion caused by spin injection from the ferromagnetic Konbu phase to the paramagnetic metal contact. The change of the summation of spin entropy ($\Delta\sigma_{\text{Spin}} > 0$) and lattice entropy ($\Delta\sigma_{\text{Lattice}} < 0$) is zero. Continuous spin injection and spin-entropy expansion take the heat from the lattice system.

Comparing the physical values of the Peltier coefficient in Eq. (1.2) and the spin-entropy expansion cooling term in Eq. (1.3) at 300 K, we should emphasize that the spin-entropy expansion cooling term is almost two times larger than the conventional Peltier effect.

Equation (1.1) indicates that heating and cooling compensate each other at $I = [\Pi + T\Delta\sigma]/R_0(\equiv I_p)$. Here, R_0 is the zero-current resistance. As shown in Fig. 1.33, I_p is defined as the cross-point of the $R - I$ curves at $\Delta R = 0$. Equation (1) leads to the product of RI_p equaling $[\Pi + T\Delta\sigma]$. $[\Pi + T\sigma]$ is determined

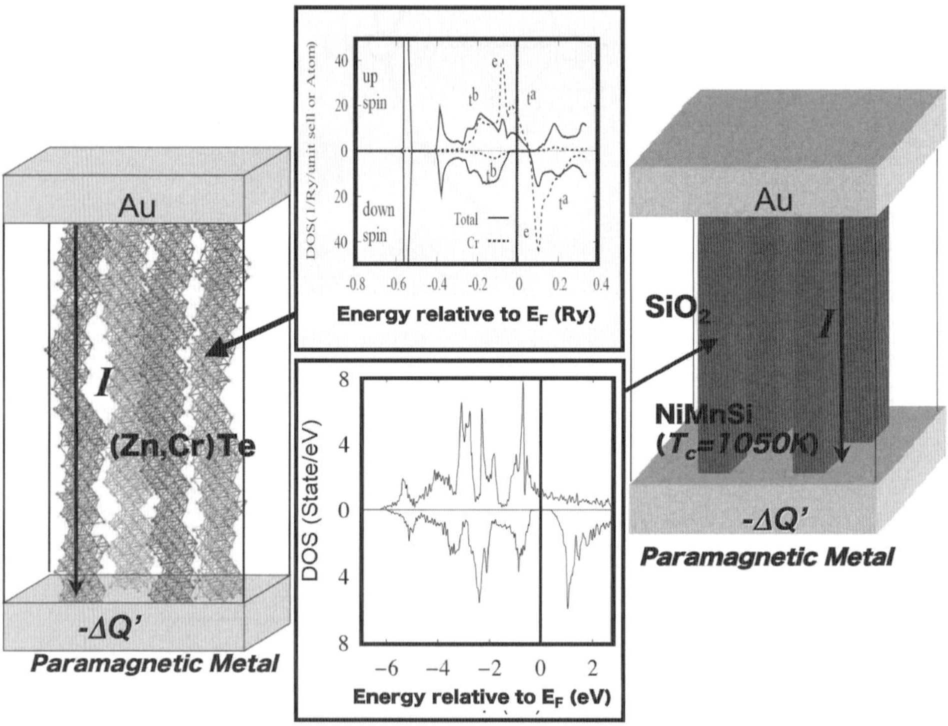

Figure 1.35. The designed thermoelectric-cooling nano-superstructure by adiabatic spin-entropy expansion from the half-Heusler ferromagnets, NiMnSi, ($T_C = 1050$ K) to paramagnetic metals, Au, in the quasi-one-dimensional nano-superstructures. The spaces between the nano-superstructure are made by SiO_2. The inserted density of states (DOS) indicates the half-metallicity (100% spin-polarized ferromagnets). The structure and magnetic properties are predicted through the calculation of the phase stability, equilibrium lattice constant (10.3291 au), electronic structure, magnetic exchange interaction (J_{ij}) and Curie temperature of $T_C = 1050$ K is calculated by Monte Carlo simulation. Applying thehigh density of charge and spin current from upper- to lower-side between the Au contacts, the upper Au part is heated up and the lower Au part is cooled down.

The *Konbu phase* sandwiched by the Au contacts is designed by self-organized two-dimensional (2D) spinodal nano-decomposition in the layer-by-layer crystal growth simulation Monte Carlo method based upon the *ab initio* calculation of chemical pair interactions between Cr–Cr, Zn–Cr, and Zn–Zn pairs. Averaged Cr concentration is 20% in (Zn, Cr)Te. Growth-position control of the *Konbu phase* is achieved by the seeding of the periodic 13 Cr atoms on the Au contact and substrate. The pink atom corresponds to Cr. The regions between the nano-superstructure columns are pure ZnTe. The inserted density of states (DOS) of the high Cr concentration region caused by 2D spinodal nano-decomposition indicates the half-metallicity (100% spin-polarized ferromagnets).

by the combination of the two materials at the Peltier-cooled Co/Au interface and the "natural value" from adiabatic spin-entropy expansion ($T\Delta\sigma = Tk_B \ln(2) = T \cdot 60\,\mu V/K$). Therefore, these are independent of the size or the resistance of the CPP nano-superstructure. Based upon the experimental data[81] for CPP Co/Au interfaces from different experimental groups, the characteristic experimental values of R_0I_p are 22.5 meV,[81] 27.9 meV,[82] 18.4 meV,[83] and 23.0 meV,[84] and these experimental values are very good agreement with the theoretically calculated value of 27.6 meV [= 9.8 meV (*conventional Peltier effect term*) + 17.8 meV (*spin-entropy-current injection-cooling term*)] at 300 K. Based upon our new mechanism and the above experimental data analyses, the thermoelectric-cooling power (RI_p^2) is eight times [$(27.6\,\text{meV}/9.8\,\text{meV})^2 = 8$] larger than the conventional Peltier effect alone. Therefore, we conclude that the major thermoelectric-cooling mechanism in the nano-superstructures of the CPP Co/Au interface is the adiabatic spin-entropy expansion mechanism, and the conventional Peltier effect alone makes a minor contribution to the thermoelectric cooling in CPP nano-superstructures. The thermoelectric-cooling power per unit area of CPP element ($10^5 \sim 10^6\,W/cm^2$)[81–84] is much larger than that of conventional thermoelectric materials ($\sim 5\,W/cm^2$).

1.4 ZnO AND GaN-BASED DMS BY SIC-LDA

The Kohn–Sham approach to the density-functional theory (DFT-KS) with the local-density approximation (LDA) is a surprisingly successful method to investigate electronic structures of various classes of materials. However, the LDA is known to have systematic failings, such as the overbinding error for solids and molecules, and the underestimation of band-gaps for semiconductors. In the calculations for dilute magnetic semiconductors (DMS), there is concern that the calculated electronic structures may contain errors caused by the LDA. In fact, if one compares the density of states calculated in the LDA with experimental photoemission spectroscopy data, discrepancies are found particularly in the width of the band-gap of the host semiconductors and the eigenvalues of the *d* orbitals of the transition-metal magnetic impurity atoms. For accurate investigation of DMS systems, it is necessary to use an approach which improves on such errors introduced by the LDA.

In this section, we make a simple correction to the DFT-KS-LDA calculation scheme. The method we employ is called the pseudopotential-like self-interaction correction (pseudo-SIC) method, which was originally developed by Filippetti and Spaldin.[85] Although the original pseudo-SIC method is implemented in a plane-wave pseudopotential code, we have modified and combined it with a code based on the Korringa–Kohn–Rostoker (KKR) method.[86] This is because the KKR method combined with the coherent-potential approximation (CPA) is very useful for investigating disordered systems like DMS systems.

In the next section, the formalism of our modified pseudo-SIC method is introduced. The calculation results for wide band-gap DMS (ZnO- and GaN-based) are given in Secs. 1.4.2 and 1.4.3, respectively.

1.4.1 Pseudo-Self-Interaction-Corrected Local Density Approximation (SIC-LDA) Method

In the DFT-KS-LDA scheme, the states considered are those of the non-interacting particles moving in a one-body effective potential. The Coulomb interaction acting among the real electrons is taken into account through the effective potential, which is given by the variation of the electron–electron interaction energy as

$$E_{ee}[n] = U[n] + E_{\mathrm{XC}}[n]. \tag{1.4}$$

Here, the first term is the Hartree (electrostatic Coulomb interaction) term,

$$U[n] = \frac{1}{2}\int\int \frac{n(\vec{r})n(\vec{r})}{|\vec{r}-\vec{r}'|} d^3r d^3r', \tag{1.5}$$

and the second term is the exchange-correlation term. Since the exact description as a functional of local densities is not known, the exchange-correlation term is replaced by an approximated functional, which is often obtained by using the LDA.

The exact electron–electron interaction energy E_{ee} with the exchange-energy E_{XC} must give exactly zero energy whenever the density $n(\boldsymbol{r})$ represents a density of one-electron states. In other words, the Hartree term and the exchange-correlation term for one-electron states must cancel each other out. In actual calculations, however, since the approximated LDA exchange-correlation energy $E_{\mathrm{XC}}^{\mathrm{LDA}}[n]$ is generally used, this cancellation condition is only partially complete and spurious interaction energy is produced. This unphysical energy is called the self-interaction error (SIE). The SIE for an orbital labeled by index i, of which the charge density is given by $n_i(\boldsymbol{r})$, becomes

$$\delta_i = U[n_i] - E_{\mathrm{XC}}^{\mathrm{LDA}}[n_i]. \tag{1.6}$$

In the original self-interaction correction (SIC) method by Perdew and Zunger,[87] the unphysical SIE of all the occupied orbitals are simply subtracted from $E_{ee}[n]$:

$$E_{ee}^{\mathrm{SIC}}[n] \equiv U[n] + E_{\mathrm{XC}}^{\mathrm{LDA}}[n] - \sum_{i}^{\mathrm{occ}} \delta_i. \tag{1.7}$$

The SIC total energy is then minimized by solving the following equations:

$$[-\nabla^2 + v_i^{\mathrm{SIC}}]\psi_i(\vec{r}) = \epsilon_i \psi_i(\vec{r}), \tag{1.8}$$

$$v_i^{\mathrm{SIC}} \equiv v^{\mathrm{LDA}}[n] - (u[n_i] + v_{\mathrm{XC}}^{\mathrm{LDA}}[n_i]), \tag{1.9}$$

where $V^{\mathrm{LDA}}[n]$ is the LDA effective potential, $u[n]$ the Hartree potential (given by the variation of U), and $v_{\mathrm{XC}}^{\mathrm{LDA}}[n]$ the LDA exchange-correlation potential (given by the variation of $E_{\mathrm{XC}}^{\mathrm{LDA}}$).

The straightforward application of the SIC method to extended systems, however, requires large computational efforts. Furthermore, since the screening effect on the Coulomb interaction is not taken into account, the full correction by the SIC method generally results in overcorrection e.g., the band-gaps become too large and the eigenvalues of localized orbitals become too low in energy. In the pseudo-SIC method, these problems are worked out by modifying the SIC potential to be expressed in terms of nonlocal projector operators and by introducing a pre-scaling factor to the SIC potential. As already mentioned, we have made modifications to the method and combined it with the KKR-CPA method. In the original implementation, the SIC potential is defined on a set of localized pseudo-atomic orbitals. In our implementation, the SIC potential is calculated by using the angular-momentum-decomposed charge density in the muffin-tin spheres of each site:

$$n_L^i(\vec{r}) = \frac{1}{\pi}\mathrm{Im}\int^{E_\mathrm{F}} G_{LL}^{ii}(\vec{r},\vec{r};\epsilon)\,d\epsilon, \tag{1.10}$$

where $L = (\ell, m)$ is the composite index of angular momentum, i the site index, and $G_{LL'}^{ii}(\vec{r},\vec{r};\epsilon)$ are the Green's functions. The corresponding SIE for an angular momentum channel L at a site i is given as

$$\delta_i = U[\tilde{n}_L^i] + E_{\mathrm{XC}}^{\mathrm{LDA}}[\tilde{n}_L^i]. \tag{1.11}$$

Here, $\tilde{n}_L^i$ is the normalized orbital charge density of the orbital:

$$\tilde{n}_L^i = \frac{n_L^i}{p_L^i}, \tag{1.12}$$

where p_L^i is the orbital occupancy

$$p_L^i \equiv \int_{\mathrm{sphere}} n_L^i(\vec{r})d^3r. \tag{1.13}$$

The SIC potential becomes

$$v_L^{\mathrm{PSIC},i} \equiv v^{\mathrm{LDA}}[n] - \frac{1}{2}p_L^i(u[\tilde{n}_L^i] + v_{\mathrm{XC}}^{\mathrm{LDA}}[\tilde{n}_L^i]), \tag{1.14}$$

where the pre-factor $1/2$ comes from the same philosophy as in the original implementation of the pseudo-SIC method. The total energy becomes the LDA total energy with SIE correction for the electron–electron interaction energy and the double-counting of the effective potential energy:

$$E^{\mathrm{PSIC}} = E^{\mathrm{LDA}} + \frac{1}{2}\sum_i\sum_L\int n_L^i(\vec{r})v_L^{\mathrm{PSIC},i}(\vec{r})d^3r - \sum_i \delta_i. \tag{1.15}$$

1.4.2 ZnO-Based DMSs in SCI-LDA Versus LDA

The electronic structures of ZnO-based DMS are now calculated, where V, Cr, Mn, Fe, Co and Ni are doped as the magnetic impurity. The lattice parameters are fixed to those of the experimental lattice constant of wurtzite-ZnO ($a = 3.249$ Å, $c = 5.204$ Å and $u = 0.3821$).[88] The concentration of transition-metal is 5% and the substitutional disorder is dealt with by CPA. All the magnetic impurity atoms are assumed to replace the host cation (zinc) atoms.

The main feature of the band structure of the host ZnO consists of the O-2p valence band, the Zn-4s conduction band and the Zn-3d semicore-like band. In experiments, the band-gap width of ZnO is 3.4 eV and the Zn-3d semicore band is observed at ~8 eV below the top of the valence band.[89] The calculated density of states (DOS) of V-, Cr-, and Mn-doped ZnO are summarized in Fig. 1.36, and those of Fe-, Co-, and Ni-doped ZnO are in Fig. 1.37. In the LDA total DOS [the upper panels of Figs. 1.36(a), (c), (e) and Figs. 1.37(a), (c), (e)], the band-gap width is ~1 eV and the Zn-3d state is located at 5 ~ 6 eV below the top of the valence band. In the SIC total DOS [the upper panels of Figs. 1.36(b), (d), (f) and Figs. 1.37(b), (d), (f)], the discrepancies are substantially improved. The band-gap width becomes ~2.5 eV, which, although still smaller than the experimental value by ~1 eV, is wider and closer to the experimental value than the LDA value. The Zn-3d semicore band is now correctly located at 7 ~ 8 eV below the host valence band.

Now let us consider the band structure of ZnO-based DMS. The calculation results (in LDA and SIC) and the experimental spectra are compared for several materials for which the photoemission spectroscopy data is available, namely, V-, Mn- and Co-doped ZnO.

The partial DOS of V-3d states in V-doped ZnO is shown in the lower panel of Fig. 1.36(a) for the LDA results, and the lower panel of Fig. 1.36(b) for the SIC results. In both results, the V-3d states appear near the Fermi level. Due to the crystal field of the near-tetrahedrally surrounding ligands, the states show tetrahedral-like splitting. In the LDA partial DOS [Fig. 1.36(a)] the clear splitting between the non-bonding e state and the anti-bonding t_a state is shown. In the majority spin channel, the fully occupied e state is located immediately below the Fermi level and the almost empty t_a state is located above the Fermi level, while the states in the minority spin channel are empty. In the SIC partial DOS, Fig. 1.36(b), the fully occupied e state is located lower in energy, while the other empty states are shifted to the higher energy side. Since SIC produces large splitting between occupied and unoccupied states, the partially occupied t_a state shows another splitting. In photoemission spectroscopy experiment,[90] the peak of the V state is found near the position of the SIC e state, which is ~1.8 eV below the Fermi level. The half-occupied state on the Fermi level is, however, not observed in the experiment. This implies a slight shifting of the Fermi level in experimental samples since ZnO shows n-type conductivity due to the electrons generated by intrinsic defects.

The LDA and SIC DOS for Mn-doped ZnO are shown in Fig. 1.36(e) and in Fig. 1.36(f), respectively. The Mn-d states have a d^5 configuration such that, both

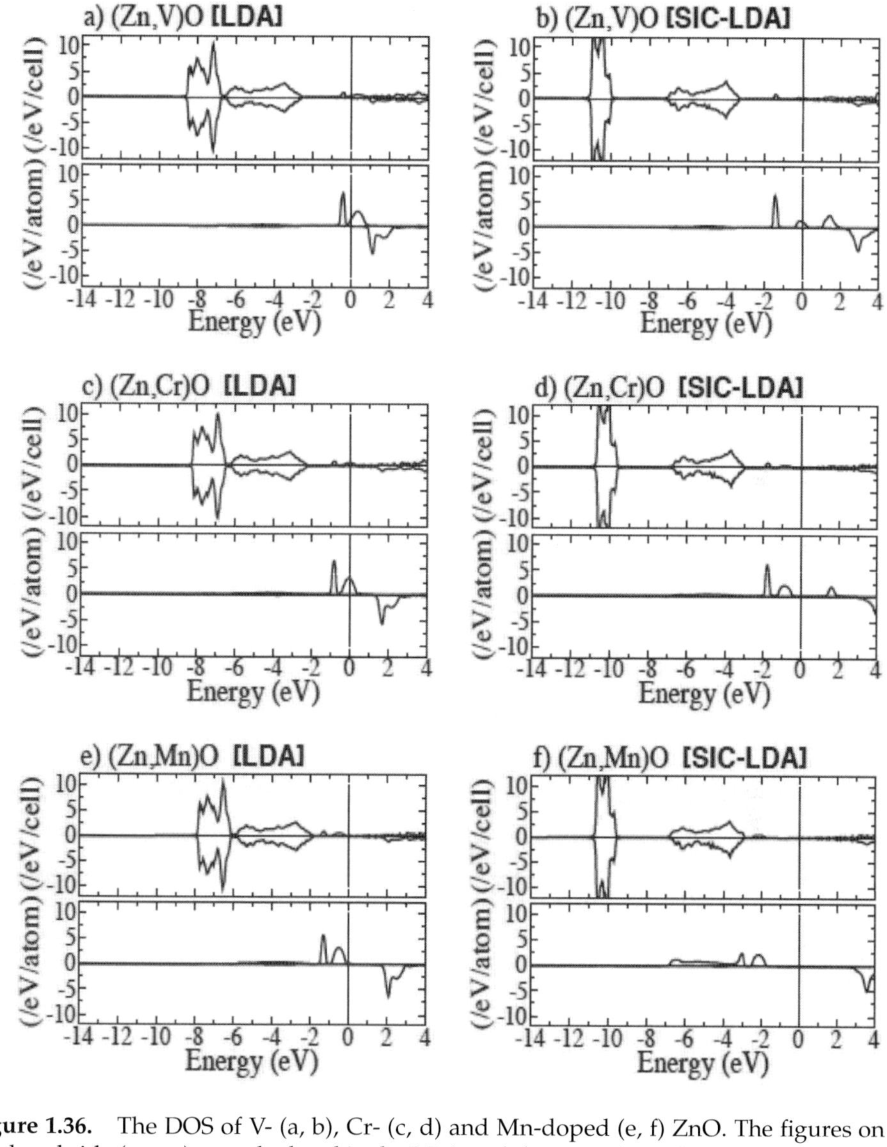

Figure 1.36. The DOS of V- (a, b), Cr- (c, d) and Mn-doped (e, f) ZnO. The figures on the left-hand side (a, c, e) are calculated in the LDA and those on the right-hand side (b, d, f) are in the SIC-LDA. For all figures, the upper panels show the total DOS and the lower panels the partial DOS of transition-metal atoms.

in LDA and SIC, all the majority spin states are occupied and the minority spin states are empty. In one experiment,[91] the Mn-*d* states are found to be strongly hybridized with the host valence band and show a broad spectral structure. The SIC result reproduces well the broad feature of the Mn-*d* states, while in the LDA result the Mn-*d* states are wrongly predicted to be mid-gap states.

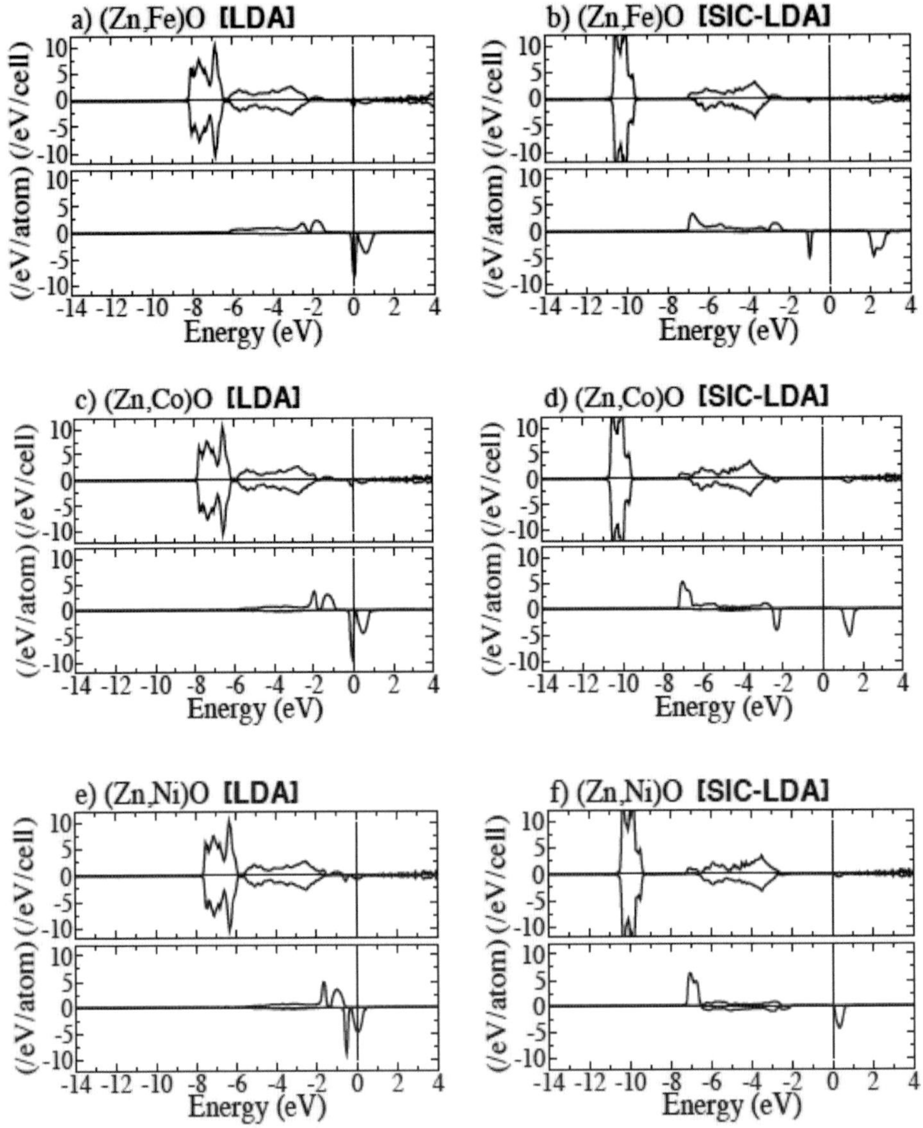

Figure 1.37. The DOS of Fe- (a, b), Co- (c, d) and Ni-doped (e, f) ZnO. The figures on the left-hand side (a, c, e) are calculated in the LDA and those on the right-hand side (b, d, f) are in the SIC-LDA. For all figures, the upper panels show the total DOS and the lower panels the partial DOS of transition-metal atoms.

In another experiment,[92] the main peak of the Co-d state is observed at $\sim$3 eV below the Fermi level and no spectral weight of the Co-d state is found in the energy range between the main peak and the Fermi level. The observation is not reproduced by the LDA result [Fig. 1.37(c)] because all the Co-d states in the LDA result appear above the host valence band, where no state is found in experiment.

On the other hand, in the SIC results [Fig. 1.37(d)], the main peak of the SIC Co-*d* partial DOS is found at the bottom of the host valence band and another large peak is found at the top of the valence band, which is ~2.5 eV below the Fermi level. The biggest difference from the LDA result is that no Co-*d* state is found near the Fermi level in the SIC result.

In the LDA results, the main structures of the transition-metal states are found above the host valence band for all the cases considered here. As a result, the *d* state appears quite near to the Fermi level. Therefore, the LDA calculations predict that the ferromagnetic ground states are strongly stabilized through the double-exchange interaction mechanism. On the other hand, in the presented calculations using the pseudo-SIC method, the transition-metal *d* states are shifted to the lower energy side and, except for V-doped ZnO, no *d* state is found near the Fermi level. For V-, Mn- and Co-doped ZnO, the SIC calculation is found to reproduce the experimental band structure data quite well. However, since no *d* state at the Fermi level is found, ferromagnetism by the double-exchange mechanism is expected to be realized.

1.4.3 Mn-Doped GaN in SIC-LDA Versus LDA

We also calculate the electronic structure and magnetic properties of Mn-doped GaN, which is one of the most promising high-Curie-temperature DMS materials. In the calculations, we use cubic zinc blende structure with the lattice parameter set to $a = 4.50$ Å.[93]

The band structure of GaN is similar to that of ZnO. The valence band mainly comes from the N-2p state and the conduction band is from the Ga-4s state. The band-gap width is 3.4 eV. Although the Zn-3d state in ZnO is found just below the valence band, the Ga-3d state in GaN is located much lower in energy. In a photoemission spectroscopy experiment,[91] the main peak of the Mn-*d* partial DOS in Mn-doped GaN is found to be located at 5 eV below the Fermi level and another small peak is found at 2 eV below the Fermi level.

In Fig. 1.38, the LDA calculation result is shown. The Mn-*d* state is found in the host band-gap. The clear tetrahedral-like splitting is shown. The fully occupied *e*-state is located at 1.5 eV below the Fermi level, which is the largest peak of the Mn-*d* state. This is near the 2 eV peak observed in experiment; however, the experimental 5 eV peak is not reproduced in the LDA result. The half-occupied *t* state is located at the Fermi level.

In the SIC result, shown in Fig. 1.39, the Mn-*d* state is shifted to the lower energy side. Due to the strong hybridization with the valence band, the Mn-*d* state becomes broader in shape. The main peak and the second largest peak are correctly found near 5 eV and 2 eV below the Fermi level. Despite the significant downward shift of the fully occupied *e* state from the LDA result, the partially occupied *t* state does not change so much. The Fermi level lies on the *t* state as in the LDA results. Thus, the ferromagnetic ground state should be stabilized by the double-exchange mechanism.

The magnetic exchange interaction, J_{ij}, of Mn-doped GaN is shown in Fig. 1.40. The figure shows the magnetic exchange interaction Mn–Mn pairs as a function of the distance of the pair. The interaction between the first-nearest neighbor pairs is strongly ferromagnetic, while as the distance is increased, the interaction is rapidly damped. In order to see the distance dependence more clearly, the interaction between only those pairs that are aligned along the [110] direction is plotted in the inset. The values are multiplied by the RKKY factor $(r/a)^3$, where a is the lattice constant. [110] is the direction in which the Ga–N–Ga bonds are aligned. The curves show the obvious short-range feature, which is consistent with the fact that the calculated DOS has a double-exchange-like structure. By applying

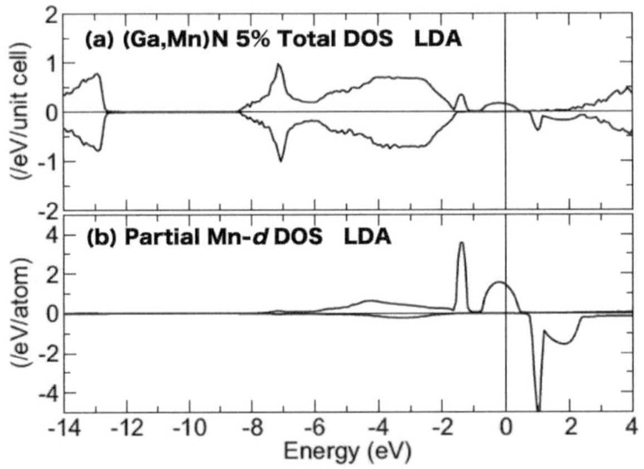

Figure 1.38. The DOS of Mn-doped GaN calculated in the LDA. The upper panel shows the total DOS and the lower panel shows the partial DOS of Mn-*d* states. The energy of zero corresponds to the Fermi level. The Mn concentration is 5%.

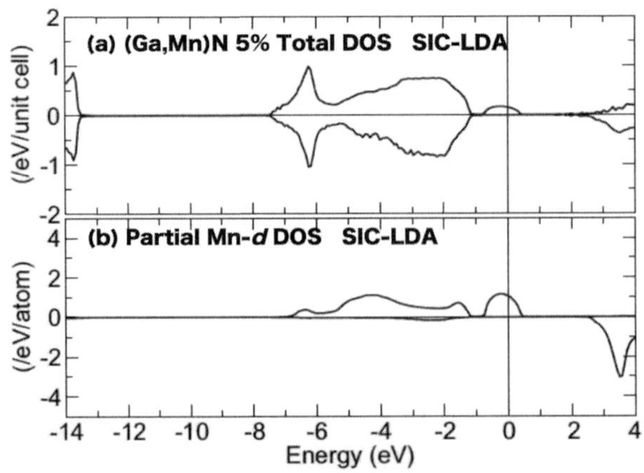

Figure 1.39. The DOS of Mn-doped GaN calculated in the SIC-LDA. The upper panel shows the total DOS and the lower panel shows the partial DOS of Mn-*d* states. The energy of zero corresponds to the Fermi level. The Mn concentration is 5%.

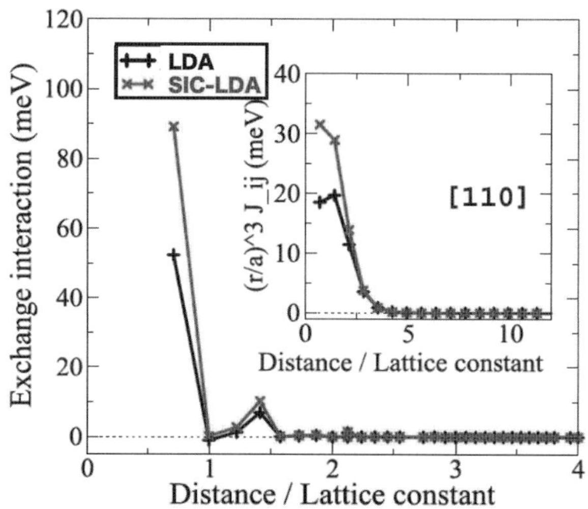

Figure 1.40. Calculated magnetic exchange interactions of Mn-doped GaN. The LDA values (black) and the SIC-LDA (red) are plotted as a function of distance between pairs of magnetic impurity atoms. The inset shows the dependence on distance by plotting only the pairs along the [110] direction and multiplying by the RKKY factor.

the SIC, the magnetic exchange interaction is enhanced. This is considered to be due to the suppression of the anti-ferromagnetic super-exchange interaction by large exchange splitting. Meanwhile, the distance dependence of the interaction is almost the same with the LDA result, i.e., it is quite short ranged. This is because the anti-bonding *t* state on which the Fermi level lies is not changed by SIC.

In Fig. 1.41, calculated Curie temperature is shown as a function of the Mn concentration. The experimental values from several studies[94–98] are also plotted. Due to the short-range feature of the magnetic exchange interaction, the Curie temperature is very low especially for the low Mn concentration region. Since the magnetic exchange interaction is larger, the values of the Curie temperature calculated in the SIC-LDA are higher than the LDA values. However, it does not reach 300 K even at 15% Mn concentration. The experimental values of Curie temperature are widely distributed from 0 K (paramagnetic) to as high as nearly 1000 K. The high values of Curie temperature, unfortunately, are not reproduced by the presented calculations, even if the LDA error is corrected by SIC. The reason why the calculated Curie temperature of Mn-doped GaN is low should be attributed to the short-range nature of the exchange interaction in Mn-doped GaN.

1.4.4 Calculated Density of States (DOS) by LDA and SIC-LDA Versus X-ray Photoemission Spectroscopy (XPS)

We next calculate the density of states (DOS) in GaAs, GaN, and ZnO-based DMS in the LDA and SIC-LDA, and compare the DOS with the imaginary part of single-particle Green's function, which is experimentally observed in X-ray

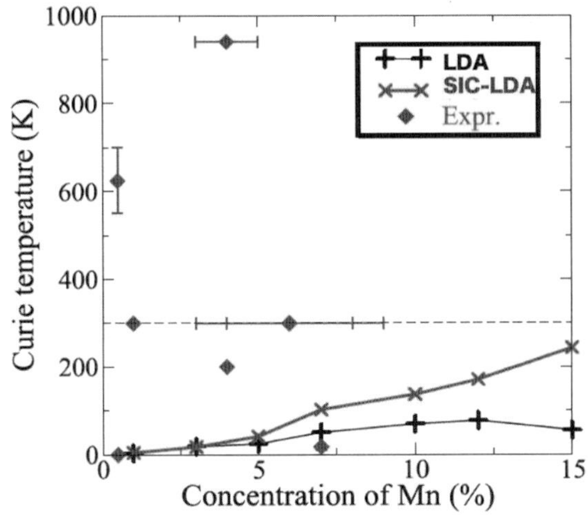

Figure 1.41. Calculated Curie temperature of Mn-doped GaN. The LDA values (black) and the SIC-LDA (red) are plotted. The experimental values from the literature[94–98] are also plotted (green squares).

photoemission experiments. In Figs. 1.42–46, the total DOS per unit cell and the partial DOS of 3*d*-states per TM atom at the TM site obtained by using the LDA and the SIC-LDA are shown for the ferromagnetic state, compared with the results from X-ray photoemission spectroscopy (XPS).[90–92] We also show the calculated exchange interaction as a function of distance in order to discuss the origin of the magnetic interactions.

In (Ga, Mn)As and (Ga, Mn)N, a smaller exchange splitting is always obtained in the LDA; however, the SIC-LDA gives a larger exchange splitting consistent with XPS data both for (Ga, Mn)As and (Ga, Mn)N (see Figs. 1.42 and 1.43). In (Ga, Mn)As, 3*d*-states at the Fermi level are suppressed in the SIC-LDA due to the larger exchange splitting, and Zener's *p–d* exchange mechanism is dominant with long-range ferromagnetic interactions [Fig. 1.42(c)]. However, Zener's double-exchange mechanism is dominant in (Ga, Mn)N due to the short-range ferromagnetic interaction originated from a partially occupied deep 3*d*-impurity band for both SIC-LDA and LDA.

Next, the calculated total and partial V-3*d* density of states of ferromagnetic (Zn, V)O with 5% V-doping obtained by the LDA and the SIC-LDA are compared with the results from XPS obtained by Ishida *et al.*[90] in Fig. 1.44. In the SIC-LDA, the energy positions of the V deep-impurity band and the Zn-3*d* state in ZnO are in good agreement with the XPS, in contrast to the LDA. Both the LDA and the SIC-LDA indicate that the partially-occupied deep-impurity band in the band-gap stabilizes the ferromagnetism due to Zener's double-exchange mechanism. This ferromagnetic interaction is short-ranged for both the LDA and SIC-LDA, and therefore, we can expect low-T_C due to the small percolation path in the dilute concentrations.

The total and partial Mn-3*d* density of states of (Zn, Mn)O with 5% Mn-doping calculated by using the LDA and the SIC-LDA are next compared with the results of XPS obtained by Mizokawa *et al.*[91] in Fig. 1.45. The agreement of the deep-impurity bands in the valence band and the Zn-3*d* state observed using the XPS and the SIC-LDA is reasonable. Since the Fermi level is located in the band-gap, we have no partially occupied deep-impurity band to stabilize the ferromagnetism caused by Zener's double-exchange mechanism. The anti-ferromagnetic super-exchange interaction for the d^5 configuration dominates the anti-ferromagnetic spin-glass (or paramagnetic) state. The SIC-LDA is quantitatively consistent with the XPS, in contrast to the LDA.

The total and partial Co-3*d* density of states of ferromagnetic (Zn, Co)O with 5% Co-doping calculated by using the LDA and the SIC-LDA (PSIC-LDA) are now compared with the results XPS obtained by Kobayashi *et al.*[92] in Fig 1.46. Agreement between the SIC-LDA and the PES is also reasonable for the Zn-3*d* state and the satellite of two-hole bound states with resonance in the valence band. Since the Fermi level is located in the band-gap, we have no partially-occupied deep-impurity band to stabilize the ferromagnetism caused by Zener's double-exchange mechanism. However, the ferromagnetic super-exchange interaction for the d^7 configuration dominates the ferromagnetic interaction. The SIC-LDA is quantitatively consistent with the XPS, in contrast to the LDA. Upon electron doping by oxygen vacancy (double donor), we can expect Zener's ferromagnetic double-exchange interaction due to the shift upwards of the Fermi level. However, we can only expect low T_C due to the short-range ferromagnetic interaction by double-exchange interaction. We can design high-T_C or

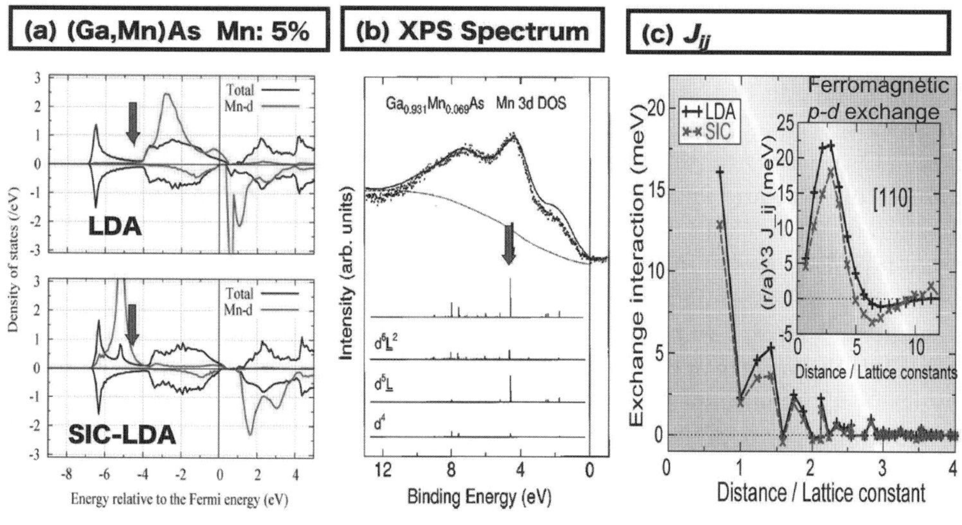

Figure 1.42. (a) Calculated total and partial 3*d* density of states in (Ga, Mn)As, (b) observed X-ray photoemission experimental data (the arrow indicates 3*d* states), and (c) calculated exchange interaction as a function of distance by the magnetic force theorem.

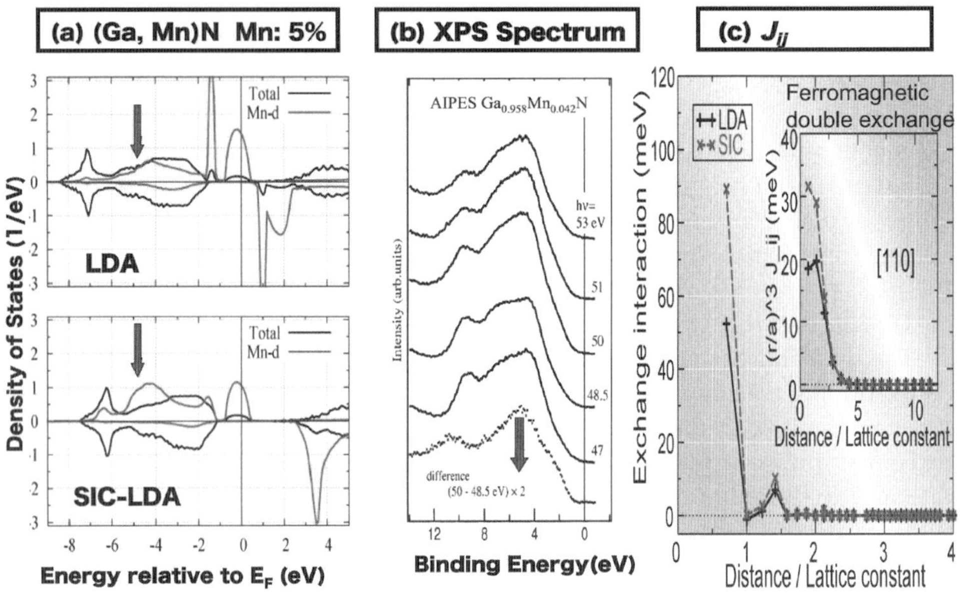

Figure 1.43. (a) Calculated total and partial 3*d* density of states in (Ga, Mn)N, (b) observed X-ray photoemission experimental data (the arrow indicates 3*d* states), and (c) calculated exchange interaction as a function of distance by the magnetic force theorem.

high-T_B (blocking temperature) by controlling spinodal nano-decomposition was discussed in Sec.1.3.

1.5 TiO_2 BASED DMS BY LDA VERSUS SIC-LDA

Titanium dioxide TiO_2 is interesting due to its many properties and occurs in several forms, such as rutile, anatase and brookite. In particular rutile- and anatase-type TiO_2 phases are, relatively, the most stable phases. However, the relative stability of rutile and anatase under ambient conditions is a controversial topic in the literature, from both the experimental and theoretical points of view.[99–102] The rutile-type TiO_2 has a tetragonal crystal structure ($a = 4.59$ Å and $c = 2.96$ Å) with a band-gap energy of 3 eV while the anatase phase also crystallizes in a tetragonal structure ($a = 3.78$ Å and $c = 9.52$ Å) with a band-gap energy of 3.2 eV.[103]

While the surface of the anatase phase is well known for an efficient photocatalytic effect,[104] the rutile phase has a variety of uses as a nanoscale insulator, such as ultrathin gate oxide field-effect transistors,[105,106] dielectric layers in capacitors for dynamic random access memory,[107] and in dye-sensitized TiO_2 solar cells with a good light power conversion efficiency.[108,109]

TiO_2 thin films can be produced by several methods including the sol-gel process,[110] different froms of sputtering form metallic[111] or ceramic[112] targets, electron beam evaporation,[113] and pulsed laser deposition.[114]

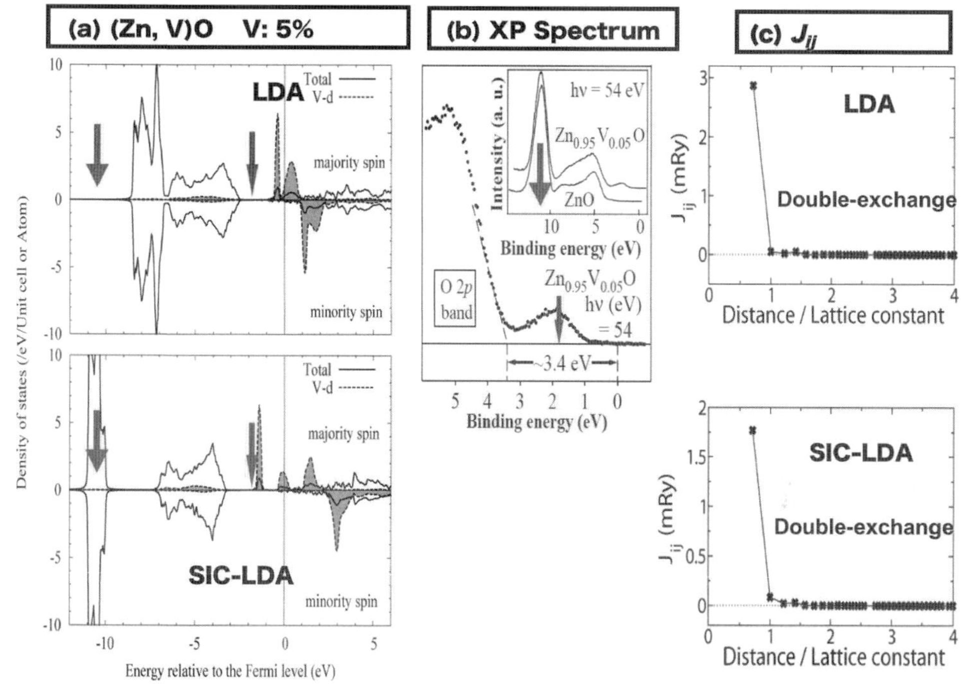

Figure 1.44. (a) Calculated total and partial 3*d* density of states in (Zn, V)O, (b) observed X-ray photoemission experimental data (the arrow indicates 3*d* states of Zn and V), and (c) calculated exchange interaction as a function of distance by magnetic force theorem.

In this study we focus on the rutile-type TiO_2 (Fig. 1.47). The material belongs to space group $P4_2/mnm$. The titanium atoms are surrounded by six oxygen atoms so that edge-shared local octahedral structures are formed.

1.5.1 Electronic Structure of (Ti, Co)O_2

The discovery of ferromagnetism at room temperature in Co-doped TiO_2 (rutile and anatase)[115,116] triggered subsequent intensive studies of various oxide semiconductors doped with magnetic ions[116,117] and was the first step in the interest in oxide-based DMS. Since the breakthrough, a number of Co-doped TiO_2 films have been reported with various physical properties. As these studies shows, while the Co^{2+} oxidation state is observed in both rutile and anatase $Ti_{1-x}Co_xO_{2-\delta}$,[118–121] there is widespread disagreement over whether the spin state is a high-spin[120–124] or a low-spin[114,115,125,126] configuration. For example, Mamiya *et al.* reported a combined experimental and theoretical Co $L_{2,3}$-edge soft X-ray magnetic circular dichroism (XMCD) study of rutile-type $Ti_{0.97}Co_{0.03}O_{2-\delta}$ as-deposited films.[121] They observed clear multiplet features at the Co $L_{2,3}$ edges in the XMCD spectrum corresponding to those in X-ray absorption spectra (XAS) of $Ti_{0.97}Co_{0.03}O_{2-\delta}$. The experimentally observed XMCD multiplet features agree qualitatively well with the results of a full atomic-multiplet calculation for

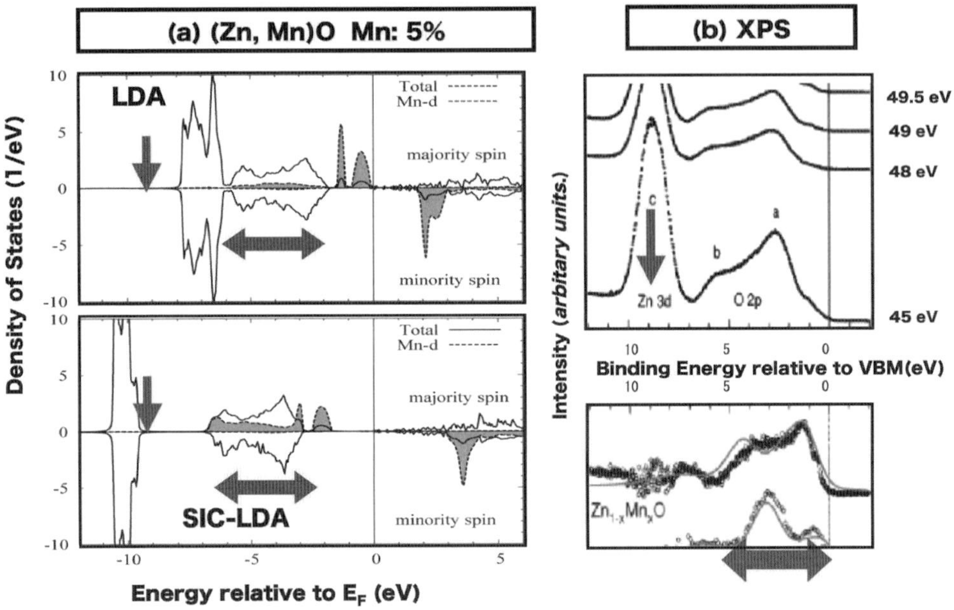

Figure 1.45. (a) Calculated total and partial 3*d* density of states in (Zn, Mn)O, (b) observed X-ray photoemission experimental data (the arrow indicates 3*d* states of Zn and Mn).

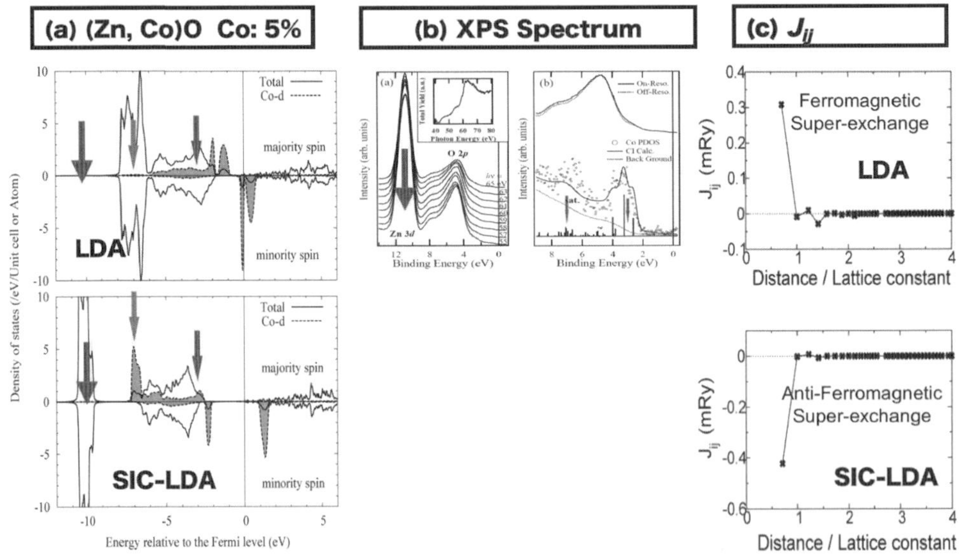

Figure 1.46. (a) Calculated total and partial 3*d* density of states in (Zn, Co)O, (b) observed X-ray photoemission experimental data (the arrow indicates 3*d* states of Zn and Co), and (c) calculated exchange interaction as a function of distance by the magnetic force theorem.

high-spin Co^{2+} ions under a D_{2h}-symmetry crystal field around Ti sites in rutile TiO_2. Thus, their observations strongly indicate intrinsic ferromagnetism arising from Co^{2+} ions substituting the Ti^{4+} ions.

From the LDA calculation point of view, however, an enhancement of the average value of the local magnetic moment relative to the Co^{2+} low-spin state with the presence of substitutional Co has been reported.[127] Thus, the experimental results do not agree with the calculated results.

The main purpose of this study is to explain from first-principles calculations the origin of the discrepancy regarding spin state between the theoretical predictions and the experimental observations. It is well known that LDA calculations underestimate the electron correlation effects, and the band-gap energy of semiconductors and occupied d states are systematically predicted at too high

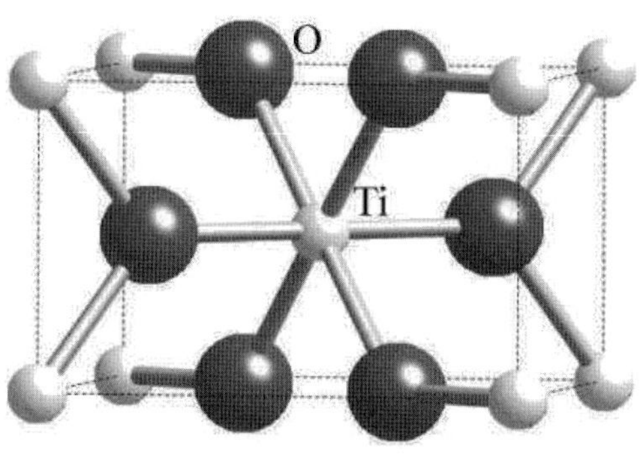

Figure 1.47. The crystal structure of rutile-type TiO_2 material. Small white and large black balls represent Ti and O, respectively.

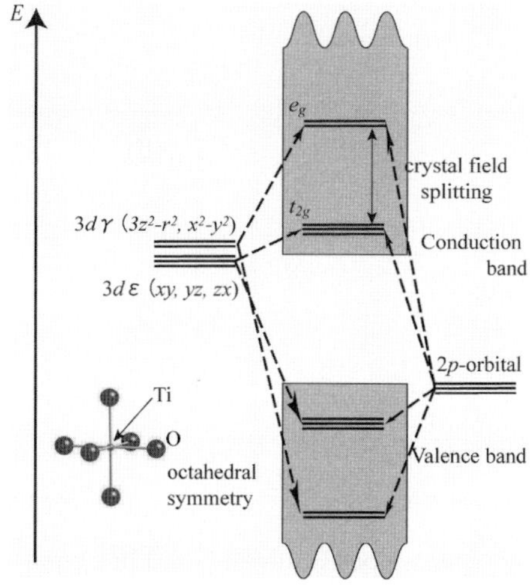

Figure 1.48. Schematic crystal field splitting diagram under octahedral coordination of host TiO_2.

energies. As a result, the hybridization between d states and ligand p states is generally overestimated in the LDA. One major source of these shortcomings of the LDA is the presence of self-interaction. We therefore carry out the calculations on TiO_2 (rutile)-based DMS in the framework of SIC-LDA. These results are compared with the ones calculated within the standard LDA.

1.5.2 Electronic Structure of Host TiO_2: LDA Versus SIC-LDA

The schematic crystal field splitting under octahedral symmetry of host TiO_2 is illustrated in Fig. 1.48. Under octahedral coordination in TiO_2 Ti $3d$-orbitals hybridize with the ligand O p-orbitals, where octahedral coordination means that distorted octahedral coordination in TiO_2 is regarded as regular octahedral coordination in order to simplify our discussion. As shown in the figure, therefore, the Ti $3d$-orbitals are split into doublet e_g and triplet t_{2g} orbitals due to the crystal field splitting under octahedral symmetry, where e_g orbitals are composed of $d_{3z^2-r^2}$ and $d_{x^2-y^2}(3d\gamma)$ and t_{2g} orbitals are composed of d_{xy}, d_{yz} and d_{zx} $(3d\varepsilon)$. This picture is revealed in the electronic structure of TiO_2.

In order to investigate the electronic structure in pure rutile TiO_2, the DOS calculated by the LDA and the SIC-LDA are shown in Figs. 1.49(a) and (b), respectively. As shown in the figures, both of the electronic structures in TiO_2 are characterized by a strong p–d hybridization between Ti-$3d$ and O-$2p$ orbitals as already mentioned. We found the apparent differences of band-gap energies between the LDA and the SIC-LDA. While the experimental band-gap energy is 3.0 eV, the calculated band-gap energy within the LDA is ~1.8 eV due to the underestimation caused by the LDA error. On the other hand, that within SIC-LDA is ~3.0 eV due to the shift of the valence bands (i.e., slightly localized O-$2p$ bands) to lower energy. Hence, the band-gap energy in the framework of the SIC-LDA agrees well with the experimental value.

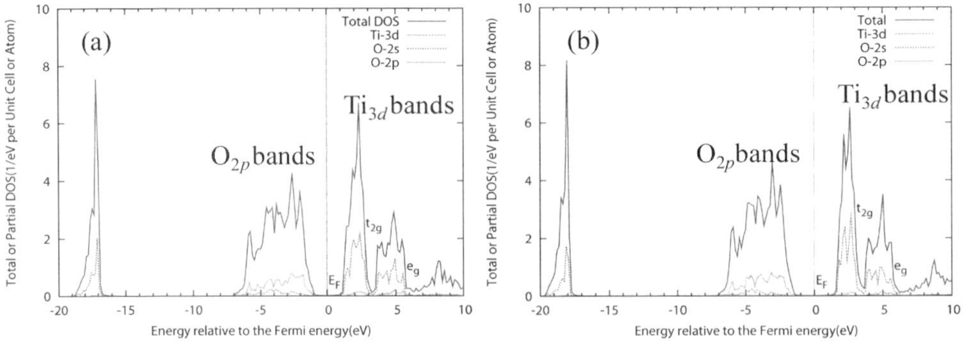

Figure 1.49. Total density of states per unit cell (solid line), partial density of Ti-$3d$ states per atom (long-dashed line), partial density of O-$2s$ states per atom (short-dashed line) and partial density of O-$2p$ states per atom (dotted line) in rutile TiO_2 in the frameworks of (a) the LDA and (b) the SIC-LDA. The horizontal axis denotes energy relative to the Fermi energy.

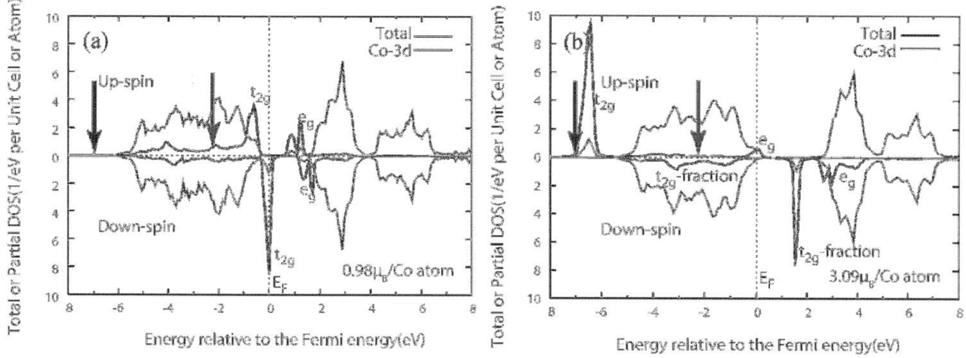

Figure 1.50. Calculated DOS in (Ti, Co)O_2 within (a) the standard LDA and (b) SIC-LDA without oxygen vacancies in the ferromagnetic state. The black line indicates the total DOS per unit cell and the red line denotes the partial density of *d* states at Co site per Co atom. Co concentration is 5% for each case. The horizontal axis is energy relative to the Fermi energy. The calculated magnetic moment per Co atom is given in the bottom right of each figure. The experimental peak positions of Co (2.2 and 7.0 eV[22]) 3*d* states are indicated by red and blue arrows, respectively.

1.5.3 Electronic Structure of (Ti, Co)O_2 and n-type (Ti, Co)O_2 with Oxygen Vacancy (Double Donor): LDA Versus SIC-LDA

Figures 1.50(a) and 1.50(b) show the calculated DOS of the ($Ti_{0.95}$, $Co_{0.05}$)O_2 DMS without oxygen vacancies in the ferromagnetic state within the standard LDA and SIC-LDA, respectively. As shown in the figures, highly localized Co-*d* states appear in the band-gap. Generally speaking, considering the formal electron configurations, the d^5 electron configurations are realized for Co^{4+}. In the case of the LDA [Fig. 1.50(a)], Fermi energy appears inside t_{2g} states for the down-spin side, a low-spin state is realized and a half-metallic density of states is also predicted, such that the formal electron configuration leads to magnetic moments of 1 μ_B per Co atom. In fact the calculated magnetic moment per Co atom is 0.98 μ_B. In the case of the SIC-LDA [Fig. 1.50(b)], on the other hand, the t_{2g} states for the down-spin side are split, leading to a formation of t_{2g} fractions, and Fermi energy appears inside both the e_g states for the up-spin side and the t_{2g} fraction for the down-spin side. Thus, an intermediate spin state is realized, and the calculated magnetic moment per Co atom is 1.50 μ_B.

From an experimental point of view, as shown in Figs. 1.50(a) and (b), the peak of the partial density of Co-*d* states does not agree with the photoemission spectra reported by Quilty *et al.* (Fig. 1.51[120]). As a result, under the LDA and SIC-LDA treatment we find that the calculated DOS without oxygen vacancies give an wrong description compared to experimental results.

In order to explain the discrepancy between the calculated DOS without oxygen vacancies and the experimental photoemission spectra, oxygen vacancies per Co ion are doped to the system so that Co ions in the +2 formal oxidation state Co^{2+} (according to experimental evidence[118–121]) are produced. Figures 1.52(a)

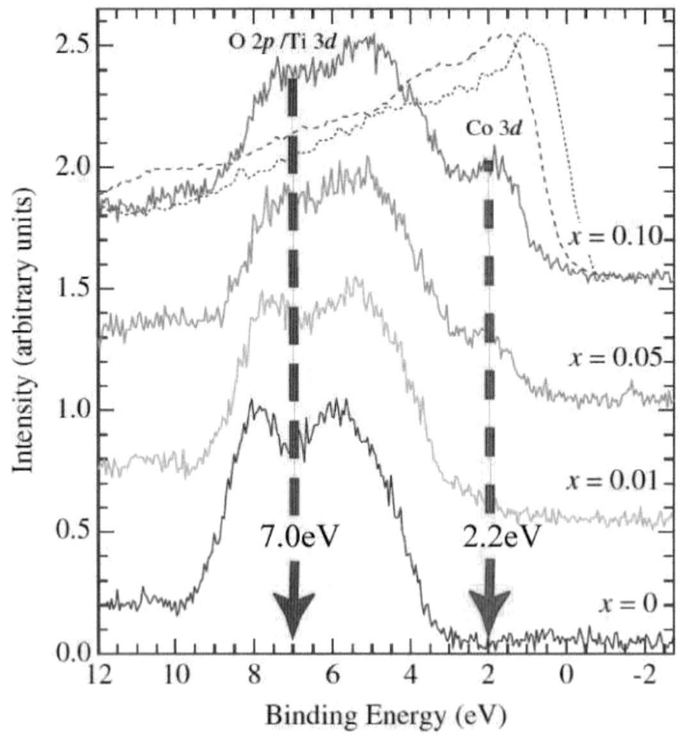

Figure 1.51. Valence band X-ray photoemission spectra in (Ti, Co)$O_{2-\delta}$ as a function of Co concentration x. Included for comparison are valence band spectra of metallic Co (dotted line) and CoO (dashed line). The red and blue arrows indicate the main peaks of Co-3d states (2.2 and 7.0 eV, respectively).

and 1.52(b) show DOS in ($Ti_{0.95}$, $Co_{0.05}$)$O_{1.95}$ with 2.5% oxygen vacancies within the LDA and the SIC-LDA, respectively. As is shown in the figures, it is found that these calculated DOS with oxygen vacancies show apparent differences from the previous DOS without oxygen vacancies. We find these differences in the energetic position of Co-d states and the ground spin state. In the case of Fig. 1.52(a), the peaks of the partial density of Co-d states appear in the band-gap (t_{2g}) and conduction bands (e_g), and the non-magnetic state is more stable than the ferromagnetic state. Obviously, the description within the LDA fails because of the LDA error. On the other hand, in the case of Fig. 1.52(b), it is found that within SIC-LDA the spin state is a high-spin state possessing a magnetic moment of 3.137μ_B per Co ion [as shown in the bottom right of Fig. 1.52(b)] and the DOS is consistent with XPS spectra (Fig. 1.51[120]) due to the energy shift of the t_{2g} and e_g states to the lower energy region by a shift in Fermi energy. This situation is schematically shown in Fig. 1.53. The left-hand side illustration with 3μ_B per Co atom in Fig. 1.53 corresponds to Fig. 1.52(b).

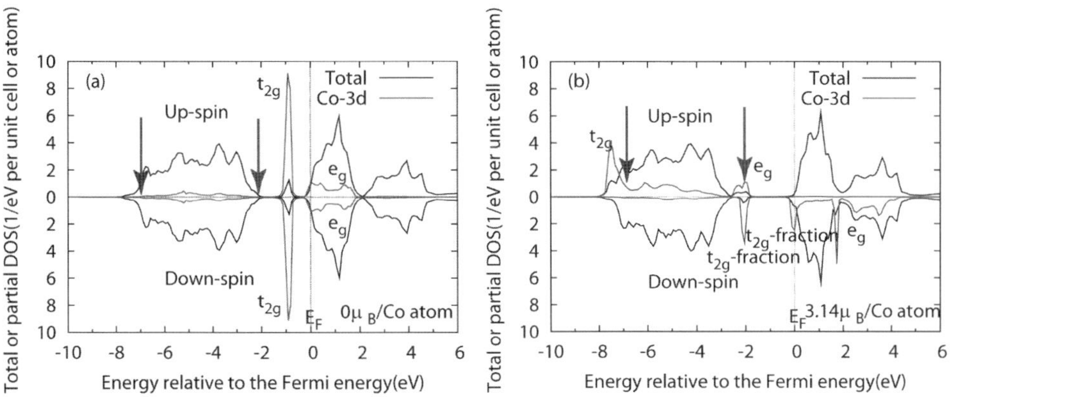

Figure 1.52. Calculated DOS in $(Ti_{0.95}, Co_{0.05})O_{1.95}$ with 2.5% oxygen vacancies within (a) the standard LDA and (b) the SIC-LDA in the ferromagnetic state. The red line indicates the total DOS per unit cell and the blue line means partial density of *d* states at Co site per Co atom. Co concentration is 5% for each case. The horizontal axis denotes energy relative to the Fermi energy. Calculated magnetic moment per Co atom is given in the bottom right of (b). The experimental peak positions of Co (2.2 and 7.0 eV[22]) 3*d* states are indicated by red and blue arrows, respectively.

1.5.4 Conclusion

We have investigated the electronic structure in TiO_2 (rutile) and $(Ti, Co)O_2$ from first-principles calculations to explain the origin of the discrepancy in spin state between the theoretical low-spin predictions and the experimental high-spin observations. Since the self-interacion error in the LDA is significant, in particular for oxides with octahedral local structure such as TiO_2, these systems were calculated within the SIC-LDA. These results have been compared with those calculated within the standard LDA. It is found that the calculated band-gap energy and energetic position of Ti-3*d* states are evidently different in the LDA and the SIC-LDA. In the case of $(Ti_{0.95}, Co_{0.05})O_2$, both the LDA and SIC-LDA give poor descriptions. In order to obtain more realistic results, we calculated the electronic structure of $(Ti_{0.95}, Co_{0.05})O_{1.95}$ with 2.5% oxygen vacancies within both the LDA and SIC-LDA such that the Co ions realize Co^{2+}. As a result, within the LDA a non-magnetic state is predicted, while, within the SIC-LDA the DOS of $(Ti_{0.95}, Co_{0.05})O_{1.95}$ shows a high-spin state. The result agrees well with experimental spin states and XPS spectra.

In conclusion it is found that calculated spin state in $(Ti, Co)O_2$ shows a high-spin state given the existence of oxygen vacancies, and the SIC-LDA method is indispensable for describing the electronic structure of $(Ti, Co)O_2$ because of the strong hybridization between Co-*d* and O-*p* orbitals under octahedral coordination.

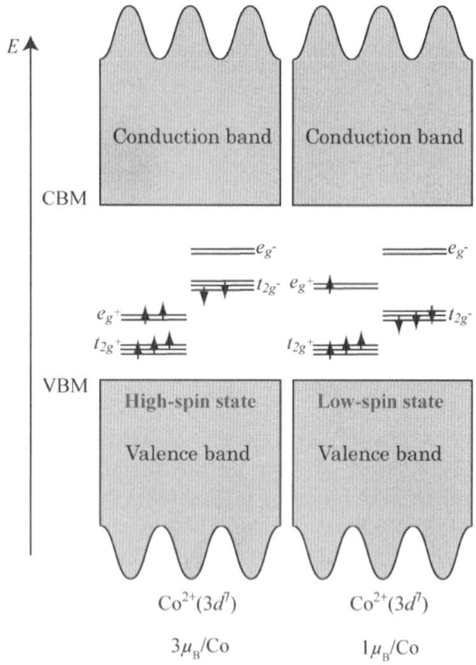

Figure 1.53. Schematic splitting of d levels in $(Ti_{0.95}, Co_{0.05})O_{1.95}$ with 2.5% oxygen vacancies and possible occupation schemes for a Co^{2+} $(3d^7 4s^0)$ atom. On the left-hand side is a high-spin configuration with $3\mu_B$ per Co atom and on the right is a low-spin configuration with $1\mu_B$ per Co atom. VBM and CBM indicate the valence band maximum and conduction band minimum, respectively.

1.6 A NEW CLASS OF DILUTE MAGNETIC SEMICONDUCTORS WITHOUT TRANSITION METAL ELEMENTS

In addition to incorporating transition metals (TM) into various semiconductors to realize ferromagnetic dilute magnetic semiconductors (DMS) for spintronics devices, recently a novel class of magnetic materials incorporating non-transition-metal atoms with/without lattice defects has attracted great attention among both theoretical and experimental scientists.

The magnetism in DMS is caused by the magnetic network of finite magnetic moments localized at sites or interstitial shells of lattices. Consequently, in order to fabricate a DMS, one can choose a certain atom to be incorporated into the host materials in such a way that the finite magnetic moments appear at lattice sites or other places in the host materials and an impurity band is formed in the band-gap. Following this treatment, DMS can be realized not only by incorporating the typical TM, but also by substituting for some atoms in the host material with a non-TM. In the materials in which the magnetism is induced by incorporating nonmagnetic impurities, the substitutional ions may have a nonzero magnetic moment and the $2p$-electrons of these ions, rather than the $3d$-electrons, play an essential role in introducing magnetism in the host materials. The $2p$-electrons can form an impurity band in the deep-band-gap, and ferromagnetism (FM) can be introduced if the Fermi level falls into this impurity band. With regard to this issue, studies have reported the magnetism caused by cation vacancies in MgO,[128] Ca vacancies in CaO,[129,130] Hf vacancies in HfO_2 with T_C exceeding 500 K,[131–134] and by hydrogen in graphite[135,136] and in C_{60}-based polymers,[137–140] etc. The magnetism induced by substituting for anions in various oxides by nonmagnetic impurities such as C and N has also been predicted, e.g., the FM caused by N and C in alkaline-earth-metal-oxides[130,141–143] and the half-metallicity and FM induced by N in α-quartz-SiO_2.[144] Furthermore, our recent research has also show the role of the $2p$-like impurity band formed by a non-TM dopant in the stabilization of FM.[145–147]

However, it seems that FM in such materials is more difficult to realize than TM-doped DMS because the exchange splitting of $2p$ spins is actually smaller than that of $3d$ spins of TM atoms. In order to realize half-metallic FM without TM impurities, the following minimum conditions must be satisfied:

(i) First, the impurity bands formed by impurities must be under Stoner's condition to allow impurities to have magnetic moments: $U > W$, where U and W stand for the effective correlation energy and the width of the mid-gap impurity band in which the Fermi level lies, respectively. The effective correlation energy U is determined through the total energies E: $U = E(N+1) + E(N-1) - 2E(N)$, where $E(N)$ is the total energy of an N-electron system.
(ii) Second, the condition for stabilizing the half-metallic FM due to the double-exchange mechanism must be maintained. This condition forces impurity bands formed by highly correlated itinerant electron states to be partially occupied.

In this section, we consider the prospects of FM and half-metallicity introduced by C or N into host insulators such as alkaline earth metal oxides, SiO_2, MgS and MgSe is given.

1.6.1 Deep Impurity Band Based Dilute Ferromagnetic Semiconductors without Transition Metal Impurities: SIC-LDA Versus LDA

It is well known that the LDA and the mean field approximation (MFA) cannot be successful for describing many materials, especially for strongly correlated systems. The LDA often overestimates the hybridization between electron states due to the underestimation of the band-gap energies of semiconductors. Moreover, the MFA often predicts T_C an excessively high value for T_C even if the substitutional concentration is lower than the percolation limit.[148] To overcome these inaccuracies, the SIC-LDA (which improves the LDA by taking the self-interaction correction into account[149–151]) and Monte Carlo simulation are employed. After

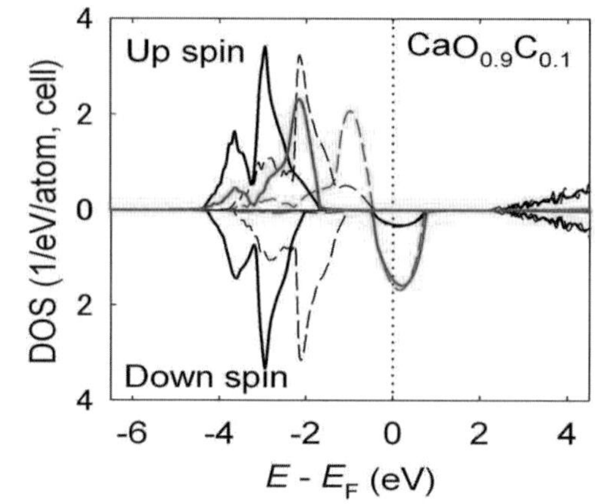

Figure 1.54. DOS of $CaO_{0.9}C_{0.1}$ by the LDA and SIC-LDA. The red lines illustrate C's $2p$ partial DOS and the black lines depict the total DOS. Dashed lines correspond to the LDA results, and solid lines to the SIC-LDA. The valence band, which originates from O's $2p$ electrons, shifts down to the lower energies to widen the band-gap in the SIC-LDA calculation.

calculating the electronic structure, the formula derived by Liechtenstein *et al.*[152] is used to calculate the exchange interaction J_{ij} between the two impurities at sites i and j in the ferromagnetic coherent potential medium. Finally, Monte Carlo simulation (MCS) is employed to estimate T_C.

A discussion about the origin of ferromagnetism without transition-metal elements based on calculation within the LDA framework is given in Refs. 141 and 142. Here, we concentrate on the results obtained with the SIC-LDA. For the sake of comparison, some results obtained by the standard LDA are also presented.

In order to compare the DOS obtained by the standard LDA with the SIC-LDA we plot the DOS of the typical case of alkaline earth metal oxides, $CaO_{1-x}C_x$ at $x = 0.10$, in Fig. 1.54.

It is found that taking the self-interaction of electrons into account, the band-gap of CaO becomes wider than that of the standard LDA. While the position of minority spin states of C's $2p$ electrons remains unchanged, the majority spin states calculated within the SIC-LDA shifts about 1.1 eV in comparison with the standard LDA, leading to a higher localization of SIC-LDA $2p$ states than for the LDA. In addition, the local magnetic moment of C increases from $1.242\mu_B$ (LDA) to $1.482\mu_B$ (SIC-LDA).[142] Since the $2p$ states of the majority spins shift down to lower energies, the exchange splitting in the SIC-LDA increases approximately two-fold as compared to that in the LDA, resulting in a possibility of the suppression of the super-exchange interaction and enhancement of the ferromagnetic Zener double-exchange mechanism. Correspondingly, the exchange coupling constant and T_C calculated within the SIC-LDA are shown to be much larger than in the case of the LDA, as seen later.

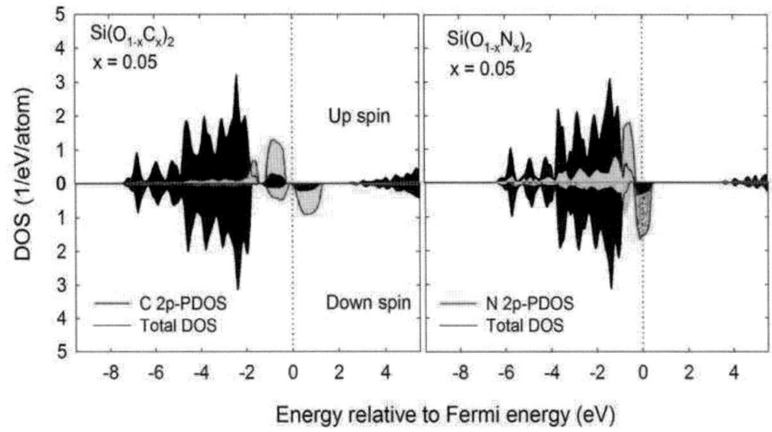

Figure 1.55. Total DOS (black) and $2p$ partial DOS per impurity atom at O sites (red) for two cases of substitutional non-TM: $Si(O_{1-x}C_x)_2$ (left figure) and $Si(O_{1-x}N_x)_2$ (right figure) at $x = 0.05$. The DOS is calculated within the LDA framework.

1.6.2 SiO_2-Based DMS

1.6.2.1 Density of states

Substituting O in SiO_2 by non-TM atoms such as C or N, one can also introduce local magnetic moments into lattice sites occupied by impurities. Correspondingly, magnetism might arise in the hosts if local magnetic moments interact with one another and form a magnetic moment net. Figure 1.55 demonstrates two cases of substitutional non-TM impurities: C ions introducing anti-FM, and N ions causing half-metallic FM in the host SiO_2. It is easy to see that Stoner's condition is satisfied in both of these cases. The $2p$ states of C are located deep within the band-gap, while the $2p$ states of N are close to the valence band of the host SiO_2. On the other hand, the Fermi level (E_F) of $Si(O_{1-x}N_x)_2$ lies near the center of the $2p$ minority spin band, leading to a half-full occupation of N's $2p$ states, whereas E_F of $Si(O_{1-x}C_x)_2$ falls between two impurity bands formed by minority spins; hence, the anti-ferromagnetic super-exchange mechanism should be dominant. For $Si(O_{1-x}N_x)_2$, the impurity band formed in the band-gap will be broadened as x increases. As a result, FM due to the double-exchange mechanism in $Si(O_{1-x}N_x)_2$ becomes more stabilized at higher x.

1.6.2.2 Stability of ferromagnetism

Figure 1.56 depicts the dependence of the total energy difference ΔE on the impurity concentration. ΔE determines the stabilization of magnetic phases in the DMS and is defined by $\Delta E = E_{DLM} - E_{FER}$, where E_{DLM} and E_{FER} are the total energies of disordered local moment and ferromagnetic states, respectively. ΔE corresponds to the fact that the ferromagnetic phase is more stabilized than the anti-ferromagnetic phase, and *vice versa*, and larger ΔE indicates a more stabilized ferromagnetic state. As seen from the figure, N introduces FM into the host material with very large positive ΔE, whereas C causes anti-ferromagnetic behavior. Consistent with Fig. 1.55, ΔE of the C-doped case is negative and it decreases as $-x^2$ with increasing impurity concentration due to the anti-ferromagnetic super-exchange interaction. In contrast, $Si(O_{1-x}N_x)_2$ is half-metallic and ferromagnetic due to Zener's double-exchange mechanism. ΔE depends on the substitutional concentration as $x^{3/2}$. Therefore, FM at high temperatures for high x can be expected to be achieved in $Si(O_{1-x}N_x)_2$.

1.6.3 Carbon Doped Alkaline Earth Metal Oxides

The substitution of O with C in alkaline earth metal oxides is treated randomly. For convenience, the lattice constants of alkaline earth metal oxide based DMS are fixed to the values of undoped crystals[153] and no distortion in lattice structures is assumed.

1.6.3.1 Half-metallicity

Figure 1.57 depicts DOS of $AO_{1-x}C_x$ (A = Mg, Ca, Sr and Ba) calculated within the SIC-LDA at $x = 0.10$. It is easy to see the half-metallic behavior of all the compounds. However, the exchange mechanism causing FM is somewhat different.

Figure 1.57(a) illustrates the DOS of $MgO_{0.9}C_{0.1}$. $MgO_{0.9}C_{0.1}$ has the largest band-gap energy due to having the smallest lattice constant. The $2p$ states of carbon are located near the top of the valence band which originates from anion p-states, resulting in the strong hybridization of $2p$ electron wave functions. The majority spin states hybridize with $2p$ states of oxygen, leading to the appearance

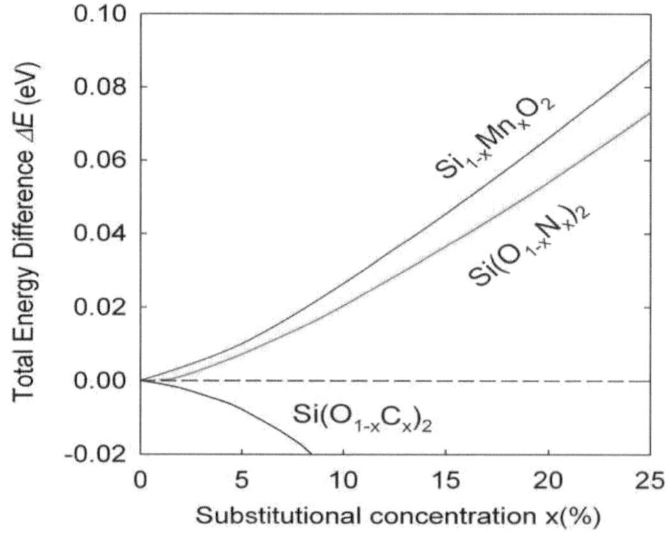

Figure 1.56. ΔE versus substitutional impurity concentration for $Si(O_{1-x}N_x)_2$ (red line), $Si(O_{1-x}C_x)_2$ (black line). ΔE of $Si_{1-x}Mn_xO_2$ (blue line)[144] is also given for comparison.

of an impurity band which connects to the top of the valence band and causes a narrower majority spin band-gap. The Fermi level falls into the impurity band formed by minority spins. This band is broadened with a half-width of about 1.5 eV. The exchange splitting in $MgO_{0.9}C_{0.1}$ approximates to 2.14 eV and is the smallest compared with the three remaining materials. This band can be broadened more strongly with increasing C concentration. As a result, the anti-ferromagnetic super-exchange interaction can easily arise if C concentration increases, and majority and minority spins can be compensated by one another. This means that FM might be suppressed at higher C concentrations. In addition, our calculations for higher x show that FM in $MgO_{1-x}C_x$ is most stabilized at $x \sim 0.10$.[142]

Figures 1.57(b)–1.57(d) show the DOS of $CaO_{0.9}C_{0.1}$, $SrO_{0.9}C_{0.1}$, and $BaO_{0.9}C_{0.1}$, respectively. As seen from the figures, the band-gap becomes narrower, and the localization of the $2p$ states becomes stronger, with increasing distance between atoms. The majority spin states are located in the valence band and cause a small broadening of the band. The Fermi level lies in the impurity band formed by minority spins with one third of this impurity band being occupied by electrons. The exchange splitting in these materials is approximately 2.3 eV, but the C's $2p$ states in $BaO_{0.9}C_{0.1}$ are the most localized. This change in the localization with respect to the lattice constant is also indicated by the increase in the local magnetic moment of C.[142] Therefore, one can expect FM to be stabilized by the predominant double-exchange mechanism in these materials even at higher C concentrations.

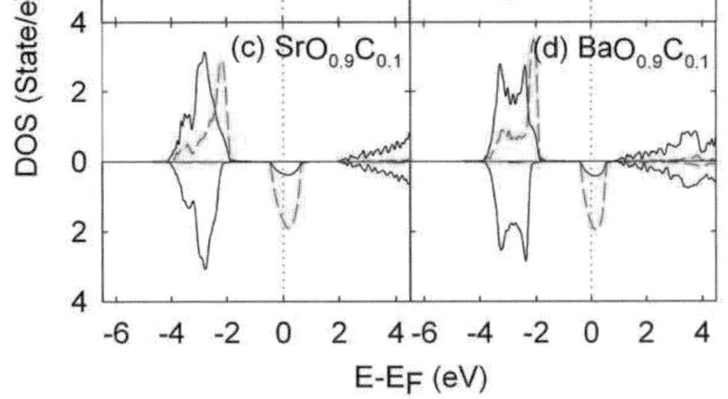

Figure 1.57. DOS of (a) $MgO_{1-x}C_x$, (b) $CaO_{1-x}C_x$, (c) $SrO_{1-x}C_x$ and (d) $BaO_{1-x}C_x$.

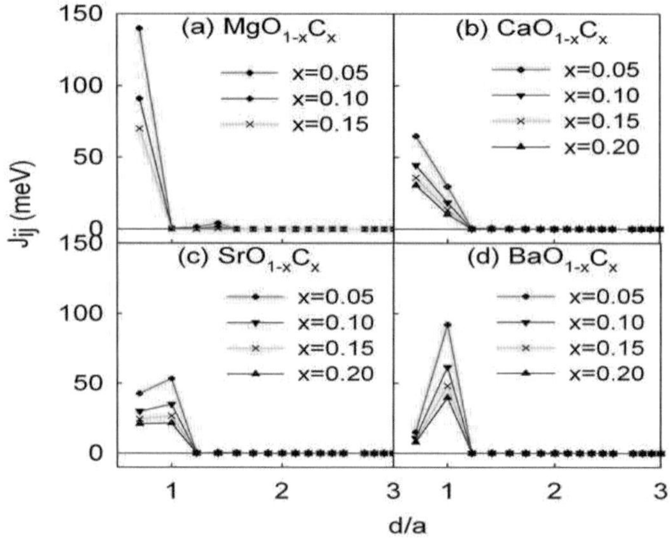

Figure 1.58. Magnetic exchange interaction between C atoms.

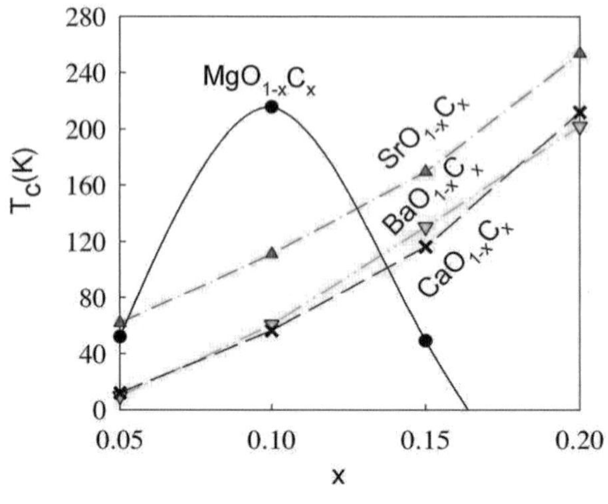

Figure 1.59. T_C versus substitutional impurity concentration for $MgO_{1-x}C_x$ (black line), $SrO_{1-x}C_x$ (red line), $BaO_{1-x}C_x$ (blue line) and $CaO_{1-x}C_x$ (green line).

1.6.3.2 Magnetic exchange interaction

The exchange coupling constants J_{ij} of C atoms in $MgO_{1-x}C_x$, $CaO_{1-x}C_x$, $SrO_{1-x}C_x$, and $BaO_{1-x}C_x$ are shown in Fig. 1.58. J_{ij} is calculated in the ferromagnetic CPA medium within the SIC-LDA.

As seen from Fig. 1.58(a), the exchange interaction of the first nearest neighbors in $MgO_{1-x}C_x$ is the strongest. The interaction strength is very strong (SIC-LDA: J_{ij} is about 140 meV at 5% C concentration) and most of the contributions come from the first nearest-neighbor interaction J_{01}. As x increases, J_{0j} calculated by the LDA sharply decreases and becomes negative for $j = 3$ and 4.[142] In contrast, the SIC-LDA gives a stronger exchange interaction, J_{ij}, especially for the first nearest-neighbor pairs. Except J_{01} and J_{04} (at $x = 10\%$), contributions from other J_{ij} are very small and can be ignored. Thus, the exchange interaction in $MgO_{1-x}C_x$ can be considered a typical case of short-ranged interactions. In addition, the exchange interaction is considerably suppressed as the C concentration increases. At $x = 0.20$, the exchange interaction becomes very small and FM is fully suppressed. For convenience, hereafter we will consider the results obtained within the SIC-LDA framework.

Interestingly, the larger the lattice constant, the greater the contribution from the second-nearest neighbors (J_{02}). For $CaO_{1-x}C_x$, J_{01} is approximately two times smaller than that in $MgO_{1-x}C_x$, but the contributions from the second nearest neighbors, J_{02}, is much stronger than for $MgO_{1-x}C_x$. For $SrO_{1-x}C_x$ and $BaO_{1-x}C_x$, contributions of the second nearest-neighbor pairs are predominant. In these former materials, the exchange interactions of the second nearest neighbors are more important than the first nearest neighbors. This can be caused by the sufficiently large distances between the first nearest neighbors, while the interactions between the second nearest atoms easily occur through the mediating atoms such as Sr or Ba, which have large ionic radii. As discussed above, the SIC-LDA enhances the ferromagnetic exchange interaction; hence, it gives J_{ij} larger than for the LDA. On the other hand, both the LDA and SIC-LDA calculations show the fact that the larger the lattice constant, the greater the contributions from the second nearest neighbors in the introduction of FM into host materials. The exchange interaction is strong but very short ranged, and the ferromagnetic double-exchange mechanism is expected to be predominant for all materials in question. Except $MgO_{1-x}C_x$, FM in $CaO_{1-x}C_x$, $SrO_{1-x}C_x$, and $BaO_{1-x}C_x$ can be more stabilized at higher x.

1.6.3.3 *Curie temperature* T_C

Using Monte Carlo simulation we calculate the Curie temperature T_C. First, we apply the Metropolis algorithm[154] to calculate the thermal average of the magnetization M and its powers. Then, the cumulant crossing method proposed by Binder[154] is employed and the fourth order cumulant U_4 is calculated as a function of temperature for different cell sizes ($14 \times 14 \times 14$, $16 \times 16 \times 16$, and $18 \times 18 \times 18$ conventional FCC cells) to find the universal fix-point at T_C. We estimate T_C for four values of the substitutional C concentrations: $x = 0.05, 0.10, 0.15$, and 0.20. The results obtained are shown in Fig. 1.59. T_C in this figure is calculated by using the set of J_{ij} obtained by employing the SIC-LDA. It should be clarified that the calculation of T_C based on the set of LDA J_{ij} gives T_C considerably smaller than the SIC-LDA, as discussed above. Figure 1.59 shows that FM in $MgO_{1-x}C_x$ is the most stabilized at $x \sim 0.10$ and can be suppressed at higher C concentrations due to the anti-ferromagnetic super-exchange interaction and by the majority and

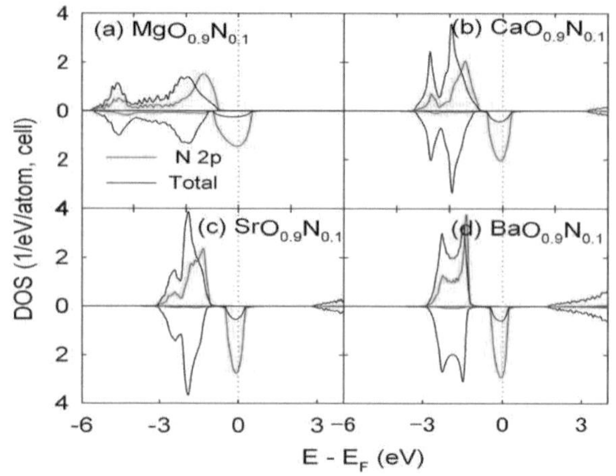

Figure 1.60. (a) DOS of $MgO_{1-x}N_x$, (b) $CaO_{1-x}N_x$, (c) $SrO_{1-x}N_x$, and (d) $BaO_{1-x}N_x$.

minority spins compensating one another. T_C of $MgO_{1-x}C_x$ (black line) increases with x in the range $0.05 < x < 0.1$ and has a peak of 216 K at $x \sim 0.10$, and then it drops sharply as a function of $-x$ at higher x. As shown in Fig. 1.57(a), the impurity band formed by $2p$ electrons of C is strongly broadened. The bandwidth W of the impurity band in the gap can be much larger than the effective correlation energy $U = E(N+1) + E(N-1) - 2E(N)$, and Stoner's condition for the existence of FM might be violated.[142] Correspondingly, FM might be fully suppressed at C concentrations higher than 16% because of the strong broadening of the impurity bands of both spin directions, resulting in the two spin channels compensating one another. In contrast to $MgO_{1-x}C_x$, the impurity states in $CaO_{1-x}C_x$, $SrO_{1-x}C_x$, and $BaO_{1-x}C_x$ are more localized and the bandwidths of impurity bands formed by minority spins in the gap are sufficiently small to satisfy Stoner's condition. Hence, FM is stabilized by a predominant ferromagnetic double-exchange mechanism and is more stabilized as x increases. Although the exchange interaction J_{01} between the first nearest-neighbor pairs considerably decreases with increasing lattice constant, owing to J_{02} contributions from the second nearest neighbors, T_C monotonously increases with x and can gain a value higher than room temperature if the C concentration becomes high enough. However, a question that arises here is the solubility of C in the materials.

1.6.4 Nitrogen Doped Alkaline Earth Metal Oxides

Similar to C-doped DMS, we also consider the role of N's $2p$ electrons in introducing FM in semiconductors.

1.6.4.1 Half-metallicity

As seen from Fig. 1.60, majority spin states are localized in the valence band. N's $2p$ electron states in the gap are more localized than C's $2p$ states. As a result, FM in these DMS can be weaker than that of the C-doped case. Since N introduces only one hole on the anion site N, two-thirds of the $2p$ states in the gap are occupied while one-third of the $2p$ states of C in the gap are occupied (see Fig. 1.57). In contrast to $MgO_{1-x}C_x$, FM in $MgO_{1-x}N_x$ is weaker at fixed concentration $x < 0.16$ but becomes more stabilized at $x > 0.16$. Thus, FM in these N-doped materials can be expected to be stable at all values of x, except where x is lower than the percolation limit.

1.6.4.2 Magnetic exchange interaction

Figure 1.61 shows the exchange coupling constant J_{ij} within the SIC-LDA for N atoms substituting O of alkaline earth metal oxides. Since the dominant contributions to the stabilization of FM comes from J_{01}, and J_{01} of $MgO_{1-x}N_x$ is approximately two times smaller than that of $MgO_{1-x}C_x$, T_C of $MgO_{1-x}N_x$ at dilute x will be much smaller than $MgO_{1-x}C_x$. However, the exchange interaction of N atoms is ferromagnetic at higher x; hence, FM is stable even at high N concentrations. A similar behavior to that shown in C-doped alkaline earth metal oxides is seen in the remaining compounds: FM introduced by N atoms is stabilized by the ferromagnetic double-exchange mechanism and contributions from second nearest neighbors is more important and becomes dominant with increasing lattice constant. However, because J_{ij} is much smaller than that for C atoms, T_C in this case could be much smaller than $AO_{1-x}C_x$ and is lower than room temperature even at higher x.

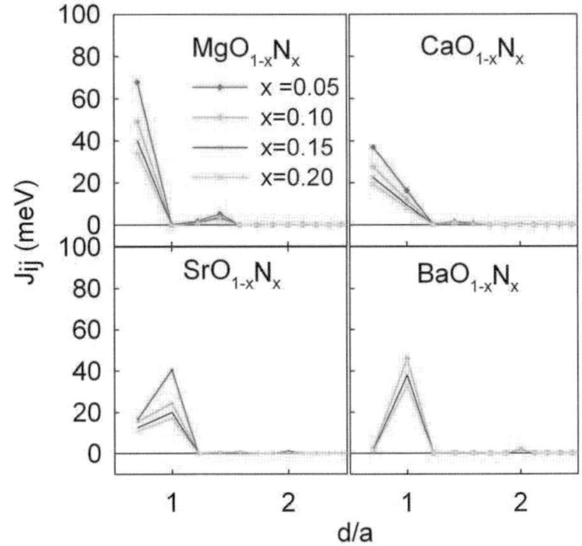

Figure 1.61. Magnetic exchange interaction between N atoms.

1.6.5 Nitrogen Doped MgS and MgSe

$MgS_{1-x}N_x$ and $MgSe_{1-x}N_x$ are quite similar to $MgO_{1-x}C_x$ in that the second nearest-neighbor pairs give a negligible contribution to the stabilization of FM. While FM in $AO_{1-x}C_x$ and $AO_{1-x}N_x$ (A = Ca, Sr, Ba) is stabilized by the first and second nearest-neighbor pairs, owing not only to the first nearest neighbors but also the third and fourth nearest-neighbor pairs, FM in $MgS_{1-x}N_x$ and $MgSe_{1-x}N_x$ can be more enhanced with increasing N concentration.

1.6.5.1 Half-metallicity

The p-PDOS calculated by performing the pseudo-potential method is shown in Fig. 1.62 for $MgS_{1-x}N_x$ and $MgS_{1-x}N_x$. A 32-atom super-cell with eight nitrogen atoms substituting for eight oxygen atoms is used throughout our DOS calculation. Unlike the DOS calculated in a CPA medium, the $2p$ states in the super-cell calculation are split into the two main parts for both spin channels, and consequently become more localized. Not only $2p$ states of N atoms but also $3p$ ($4p$) states of S (Se) atoms are polarized. The $2p$-like band formed in the gap is caused by the hybridization of $2p$ electron states of N and S (or Se). E_F falls into the minority band in the gap. The PDOS picture shows that FM can be stabilized by the ferromagnetic double-exchange mechanism.

1.6.5.2 Magnetic exchange interaction and Curie temperature

Figure 1.63 shows the exchange coupling constant J_{ij} calculated within the SIC-LDA in a ferromagnetic CPA medium. The exchange interaction between N atoms in N-doped MgS and MgSe is ferromagnetic and stronger than in N-doped alkaline earth metal oxides, and FM is stabilized not only by the first nearest-neighbor pairs but also by the third and fourth nearest-neighbor pairs. In particular, the exchange

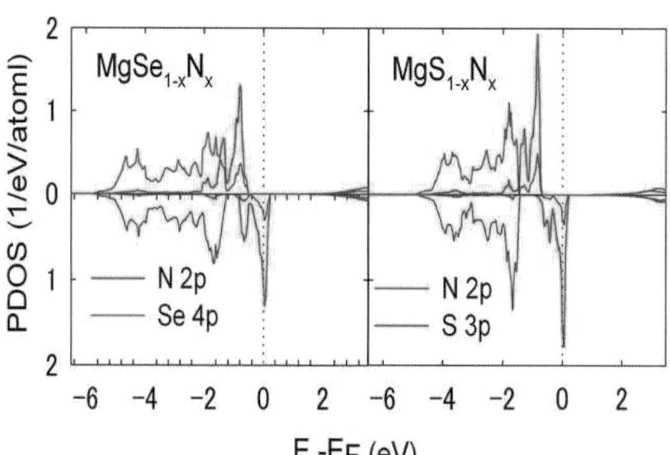

Figure 1.62. Pseudo-potential p-PDOS in $MgS_{1-x}N_x$ and $MgSe_{1-x}N_x$.

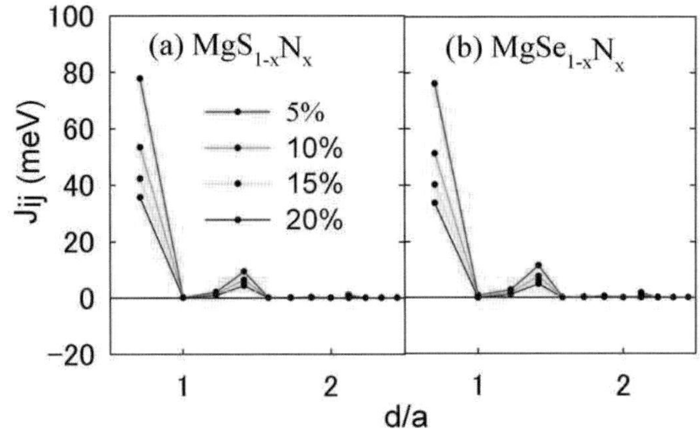

Figure 1.63. Magnetic exchange interaction between N atoms.

interaction remains ferromagnetic and rather strong at high N concentration in comparison with $MgO_{1-x}C_x$. Although J_{01} of $MgS_{1-x}N_x$ is slightly larger than $MgSe_{1-x}N_x$, J_{0j} (with $j > 1$) of $MgS_{1-x}N_x$ is smaller, particularly for J_{04}. Using the set of J_{ij} as the input data, we employ the MCS to evaluate T_C.

The concentration dependence of T_C for $MgS_{1-x}N_x$ and $MgSe_{1-x}N_x$ is shown in Fig. 1.64. The solid circles and triangles indicate the calculated T_C of $MgS_{1-x}N_x$ and $MgSe_{1-x}N_x$, respectively. Solid (red) and dashed (black) lines are also plotted to guide the eye. T_C of both of DMS sharply increases with N concentration. The percolation threshold x_0 is evaluated as approximately equal to 0.034 for $MgSe_{1-x}N_x$ and to 0.041 for $MgS_{1-x}N_x$. The greater localization of the $2p$ wave function and the smaller J_{0j} with $j > 1$ in $MgS_{1-x}N_x$ in comparison with $MgSe_{1-x}N_x$ are reflected in the magnitude of T_C. Given J_{0j} with $j > 1$, T_C of $MgSe_{1-x}N_x$ is higher than that of $MgS_{1-x}N_x$.

1.6.6 Conclusion

We have discussed the ferromagnetism introduced by the incorporation of non-transition metal elements into SiO_2, alkaline earth metal oxides, MgS and MgSe. The electronic structures and exchange coupling constants are calculated by applying the LDA and SIC-LDA. The dominant exchange mechanism and the role of the exchange interaction between nearest-neighbor pairs in these DMS have been discussed through the calculation of the DOS and the exchange coupling constant. The stabilization of ferromagnetism in these DMS is considered via the estimation of different total energy and T_C. T_C has been evaluated by Monte Carlo simulation for C-doped alkaline earth metal oxides, N-doped MgS and N-doped MgSe.

Our first-principles calculations show that carbon atoms can introduce half-metallic ferromagnetism, which is stabilized by the ferromagnetic double-exchange mechanism, into alkaline earth metal oxides whereas it causes

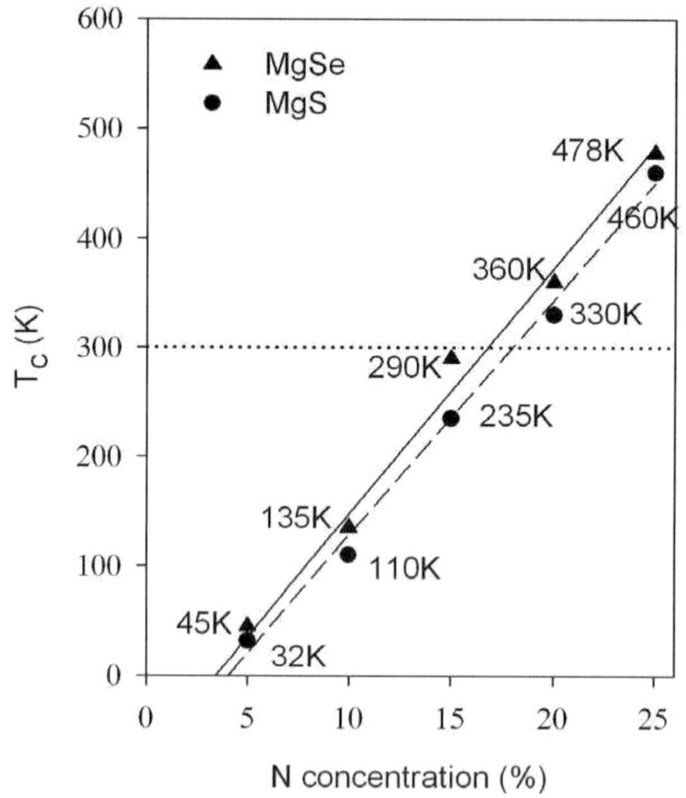

Figure 1.64. Curie temperature of $MgS_{1-x}N_x$ and $MgSe_{1-x}N_x$.

anti-ferromagnetic behavior in SiO_2 due to the predominant anti-ferromagnetic super-exchange interaction between C atoms. Except in $MgO_{1-x}C_x$, FM in other C-doped alkaline earth metal oxides is more stabilized by increasing C concentration. The substitution of anions in SiO_2, alkaline earth metal oxides, MgS and MgSe by N atoms can induce half-metallicity and FM due to the predominant double-exchange mechanism. In alkaline earth metal oxides based DMS, J_{02} becomes larger the distance between substitutional atoms is increased. Consequently, the exchange interaction of second nearest-neighbor pairs is more important than the first nearest-neighbor pairs in SrO based DMS and it becomes dominant in BaO based DMS. While the FM in alkaline earth metal oxides based DMS is stabilized by the exchange interaction of the first and second nearest-neighbor pairs, the exchange interaction range of N atoms in MgS and MgSe is rather longer and the contribution of J_{0j} with $j > 1$ becomes very important.

Thus, without transition metals, one can introduce ferromagnetism into various insulators. However, in order to realize ferromagnetism above room temperature without transition metal elements we need to incorporate N (or C) atoms at a sufficiently high concentration. The solubility problem of impurities such as C and N in these host materials is the open question that must be addressed.

1.7 SUMMARY

Based upon *ab initio* electronic structure calculations, we have considered computational nano-materials design for the realization of future semiconductor nano-spintronics. In Sec. 1.1, we discussed why we need semiconductor nano-spintronics based on the present status and future limitations of Si-CMOS technology according to Gordon Moore's law. We showed how we can design and realize semiconductor nano-spintronics based on a combination of top-down (nano-lithography) and bottom-up (self-organization) nanotechnologies. In order to go beyond the limitations of Si-CMOS technologies, we propose a possible new methodology to realize a new design-based class of nano-spintronics with high Curie temperature ($T_C > 1000\,K$) or high blocking (B) temperature (T_B): high-density (Tbit/(inch)^2), high-speed-switching (THz), and energy-saving (non-volatile) spintronics devices. For the realization of semiconductor nano-spintronics, we propose a colossal magnetic response controlled by electric field or photonic excitation, and colossal-thermal management (spincaloritronics) nano-spintronics devices.

In Sec. 1.2, we proposed Zener's *p–d* exchange interaction as a ferromagnetic mechanism in the (Ga, Mn)Sb and (Ga, Mn)As, where the Mn-3*d* level is lower than the Sb-5*p* or As-4*p* levels and the 3*d* state becomes a bonding state in the deep valence band, and the *p*-state is pushed up into the band-gap as an anti-bonding state. Negatively polarized *p*-states relative to Mn-3*d* states stabilize the ferromagnetic states in Zener's *p–d* exchange interactions. On the other hand, Zener's double-exchange mechanism dominates the ferromagnetism in (Ga, Mn)N, (Ga, Cr)N, II–VI-based DMS such as (Zn, Cr)Te, n-type-doped (Zn, Co)O, and (Zn, V)O.

We proposed a design methodology to obtain accurate (*almost exact*) T_C by combining the *ab initio* calculation of the exchange interaction as a function of distance based on the atomic force theorem with Monte Carlo simulation. The calculated T_C by Monte Carlo simulation is in good agreement with experimental data. However, the homogeneous DMS indicates a low T_C due to the percolation problem given a low concentration of 3*d* transition metal impurities. Hence, it is very difficult to realize a high-T_C DMS for low concentrations of transition metal impurities.

In an effort to realize a high Curie temperature (T_C) in DMS ($T_C > 1000\,K$), in Secs. 1.2 and 1.3 we discuss the electronic structure and ferromagnetic mechanism in III–V and II–VI-based DMS, spinodal nano-decomposition, and blocking phenomena in semiconductor nano-magnets based on the local density approximation (LDA).

In Sec. 1.3, we discuss spinodal nano-decomposition in inhomogeneous DMS and a new fabrication method of 100 Tbits/inch^2 for semiconductor nano-magnets using two-dimensional and three-dimensional spinodal nano-decomposition by self-organization, and for a colossal magnetic response controlled by electric field or photonic excitation. We have shown that two-dimensional spinodal nano-decomposition realized a self-organized quasi-one-dimensional nano-wire

(*Konbu phase*) and three-dimensional spinodal nano-decomposition realizes a quantum dot (*Dairiseki phase*). Based on Monte Carlo simulation of crystal growth using the Ising model mapped by *ab initio* calculations of the effective chemical pair interaction, we have simulated spinodal nano-decomposition, position control by nanoscale seeding, and shape control by changing the atomic vapor pressure during the crystal growth. We have shown that there is a high blocking temperature caused by the magneto-crystal anisotropy or shape anisotropy in the *Konbu phase* or *Dairiseki phase*. These predicted nano-superstructures in the semiconductor matrixes are experimentally observed in the recent MBE or MOCVD experiments.

Section 1.4 is devoted to ZnO- and GaN-based DMS using the self-interaction corrected LDA (SIC-LDA), which is very important for highly correlated systems such as DMS. We compared the calculated density of states using the SIC-LDA and LDA with results of X-ray photoemission spectroscopy (XPS), and obtained an almost perfect agreement between the SIC-LDA and XPS. Based on the SIC-LDA results, we have confirmed that Zener's $p-d$ exchange mechanism is dominant in (Ga, Mn)As and Zener's double–exchange mechanism is dominant in (Ga, Mn)N, (Zn, Cr)Te, (Zn, V)O, and n-type-doped (Zn, Co)O.

In Sec. 1.5, we reported the calculated electronic structure for TiO_2-based. DMS using the SIC-LDA. Without carrier doping in (Ti, Co)O_2, the LDA predicted low-spin ground states, while the SIC-LDA predicted high-spin ground states; however, agreement with the XPS data was poor for both the LDA and the SIC-LDA, suggesting the role of oxygen vacancies. Based on an oxygen vacancy of 2.5% and Co (5%) co-doping, we obtained high-spin ferromagnetic states with n-type materials and achieved almost perfect agreement between the SIC-LDA and XPS data. In transition metal oxides such as $3d$ transition metal-doped TiO_2, the LDA is very poor at describing the correct electronic structure and we need to use the SIC-LDA calculation for TiO_2-based DMS.

In Sec. 1.6, we discussed the materials design of a new, oxide-based class of ferromagnetic DMS without transition-atom impurities by controlling the deep-impurity-band width (W) and the impurity-correlation energy (U). We compared the calculated density of states in the LDA and SIC-LDA in the wide band-gap semiconductor-based DMS with C or N-doping in CaO, MgO, etc.

ACKNOWLEDGMENTS

This research was partially supported by a Grant-in-Aid for Scientific Research in Priority Areas "Quantum Simulators and Quantum Design" (No. 17064014), a Grand-in-Aid for Scientific Research for young researchers, Japan Science and Technology Agency (CREST), New Energy and Industrial Technology Development Organization Nanotech, the 21st Century Center of Excellence, and the Japan Society for the Promotion of Science core-to-core program "Computational Nano-materials Design." This research was also supported in part by the National

Science Foundation under Grant No. PHY99-07949 (Kavli Institute for Theoretical Physics SPINTRONICS06 program, University of California, Santa Barbara), by the Kansai Research Foundation for technology promotion (KRF), the Foundation of C&C Promotion, the Murata Science Foundation, and the Inoue Foundation for Science (IFS).

References

[1] S. A. Wolf *et al.*, *Science* **294**, 1488 (2001).
[2] H. Akai and P. H. Dederichs, *Phys. Rev. B* **47**, 8739 (1993).
[3] H. Akai, *Phys. Rev. Lett.* **81**, 3002 (1998).
[4] M. Toyoda, H. Akai, K. Sato and H. Katayama-Yoshida, *Physica B* **376**, 647 (2006).
[5] M. Toyoda, H. Akai, K. Sato and H. Katayama-Yoshida, *Phys. Status Solidi C* **3**, 4155 (2006).
[6] V. A. Dinh, M. Toyoda, K. Sato and H. Katayama-Yoshida, *J. Phys. Soc. Japan*, **75**, 093705 (2006).
[7] H. Shiba, *Prog. Theor. Phys.* **46**, 77 (1971).
[8] K. Sato and H. Katayama-Yoshida, *Jpn. J. Appl. Phys.* **39**, L555 (2000).
[9] K. Sato and H. Katayama-Yoshida, *Jpn. J. Appl. Phys.* **40**, L485 (2001).
[10] K. Sato and H. Katayama-Yoshida, *Jpn. J. Appl. Phys.* **40**, L334 (2001).
[11] K. Sato and H. Katayama-Yoshida, *Jpn. J. Appl. Phys.* **40**, L651 (2001).
[12] K. Sato and H. Katayama-Yoshida, *Semicond. Sci. Technol.* **17**, 367 (2002); T. Fukushima, K. Sato and H. Katayama-Yoshida, *Jpn. J. Appl. Phys.* **43**, L1416 (2004).
[13] K. Sato, P. H. Dederichs and H. Katayama-Yoshida, *Europhys. Lett.* **61**, 403 (2003).
[14] K. Sato, P. H. Dederichs, H. Katayama-Yoshida and J. Kudrnovsky, *J. Phys. Cond. Matt.* **16**, S5491 (2004).
[15] J. Kanamori, *J. Phys. Chem. Solids* **10**, 87 (1959).
[16] J. B. Goodenough, *Phys. Rev.* **100**, 564 (1955).
[17] K. Sato, W. Schweika, P. H. Dederichs and H. Katayama-Yoshida, *Phys. Rev. B* **70**, 201202(R) (2004).
[18] L. Bergqvist, O. Eriksson, J. Kudrnovsk'y, V. Drchal, P. Korzhavyi and I. Turek, *Phys. Rev. Lett.* **93**, 137202 (2004).
[19] K. Sato, H. Katayama-Yoshida and P. H. Dederichs, *Jpn. J. Appl. Phys.* **44**, L948 (2005).
[20] T. Fukushima, K. Sato, H. Katayama-Yoshida and P. H. Dederichs, *Jpn. J. Appl. Phys.* **45**, L416 (2006).
[21] H. Katayama-Yoshida, K. Sato, T. Fukushima, M. Toyoda, H. Kizak V. A. Dinh and P. H. Dederichs, *J. Magn. Magn. Mater.* **310**, 2070 (2007).
[22] H. Katayama-Yoshida, K. Sato, T. Fukushima, M. Toyoda, H. Kizaki, V. A. Dinh and P. H. Dederichs, *Phys. Status Solidi A* **204**, 15 (2007).
[23] T. Fukushima, K. Sato and H. Katayama-Yoshida, *Phys. Stat. Sol. A* **203** (2006) 2751.
[24] T. Fukushima, K. Sato, H. Katayama-Yoshida and P. H. Dederichs, *Phys. Stat. Sol. C* **3**, 4139 (2006).
[25] K. Sato, T. Fukushima and H. Katayama-Yoshida, *Jpn. J. Appl. Phys.* **46**, L682 (2007).
[26] D. Stauffer and A. Aharony, *Introduction to Percolation Theory*, 2nd Edn. (Taylor and Francis, London, 1994).

[27] H. Kizaki, K. Sato, A. Yanase and H. Katayama-Yoshida, *Jpn. J. Appl. Phys.* **44**, L1187 (2005).

[28] V. A. Dinh, K. Sato and H. Katayama-Yoshida, *Solid State Commun.* **136**, 1 (2005).

[29] K. Kenmochi, V. A. Dinh, K. Sato, A. Yanase and H. Katayama-Yoshida, *J. Phys. Soc. Jpn.* **73**, 2952 (2004).

[30] H. Katayama-Yoshida, T. Fukushima and K. Sato, *Jpn. J. Appl. Phys.* **46**, L777 (2007).

[31] T. Dietl, *Semicond. Sci. Technol.* **17**, 377 (2002).

[32] J. Koenig, J. Schliemann, T. Jungwirth and A. H. MacDonald, in *Electronic Structure and Magnetism of Complex Materials*, eds. D. J. Singh and D. A. Papaconstantopoulos (Springer, New York, 2002), p. 163.

[33] T. Jungwirth, J. Sinova, J. Ma'sek, J. Ku'cera and A. H. MacDonald, *Rev. Mod. Phys.* **78**, 809 (2006).

[34] J. Kanamori and K. Terakura, *J. Phys. Soc. Jpn.* **70**, 1433 (2001).

[35] T. Yamamoto and H. Katayama-Yoshida, *Jpn. J. Appl. Phys.* **36**, L180 (1997).

[36] K. Sato and H. Katayama-Yoshida, *Jpn. J. Appl. Phys.* **46**, L1120 (2007).

[37] H. Akai and P. H. Dederichs, *Phys. Rev. B* **47**, 8739 (1993).

[38] H. Akai, http://sham.phys.sci.osaka-u.ac.jp/kkr/

[39] K. Sato and H. Katayama-Yoshida, *Semicond. Sci. Technol.* **17**, 365 (2002).

[40] K. Sato, P. H. Dederichs and H. Katayama-Yoshida, *Europhys. Lett.* **61**, 403 (2003).

[41] A. I. Liechtenstein, M. I. Katsnelson, V. P. Antropov and V. A. Gubanov, *J. Magn. Magn. Mater.* **67**, 65 (1987).

[42] L. Bergqvist, O. Eriksson, J. Kudrnovsky, V. Drchal, P. Korzhavyi and I. Turek, *Phys. Rev. Lett.* **93**, 137202 (2004).

[43] K. Sato, W. Schweika, P. H. Dederichs and H. Katayama-Yoshida, *Phys. Rev. B* **70**, R201202 (2004).

[44] T. Fukushima, K. Sato and H. Katayama-Yoshida, *Jpn. J. Appl. Phys.* **43**, L1416 (2004).

[45] K. Sato, P. H. Dederichs and H. Katayama-Yoshida, *J. Phys. Soc. Jpn.* **76**, 24717 (2007).

[46] K. Sato, H. Katayama-Yoshida and P. H. Dederichs, *Jpn. J. Appl. Phys.* **44**, L948 (2005).

[47] T. Fukushima, K. Sato, H. Katayama-Yoshida and P. H. Dederichs, *Jpn. J. Appl. Phys.* **45**, L416 (2006).

[48] K. Sato, T. Fukushima and H. Katayama-Yoshida, *Jpn. J. Appl. Phys.* **46**, L682 (2007).

[49] K. Sato and H. Katayama-Yoshida, *Jpn. J. Appl. Phys.* **46**, L1120 (2007).

[50] H. Akai, *Phys. Rev. Lett.* **81**, 3002 (1998).

[51] K. Sato, P. H. Dederichs, H. Katayama-Yoshida and J. Kudrnovsky, *J. Phys. Cond. Matt.* **16**, S5491 (2004).

[52] B. Belhadji, L. Bergqvist, R. Zeller, P. H. Dederichs, K. Sato and H. Katayama-Yoshida, *J. Phys. Cond. Matt.* **19**, 436227 (2007).

[53] T. Dietl, *Semicond. Sci. Technol.* **17**, 377 (2002).

[54] J. Kanamori and K. Terakura, *J. Phys. Soc. Jpn.* **70**, 1433 (2001).

[55] G. M. Dalpian, S. H. Wei, X. G. Gong, A. J. R. da Silva and A. Fazzio, *Solid State Commun.* **138**, 353 (2006).

[56] D. Stauffer and A. Aharony, *Introduction to Percolation Theory*, 2nd Edn. (Taylor and Francis, Philadelphia, 1991).

[57] D. P. Landau and K. Binder, *A Guide to Monte Carlo Simulations in Statistical Physics* (Cambridge University Press, Cambridge, 2000).

[58] D. A. Dimitrov and G. M. Wysin, *Phys. Rev. B* **54**, 9237 (1996).

[59] S. Sonoda, S. Shimizu, T. Sasaki, Y. Yamamoto and H. Hori, *J. Cryst. Growth* **237**, 1358 (2002).

[60] M. Hashimoto, Y.-Z. Zhou, M. Kanamura and H. Asahi, *Solid. State Commun.* **122**, 37 (2002).

[61] H. Saeki, H. Tabata and T. Kawai, *Solid State Commun.* **120**, 439 (2001).

[62] H. X. Liu, S. Y. Wu, R. K. Singh, L. Gu, D. J. Smith, N. R. Dilley, L. Montes, M. B. Simmonds and N. Newman, *Appl. Phys. Lett.* **85**, 4076 (2004).

[63] H. Saito, V. Zayets, S. Yamagata and K. Ando, *Phys. Rev. Lett.* **90**, 207202 (2003).

[64] N. Ozaki, N. Nishizawa, K.-T. Nam, S. Kuroda and K. Takita, *Phys. Stat. Sol. C* **1**, 957 (2004).

[65] H. Ohno, A. Shen, F. Matsukura, A. Oiwa, A. Endo, S. Katsumoto and Y. Iye, *Appl. Phys. Lett.* **69**, 363 (1998).

[66] S. Kuroda, N. Nishizawa, K. Takita, M. Mitome, Y. Bando, K. Osuch and T. Dietl, *Nature Mater.* **6**, 440 (2007).

[67] M. Jamet, A. Barski, T.Devillers, V. Poydenot, R. Dujardin, P. Bayle-Guillemaud, J. Rothman, E. Bellet-Amalric, A. Marty, J. Cibert, R. Mattana and S. Tatarenko, *Nature Mater.* **5**, 653 (2006).

[68] K. Sato and H. Katayama-Yoshida, *J. Jpn. Appl. Phys.* **46**, L1120 (2007).

[69] F. Ducastelle and F. Gautier, *J. Phys.* **F6** 2039 (1976)

[70] L. Gu, S. Y. Wu, H. X. Liu, R. K. Singh, N. Newman and D. J. Smith, *J. Magn. Magn. Matt.* **290**, 1395 (2005).

[71] M. Tay, Y. Wu, G. C. Han, T. C. Chong, Y. K. Zheng, S. J. Wang, Y. Chen and X. Pan, *J. Appl. Phys.* **100**, 063910 (2006).

[72] A. Bonanni, M. Kiecana, C. Simbrunner, T. Li, M. Sawicki, M. Wegscheider, M. Quast, H. Przybylinska, A. Navarro-Quezada, R. Jakiela, A. Wolos, W. Jantsch and T. Dietl, *Phys. Rev. B* **75**, 125210 (2007).

[73] G. Bouzerar, T. Ziman and J. Kudrnovsky, *Europhys. Lett.* **69**, 812 (2005).

[74] S. Hilbert and W. Nolting, *Phys. Rev. B* **70**, 165203 (2004).

[75] T. Fukushima, K. Sato and H. Katayama-Yoshida, *Jpn. J. Appl. Phys.* **45**, L416 (2006).

[76] N. D. Mermin and H. Wagner, *Phys. Rev. Lett.* **17**, 1133 (1966).

[77] T. Yamamoto and H. Katayama-Yoshida, *Jpn. J. Appl. Phys.* **36**, L180 (1997).

[78] K. Sato and H. Katayama-Yoshida, *Jpn. J. Appl. Phys.* **46**, L1120 (2007).

[79] H. Katayama-Yoshida, T. Fukushima, V. A. Dinh and K. Sato, *Jpn. J. Appl. Phys.* **46**, L777 (2007).

[80] V. A. Dinh, K. Sato and H. Katayama-Yoshida, *J. Phys. Soc. Jpn.* **77**, 014705 (2008).

[81] A. Fukushima, K. Yagami, A. A. Tulapurkar, Y. Suzuki, H. Kubota, A. Yamamoto and S. Yuasa, *Jpn. J. Appl. Phys.* **44**, L12 (2005).

[82] F. J. Albert, J. A. Katine, R. A. Buhrman and D. C. Ralph, *Phys. Rev. Lett.* **77**, 3809 (2000).

[83] S. I. Kiselev, C. J. Sankey, I. N. Krivorotov, N. C. Emely, R. J. Schoelkoph, R. A. Buhrman and D. C. Ralph, *Nature* **425**, 380 (2003).

[84] R. H. Koch, J. A. Katine and J. Z. Sun, *Phys. Rev. Lett.* **92**, 88302 (2004).

[85] A. Filippetti and N. A. Spalding, *Phys. Rev. B* **67**, 125109 (2003).

[86] H. Akai and P. H. Dederichs, *Phys. Rev. B* **47**, 8739 (1993).

[87] J. P. Perdew and A. Zunger, *Phys. Rev. B* **23**, 5048 (1981).

[88] K. Kihara and G. Donnay, *Can. Mineral.* **23**, 647 (1985).

[89] W. Göpel, Pollmann, I. Ivanov and B. Reihl, *Phys. Rev. B* **26**, 3144 (1982).

[90] Y. Ishida, J. I. Hwang, M. Kobayashi, A. Fujimori, H. Saeki, H. Tabata and T. Kawai, *Physica B Cond. Matt.* **351** 304 (2004).

[91] T. Mizokawa, T. Nambu, A. Fujimori, T. Fukumura and M. Kawasaki, *Phys. Rev. B* **65**, 085209 (2002).

[92] M. Kobayashi, Y. Ishida, J. I. Hwang, T. Mizokawa, A. Fujimori, K. Mamiya, J. Okamoto, Y. Takeda, T. Okane, Y. Saitoh, Y. Muramatsu, A. Tanaka, H. Saeki, H. Tabata and T. Kawai, *Phys. Rev. B* **72**, 201201(R) (2005).

[93] K. Uesugi, N. Morooka and I. Suemune, *Appl. Phys. Lett.* **74**, 1254 (1999).
[94] M. L. Reed, N. A. El-Masry, H. H. Stadelmaier, M. K. Ritums, M. J. Reed, C. A. Parker, J. C. Roberts and S. M. Bedair, *Appl. Phys. Lett.* **79**, 3473 (2001).
[95] G. T. Thaler, M. E. Overberg, B. Gila, R. Frazier, C. R. Abernathy, S. J. Pearton, J. S. Lee, S. Y. Lee, Y. D. Park, Z. G. Khim, J. Kim and F. Ren, *Appl. Phys. Lett.* **80**, 3964 (2002).
[96] N. Theodoropoulou, A. F. Hebard, M. E. Overberg, C. R. Abernathy, S. J. Pearton, S. N. G. Chu and R. G. Wilson, *Appl. Phys. Lett.* **78**, 3475 (2001).
[97] M. E. Overberg, C. R. Abernathy, S. J. Pearton, N. A. Theodoropoulou, K. T. McCarthy and A. F. Hebard, *Appl. Phys. Lett.* **79**, 1312 (2001).
[98] K. H. Ploog, S. Dhar and A. Trampert, *J. Vac. Sci. Technol. B* **21**, 1756 (2003).
[99] J. Muscut, V. Swamy and N. Harrison, *Phys. Rev. B* **65**, 224112 (2002).
[100] M. Calatayud, P. Mori-Sánchez, A. Beltran, A. Pendás, E. Francisco, J. Andrés and J. Recio, *Phys. Rev. B* **64**, 184113 (2001).
[101] M. Ranade, A. Navrotsky, H. Zhang, J. Banfield, S. Elder, A. Zaban, P. Borse, S. Kulkarni, G. Doran and H. Whitfeld, *Proc. Natl. Acad. Sci USA* **99**, 6476 (2002).
[102] J. Banfield, B. Bischoff and M. Anderson, *Chem. Geol.* **110**, 211 (1993).
[103] H. Tang, H. Berger, P. E. Schmid and F. Levy, *Sol. Stat. Commun.* **92**, 267 (1994).
[104] J. Augustynski, *J. Electronchim. Acta* **38**, 43 (1993).
[105] S. A. Campbell *et al.*, *IBM J. Res. Dev.* **43**, 383 (1999).
[106] C. K. Mati *et al.*, *Microelectron. Eng.* **72**, 253 (2004).
[107] S. K. Kim *et al.*, *Appl. Phys. Lett.* **85**, 4112 (2004).
[108] B. O'Regan and M. Gratzel, *Nature* **353**, 737 (1991).
[109] P. J. Sevastián, A. Olea, J. Campos, J. A. Toledo and S. A. Ganboa, *Solar Energy Mater. Solar Cells* **81**, 349 (2004).
[110] S. M. Tracey, S. N. B. Hodgson, A. K. Ray and Z. Ghassemlooy, *J. Mater. Process. Technol.* **77**, 86 (1998).
[111] T. M. Viseu, B. Almeida, M. Stchakovsky, B. Drevillon, M. I. C. Ferreira and J. B. Sousa, *Thin. Sol. Films* **401**, 216 (2001).
[112] G. H. Li, L. Yang, Y. X. Jin and L. D. Zhang, *Thin. Sol. Films* **368**, 163 (2000).
[113] T. S. Yang, C. B. Shiu and M. S. Wong, *Surf. Sci.* **548**, 75 (2004).
[114] Y. Matsumoto, M. Murakami, T. Shono, T. Hasegawa, T. Fukumura, M. Kawasaki, P. Ahmet, T. Chikyow, S. Koshihara and H. Koinuma, *Science* **291**, 854 (2001).
[115] Y. Matsumoto, R. Takahashi, M. Murakami, T. Koida, X.-J. Fan, T. Hasewaga, T. Fukumura, M. Kawasaki, S.-Y. Koshihara and H. Koinuma, *Jpn. J. Appl. Phys.* **40**, L1204 (2001).
[116] S. B. Ogale *et al.*, *Phys. Rev. Lett.* **91**, 077205 (2003).
[117] P. Sharma, A. Gupta, K. V. Rao, F. J. Owens, R. Sharma, R. Ahuja, J. M. O. Guillen, B. Johansson and G. A. Gehring, *Nat. Mater.* **2**, 673 (2003).
[118] S. A. Chambers, S. M. Heald and T. Droubay, *Phys. Rev. B.* **67**, 100401 (2003).
[119] M. L. Cui, J. Zhu, X. Y. Zhong, Y. G. Zhao and X. F. Duan, *Appl. Phys. Lett.* **85**, 1698 (2004).
[120] J. W. Quilty, A. Shibata, J. Y. Son, K. Takubo, T. Mizokawa, H. Toyosaki, T. Fukumura and M. Kawasaki, *Phys. Rev. Lett.* **96**, 027202 (2006).
[121] K. Mamiya, K. Koide, A. Fujimori, H. Takano, H. Manaka, A. Tanaka, H. Toyosaki, T. Fukumura and M Kawasaki, *Appl. Phys. Lett.* **89**, 062506 (2006).
[122] M. S. Park, S. K. Kwon and B. I. Min, *Phys. Rev. B* **65**, 161201R (2002).
[123] J.-Y. Kim *et al.*, *Phys. Rev. Lett.* **90**, 017401 (2003).
[124] H. Weng, X. Yang, J. Dong, H. Mizuseki, M. Kawasaki and Y. Kawazoe, *Phys. Rev. B* **69**, 125219 (2004).
[125] S. A. Chambers *et al.*, *Appl. Phys. Lett.* **79**, 3467 (2001).

[126] X. F. Yao, T. J. Zhou, Y. X. Gai, T. C. Chong and J. P. Wang, *J. Appl. Phys.* **95**, 7375 (2004).

[127] J. M. Sullivan and S. C. Erwin, *Phys. Rev. B* **67**, 144415 (2003).

[128] L. E. Halliburton, D. L. Cowan, W. B. J. Blake and J. E. Wertz, *Phys. Rev. B* **8**, 1610 (1973).

[129] I. S. Elfimov, S. Yunoki and G. A. Sawazky, *Phys. Rev. Lett.* **89**, 216403 (2002).

[130] K. Kenmochi, M. Seikei, K. Sato, A. Yanase and H. Katayama-Yoshida, *Jpn. J. Appl. Phys.* **43**, L934 (2004).

[131] M. Venkatesan, C. B. Fitzgerand and J. M. D. Coye, *Nature* (*London*) **430**, 630 (2004).

[132] J. M. D. Coye, M. Venkatesan, P. Stamenov, C. B. Fitzgerand and L. S. Dorneles, *Phys. Rev. B* **72**, 024450 (2005).

[133] C. P. Pemmaraju and S. Sanvito, *Phys. Rev. Lett.* **94**, 217205 (2005).

[134] J. Osorio-Guillen, S. Lany, S. V. Barabash and A. Zunger, *Phys. Rev. Lett.* **96**, 107203 (2006).

[135] K. Kusakabe and M. Maruyama, *Phys. Rev. B* **67**, 092406 (2003).

[136] P. Esquinazi, D. Spemann, R. Hohne, A. Setzer, K. H. Han and T. Butz, *Phys. Rev. Lett.* **91**, 227201 (2003).

[137] T. L. Makarova, B. Sundquist, R. Hohne, P. Esquinazi, Y. Kopelevich, P. Scharff, V. A. Davidov, L. S. Kashevarova and A. V. Rakhmanina, *Nature* (*London*) **413**, 716 (2001).

[138] P. Esquinazi, A. Setzer, R. Hohne, C. Semmelhack, Y. Kopelevich, D. Spemann, T. Butz, B. Kohlstrunk and M. Losche, *Phys. Rev. B* **66**, 024429 (2002).

[139] B. Narymbetov, A. Omerzu, V. V. Kabanov, M. Tokumoto, H. Kobayashi and D. Michailovic, *Nature* (*London*) **407**, 883 (2000).

[140] Y. Ma, P. O. Lehtinen, A. S. Foster and R. M. Nieminen, *Phys. Rev. B* **72**, 085451 (2005).

[141] K. Kenmochi, V. A. Dinh, K. Sato, A. Yanase and H. Katayama-Yoshida, *J. Phys. Soc. Jpn.* **73**, 2952 (2004).

[142] V. A. Dinh, M. Toyoda, K. Sato and H. Katayama-Yoshida, *J. Phys. Soc. Jpn.* **75**, 093705 (2006).

[143] I. R. Shein, A. N. Enyashin, A. L. Ivanovskii, *Phys. Rev. B* **75**, 245404 (2007).

[144] V. A. Dinh, K. Sato and H. Katayama-Yoshida, *Solid State Commun.* **136**, 1 (2005).

[145] V. A. Dinh, K. Sato and H. Katayama-Yoshida, *Jpn. J. Appl. Phys.* **42**, L888 (2003).

[146] V. A. Dinh, K. Sato and H. Katayama-Yoshida, *J. Phys. Cond. Matt.* **16**, S5705 (2004).

[147] V. A. Dinh, K. Sato and H. Katayama-Yoshida, *J. Supercond. Incorp. Novel Mag.* **18**, 47 (2005).

[148] K. Sato, P. H. Dederichs and H. Katayama-Yoshida, *Phys. Rev. B* **70**, 201202 (2004).

[149] V. A. Asimov, J. Zaanen and O. K. Andersen, *Phys. Rev. B* **44**, 943 (1991).

[150] A. Svane and O. Gunnarsson, *Phys. Rev. Lett.* **65**, 1148 (1990).

[151] A. Filippetti and N. A. Spaldin, *Phys. Rev. B* **67**, 125109 (2003).

[152] A. I. Liechtenstein, M. I. Katsnelson, V. P. Antropov and V. A. Gubanov, *J. Magn. Magn. Matt.* **67**, 65 (1987).

[153] W. G. Wyckoff, *Crystal Stuctures*, Vol. 1, 2nd Ed. (Wiley Interscience, New York, 1974).

[154] K. Binder and D. W. Heermann, *Monte Carlo Simulation in Statistical Physics* (Springer, Berlin, 2002).

Chapter Two

Transition Metal Doped ZnO

David Norton* and Stephen J. Pearton
Department of Materials Science and Engineering, University of Florida, Gainesville, FL 32611, USA
**dnort@mse.ufl.edu*

Mathew P. Ivill
US Army Research Laboratory, Aberdeen Proving Ground, MD 21005, USA

Fan Ren
Department of Chemical Engineering, University of Florida, Gainesville, FL 32611, USA

The properties of transition metal doped ZnO thin films are discussed. A review of current theories is presented, along with experimental results for the two most studied transition metal dopants, namely Mn and Co.

2.1 INTRODUCTION

The nascent field of spin electronics (or spintronics) represents the convergence of two fundamental properties of the electron — charge and spin — to form the basis for a new class of device design.[1–10] Operating by the manipulation, transport, and detection of charge carrier spins, spintronics is expected to improve upon traditional electronic and photonic devices, allowing for enhancements in the form of reduced power consumption, faster device operation, and new forms of information computation. Spintronics may lead to devices such as spin-polarized LEDs, spin-FETs, and spin-based qubits for quantum computers. Increased functionalities are also expected, such as integrated magnetic/electronic operations on the same chip. One aspect of spintronics of particular interest is the creation and control of spin-polarized currents in semiconducting material. Ferromagnetic semiconductors allow the "tools" of conventional semiconductor technology to be utilized, such as heterostructures and *pn* junctions, providing a convenient platform for a wide variety of device design. In order to

Handbook of Spintronic Semiconductors by W M Chen & I A Buyanova

www.panstanford.com
978-981-4267-36-6

fully realize semiconductor-based spintronics, significant challenges related to the lifetime, control, and detection of spin polarized carriers in semiconductors must be addressed.

Semiconducting behavior in ferromagnetic material was first uncovered with the discovery of the europium chalcogenides (EuO) and chromium spinels ($CdCr_2S_4$, $CdCr_2Se_4$). These materials are true ferromagnetic semiconductors in the sense that they have magnetic atoms intrinsic to the crystal sublattice. The elegant interplay between band electrons and localized magnetic ions in these materials brought about extensive research and scientific interest into the field. However, these materials have been limited to the realm of academic research for several reasons.[11–13] First of all, their crystal structures are incompatible with conventional semiconductors, like Si and GaAs, making their integration with contemporary electronics difficult. The synthesis of these materials is also cumbersome and hard to reproduce, making industrial production of the crystals expensive. And lastly, low ferromagnetic ordering temperatures ($T_C < 100$ K) make them less attractive for practical applications. Materials that can retain their ferromagnetism above room temperature are crucial to the practical application of spintronic devices.

Advancing in a slightly different direction, contemporary work has focused on making non-magnetic semiconductors magnetic by doping them with small amounts (typically a few percent) of magnetic atoms. This class of materials has attracted renewed interest in the development of magnetic semiconductors. Such compounds are known as dilute magnetic semiconductors, or DMS, because of the dilute concentrations of magnetic impurities. Notice these materials are fundamentally different than the Eu chalcogenides and Cr spinels since the magnetic atoms are artificially added into the lattice; the magnetic atoms are not a part of the periodic crystal structure of the parent material.

Early studies of DMS materials began with Mn-doped II-VI alloys of the form $A^{II}\ B^{VI}_{1-x}Mn_x$ (where A^{II} = Zn, Cd, Hg and B^{VI} = S, Se, Te). These materials were heavily studied in the 1980s and have been comprehensively reviewed by Furdyna.[14] It is worthwhile reviewing some aspects of these materials since ZnO also belongs to the II-VI family of semiconductors. The ternary structure of these compounds makes them amenable to tuning the lattice and band parameters by varying alloy composition, making them an attractive candidate for the preparation of heterostructure devices. The alloys crystallize into either the zinc-blende or wurtzite structure and are formed by sp^3 tetrahedral bonding, incorporating the valence s-electrons from the group II metal and the p-electrons from the group VI element. Elemental Mn has a half-filled $3d$-shell and two valence ($4s^2$) electrons. Manganese atoms may substitute on the group II sites as Mn^{+2} by giving up these two valence electrons. High solubilities of Mn in the host materials while maintaining the zinc-blende or wurtzite structures are possible, which is thought to arise from the chemical similarity of Mn^{+2} to the group II element.[14] Since the $3d$-shell of Mn is half-filled, it requires substantial energy to add an electron; this makes the $3d^5$ orbit act chemically similar to a $3d^{10}$ orbit. The magnetic properties of these alloys are dictated by the exchange interactions between local atomic moments (provided by the Mn) and the *sp*-band electrons, and have dramatic effects on the optical and electrical properties of the material, such as giant Faraday rotation and bound magnetic polaron formation. Driven mostly by superexchange mechanisms — an indirect exchange interaction mediated through the anion — these systems exhibit high temperature paramagnetism, low temperature spin-glass phase, and type III antiferromagnetic ordering.[14] Neutron diffraction studies show that the antiferromagnetic ordering of these structures is limited to short ranges, implying that the magnetic ordering is confined to the formation of small cluster regions.[14] However, ferromagnetic ordering has been achieved in low-dimensional CdMnTe quantum wells driven by hole-mediated exchange, albeit with low Curie temperature ($T_C < 2$ K).[15] An additional obstacle to the practical applicability of II-VI material is the capability of doping the material both *n*-type and *p*-type (bipolar doping). Again, these materials were not considered practical since the materials that did show ferromagnetism were restricted to very low ordering temperatures.

In the early 1990s, a technological advancement in DMS materials occurred with the discovery of ferromagnetism up to ~35 K in Mn-doped InAs.[13,16–18] InAs is an established III-V compound semiconductor material. Transition metal species are known to have very low solubility in host III-V materials, but the problem was overcome by non-equilibrium epitaxial growth using low temperature molecular beam epitaxy. III-V materials find widespread application in the electronics industries as high-speed digital devices, visible and infra-red light-emitting diodes and lasers, and magnetic sensors. The demonstration of ferromagnetism in InMnAs offered the intriguing opportunity to study spin-based phenomena in these well-established semiconductor devices.

Eventually, the success of MBE grown InMnAs led to the development of Mn-doped GaAs DMS. Segregation of Mn secondary phases, namely the MnAs phase, was suppressed using low temperature growth ($T_g = 250°$C).[12] Mn acts as an acceptor dopant when substituted on the group III sites leading to high hole concentrations which, as explained later, is a necessity for ferromagnetism in the material. The coupling of the charge- and spin-based processes have been repeatedly proven in GaMnAs, including the realization of spin-polarized light emission[19] and electrical and optical control over the ferromagnetism.[20] Unfortunately, GaMnAs is limited by its low Curie temperature of 170 K (well below room temperature). Raising the Curie temperature has been a prime challenge for GaMnAs based DMS.

2.2 DMS THEORY: THE PHYSICAL ORIGINS OF FERROMAGNETISM IN DMS

Understanding the physical mechanism behind magnetic ordering in DMS materials is an essential ingredient to their further development. Indeed, if both a conceptual and quantitative foundation to the origin of ferromagnetism in ZnO is developed, they may provide the direction necessary to a successful recipe for the fabrication of higher T_c materials with useful spin polarization. At the present time, however, there is an incomplete understanding of the origin of ferromagnetism in TM-doped ZnO and the subject remains an issue of active debate. This section will discuss the contemporary theories on the subject and the path they provide for subsequent research.

2.2.1 Dietl's Mean-Field Theory

The motivation for studying ZnO for spintronics began with the work of Dietl *et al.*[21] Dietl and co-workers employed a mean-field model of ferromagnetism, as originally described by Zener, to the case of III-V and II-IV compound semiconductors to predict the Curie temperatures for various Mn-doped DMS semiconductors and oxides. Zener's original model proposed that ferromagnetism is driven by the exchange interaction between carriers and localized moments. Zener's early model was neglected because it was found unsuitable to describe the magnetism of transition metals.[21] However, Dietl found that it could be used to accurately predict the ferromagnetic Curie temperatures of GaMnAs and ZnMnTe.

The model assumes that the ferromagnetic exchange interactions occur between localized spins doped into the semiconductor matrix and are mediated by charge carriers. These spins are assumed to be randomly distributed throughout the host semiconductor lattice. Specifically, the doped Mn ions reside on group II or III sites and provide the localized spins. Conceptually, the effect may be envisaged as a feedback between the magnetic spins and the carriers by a spin-spin process; the localized magnetic spin induces carrier polarization which then induces magnetic polarization (Fig. 2.1).[10,22] The model suggests that high values of T_C are obtainable in p-type material, while the T_C of n-type material should be constrained to lower temperatures. This can be attributed to both the large p–d exchange integral ($N_o\beta$) and density of states of the valence band, while the conduction band's s–d exchange ($N_o\alpha$) and density of states are significantly smaller.[23] Note that in the case of III-V semiconductors, Mn also acts as an acceptor dopant whereas it substitutes isovalently in the II-VI semiconductors. Dietl's calculations are summarized in Fig. 2.2 as a function of the semiconductor band-gap (based on a Mn concentration of 5% and a hole concentration of 3.5×10^{20} cm^{-3}). Of particular relevance are the predicted Curie temperatures in excess of room temperature for GaN and ZnO.

Diet's theory has proven useful in understanding the experimental results for GaMnAs. However, it does not appear to be consistent with the experimental results of transition metal doped wide band-gap semiconductors, such as the predictions for GaN and ZnO. This stems from several reasons, including the difficulty in experimentally preparing p-type ZnO material and the many observations of ferromagnetism in n-type ZnO DMS. Nevertheless, Dietl's original theory has led to multiple experimental and computational studies of transition metal doping in ZnO.

2.2.2 First-Principles Design: DFT Calculations

Sato and Katayama-Yoshida have employed first-principles design to investigate ferromagnetism in both semiconductor and oxide spintronics.[23–27] The magnetic stability of transition metal doped ZnO was calculated using density functional theory (DFT) within the framework of the local density approximation (LDA). The random distribution of transitional metal ions over the lattice (creating disorder in the alloy) was inherently included in the calculations by the coherent potential approximation (CPA). Magnetic stability was calculated by comparing the total energy difference between the ferromagnetic and spin-glass state, the lower of the two representing the ground state of the system. In the case of Mn, their results are consistent with Dietl's theory that the ferromagnetic state is stabilized with the addition of hole doping, and without holes the spin-glass state is favored. However, V, Cr, Fe, Co, and Ni impurities were predicted to be ferromagnetic without the need of additional charge carriers. Electron doping further stabilized the ferromagnetic state in these alloys. Their work also points to a contribution of d states at the Fermi level hinting at some delocalization of these d states. It was suggested that this could lead to ferromagnetic ordering through a double-exchange interaction in which ferromagnetic alignment is stabilized by the hopping of $3d$ electrons between neighboring TM sites. This mechanism is driven by partially unoccupied up-spin (or down-spin) states in the $3d$ band and is therefore not possible in the case of Mn, which exhibits a half-filled $3d$ band. In the case of p-type doping, however, the transference of weakly-bound $3d$ electrons between Mn ions may be mediated by the presence of holes. The valence band p-states hybridize with the $3d$ states of Mn and itinerant holes can retain their d-like character. This was suggested to stabilize the ferromagnetic phase for Mn doping.

2.2.3 Ferromagnetism in a Localized Carrier Regime

An additional theoretical approach considers whether ferromagnetic ordering between the localized spins can originate from localized carriers. Ferromagnetism in this localized carrier regime can be explained through the formation of bound magnetic polarons (BMPs).[28–30] A BMP is a quasi-particle comprised of the localized carrier and the magnetic atoms encompassed within its radius (Fig. 2.3).

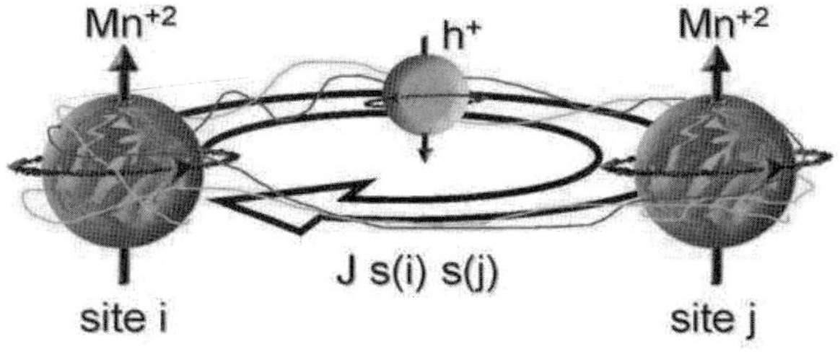

Figure 2.1. Schematic representation of magnetic exchange between two Mn ions mediated by a delocalized hole. Adapted from Ref. 10.

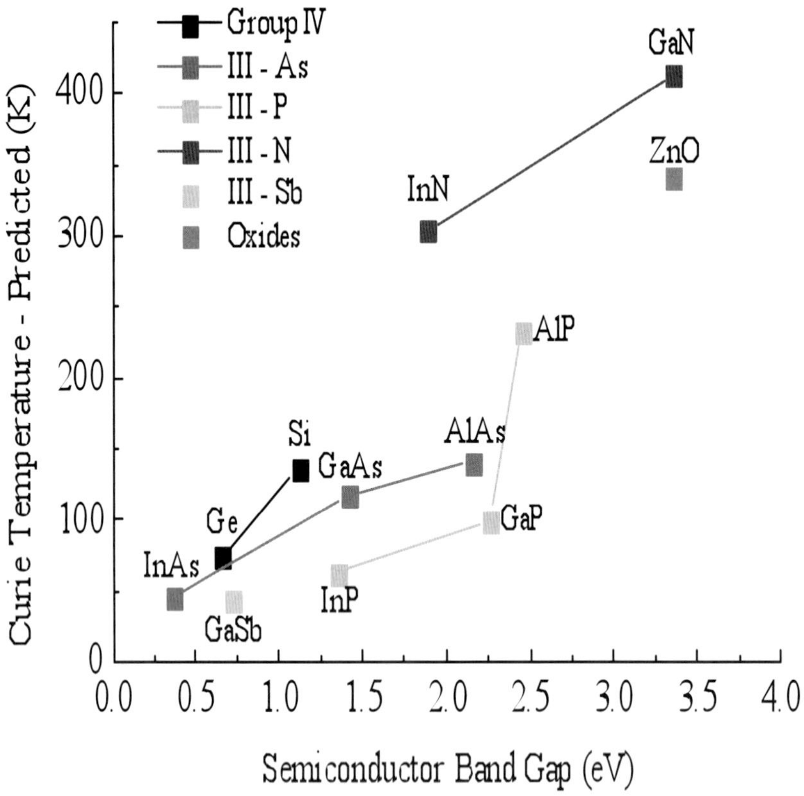

Figure 2.2. Predicted Curie temperatures based on Dietl's calculations[21] for 3% Mn and a hole concentration of 3.5×10^{20} cm^{-3} (after Ref. 24).

The localized carrier is bound to its associated defect (such as a donor atom if the carrier is an electron) in a hydrogenic orbital of radius, $r_h = \varepsilon(m/m^*)a_o$, where ε is the high frequency dielectric constant, m^* is the effective mass, and a_o is the Bohr radius (53 pm).[31] This radius can be large (~8 Å in ZnO) extending over several lattice constants, and can encompass a number of magnetic dopants depending on their concentration. The exchange interaction between the bound carrier and the magnetic moments tends to align the moments parallel to one another inside the BMP. At high temperatures, the BMPs may be isolated from one another. However, as the temperature is lowered, the BMP radius grows and the individual BMPs begin to overlap. Overlapping BMPs become correlated and their spins align, producing long-range ferromagnetic interactions.[29] At a critical temperature, the overlapping BMPs are percolated throughout the sample and the transition to ferromagnetism occurs. The BMP model is equally applicable to n-type or p-type material.[30]

The BMP model allows for ferromagnetism in an insulating or semi-insulating regime. This is especially attractive in the case of ZnO where high p-type doping, as required by Dietl's model, is inherently difficult.

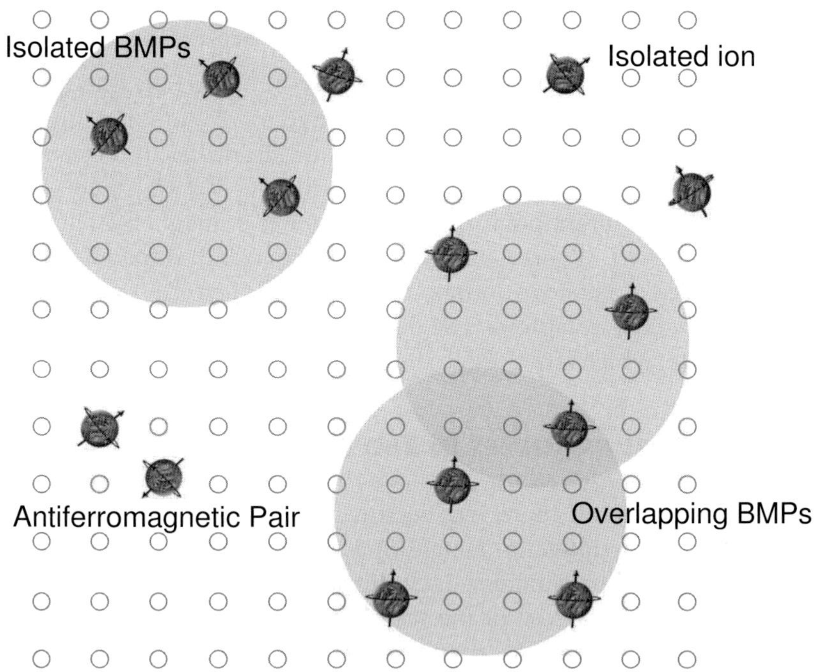

Figure 2.3. Illustration of bound magnetic polarons. An electron bound within its hydrogenic orbital couples to magnetic impurities causing them to align parallel to one another inside the orbit radius (adapted from Ref. 31).

2.2.4 Ferromagnetism in a Spin-Split Conduction Band

Coey *et al.* have proposed another model for ferromagnetism in DMS materials based on a spin-split donor impurity band.[31] The model is consistent with the observed magnetization for *n*-type transition metal doped ZnO. In this model, donor defects (which could arise from either oxygen vacancies or zinc interstitials in the case of ZnO) overlap at large concentrations to form an impurity band. The impurity band can interact with local magnetic moments through the formation of bound magnetic polarons. Within each BMP the bound carrier interacts with the magnetic dopants inside its radius aligning the spins of the magnetic dopants parallel to one another. Ferromagnetism is achieved when the BMPs overlap to form a continuous chain throughout the material, thus percolating ferromagnetism in the DMS. However, Coey showed that in this model, to achieve a high T_C, a fraction of the polaronic charge must delocalize (or hybridize) onto each magnetic dopant. In a band scheme, this occurs when the impurity band overlaps with unoccupied *d* levels of the magnetic dopant. It was shown that for Sc, Ti, and V, the spin-up states of the 3*d* TM metal are aligned with the impurity levels, resulting in significant alignment. Similarly for Fe, Co, and Ni doping, the spin-down states perform the same function. Interestingly, Mn and Cr doping would not lead to strong magnetization due to small hybridization.

Within the framework of Coey's model, Kittilstved *et al.* have performed detailed spectroscopic experiments on cobalt-doped ZnO.[32] Their results show that the singly ionized Co^{+} state lies close to the conduction band, similar in energy to a shallow donor state. Since the energies are similar, charge transfer can occur between the cobalt atoms and the donor impurities, thus leading to the hybridization necessary for ferromagnetism. Kittilstved *et al.* have also shown that this leads to an inherent polarity difference for ferromagnetism in cobalt and manganese-doped ZnO. Whereas ferromagnetism in cobalt-doped ZnO is closely tied to the presence of shallow donors, manganese-doped ZnO is closely tied to the presence of shallow acceptors. The difference lies in the location of the singly ionized Mn^{+3} state, which sits close to the valence band in ZnO.

2.3 EXPERIMENTAL PROGRESS IN ZnO DMS

At the experimental level, there has been a wide distribution in the magnetic properties reported for transition metal doped ZnO. Experiments have now covered a broad range of parameters, including various transition metal dopants (every element in the first row of the transition metal series has now been surveyed), compositional variations, preparation techniques and growth conditions, and post-growth processing. The observed results are often conflicting and non-reproducible between research groups.

The discrepancy in the observed properties likely stems from differrent growth techniques, synthesis conditions, and insufficient characterization. Most of the difficulties arise in determining if the material is a true DMS (TM atoms randomly substituting Zn lattice sites) or if ferromagnetism originates from TM clustering or dopant-induced secondary phases. In any case, the results indicate that the underlying mechanisms of ferromagnetism in transition metal doped ZnO are quite sensitive to growth conditions and must be clearly delineated by careful analysis.

Describing all the experimental trials in ZnO DMS over the past several years would be tedious and overwhelming. There are already several reviews covering the subject.[33–36] Instead, to provide a flavor of the experimental progress, a brief summary of the two most widely examined dopants in ZnO, namely Mn and Co, is provided.

2.3.1 *Mn-Doped ZnO*

Fukumura and co-workers were some of the first to study Mn-doped ZnO DMS using PLD.[37] A large solubility of 35% Mn was achieved while retaining the wurtzite structure of ZnO (reminiscent of the earlier studies on II-Mn-VI compounds discussed earlier). This is over the thermodynamic solid-solubility limit of Mn in ZnO and is a testament to the non-equilibrium conditions obtainable by thin film growth. They later showed the heavily doped alloy to exhibit spin-glass behavior with a spin-freezing temperature of $\sim$13 K due to strong antiferromagnetic exchange coupling between neighboring Mn atoms.[38] The high solubility of Mn achieved in ZnO motivated other experimental efforts into the synthesis of

ZnMnO. While some groups reported ferromagnetism, others observed antiferromagnetic, spin-glass, or paramagnetic behavior (for example, see Ref. 36).

Sharma and co-workers were the first to report ferromagnetism above room temperature in dilute Mn-doped ZnO bulk and thin film samples.[39] Bulk pellets with a nominal concentration of 2 at% Mn (EDS showed the actual concentration to be much lower at ~0.3 at%) sintered below 700°C were found to have a Curie temperature of 420 K. Additionally, thin films deposited by PLD with 2.2 at% Mn were shown to exhibit ferromagnetism at room temperature. However, using similar preparation techniques to Sharma *et al.*, Kundaliya and co-workers[40] demonstrated that the observed high-temperature ferromagnetism resulted from a metastable phase (oxygen-vacancy-stabilized $Mn_{2-x}Zn_xO_{3-\delta}$), and not from the proposed carrier-mediated ferromagnetism between Mn atoms. Likewise, there has also been a discrepancy in the reported overall distribution of Mn in ZnO. For example, a homogenous distribution of Mn was observed by Cheng and Chien,[41] while Jin *et al.*[42] found clustering of Mn atoms. Clearly, thorough characterization of the chemical structure is needed to fully appreciate and understand the origin of ferromagnetism in these materials.

Original theories motivating the search for ferromagnetism in ZnO DMS emphasize the importance of holes in mediating the exchange interaction between doped Mn atoms. Dietl's mean-field calculations predict that room temperature ferromagnetism is possible in Mn-doped ZnO that is heavily doped with holes, while carrier-mediated ferromagnetism in *n*-type material should be limited to lower temperatures. The work by Kittilstved and co-workers[32] demonstrated that the Mn^{+3} charge transfer energy lies close to the valence band, similar in energy to ZnO acceptor states. It is thought that this can lead to large hybridization necessary to support ferromagnetic ordering.

The synthesis and magnetic properties of Mn-doped ZnO epitaxial films co-doped for carrier density modulation has also been considered.[43] Co-doping allows independent control over the magnetic and electronic properties of the material by doping for each separately. In II-VI materials, Mn^{+2} is isovalent and should not directly introduce carriers. By co-doping II-VI semiconductors, Mn provides the localized spins while an additional dopant can be used to control the carrier concentration. This provides a convenient platform to study the effects of carrier concentration on the observed magnetic properties in ZnO DMS. Consider co-doping experiments with Sn. As a group IV cation, Sn can exist in either the 4+ or 2+ valence, although the 4+ valence is most common. As such, it can serve either as a doubly ionized donor or isoelectronic impurity. For the ZnO films deposited in this work, Sn behaves as a donor.

Epitaxial Mn, Sn-doped ZnO films were grown by conventional pulsed-laser deposition. The targets were fabricated with a nominal composition of 3 at% Mn and 0, 0.1, 0.01, and 0.001 at% Sn. Films were deposited onto single-crystal, c-plane oriented sapphire substrates. Film growth was conducted over a temperature range of 400–600°C.

The Mn valence state in the ZnO lattice was investigated using X-ray photoemission spectroscopy (XPS). Figure 2.4 shows the core level XPS spectra for a ZnO film doped with 3% Mn and 0.01% Sn. The data was charge corrected by shifting the O 1*s* peak to 530.1 eV. The film was sputtered with Ar^+ for 4 min to remove surface contamination. The Mn 2*p*3/2 binding energy is 640.7 eV. This is consistent with values assigned to Mn^{+2} in ZnO.[44,45] There is a ~6 eV energy difference between the Mn 2*p*3/2 and its higher binding energy satellite peak. The binding energy and satellite peak are consistent with that reported for single crystal MnO by Langell *et al.*[46] They found the satellite structure to be particularly sensitive to the oxide stoichiometry. In the case of Mn_2O_3 (the Mn^{+3} valence) the binding energy was higher at 641.1 eV and the satellite structure tended to decrease for the higher oxide phases.[46] The binding energies for metallic Mn and Mn^{+4} sit close to the Mn^{+2} value, but the energy for Mn has been seen at 637.7 eV and that for Mn^{+4} at 642.4 eV in ZnO.[44] The binding energy and satellite structure for the measured film suggests that most of the Mn doped into the ZnO is in the +2 valence state.

The room-temperature resistivity for the Mn-doped ZnO films with varying concentrations of Sn was determined using a four-point van der Pauw geometry. Defect chemistry calculations for Mn-doped ZnO indicate that Mn^{+2} forms a donor level ~2.0 eV below the conduction-band edge.[47] Defect chemistry calculations also indicate a reduction in Zn interstitials with Mn doping. Zn interstitials are generally accepted as the primary shallow donor defects in nominally undoped ZnO. This will yield an increase in resistivity for Mn-doped films as compared to undoped material.[47–49] The Mn-doped ZnO films with no Sn exhibit a resistivity on the order of $10^2 \Omega$ cm with a carrier density of mid-$10^{16}/cm^3$. This carrier density is substantially lower than that seen for undoped epitaxial films and is consistent with the reduction of shallow donors. Limited results on the doping behavior of Sn in ZnO indicate that it introduces a donor state,[50–54] although in some II-VI compound semiconductors, Sn is an amphoteric dopant, substituting on either the II or VI site.[55,56] For ZnO, the expectation is that Sn will substitute on the Zn site due to a close match in ionic radii between Zn^{+2} (0.074 nm) and Sn^{+4} (0.069 nm). For the epitaxial films considered in this work, Sn behaves as a donor. The resistivity of the films drops rapidly with Sn doping with a minimum of 0.185 Ω cm for an Sn concentration of 0.1%. Hall measurements indicate that the films are increasingly *n*-type with Sn doping up to 0.1 at%.

The magnetic properties of the films were measured using a SQUID magnetometer. The diamagnetic responses of the substrate and host semiconductor were subtracted from the magnetization plots. The primary focus of the measurements was to determine how the magnetic properties of the films changed as a function of electron density as controlled by Sn concentration. All the *M* versus *H* loops show hysteretic behavior with a finite coercivity and loop closure.

As noted earlier, increasing Sn concentration increases electron density and conductivity. Figure 2.5 shows the room-temperature magnetization versus field behavior for the ZnO samples containing 3% Mn and Sn contents of 0%, 0.1%, 0.01%, and 0.001%. Magnetization is given as the magnetic moment per Mn dopant

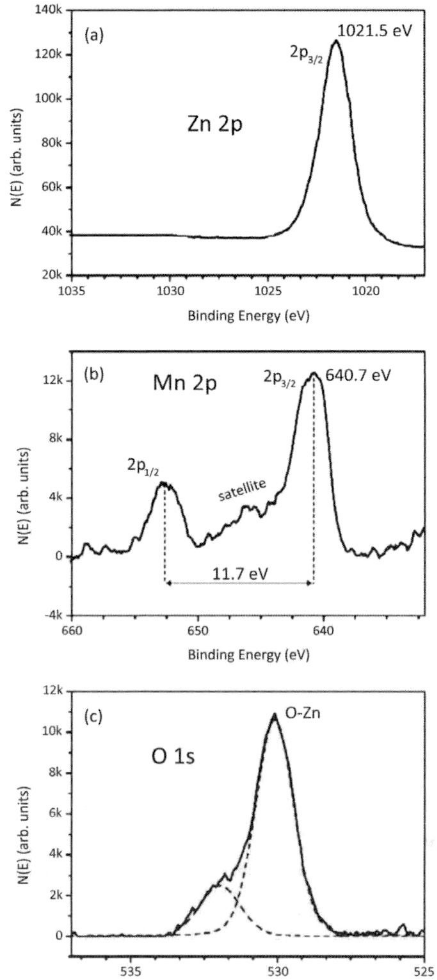

Figure 2.4. XPS spectra for ZnO:3% Mn film co-doped with 0.01% Sn. (a) Zn $2p3/2$ spectrum, (b) Mn $2p$ spectrum, and (c) O $1s$ spectrum. The film was sputtered in Ar for 4 min and charge corrected to the O $1s$ peak.

ion. Initially, there is an increase in magnetization with minimal Sn doping. However, with increasing Sn doping, there is an inverse correlation between the Sn content and saturation magnetization: as the electron density increases with Sn doping the magnetization decreases. The maximum magnetization corresponds to a magnetic moment per Mn ion of $\sim$0.5 μB/Mn. This is consistent with the bound magnetic polaron model in which only a fraction of the Mn ions are expected to order ferromagnetically due to competing super-exchange antiferromagnetic interactions between neighboring Mn ions.[57] The inverse correlation of saturation magnetization with electron density is interesting and provides some insight into the mechanism for ferromagnetism in Mn-doped ZnO. Overlap of the Mn d states with the valence band suggests that holes are necessary in order to induce ferromagnetic order. For semi-insulating films to exhibit ferromagnetism, the bound magnetic polaron model provides a mechanism whereby holes that are localized at or near the Mn ions are responsible for mediating ferromagnetism. The addition of electrons to the system will move the Fermi energy level up in the band-gap, resulting in a decrease in hole density and a reduction in magnetization. This is consistent with Kittilstved and co-worker's observation where ferromagnetism was induced when the holes from the acceptor states hybridize with the charge transfer state of Mn.

The magnetic properties of Mn-doped ZnO codoped with P, a known *p*-type dopant in ZnO were also considered.[58] The behavior of phosphorus in ZnO epitaxial films both as-deposited and upon annealing has been reported in detail elsewhere.[59] For the as-deposited films, the inclusion of phosphorus yields a significant increase in electron density, resulting in ZnO that is highly conductive and *n*-type. The shallow donor behavior in the as-deposited films is inconsistent with P substitution on the O site, and presumably originates from either substitution on the Zn site or the formation of a phosphorus-bearing complex. Previous work has shown that the defect-related carrier density in nominally undoped ZnO can be reduced via high temperature annealing in oxygen or air. In the case of undoped material, the reduction in donor density is presumed due to either a reduction in oxygen vacancies, Zn interstitials, or perhaps out-diffusion of hydrogen that is incorporated in the ZnO lattice during synthesis. In order to reduce electron density, annealing in oxygen can be performed. As the films are annealed at increasing temperatures, the resistivity of the phosphorus-doped films increases rapidly.

The magnetic properties of the films were measured using a Quantum Design SQUID magnetometer. The primary focus of the measurements was to determine how the magnetic properties of the films changed as a function of electron density as controlled by P doping. Figure 2.6 shows the room temperature magnetization as a function of applied magnetic field for epitaxial ZnO:3% Mn films both without and with P co-doping. For the Mn-doped film with no P, saturation in the

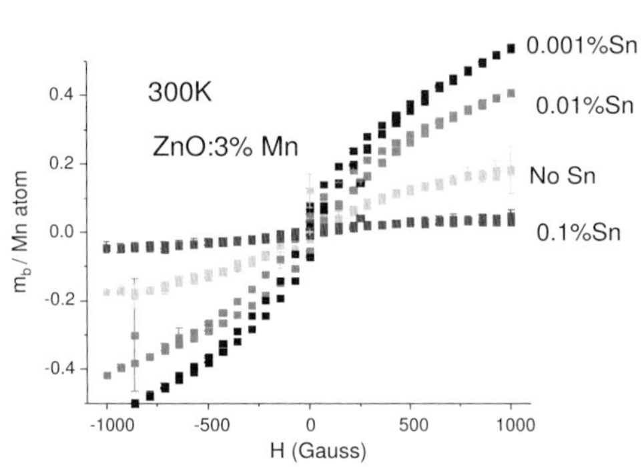

Figure 2.5. Magnetization measured at 300 K for epitaxial ZnO:3% Mn films that are co-doped with 0.001% Sn, 0.01% Sn, 0.1% Sn, and no Sn. There appears to be an inverse correlation of the Sn content with the saturation magnetization.

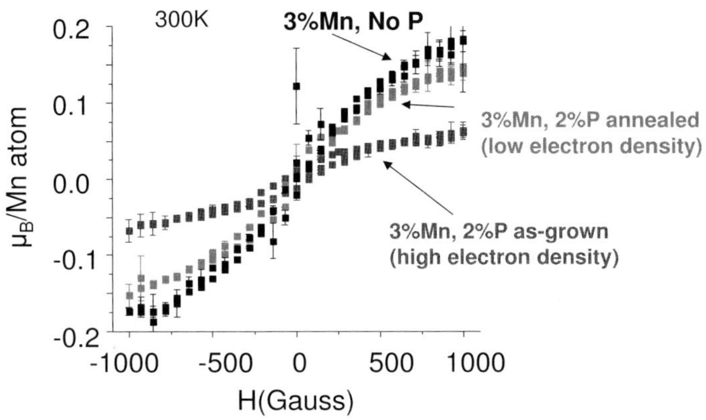

Figure 2.6. Room temperature SQUID measurements for epitaxial ZnO:3% Mn, 2% P films before and after annealing. Also shown is a ZnO:3% Mn film with no P.

magnetization is observed, but with little evidence for hysteresis in the *M* versus *H* curves. The as-deposited ZnO film doped with both Mn and P showed a reduction in magnetization/Mn ion. This is consistent with the proposed models for Mn-doped ZnO where ferromagnetic ordering is not favored by electron doping. Most interesting is the saturation magnetization behavior as the P doped samples are annealed. As noted earlier, increasing P concentration in as-deposited films initially increases electron density and conductivity. Consequently there is a decrease in magnetization with intial P doping due to the increase in electron density. After annealing, however, the electron density is decreased causing an increase in the magnetization. Similar results are seen at 10 K. This behavior is analagous to the inverse correlation of saturation magnetization with electron density observed in the Sn co-doping experiments. The important observation is that the activation of acceptor states for hole formation correlates with an enhancement in ferromagnetism. The holes may be delocalized, but with low mobility, thus yielding low conductivity. In this case, the carrier mediated mechanism may suffice without the need of invoking bound polarons as inherent to the ferromagnetic ordering. In either case, the addition of electrons to the system will move the Fermi energy level up in the band-gap resulting in a decrease in hole density and a reduction in magnetization. This appears consistent with early work on trivalent doped (Zn, Mn)O where no ferromagnetism was observed for heavily *n*-type films. It may also explain the discrepancy from other studies of Mn-doped ZnO films in which the intrinsic defect-mediated donor states are high in density. It should be noted that the amount of magnetization in the material remains relatively low at all temperatures.

The results of this study are consistent with studies on the carrier type dependence in Co- and Mn-doped ZnO nanocrystalline films.[32,60] In that case, ferromagnetism was observed in Mn-doped ZnO nanocrystals only when nitrogen, a group V acceptor dopant, was introduced during the synthesis process. Based on this and other properties, it was concluded that ferromagnetism in ZnO is closely tied to the charge transfer electronic structure of the transition metal dopant. For Mn, ferromagnetism is induced when the holes from the acceptor ion delocalize onto Mn^{2+}. Our results are consistent with this conclusion. However, one of the challenges in understanding these and other similar results stems from the low carrier concentration in the sample, negating an explanation based on the mean-field approximation.

2.3.2 Co-Doped ZnO

One of the initial works on cobalt-doped ZnO DMS was by Ueda *et al.*[61] They found the material to be ferromagnetic above 280 K with 5–25% Co and 1% Al (added as an n-type dopant) without the observation of secondary phases. Differences in the magnetization were attributed to differences in the conductivity; films with higher carrier concentrations ($\sim 10^{20}\,cm^{-3}$) showed ferromagnetic features with higher M_s and T_C. Since then, additional experimental studies have investigated the properties and origin of ferromagnetism in cobalt-doped ZnO. Again, the results are conflicting with reports of ferromagnetism in phase pure films,[62,63] ferromagnetism from clusters,[64] and no observed ferromagnetism.[65]

The first report of reversible (controlled) switching of ferromagnetism in a DMS at 300 K was demonstrated by Schwartz and Gamelin in cobalt-doped ZnO.[66] The reversibility was mediated by the incorporation and removal of Zn interstitials. The Zn interstitial (Zn_i) is a known n-type dopant that produces a shallow donor level below the conduction band. Diffusing Zn_i into the lattice lowers the conductivity and activates room temperature ferromagnetism. Removing Zn_i, by heating in air, returned the films to an insulating state and subsequently quenched the ferromagnetism. The process was reversible over many cycles. This reversibility is evidence that free carriers activate ferromagnetism in cobalt-doped ZnO. The process was observed in both MOCVD grown films and ZnO:Co nanoparticle films prepared by spin coating. Strong hybridization of Zn_i donor states with Co^{+2} states near the conduction band (which, as explained earlier, is theoretically believed to cause ferromagnetism) was used to explain the magnetic ordering. Conduction electrons, derived from the Zn_i donors, delocalize over several Co^{+2} ions and ferromagnetically align their spins through a double-exchange interaction.

Importantly, from the same group, Kittilstved was able to demonstrate a chemical polarity difference between the ferromagnetism in ZnCoO and ZnMnO.[60] Specifically, p-type ZnMnO led to ferromagnetism, while ferromagnetism in ZnCoO was activated by n-type doping. Doping of the ZnMnO was done on a local level by N-capping ZnMnO nanoparticles with amines. ZnCoO nanoparticle films were made n-type by capping with oxygen. Reversing the capping layers, ZnCoO:N and ZnMnO:O, led to the disappearance of ferromagnetism in both sets of films. Optical absorption, MCD, and photoconductivity measurements were employed to understand this inherent polarity difference.[32] For n-type ZnCoO, the authors showed that a resonance in the charge transfer ($Co^{+1} \rightarrow Co^{+2} + e_{CB}$,

$\Delta E \approx 0.27$ eV) and donor state energies can lead to a large hybridization necessary for ferromagnetism. For ZnMnO, a similar resonance was observed but derived from the Mn^{+3} state close to the valence band ($Mn^{+3} \rightarrow Mn^{+2} + h^{+}$, $\Delta E \approx 0.22$ eV) with acceptor state energies.

In recent work, the structure and magnetic behavior of cobalt-doped ZnO thin films have been examined for cobalt concentration varied over a wide range.[67] Cobalt-doped ZnO epitaxial films were deposited via pulsed laser deposition (PLD)[68] onto c-plane oriented sapphire substrates. The targets were mixed to give proportions of $Zn_{1-x}Co_xO$ with $x = 0.00, 0.02, 0.05, 0.10, 0.15$, and 0.30. A temperature range of 400–600°C and oxygen pressures up to 0.02 mTorr were used in the experiments.

Crystal structure and phase analysis were characterized using X-ray diffraction (XRD) in Bragg–Brentano geometry. From the θ–2θ X-ray diffraction patterns in Fig. 2.7, there is a small peak noticeable in some of the Co-doped films around $2\theta = 44.4$ degrees. This peak is consistently recorded at 15% and 30% cobalt and occasionally seen in films with lower cobalt concentrations, such as the film with 5% Co. The peak intensity is low (only a few hundred counts above background) and does not correspond to any ZnO or substrate peaks. Both the small intensity and 2θ position make identification of the peak difficult using XRD as there are several cobalt containing phases with similar 2θ values of around 44.4 degrees, including the spinel family of cobalt oxides and cobalt metal.[69] A longer scan (step size = 0.005°, time/step = 5.0 sec) around this peak, performed on the film with 30% Co, shows one broad Gaussian peak and does not show signs of asymmetry or overlapping peaks. Determination of the phase responsible for the peak is critical since the presence of ferromagnetic cobalt metal could contribute to the magnetic signature of the films. The cubic and spinel cobalt oxides are antiferromagnetic, though some papers report that small nanocluster powders of cobalt oxides are ferromagnetic due to uncompensated surface spins.[70,71]

High resolution XRD and TEM were used to characterize the secondary phase observed in the powder XRD scans. Cross-sectional TEM was used to more precisely delineate the nature and location of the extra phase as shown in Fig. 2.8. Most of the precipitation occurs near the film/substrate interface in the form of small (~5 nm) particulates. However, for most of the film the cobalt dopant appears to reside in the ZnO lattice without precipitation. TEM diffraction patterns of the film and nanoprecipitates are shown in Fig. 2.9. The particles appear to be oriented with the lattice and have d-spacings of $d_{002} = 0.20$ nm and $d_{-210} = 0.13$ nm. This is consistent with metallic cobalt, which can exist in either a hexagonal or face-centered cubic structure. Oriented cobalt nanoparticles have also been observed in ion-implanted cobalt-doped ZnO.[72]

At higher growth temperatures, cobalt precipitates out of the lattice as CoO. At 500°C the CoO phase begins to form along with the cobalt metal. At 600°C the cobalt metal phase disappears and the CoO phase becomes prominent. Thermodynamically, CoO is the most stable phase at these temperatures and pressure. This can be seen from the thermodynamic predominance diagram for cobalt oxide.[67]

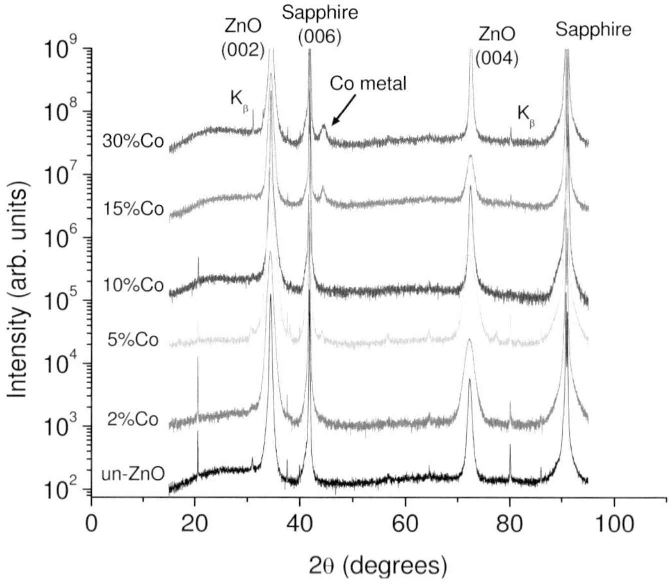

Figure 2.7. XRD scans for a series of ZnCoO films grown in vacuum at 400°C. The films are predominately *c*-axis oriented ZnO. At high cobalt concentrations, cobalt metal precipitation appears (corresponding to the peak near $2\theta = 44.4°$).

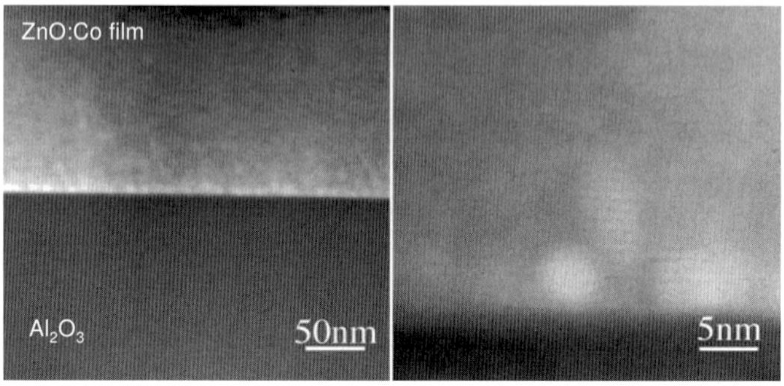

Figure 2.8. TEM micrographs of a sample doped with 30 at% Co. This film shows the presence of metallic cobalt in the XRD scan. The precipitation appears mostly at the substrate/film interface in the form of ~5 nm particles.

Interestingly, the formation of Co_3O_4 is expected at lower growth temperatures and not the metallic cobalt phase determined from TEM. This suggests the cobalt metal phase is thermodynamically metastable, but gains some structural stability from the ZnO lattice. This is consistent with the well-oriented particles observed in TEM.

The volume magnetization of the films was measured using a SQUID magnetometer. Before measuring, the backs and sides of the samples were etched in nitric acid (50% nitric/50% deionized water) to remove excess silver paint and contaminants that could contribute a spurious magnetic signal measured by the SQUID.

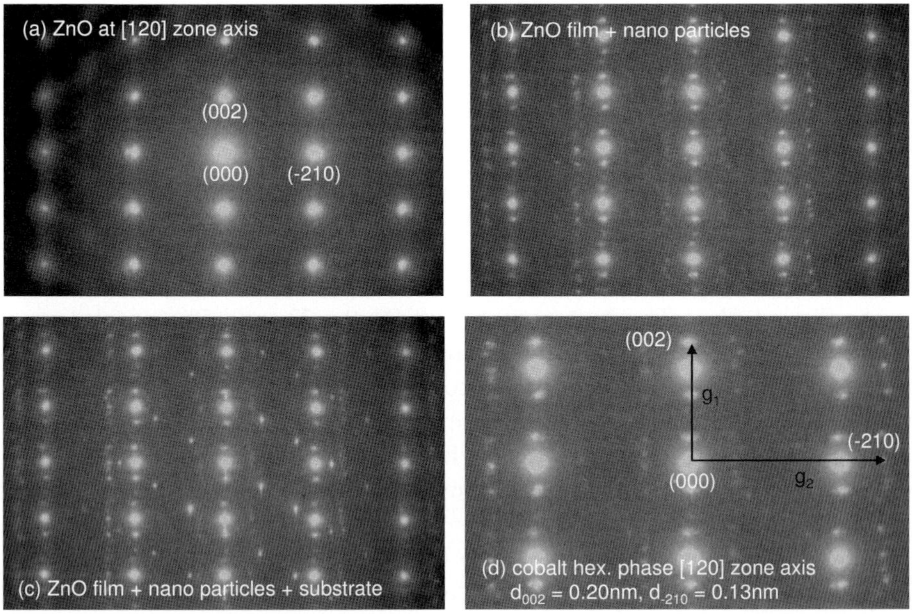

Figure 2.9. Selected area TEM diffraction patterns of a ZnO film doped with 30 at% Co grown at 400°C in vacuum: (a) wurtzite ZnO along [120] zone axis, (b) ZnO film + nanoparticles, (c) ZnO film + nanoparticles + sapphire substrate, (d) nanoparticles along [120] zone axis with $d_{002} = 0.20$ nm and $d_{-210} = 0.13$ nm, which is consistent with metallic cobalt.

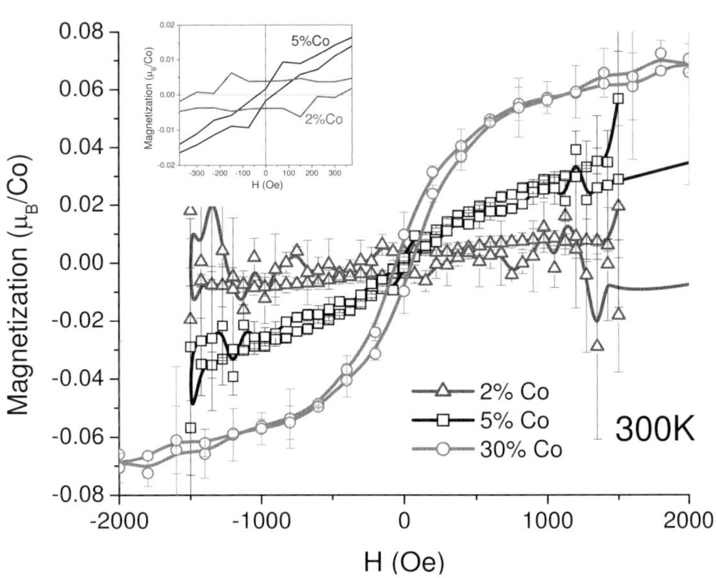

Figure 2.10. Room temperature SQUID magnetization curves for films deposited at 400°C in vacuum with 2%, 5% and 30% Co.

Each film surface was first coated in photoresist and baked for 20 min at 50°C to help protect the film during etching, and then floated on top of the nitric acid for 3 min. The photoresist was removed by rinsing in acetone. The SQUID magnetization data is normalized by the number of Bohr magnetons (μ_B) per Co atom. To convert the raw magnetization data from emu units to μ_B/Co, the cobalt atoms are assumed to occupy Zn sites and a cation density of 4.18×10^{22} cm^{-3} is used for the conversion.

Magnetization data taken at 300 K for a series of films grown at 400°C under vacuum with different amounts of cobalt are displayed in Fig. 2.10. Films grown in higher oxygen pressures (up to 20 mTorr) show very little if any magnetization. Films doped with 2% and 5% cobalt show smaller magnetization than the more heavily alloyed film. The hysteresis curves for the 2% and 5% samples are not well developed. A closer look reveals that the curves have a broadening at low field, which appears to be an additional component to the curve (inset of Fig. 2.10). This may represent a small ferromagnetic component superimposed on an overall paramagnetic state. This is consistent with other reports that the Co sublattice is paramagnetic while an additional ferromagnetic component is present, presumably from defects in the anion sublattice.[73] However, this behavior is not apparent in the 30% Co-doped film. The 30% Co film shows ferromagnetism up to 300 K with a clear hysteretic shape. The sample has a room temperature magnetization approaching 0.08 μ_β/Co at an applied field of 3 T. Additionally, a film doped with 30% Co that showed some metallic cobalt segregation (a film deposited at low base pressure) was measured and displayed a room-temperature saturation of around 0.22 μ_β/Co (not shown).

At higher pressures the ferromagnetic moment decreases considerably. Figure 2.11 shows the comparison between room temperature SQUID measurements for 30% Co-doped films prepared in vacuum and in 0.02 mTorr O_2. Neither of these films shows metallic cobalt precipitation by XRD. There is a 10-fold reduction in the saturation magnetization ($M_s \sim 0.009\,\mu_B$/Co) and the hysteresis loop closes to a coercivity of 10 Oe. The decrease in ferromagnetism in the presence of molecular oxygen is evidence that the intrinsic growth defects in ZnO play a significant role in mediating the observed magnetic properties. It is worth noting that the room temperature resistivity increases from 0.566 ohm cm when grown in vacuum to 5.579 ohm cm when grown in 0.02 mTorr O_2 indicating a change in the defect concentration.

Theoretical modeling shows that Co^{+2}–oxygen vacancy (Co–Ov) pairs are capable of producing a long-range ferromagnetic ground state in ZnO.[74] A concentration of about 6% Co–Ov pairs are estimated to place ZnO in the ferromagnetic state and produce exchange couplings large enough to produce anomalous Hall effect in transport measurements.[74] We have observed the presence of the anomalous Hall effect at room temperature in select films, including films doped with 15% and 30% cobalt.

Another factor to consider is the low magnetic saturation in these films. The high-spin moment ($3d^7$) of Co^{+2} is 3 μ_B, while that of metallic cobalt is 1.7 μ_B. The saturation magnetization of the 30% Co-doped film is rather small at 0.08 μ_B/Co. This small value, however, is in good agreement with reports on Co-doped ZnO that has been modified by vacuum annealing[75] or by Zn vapor anneals.[66] Also, the increased magnetization with cobalt concentration is in agreement with some results[76] and in contrast with others that show a decreasing magnetization with increasing cobalt concentration.[63] The exchange coupling between Co–Co nearest neighbors in ZnO are believed to be antiferromagnetic[77]; thus large concentrations of Co–Co pairs should reduce the overall moment. So there are likely competing mechanisms between regions of Co–defect pair ferromagnetism, Co–Co nearest-neighbor antiferromagnetic exchange, and isolated cobalt atoms that act paramagnetically. At large concentrations, there may be enough Co–defect pairs to produce FM, but this will be compensated by Co pairs that are antiferromagnetically aligned, which will reduce the overall moment per cobalt atom. The small observed moment in these films is likely a result of the competing mechanisms. The larger magnetization from depositing in an oxygen-deficient environment suggests oxygen vacancies are responsible, but the unintentional incorporation of hydrogen during growth under higher residual base pressures is also a possibility. It has been suggested that hydrogen could play a role in mediating ferromagnetism in transition-metal doped ZnO.[78] It is known that hydrogen easily diffuses into and out of ZnO at moderate temperatures,[79] and recent experiments suggest some correlation between magnetization and hydrogen content in transition metal doped ZnO.[80]

2.4 CONCLUSIONS AND OUTLOOK

The field of spintronics has recently attracted much attention because of its potential to provide new functionalities and enhanced performance in

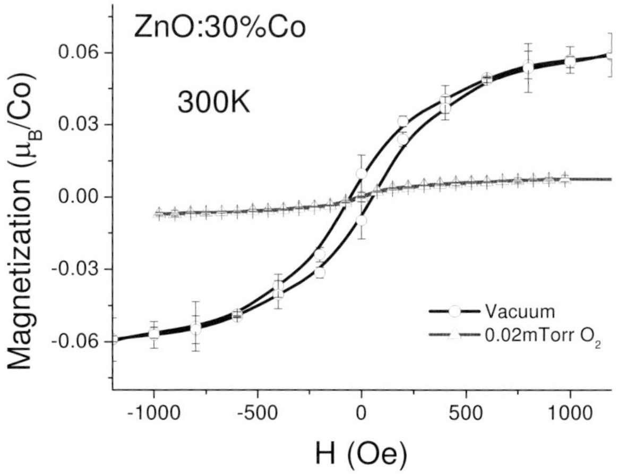

Figure 2.11. Room temperature magnetic hysteresis curves for 30% Co-doped films deposited at 400°C under vacuum and in 0.02 mTorr O_2.

conventional electronic devices. Oxide materials provide a convenient platform to study the spin-based functionality in host semiconducting material. The most pertinent open question is whether the apparent spin ordering seen in transition metal doped oxides, such as ZnO, extends to the carriers useful in transport and device applications.

References

[1] G. A. Prinz, *Physics Today* **4**, 58 (1995).
[2] G. A. Prinz, *Science* **282**, 1660 (1998).
[3] S. A. Wolf, A. Y. Chtchelkanova and D. M. Treger, *IBM J. Res. Devel.* **50**, 101 (2006).
[4] S. A. Wolf, D. Treger and A. Chtchelkanova, *MRS Bull.* **31**, 400 (2006).
[5] S. A. Wolf, D. D. Awschalom, R. A. Buhrman, J. M. Daughton, S. V. Molnar, M. L. Roukes, A. Y. Chtchelkanova and D. M. Treger, *Science* **294**, 1488 (2001).
[6] S. J. Pearton, D. P. Norton, R. Frazier, S. Y. Han, C. R. Abernathy and J. M. Zavada, *IEE Proc. Circ. Dev. Syst.* **152**, 312 (2005).
[7] H. Ohno, D. Chiba, F. Matsukura, T. Omlya, E. Abe, T. Dietl, Y. Ohno and K. Ohtani, *Nature* **408**, 944 (2000).
[8] T. Schallenberg and H. Munekata, *Appl. Phys. Lett.* **89**, 042507 (2006).
[9] A. M. Nazmul, S. Sugahara and M. Tanaka, *Phys. Rev. B* **67**, 241308 (2003).
[10] S. J. Pearton, C. R. Abernathy, D. P. Norton, A. F. Hebard, Y. D. Park, L. A. Boatner and J. D. Budai, *Mat. Sci. Eng. R* **40**, 137 (2003).
[11] C. Haas, *Crit. Rev. Solid State Mater. Sci.* **1**, 47 (1970).
[12] H. Ohno, *Science* **281**(5379), 951 (1998).
[13] G. Thaler, PhD Dissertation, University of Florida (2004).
[14] J. K. Furdyna, *J. Appl. Phys.* **64**, R29 (1988).
[15] A. Haury, A. Wasiela, A. Arnoult, J. Cibert, S. Tatarenko, T. Dietl and Y. M. d'Aubigne, *Phys. Rev. Lett.* **79**, 511 (1997).
[16] H. Ohno, H. Munekata, S. von Molnar and L. L. Chang, New iii-v diluted magnetic semiconductors (invited), *35th Annual Conference on Magnetism and Magnetic Materials*, Vol. 69, April 15 (1991), pp. 6103–6108.
[17] S. Koshihara, A. Oiwa, M. Hirasawa, S. Katsumoto, Y. Iye, C. Urano, H. Takagi and H. Munekata, *Phys. Rev. Lett.* **78**, 4617 (1997).
[18] H. Munekata, H. Ohno, S. von Molnar, A. Segmuller, L. L. Chang and L. Esaki, *Phys. Rev. Lett.* **63**, 1849 (1989).
[19] Y. Ohno, D. K. Young, B. Beschoten, F. Matsukura, H. Ohno and D. D. Awschalom, *Nature* **402**, 790 (1999).
[20] A. M. Nazmul, S. Kobayashi, S. Sugahara and M. Tanaka, *Jpn. J. Appl. Phys. Lett.* **43**, 233 (2004).
[21] T. Dietl, H. Ohno, F. Matsukura, J. Cibert and D. Ferrand, *Science* **287**, 1019 (2000).
[22] H. Ohno and F. Matsukura, *Solid State Commun.* **117**, 179 (2001).
[23] T. Dietl, *Semicond. Sci. Technol.* **17**, 377 (2002).
[24] S. J. Pearton, W. H. Heo, M. Ivill, D. P. Norton and T. Steiner, *Semicond. Sci. Technol.* 19, R59 (2004).
[25] H. Katayama-Yoshida and K. Sato, *Physica B* **327**, 337 (2003).
[26] K. Sato and H. Katayama-Yoshida, *Semicond. Sci. Technol.* **17**, 367 (2002).
[27] K. Sato and H. Katayama-Yoshida, *Jpn. J. Appl. Phys. Lett.* **39**, 555 (2000).

[28] M. Berciu and R. N. Bhatt, *Phys. Rev. Lett.* **87**, 107203 (2001).

[29] P. A. Wolff, R. N. Bhatt and A. C. Durst, *J. Appl. Phys.* **79**, 5196 (1996).

[30] A. Kaminski and S. D. Sarma, *Phys. Rev. Lett.* **88**, 247202 (2002).

[31] J. M. D. Coey, M. Venkatesan and C. B. Fitzgerald, *Nature Mater.* **4**, 173 (2005).

[32] K. R. Kittilstved, W. K. Liu and D. R. Gamelin, *Nature Mater.* **5**, 291 (2006).

[33] R. Janisch, P. Gopal and N. A. Spaldin, *J. Phys. Cond. Matt.* **17**, R657 (2005).

[34] C. Liu, F. Yun and H. Morkoc, *J. Mater. Sci. Mater. Electron.* **16**, 555 (2005).

[35] S. A. Chambers, *Surf. Sci. Rep.* **61**, 345 (2006).

[36] S. J. Pearton, D. P. Norton, M. P. Ivill, A. F. Hebard, J. M. Zavada, W. M. Chen and I. A. Buyanova, *IEEE Trans. Electron Dev.* **54**, 1040 (2007).

[37] T. Fukumura, Z. Jin, A. Ohtomo, H. Koinuma and M. Kawasaki, *Appl. Phys. Lett.* **75**, 3366 (1999).

[38] T. Fukumura, Z. Jin, M. Kawasaki, T. Shono, T. Hasegawa, S. Koshihara and H. Koinuma, *Appl. Phys. Lett.* **78**, 958 (2001).

[39] P. Sharma, A. Gupta, K. V. Rao, F. J. Owens, R. Sharma, R. Ahuja, J. M. O. Guillen, B. Johansson and G. A. Gehring, *Nature Mater.* **2**, 673 (2003).

[40] D. C. Kundaliya, S. B. Ogale, S. E. Lofland, S. Dhar, C. J. Metting, S. R. Shinde, Z. Ma, B. Varughese, K. V. Ramanujachary, L. Salamanca-Riba and T. Venkatesan, *Nature Mater.* **3**, 709 (2004).

[41] X. M. Cheng and C. L. Chien, *J. Appl. Phys.* **93**, 7876 (2003).

[42] Z.-W. Jin, Y. Z. Yoo, T. Sekiguchi, T. Chikyow, H. Ofuchi, H. Fujioka, M. Oshima and H. Koinuma, *Appl. Phys. Lett.* **83**, 39 (2003).

[43] M. Ivill, S. J. Pearton, D. P. Norton, J. Kelly and A. F. Hebard, *J. Appl. Phys.* **97**, 053904 (2005).

[44] H. T. Cao, Z. L. Pei, J. Gong, C. Sun, R. F. Huang and L. S. Wen, *J. Solid State Chem.* **177**, 1480 (2004).

[45] Z.-B. Gu, C.-S. Yuan, M.-H. Lu, J. Wang, D. Wu, S.-T. Zhang, S.-N. Zhu, Y.-Y. Zhu and Y.-F. Chen, *J. Appl. Phys.* **98**, 053908 (2005).

[46] M. A. Langell, C. W. Hutchings, G. A. Carson and M. H. Nassir, *J. Vacuum Sci. Technol. A* **14**, 1656 (1996).

[47] J. Han, P. Q. Mantas and A. M. R. Senos, *J. Euro. Ceram. Soc.* **22**, 49 (2002).

[48] Z. Zhou, K. Kato, T. Komaki, M. Yoshino, H. Yukawa, M. Morinaga and K. Morita, *J. Euro. Ceram. Soc.* **24**, 139 (2004).

[49] N. Ohashi, J. Tanaka, T. Ohgaki, H. Haneda, M. Ozawa and T. Tsurumi, *J. Mater. Res.* **17**, 1529 (2002).

[50] F. P. D., J. Morales, W. E. L., E. Andrade and M. Miki-Yoshida, *Thin Solid Films* **366**, 16 (2000).

[51] A. Bougrine, A. E. Hichou, M. Addou, J. Ebothe, A. Kachouane and M. Troyon, *Mater. Chem. Phys.* **80**, 438 (2003).

[52] J.-H. Lee and B.-O. Park, *Thin Solid Films* **426**, 94 (2003).

[53] H. Sato, T. Minami and S. Takata, *J. Vacuum Sci. Technol. A* **11**, 2975 (1993).

[54] E. Holmelund, J. Schou, S. Tougaard and N. B. Larsen, *Proceedings of COLA* 2001, Tsukuba, Japan, October 1 (Elsevier Science, 2001), pp. 467–471.

[55] I. Turkevych, R. Grill, J. Franc, P. Hoschl, E. Belas, P. Moravec, M. Fiederle and K. W. Benz, *Cryst. Res. Technol.* **38**, 288 (2003).

[56] O. Panchuk, A. Savitskiy, P. Fochuk, Y. Nykonyuk, O. Parfenyuk, L. Shcherbak, M. Ilashchuk, L. Yatsunyk and P. Feychuk, *J. Cryst. Growth* **197**, 607 (1999).

[57] T. Dietl, M. Sawicki, L. V. Khoi, J. Jaroszynski, P. Kossacki, J. Cibert, D. Ferrand, S. Tatarenko and A. Wasiela, *Physica Status Solidi (B)* **229**, 665 (2002).

[58] M. Ivill, S. J. Pearton, Y. W. Heo, J. Kelly and A. F. Hebard, *J. Appl. Phys.* **101**, 123909 (2007).

[59] Y. W. Heo, S. J. Park, K. Ip, S. J. Pearton and D. P. Norton, *Appl. Phys. Lett.* **83**, 1128 (2003).

[60] K. R. Kittilstved, N. S. Norberg and D. R. Gamelin, *Phys. Rev. Lett.* **94**, 147209 (2005).

[61] K. Ueda, H. Tabata and T. Kawai, *Appl. Phys. Lett.* **79**, 988 (2001).

[62] S. Ramachandran, A. Tiwari and J. Narayan, *Appl. Phys. Lett.* **84**, 5255 (2004).

[63] C. B. Fitzgerald, M. Venkatesan, J. G. Lunney, L. S. Dorneles and J. M. D. Coey, *Appl. Surf. Sci.* **247**, 493 (2005).

[64] J. H. Park, M. G. Kim, H. M. Jang, S. Ryu and Y. M. Kim, *Appl. Phys. Lett.* **84**, 1338 (2004).

[65] Z. Jin, T. Fukumura, M. Kawasaki, K. Ando, H. Saito, T. Sekiguchi, Y. Z. Yoo, M. Murakami, Y. Matsumoto, T. Hasegawa and H. Koinuma, *Appl. Phys. Lett.* **78**, 3824 (2001).

[66] D. A. Schwartz and D. R. Gamelin, *Adv. Mater.* **16**, 2115 (2004).

[67] M. Ivill, S. J. Pearton, S. Rawal, L. Leu, P. Sadik, R. Das, A. F. Hebard, M. Chisholm, J. D. Budai and D. P. Norton, *New J. Phys.* **10**, 065002 (2008).

[68] D. P. Norton, *Mat. Sci. Eng. R* **43**, 139 (2004).

[69] H. Zhou, L. Chen, V. Malik, C. Knies, D. M. Hofmann, K. P. Bhatti, S. Chaudhary, P. J. Klar, W. Heimbrodt, C. Klingshirn and H. Kalt, *Phys. Status Solidi A* **204**, 112. (2007).

[70] H. Zhang and X. Chen, *Nanotechnology* **16**, 2288 (2005).

[71] Y. Ichiyanag, Y. Kimishima and S. Yamada, *J. Magn. Magn. Mater.* **272–276**, E1245 (2004).

[72] D. P. Norton, M. E. Overberg, S. J. Pearton, K. Pruessner, J. D. Budai, L. A. Boatner, M. F. Chisholm, J. S. Lee, Z. G. Khim, Y. D. Park and R. G. Wilson, *Appl. Phys. Lett.* **83**, 5488 (2003).

[73] A. Barla, G. Schmerber, E. Beaurepaire, A. Dinia, H. Beiber, S. Colis, F. Scheurer, J. P. Kappler, P. Imperia, F. Nolting, F. Wilhelm, A. Rogalev, D. Muller and J. J. Grob, *Phys. Rev. B* **76**, 125201 (2007).

[74] C. D. Pemmaraju, R. Hanafin, T. Archer, H. B. Braun and S. Sanvito, *arXiv:0801.4945v1 [cond-mat.mtrl-sci]* (2008).

[75] A. C. Tuan, J. D. Bryan, A. B. Pakhomov, V. Shutthanandan, S. Thevuthasan, D. E. Mc-Cready, D. Gaspar, M. H. Engelhard, J. W. Rogers, K. Krishnan, D. R. Gamelin and S. A. Chambers, *Phys. Rev. B* **70**, 054424 (2004).

[76] S. W. Lim, D. K. Hwang and J. M. Myoung, *Solid State Commun.* **125**, 231 (2003).

[77] T. Chanier, M. Sargolzaei, I. Opahle, R. Hayn and K. Koepernik, *Phys. Rev. B* **73**, 134418 (2006).

[78] C. H. Park and D. J. Chadi, *Phys. Rev. Lett.* **94**, 127204 (2005).

[79] K. Ip, M. E. Overberg, Y. W. Heo, D. P. Norton, S. J. Pearton, C. E. Stutz, B. Luo, F. Ren, D. C. Look and J. M. Zavada, *Appl. Phys. Lett.* **82**, 385 (2003).

[80] O. D. Jayakumar, I. K. Gopalakrishnan, K. Shashikala, S. K. Kulshreshtha and C. Sudakar, *Appl. Phys. Lett.* **89**, 202507 (2006).

Chapter Three

Spintronics in III-Nitride Based Materials

Ryan P. Davies*, Cammy R. Abernathy and Stephen J. Pearton

Department of Materials Science and Engineering, University of Florida, Gainesville, FL 32611-6400, USA

**rydavies@ufl.edu*

GaN based materials have continued to be at the forefront of spintronics research due to the demonstration of room temperature ferromagnetism in these materials. A goal of this research is the fabrication of spintronic devices that may provide easy incorporation into existing GaN processing technologies, higher integration density, and less power consumption than their electronic counterparts while achieving similar speeds. The progression of the ferromagnetic metals incorporated into GaN has moved from transition metals to rare earth metals such as the lanthanides. In this chapter, we review this progression for material selection as well as methods of their incorporation and models for the mechanism for ferromagnetism in these materials.

3.1 INTRODUCTION

Spin transport electronics (spintronics) continues to be a topic of interest since the shrinkage of the current predominant semiconductor materials is leading to limited functionality at the atomic scale. Due to this limitation, scientists and engineers are researching other options in hopes of continuing the fulfillment of Moore's law. A spintronic device would utilize the electron spin and spin transport to achieve additional functionality compared to traditional electronics' sole use of the electron charge. The exploitation of electron spin in practical devices has already been demonstrated as spin valves are used as magnetic tunnel junctions in magnetic random access memory (MRAM), a platform that aims to supersede current memory technologies. Since electron spin exists in either a spin up or spin down configuration, spintronic materials would also open the door for the development of rewritable devices that are nonvolatile (can easily retain their data storage even when turned off).

Handbook of Spintronic Semiconductors by W M Chen & I A Buyanova

www.panstanford.com
978-981-4267-36-6

III-nitride materials such as GaN and AlN have established themselves as ideal material bases for certain electronic and optoelectronic devices. Studies of electron spin in GaN have laid the groundwork for the feasibility of III-nitride dilute magnetic semiconductor (DMS) materials. Spin coherence in GaN has been observed at room temperature with a lifetime in the picosecond range.[1] Spin injection has been heavily researched and difficulties have arisen in injecting spins from metals into semiconductors.[2,3] Due to these difficulties, incorporating the magnetic metal into the host material has become ideal. The optimal incorporation of the metal would result in sustaining the crystal quality of the host material so that the material can be processed into a functional device structure. This review focuses on the advancements made in the study of III-nitride DMS materials utilizing different magnetic metal impurities and different incorporation methods.

3.2 MATERIAL SELECTION FOR SPINTRONIC SEMICONDUCTORS

As with almost all electronic devices, the practical application of spintronic devices hinges on being operational at room temperature. The suitability of these devices then requires that the constituent material(s) possess a Curie temperature (T_C) above room temperature. Additional beneficial distinctions for device operation would include the ability to control the Fermi level via doping, high quality crystal growth from substrate lattice matching, and radiation hardness for specific applications. Another desirable characteristic of spintronic device materials would be the availability of a current technology base. This necessity of ferromagnetic ordering at a practical device operating temperature and easy incorporation into an existing semiconductor industry has paved the way for III-nitride DMS materials after numerous other materials have been proposed, researched, and deemed unsatisfactory. Figure 3.1 shows a schematic of how doping a semiconductor with a transition metal ion can lead to alignment of spins associated with that ion and produce ferromagnetism. In wide band-gap DMS materials there is still considerable controversy as to the mechanism that leads to the observed ferromagnetism, especially since the formation of magnetically active clusters or second phases, that are very difficult to detect in some cases, may control the magnetic properties (bottom of Fig. 3.1). This is generally considered undesirable since then the electron population is not polarized and most spin device concepts assume a polarized electron population.

First, europium chalcogenides and II–IV semiconductors were considered for magnetic semiconductors. The Eu-chalcogenides proved problematic due to this material's lattice mismatch with the Si and GaAs substrates used for crystal growth.[4] The II-IV semiconductors were difficult to dope *p*- or *n*-type and the resulting magnetization was not of a ferromagnetic nature.[4,5] Also, both of these material systems had T_Cs at approximately 100 K, well below room temperature.[6] Even though these material systems had low T_C values, Dietl *et al.* computed the T_C of GaN and ZnO doped with 5% Mn to be above room temperature.[7] As seen in

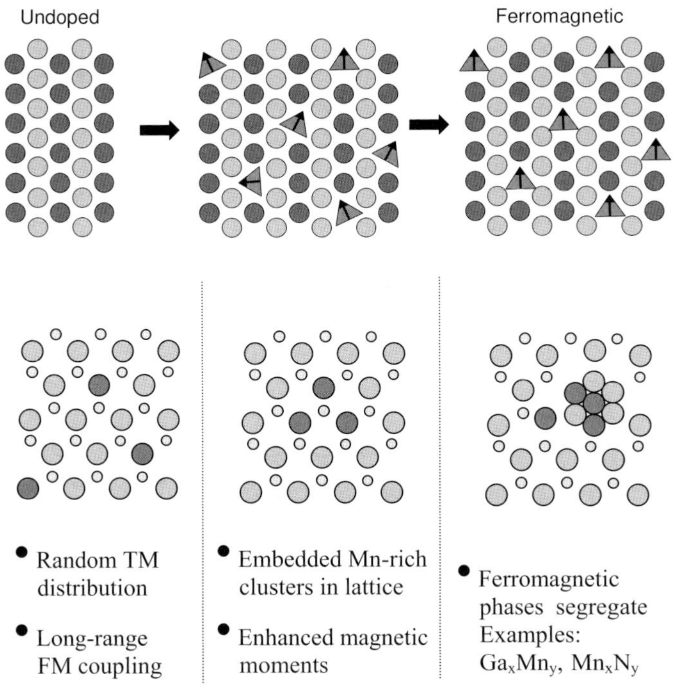

Figure 3.1. The additional of certain transition metal or rare earth ions can lead to alignment of spins and ferromagnetism in the compound semiconductor host (top). However, instead of a random distribution of these ions (bottom left), clusters (bottom center) or magnetically active second phases may form (bottom right).

Fig. 3.2, experimentally observed T_Cs for certain compound semiconductors show a correlation with band-gap. Later work has shown a better correlation with lattice constant, as depicted in Fig. 3.3. The wider band-gap semiconductors tend to have smaller lattice constants, large *p–d* hybridization and small spin-orbit interaction, and are predicted to have higher Curie temperatures. Computational data predicting above room temperature T_C values for III-V semiconductor materials doped with transition metals sparked interest in realizing a suitable material for spintronic applications. Mn-doped InAs was the first III-V material to exhibit ferromagnetic ordering due to holes as the carrier type. This material had to be grown in low temperature regimes via molecular beam epitaxy (MBE) to impede the growth of MnAs clusters and resulted in single phase *p*-type InMnAs with a T_C of 7.5 K.[8,9] Due to the more extensive knowledge base concerning GaAs and the possibility of numerous device applications, this material was strongly considered. Unfortunately, the highest T_C reported for GaMnAs was approximately 110 K[10] and the corresponding research abated due to the discovery of higher T_C values for other DMS III-V materials. Room temperature ferromagnetism was later shown in Mn-doped GaP,[11] GaN,[12–14] and AlN.[15,16]

Fueled by the computation data by Dietl *et al.*,[7] research in transition metal doped III-V materials sped forward and issues such as crystal quality, growth

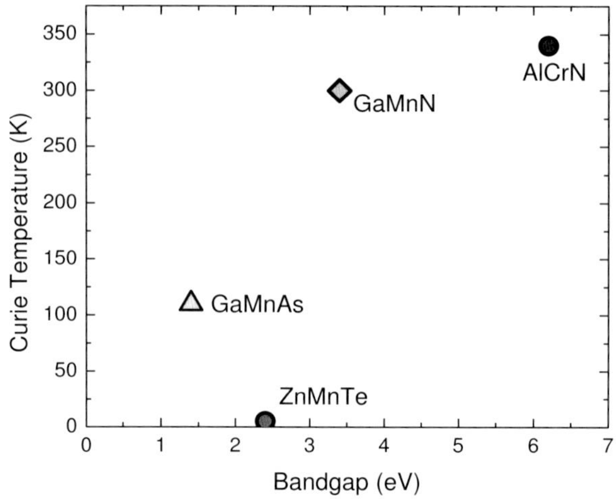

Figure 3.2. Some experimentally observed Curie temperatures (T_C) for dilute magnetic compound semiconductors.

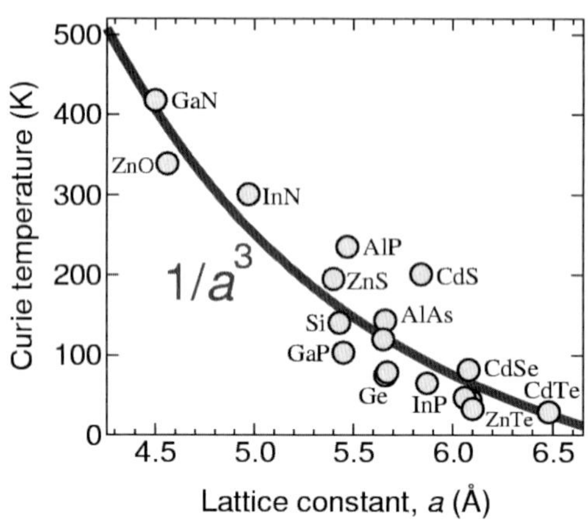

Figure 3.3. Predicted Curie temperatures as a function of lattice constant for a variety of semiconductors with transition metal doping [after S. C. Erwin (Naval Research Laboratory)]. The materials predicted to have high T_Cs have large *p–d* hybridization and small spin-orbit interaction.

conditions, and the ferromagnetic mechanism in these materials arose. When considering the III-nitride materials as the host matrix, both GaN and AlN were favored due to the extensive knowledge base arising from their suitability in optical and electronic applications such as ultraviolet LEDs and high frequency, high temperature transistors. Primarily, transition metal doped GaN was grown via MBE under non-equilibrium conditions to suppress the formation of precipitating secondary phases. These phases could form from the impurities aggregating into nanoclusters and were problematic because they could be the origin of the ferromagnetic ordering in the material.[17] One of the most studied transition metal III-nitrides is Mn-doped GaN, which has had T_Cs values ranging as low as 10 K to an estimated 940 K.[12–14,18,19] Overall, above a certain Mn atomic concentration, this material exhibits paramagnetic behavior.[12–14,18] Also, the magnetic properties of Mn-doped GaN are not thermally stable without co-doping the material with oxygen up to 10 at%.[20]

Research in DMS materials progressed towards other transition metals such as Fe, Co and Cr, with Cr emerging as an excellent candidate. Cr-doped GaN has shown ferromagnetic ordering above room temperature,[21,22] and up to 900 K.[23] The magnetic moment increases with Cr concentration until the substitutional sites in the lattice are saturated and the Cr incorporates interstitially, which leads to degradation of the magnetic properties of the material.[23] Cr-doped GaN exhibited similar ferromagnetic ordering before and after being annealed at 700°C for 1 min,[24] allowing this material to withstand high temperature device processing. Thin films of Cr-doped GaN displaying good crystal quality also showed magnetic moments up to 1.8 μ_B/Cr atom.[25]

The rare earth metals, or lanthanide series elements, are the latest magnetic dopants to receive strong consideration. Gd, Tm, Eu and Er have all been studied in GaN since their excited states lead to emission lines throughout the spectrum from the ultraviolet (UV) through the visible to the infrared (IR).[26] Utilizing these emission lines with the wide band-gap of GaN has led to research in developing these materials into tunable light emitting devices. Eu-doped GaN has been shown to exhibit room temperature ferromagnetic ordering.[27] Also, a paramagnetic-like phase has been reported to coexist with a ferromagnetic-like phase in Eu-doped GaN, which has been attributed to divalent and trivalent Eu ions.[28] M. Hashimoto *et al.* considered the Eu^{2+} ions to be responsible for the ferromagnetic behavior at room temperature but estimate that only 1.5% of the total Eu ions are divalent due to the intra-atomic *f*–*f* transitions shown by photoluminescence (PL) data.[29] The coexistence of paramagnetic and ferromagnetic ordering has also been seen in Er-doped GaN grown via gas source MBE.[30] Er-doped GaN grown via metal-organic chemical vapor deposition (MOCVD) exhibits room temperature ferromagnetism with an increasing saturation magnetization with increasing Er concentration.[31] Since the preferred valency of Gd in GaN is trivalent, this rare earth element is ideal to substitute for the trivalent Ga atoms.[32] Gd-doped GaN has shown ferromagnetism above room temperature with a Gd concentration below 10^{16} cm^{-3}.[33,34] Gd-implanted GaN and AlN have also exhibited ferromagnetism above room temperature after annealing at 900°C.[35] One of the main contributing forces to the upsurge of research of Gd in GaN was the colossal magnetic moment of up to 4,000 μ_B/Gd atom reported from magnetic measurements by Dhar *et al.*[33] The significance of this discovery is seen when comparing this colossal magnetic moment to that of bulk Gd (8 μ_B/Gd atom) and GdN (7 μ_B/Gd atom). The observation of an order of magnitude larger magnetic moment per Gd atom for implanted thin films in relation to epitaxially grown thin films has brought the relevance of defects into the discussion of the mechanisms of ferromagnetism.[36]

3.3 MODELS FOR THE MECHANISMS OF FERROMAGNETISM

Numerous models have been suggested to explain the ferromagnetic phenomena in DMS materials. As the research base supporting these materials expands, these models are refined to better explain the nature of the ferromagnetism present and the interactions of the specific components necessary for ferromagnetism. Since each model has the tendency to be rejected based on findings for other magnetic impurities in other DMS material systems, a case-by-case approach to applying a ferromagnetic mechanism becomes more suitable. This section will include a brief review of the prominent models proposed for DMS materials. The currently accepted picture for DMS ferromagnetism is that it is the local antiferromagnetic coupling between the carriers (i.e., holes in GaMnAs) and the Mn magnetic moments that leads to long-range ferromagnetic ordering of Mn local moments. The carrier system also becomes spin-polarized in the process with the carrier magnetic moment directed against the Mn magnetic ordering by virtue of the antiferromagnetic hole-Mn coupling, but the total magnetic moment of the spin-polarized carriers is extremely small since $n_c < n_i$ and $|S| > |s|$, where S and s are, respectively, the Mn and hole spin. Depending on the relative magnitudes of hole and Mn moment densities n_c and n_i, and the coupling strength J, HM describes the Kondo model or the Kondo lattice model or *s–f* (or *s–d*) Zener model or the double-exchange model or the RKKY model or (in the presence of strong disorder) the (RKKY) spin glass model. In the "simplest" mean-field model, T_C^{MF} increases monotonically with the exchange coupling, the moment density, and the carrier density as

$$T_C^{MF} \propto |J|(n_c n_i)^{1/2}.$$

The non-degenerate MFT provides a transition temperature dependence given by

$$T_C^{MF} \propto J^2 n_c^{1/3} n_i.$$

There is some emerging consensus that the appropriate picture for DMS ferromagnetism in strongly disordered insulating materials is the bound magnetic polaron percolation picture whereas at higher densities (and/or lower disorder), where the system is metallic, the dynamical mean-field picture (which becomes equivalent to the Weiss mean-field picture at low values of exchange coupling insufficient to cause the impurity band formation) is more appropriate.

3.3.1 Free Carrier Mediated Model

In one of the first models explaining ferromagnetism with relation to transition metals, Zener applied a mean-field approximation that defined the DMS material as an alloy with random metallic atoms sitting substitutionally in the host lattice.[37] In this model, the proclivity of the ferromagnetic alignment of *d* electron spins was due to the spin coupling between the incomplete *d* shell and conduction electrons.[37] Due to the neglect of the peripatetic character of the magnetic electrons and the quantum (Friedel) oscillations of the electron spin polarization around the localized spins, this model was ultimately deserted. Dietl *et al.* built on this model by pointing out that, for semiconductors, the effect of the Friedel oscillations averages to zero since the mean distance between the carriers is greater than that between spins and that due to this the Zener model becomes equivalent to the Ruderman–Kittel–Kasuya–Yosida (RKKY) interaction model.[7] The RKKY interaction model relied on the DMS material possessing a high carrier density (on the order of $10^{20}\,cm^{-3}$). Considering this model, high hole densities were demonstrated to drive a paramagnetic-ferromagnetic phase transition in II–VI DMS materials.[38] For transition metal-doped GaN, the RKKY interaction model was questioned due to the degenerate electrons or holes being hardly accessible in this wide band-gap semiconductor.[39]

3.3.2 Percolation Model

Observations such as high ferromagnetic transition temperatures occurring in non-degenerate semiconductors[40] paved the way for a new theory in DMS materials doped with magnetic atoms. The polaron percolation model theorizes that when the concentration of carriers is much smaller than that of the magnetic impurities, exchange interactions between the localized carriers and magnetic impurities lead to their mutual polarization.[41] Due to this interaction, a "bound magnetic polaron" is formed whose effective radius grows as the temperature decreases.[42] The ferromagnetic mechanism exists because, as the polarons grow and overlap, they form clusters that have all their spins aligned in the same direction.[41] The limitation of this theory lies in the necessity of the carrier concentration being much smaller than the magnetic impurity concentration, which may not be the case for some DMS materials.

3.3.3 Extension to Rare Earth Magnetic Impurities

Previously, the theoretical work had focused on the role of transition metals in DMS materials. Extending the models mentioned above to the lanthanide series becomes problematic as the additional f shell interactions further convolute an already complicated matter. Considering Gd-doped GaN, magneto-photoluminescence (PL) measurements support a percolation-like model of long-range spin polarization of the GaN matrix by Gd atoms.[33,34] Using *ab initio* band structure calculations, an electron-stabilized ferromagnetic model has been proposed.[43] According to this model, the introduction of Gd causes localized states below the conduction band due to coupling between the s and f suborbitals. Spin polarization occurs due to these new states being filled by donor electrons, which are possibly provided by the large concentration of intrinsic oxygen in the GaGdN films. Since this model mainly focuses on carriers, a more complete model is necessary based on recent DMS materials research and the presence of other species in these materials due to the host matrix and dopant incorporation.

3.3.4 Models Considering Defects

Understanding the role of defects with respect to ferromagnetic ordering in these materials is essential when you consider an electron-stabilized model. Intuitively from this model, any competition for electrons, such as defects, should decrease the magnetic ordering in the materials. Extensive studies and modeling of the role of specific defects has been pursued. In Gd-implanted GaN, the introduction of Gd might be inducing the magnetic moment of Ga and/or N interstitials due to the long-range spin polarization of these defects.[44] Considering the same phenomenon in epitaxially grown Gd:GaN, a large concentration of these defects ($\sim 10^{19}\,\mathrm{cm}^{-3}$) is required. The majority of the earlier reports of colossal magnetic moments in these materials were exhibited in films grown via MBE, a non-equilibrium method allowing for the formation of such defects during growth. The formation of these defects may arise due to the larger atomic size of Gd compared to the Ga atom that constitutes the host material. X-ray linear dichroism and X-ray magnetic circular dichroism (XMCD) measurements of Gd:GaN showed that about 85% of the Gd goes to substitutional Ga sites and a small XMCD signal is detectable for Gd, emphasizing the role of the GaN host matrix for the overall magnetic ordering.[45] Liu *et al.* noted that the strain exerted on the GaN thin films due to the lattice mismatched substrate and the atomic size differential between the large Gd atom and the relatively smaller host Ga atom leads to the actuality of a high density of vacancies.[46] Total-energy electronic-structure calculations have shown that a ferromagnetic interaction exists between Ga vacancies in the presence of Gd atoms which leads to an increase in magnetic moment with increasing Ga vacancies.[47] As noted above, these Ga vacancies are most likely formed during Gd incorporation, and the formation of Gd atom–Ga vacancies defect complexes is energetically favorable.[47] In addition to these findings, Ga vacancies located between two Gd atoms separated by about 15 Å favor ferromagnetic interaction while in the absence of vacancies these Gd atoms exhibit barely any interaction.[48] Further modeling involving a computational consideration of all the species present in DMS materials, such as carriers and defects, would greatly assist in understanding the overall ferromagnetic mechanism.

3.4 III-NITRIDE DMS MATERIALS

III-nitride materials have been studied as the host matrix for magnetic impurity incorporation due to their beneficial material properties and their existing semiconductor technology base. As shown in Fig. 3.4, the type of existing magnetic-semiconductor systems such as magnetic random access memory (MRAM) relies on a hybrid approach that uses a multilayer stack of metals to form the memory element on top of an Si circuit. With true magnetic semiconductors, one could hope to have both the memory and electronic circuit functions in the same material, leading to packing density and speed improvements. The magnetic impurity Gd has been incorporated into III-nitride materials via mainly MBE growth with more

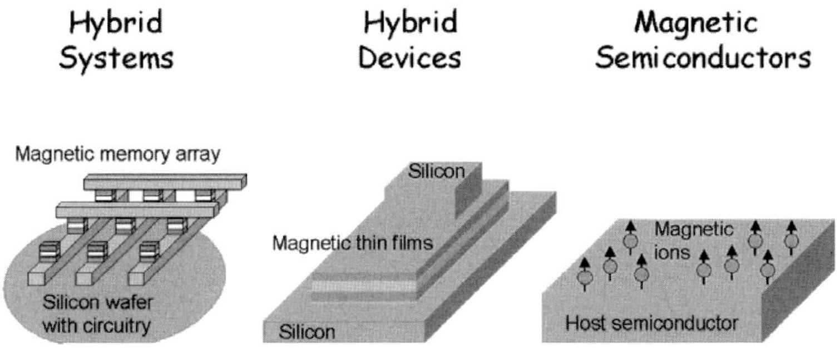

Figure 3.4. Existing technologies for combining magnetic and electronic functions use hybrid approaches. The availability of magnetic semiconductors would enable large increases in device performance.

current studies involving incorporation via MOCVD growth or ion implantation. Due to the stronger technology base for GaN, this section will focus on this III-nitride material with an additional subsection focusing on AlN.

3.4.1 Gd-Doped GaN

As mentioned above, a vast majority of the preliminary studies of Gd-doped GaN were carried out on films grown via MBE.[33,34,49–51] $Ga_{0.94}Gd_{0.06}N$ was shown to exhibit ferromagnetic behavior from 7 to 400 K.[49] Two potential sources of the ferromagnetic ordering may be either of the ferromagnetic materials elemental Gd and the rocksalt structure of GdN, which have T_C values of 307.7 and 72 K, respectively.[52] Since the Gd-doped GaN samples in this study had a T_C value that exceeded that of either elemental Gd and GdN, the hysteresis in the *M–H* curves was attributed to the DMS material.[49] Thickness studies were also performed to determine if interfacial effects may give rise to the ferromagnetic source.[34,51] The ferromagnetic behavior was seen to be characteristic of the bulk material as the measured magnetization per unit area for a given Gd concentration scaled with sample thickness.[34] Also, Fig. 3.5 shows how Gd-doped GaN samples were grown at thicknesses between 0.1 and 0.6 μm and the magnetic signal normalized by sample volume did not steadily decrease with increasing film thicknesses.[51] Gd-doped GaN samples have shown very low Gd concentrations, below the secondary ion mass spectrometry (SIMS) background level of about 10^{17} atoms/cm^3.[33,34,50] SIMS analysis also showed an oxygen content on the order of 10^{19} atoms/cm^3, similar to results seen in transition metal-doped GaN grown in the same system, and no other magnetic impurities present.[51] Figure 3.6 shows the optimum Gd cell temperature for achieving the highest magnetization under our conditions is 1050°C. The Gd-doped GaN thin film shown in Fig. 3.7 exhibits a T_C value above room temperature due to the lack of a change in hysteresis at 10 K and 300 K for magnetization versus applied field measurements and the lack of closure between the field cooled and zero field cooled lines measured at an applied field of 200 Oe.[51]

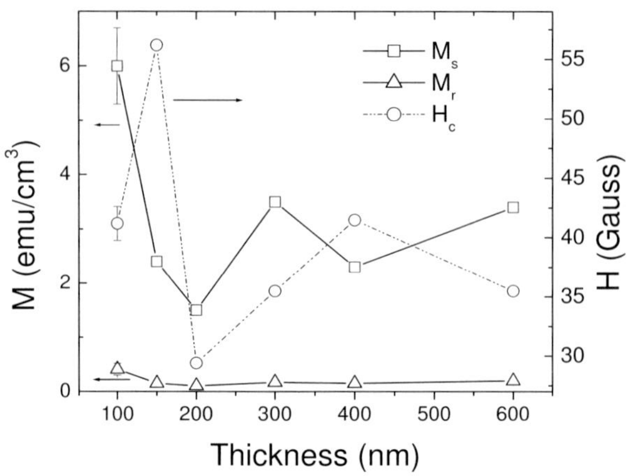

Figure 3.5. Saturation magnetization (M_S), coercivity (H_c) and remnant magnetization (M_r) versus film thickness measurements taken at 300 K. M_S was taken at an applied field of 1,000 G from magnetization versus applied field measurements. H_c and M_r values were also obtained from magnetization versus applied field measurements.[51]

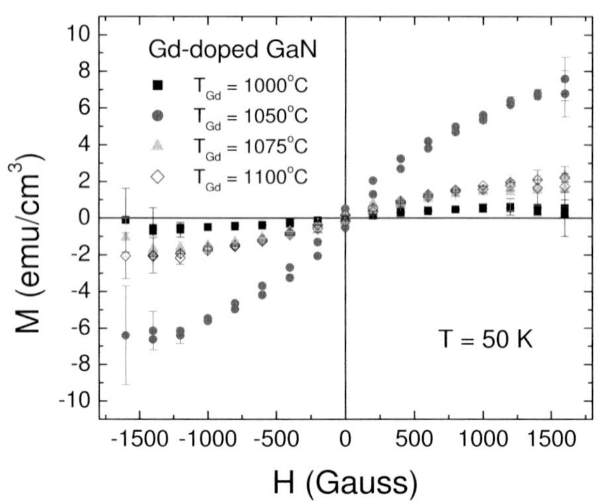

Figure 3.6. Hysteresis loops for Gd-doped GaN grown with different Gd cell temperatures [after J. K. Hite (Naval Research Laboratory)].

Gd-doped GaN samples with a Gd concentration as high as 12.5% grown at low temperatures via MBE have also shown ferromagnetic behavior.[50]

3.4.2 Gd- and Si-co-doped GaN

GaN was co-doped with both Gd and Si to study the effects that a higher carrier density and the presence of a shallow donor would have on ferromagnetism.[51,53,54] To modulate the Si incorporation and thus the carrier density, the Si cell

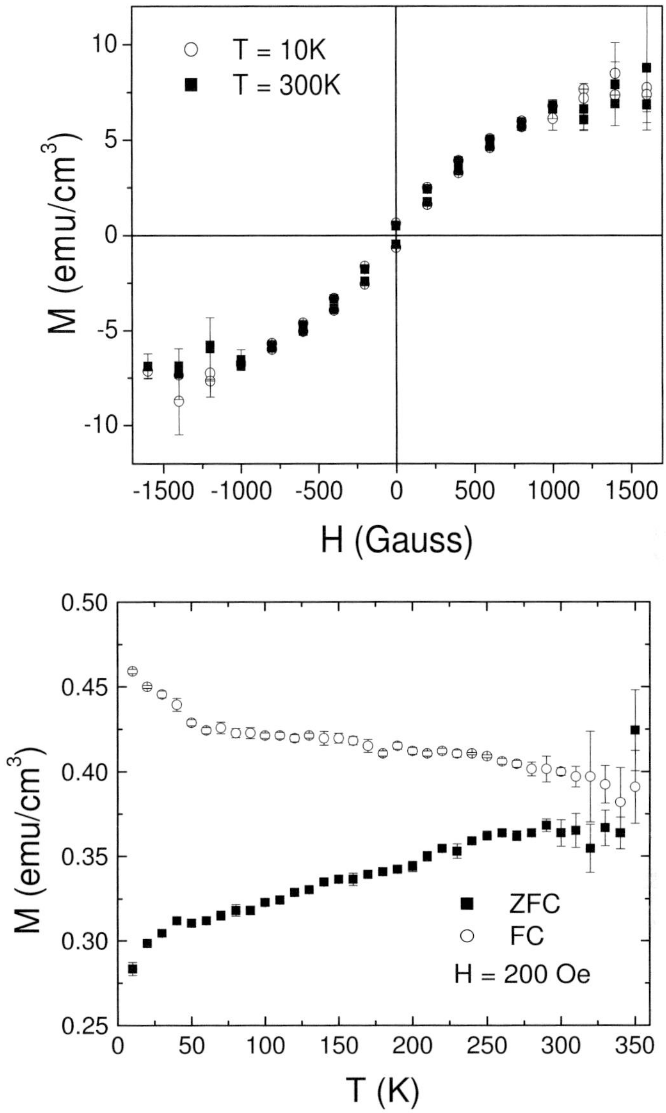

Figure 3.7. Top: Magnetization versus applied field curves at 10 K and 300 K for a 1000 Å thick GaGdN film grown at a Gd cell temperature of 1050°C. Bottom: Magnetization versus temperature curve at an applied field of 200 Oe for the same sample showing both field cooled and zero field cooled traces.[50]

temperature was varied between 1000°C and 1200°C.[51,53] As seen in Fig. 3.8, at an Si cell temperature of 1000°C, the material was quite resistive but carrier concentration still increased about an order of magnitude with each 100°C increase in Si cell temperature (up to $1.45 \times 10^{18}\,\mathrm{cm}^{-3}$).[53] As a dopant in GaN, Si behaves as a shallow donor in the wurtzite crystal structure with an ionization energy of between 0.12 and 0.20 eV.[55] Figure 3.8 also shows how the saturation magnetization (M_s)

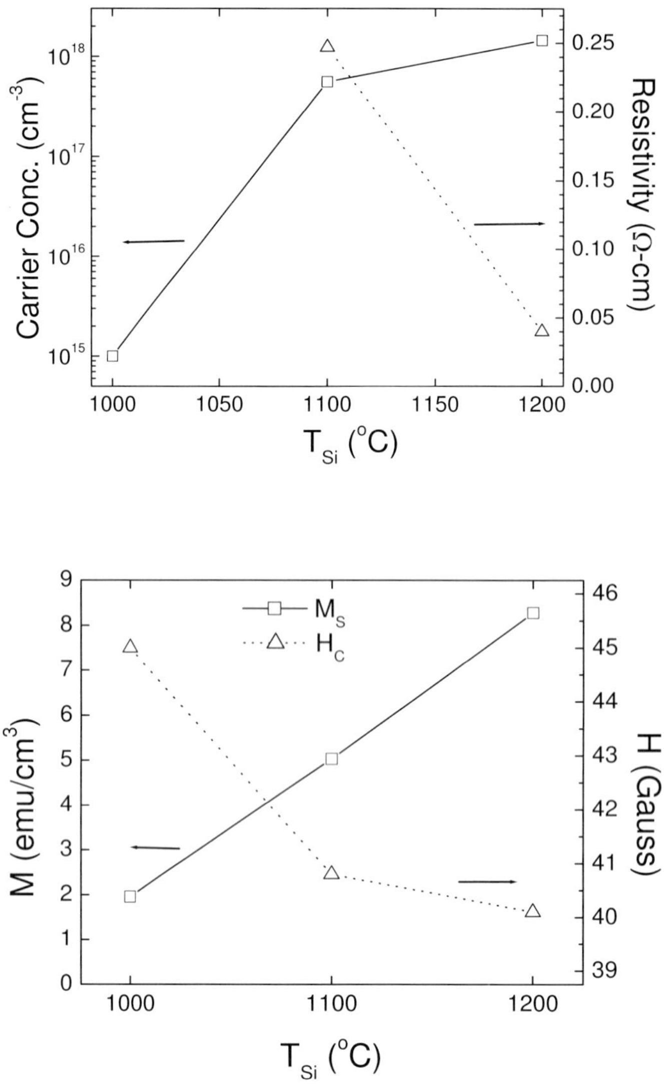

Figure 3.8. Top: Carrier concentration and resistivity of GaGdN:Si versus Si cell temperature (T_{Si}) [after J. K. Hite (Naval Research Laboratory)]. Bottom: Saturation magnetization (M_s) and coercivity (H_c) versus T_{Si} for GaGdN:Si. The Gd cell temperature was 1050°C [after J. K. Hite (Naval Research Laboratory)].

increased and the coercivity (H_c) decreased with increasing Si cell temperature.[53] The normalized magnetization measurements show that the additional carriers are having some type of effect on the ferromagnetic ordering of the material.

As described earlier, Dalpain *et al.* have postulated that the ferromagnetic phase can be stabilized by introducing electrons while the Gd atoms couple antiferromagnetically.[43] The M_s measured in the co-doped GaN thin film with the highest Si cell temperature was greater than the M_s measured in any of the

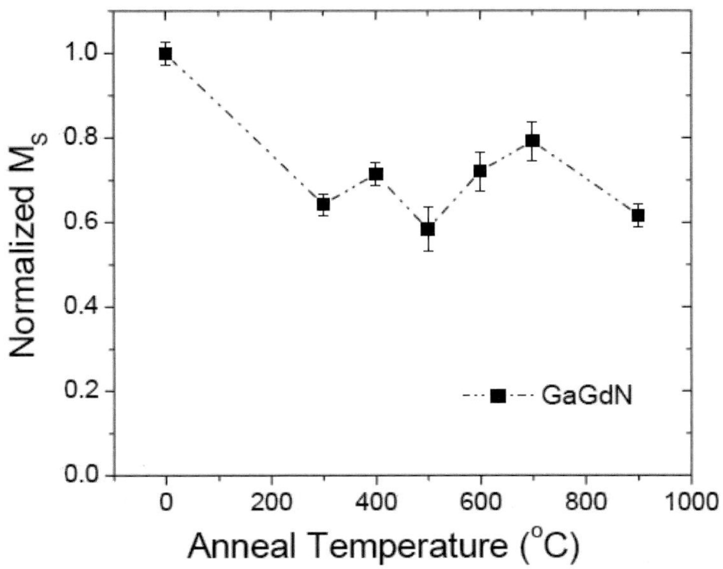

Figure 3.9. Change in magnetic parameters of Gd-doped GaN annealed at different temperatures after growth.

purely Gd-doped GaN thin films and the co-doped materials also exhibited room temperature ferromagnetism.[51,53] Also, when considering higher Gd concentrations, the M_S value (1046 emu/cm^3) of Si co-doped GaGdN was seven times larger than that of non-doped GaGdN with the same Gd concentration of 8.9% for both samples which showed clear hysteresis at room temperature.[54] Zhou *et al.* state that co-doping Gd and Si in GaN is expected to increase the shallow donor density and to strengthen the ferromagnetic interaction.[54] An additional important consideration is the stability of the material during subsequent processing, in which Ohmic formation requires annealing at temperatures as high as 800–900°C. Figure 3.9 shows that the Gd-doped GaN is stable to these temperatures. Even though the magnetization decreases with annealing, at approximately 300°C, the magnetic signal stabilizes at 70% to 75% of the as-grown value.

3.4.3 Gd-Doped AlGaN and Heterostructures

AlGaN was also explored as the host matrix for Gd doping due to the prevalent use of this material in heterostructures with GaN for such applications as high electron mobility transistors (HEMT),[56–59] UV light-emitting diodes (LED),[60,61] and UV photodetectors.[62] For the latter case, the selection of the cut-off wavelength can be achieved by changing the ratio of the group III species. Previously, transition metal-doped AlGaN was suggested as a possible candidate for such spintronic devices as polarized light emitters or spin transistors.[63] For MBE grown Gd-doped AlGaN thin films, Al cell temperatures ranging from 1000°C to 1150°C were used with the largest M_s and smallest H_c values seen at an Al cell temperature of 1025°C.[64] This optimal Gd-doped AlGaN exhibited room temperature ferromagnetism and had an M_s of ~22 emu/cm^3 and H_c of ~25 Oe compared to an M_s of ~11 emu/cm^3 and H_c of ~34 Oe for Gd-doped AlN and an M_s of ~8 emu/cm^3 and H_c of ~42 Oe for Gd-doped GaN.[64]

Gd-doped heterostructures such as GaN/AlN and AlGaN/GaN were also studied to determine the feasibility of DMS layers for future spintronic device applications. An MBE grown heterostructure of GaGdN/AlN with layer thicknesses of 90 Å/52 Å and a periodicity of 10 layers exhibited ferromagnetic behavior at room temperature while a similar heterostructure of GaGdN/AlN with layer thicknesses of 25 Å/100 Å and a periodicity of 33 layers did not.[64] Si co-doped heterostructures have also been explored due to the importance of Fermi level engineering for specific device functions. Heterostructures of GaGdN:Si/AlN with layer thicknesses of 90 Å/80 Å and periodicity of 10 layers and 180 Å/80 Å and a periodicity of five both demonstrated ferromagnetic ordering at room temperature, while the latter heterostructure showed a higher M_s.[64] As mentioned above, the AlGaN/GaN heterostructure has been widely studied and utilized in numerous device applications; therefore, integrating Gd-doped AlGaN layers into the existing technology base for this material could prove beneficial. The AlGaN:Gd/GaN heterostructure with layer thicknesses of 185 Å/87 Å and a periodicity of 10 was ferromagnetic at room temperature, but the M_s value was about half of that measured for bulk AlGaN:Gd.[64]

3.4.4 Gd-Implanted GaN

After the observation of about an order of magnitude larger magnetic moment per Gd atom for Gd-implanted GaN as compared to epitaxially grown Gd-doped GaN,[36] the interest in incorporating Gd via ion implantation increased. Gd^{3+} ions implanted into single-crystal GaN epilayers at total doses of 3–6$\times 10^{14}$ cm^{-2} and annealed at 700°C to 1000°C showed the formation of second phases such as Gd_3Ga_2, GdN and Gd from X-ray diffraction (XRD) measurements.[35] Annealing the implanted samples at 900°C proved to be the optimal temperature for total magnetization, even though the ferromagnetic ordering could be attributed to Gd precipitates.[35] Another annealing study showed that for GaN implanted with Gd at a dose of 2.4×10^{11} cm^{-2} and then annealed at 900°C, M_s was reduced as compared to a non-annealed sample.[44] Since XRD results showed the presence of Ga and N interstitials in the implanted layers and the density of these defects were reduced upon annealing, Khaderbad *et al.* speculated that Gd could be inducing magnetic moment in either or both of the interstitials and giving rise to an effective colossal magnetic moment.[44] These findings bring into question the role of defects in the ferromagnetic ordering due to the larger defect formation from the bombardment and incorporation of the large Gd ion into the GaN host matrix as compared to that from the non-equilibrium epitaxial growth of GaGdN. Further studies into the type of defect(s) present and their interactions with the implanted species in these materials would help in understanding the overall ferromagnetic mechanism.

3.4.5 Gd-Implanted AlGaN/GaN Device Structures

Gd incorporation into the AlGaN/GaN heterostructure has also been explored. At 4.2 K, the extraordinary Hall effect and anisotropic magnetoresistance were observed for a GaN/$Al_xGa_{1-x}N$ HEMT structure containing a two-dimensional electron gas implanted with a Gd dose of 3×10^{11} cm^{-2}.[65] The potential interaction of carriers and defects in this material was examined. Lo *et al.* postulated that the longitudinal resistance monotonically decreasing and not reaching a minimum with increasing magnetic field can be attributed to electron spin-orbit coupling and the possible contribution of spin-dependent scattering between electrons and the paramagnetic defects induced by Gd implantation.[65] For the same Gd implanted GaN/$Al_xGa_{1-x}N$ HEMT structure mentioned above, the measured Hall resistance decreasing with increasing temperature above 140 K could be attributed to the diamagnetic defects induced by Gd implantation since the Hall resistance for a similar non-implanted sample showed a very weak dependence on temperature that was attributed to intrinsic paramagnetic defects.[66] Further studies into the interplay between carriers and defects could provide additional insight into the possible ferromagnetic mechanism occurring in these materials.

3.4.6 Gd-Doped and Implanted AlN

AlN has also been considered as a potential host matrix for magnetic impurities due to the possibility of integrating magnetic functionality with this material's use in deep UV optoelectronics. Cathodoluminescence (CL) measurements of MBE grown Gd-doped AlN showed the luminescence from the trivalent Gd ion at 318.15 nm (3.8969 eV) in AlN.[67] AlN has also been implanted with Gd^{3+} ions to study both the ferromagnetic response[35] and UV light emission.[68] XRD results for the implanted material showed only Gd and AlN peaks and this material exhibited ferromagnetic behavior at both 5 K and 300 K.[35] UV PL measurements of Gd-implanted AlN also showed an emission at 318 nm that is characteristic of the inner $4f$ shell ${}^6P_{7/2}$–${}^8S_{7/2}$ transition of the Gd^{3+} ion.[69] A similar emission has been seen in CL studies of Gd-implanted AlN.[68] These results show the possibility of integrating the magnetic and optical properties of this material into novel device applications.

3.5 SUMMARY AND ISSUES TO BE RESOLVED

III-nitride based materials have been extensively studied as the host matrix for magnetic impurity incorporation and potential use for spintronic applications. These materials have an existing technology base which would allow for an easy integration when optimal DMS materials are developed. Some of these DMS materials have exhibited ferromagnetism through a range of temperatures up to and beyond room temperature. Transition metal dopants, such as Mn and Cr, have been incorporated into GaN and AlN in large atomic percentages to achieve the highest magnetic signal. Unfortunately, these dopants would prove deleterious for future spintronic applications due to the introduction of an impurity band which makes the discrete spin splitting difficult. Gd has been focused on due to the colossal magnetic moment that can be achieved with this dopant in very dilute limits. Magnetic measurements for GaN and AlN incorporated with Gd have shown that ferromagnetic ordering remains from low temperatures to above room temperature, but the specific ferromagnetic mechanism taking place in these materials is still undetermined.

As described earlier, numerous models have been proposed to explain the ferromagnetism seen in these DMS materials. Carriers, defects and the host matrix have all been postulated to play a role in inducing ferromagnetic ordering. Even though most models agree that the magnetic dopant is the source of the ferromagnetism, there is still much work to be done in understanding how these dopants interact with the host matrix, carriers and defects in their vicinity. Additional characterization of the materials showing room temperature ferromagnetism would provide more information to develop an operable mechanism for the specific combination of host matrix material(s), magnetic dopant, incorporation technique, carrier type and density, etc. Since a major goal of spintronics research is the development of a functional device, further work in incorporating magnetic dopants in HEMT and other device structures would show the feasibility of spin injection, spin manipulation via either gate voltage or magnetic fields from ferromagnetic contacts, and spin detection.

ACKNOWLEDGMENTS

This work was supported by Army Research Office contract W911-NF-0410296.

References

[1] B. Beschoten, E. Johnston-Halperin, D. K. Young, M. Poggio, J. E. Grimaldi, S. Keller, S. P. DenBaars, U. K. Mishra, E. L. Hu and D. D. Awschalom, *Phys. Rev. B* **63**, 121202(R) (2001).

[2] P. R. Hammar, B. R. Bennett, M. J. Yang and Mark Johnson, *Phys. Rev. Lett.* **83**, 203 (1999).

[3] F. G. Monzon, H. X. Tang and M. L. Roukes, *Phys. Rev. Lett.* **84**, 5022 (2000).

[4] H. Ohno, *J. Magn. Mag. Mater.* **200**, 110 (1999).

[5] H. Akinaga and H. Ohno, *IEEE Trans. Nanotech.* **1**, 9 (2002).

[6] T. Dietl and H. Ohno, *Physica E* **9**, 185 (2001).

[7] T. Dietl, H. Ohno, F. Matsukura, J. Cibert and D. Ferrand, *Science* **287**, 1019 (2000).

[8] H. Munekata, H. Ohno, S. von Molnar, A. Segmeuller, L. L. Chang and L. Esaki, *Phys. Rev. Lett.* **63**, 1849 (1989).

[9] H. Ohno, H. Munekata, S. von Molnar and L. L. Chang, *J. Appl. Phys.* **69**, 6103 (1991).
[10] H. Ohno, *Science* **281**, 951 (1998).
[11] M. E. Overberg, B. P. Gila, G. T. Thaler, C. R. Abernathy, S. J. Pearton, N. Theodoropoulou, K. T. McCarthy, S. Arneson, A. F. Hebard, S. N. G. Chu, R. G. Wilson, J. M. Zavada and Y. D. Park, *J. Vac. Sci. Tech. B* **20**, 969 (2002).
[12] M. Hashimoto, Y. K. Zhou, H. Tampo, M. Kanamura and H. Asahi, *J. Cryst. Growth* **252**, 499 (2003).
[13] G. Thaler, R. Frazier, B. Gila, J. Stapleton, M. Davidson, C. R. Abernathy, S. J. Pearton and C. Segre, *Appl. Phys. Lett.* **84**, 1314 (2004).
[14] G. Thaler, R. Frazier, B. Gila, J. Stapleton, M. Davidson, C. R. Abernathy, S. J. Pearton and C. Segre, *Appl. Phys. Lett.* **84**, 2578 (2004).
[15] Z. Liu, J. De Boeck, V. V. Moshchalkov and G. Borghs, *J. Magn. Mag. Mater.* **242–245**, 967 (2002).
[16] R. Frazier, G. Thaler, M. Overberg, B. Gila, C. R. Abernathy and S. J. Pearton, *Appl. Phys. Lett.* **83**, 1758 (2003).
[17] M. Van Scilfgaarde and O. N. Myrasov, *Phys. Rev. B* **63**, 233205 (2001).
[18] M. E. Overberg, C. R. Abernathy, S. J. Pearton, N. A. Theodoropoulou, K. T. McCarthy and A. F. Hebard, *Appl. Phys. Lett.* **79**, 1312 (2001).
[19] S. Sonoda, S. Shimizu, T. Sasaki, Y. Yamamoto and H. Hori, *J. Cryst. Growth* **237**, 1358 (2002).
[20] G. T. Thaler, Development of gallium nitride based dilute magnetic semiconductors for magneto-optical applications, http://purl.fcla.edu/fcla/etd/UFE0006460, PhD dissertation, University of Florida, USA (2004).
[21] Y.-K. Zhou, M. Hashimoto, M. Kanamura and H. Asahi, *J. Supercond.* **16**, 37 (2003).
[22] G. T. Thaler, R. M. Frazier, C. R. Abernathy and S. J. Pearton, *Appl. Phys. Lett.* **86**, 131901 (2005).
[23] H. X. Liu, S. Y. Wu, R. K. Singh, L. Gu, D. J. Smith, N. Newman, N. R. Dilley, L. Montes and M. B. Simmonds, *Appl. Phys. Lett.* **85**, 4076 (2004).
[24] G. T. Thaler, R. M. Frazier, C. R. Abernathy and S. J. Pearton, *Appl. Phys. Lett.* **86**, 131901 (2005).
[25] L. Gu, S. Y. Wu, H. X. Liu, R. K. Singh, N. Newman and D. J. Smith, *J. Magn. Mag. Mater.* **290–291**, 1395 (2005).
[26] J. Solé, An Introduction to the Optical Spectroscopy of Inorganic Solids (Wiley, New York, 2005), p. 202.
[27] J. Hite, G. T. Thaler, R. Khanna, C. R. Abernathy, S. J. Pearton, J. H. Park, A. J. Steckl and J. M. Zavada, *Appl. Phys. Lett.* **89**, 132119 (2006).
[28] H. Asahi, Y. K. Zhou, M. Hashimoto, M. S. Kim, X. J. Li, S. Emura and S. Hasegawa, *J. Phys. Cond. Mater.* **16**, S5555 (2004).
[29] M. Hashimoto, A. Yanase, R. Asano, H. Tanaka, H. Bang, K. Akimoto and H. Asahi, *Jpn. J. Appl. Phys.* **42**, L1112 (2003).
[30] H. Bang, J. Sawahata, M. Tsunemi, J. Seo, H. Yanagihara, E. Kita and K. Akimoto, Structural and magnetic properties of Er doped GaN, in *International Symposium on Compound Semiconductors* (IEEE, Piscataway, 2003), p. 114.
[31] J. M. Zavada, N. Nepal, C. Ugolini, J. Y. Lin, H. X. Jiang, R. Davies, J. Hite, C. R. Abernathy and S. J. Pearton, *Appl. Phys. Lett.* **91**, 054106 (2007).
[32] A. Svane, N. E. Christensen, L. Petit, Z. Szotek and W. M. Temmerman, *Phys. Rev. B* **74**, 165204 (2006).
[33] S. Dhar, O. Brandt, M. Ramsteiner, V. F. Sapega and K. H. Ploog, *Phys. Rev. Lett.* **94**, 037205 (2005).

[34] S. Dhar, L. Perez, O. Brandt, A. Trampert, K. H. Ploog, J. Keller and B. Beschoten, *Phys. Rev. B* **72**, 245203 (2005).

[35] S. Y. Han, J. Hite, G. T. Thaler, R. M. Frazier, C. R. Abernathy, S. J. Pearton, H. K. Choi, W. O. Lee, Y. D. Park, J. M. Zavada and R. Gwilliam, *Appl. Phys. Lett.* **88**, 042102 (2006).

[36] S. Dhar, T. Kammermeier, A. Ney, L. Perez, K. H. Ploog, A. Melnikov and A. D. Wieck, *Appl. Phys. Lett.* **89**, 062503 (2006).

[37] C. Zener, *Phys. Rev.* **81**, 440 (1951).

[38] T. Dietl, A. Haury and Y. Merle d'Aubigné, *Phys. Rev. B* **55**, R3347 (1997).

[39] V. I. Litvinov, *Phys. Rev. B* **72**, 195209 (2005).

[40] V. I. Litvinov and V. K. Dugaev, *Phys. Rev. Lett.* **86**, 5593 (2001).

[41] A. Kaminski and S. Das Sarma, *Phys. Rev. B* **68**, 235210 (2003).

[42] A. Kaminski and S. Das Sarma, *Phys. Rev. Lett.* **88**, 247202-1 (2002).

[43] G. M. Dalpian and S.-H. Wei, *Phys. Rev. B* **72**, 115201 (2005).

[44] M. A. Khaderbad, S. Dhar, L. Pérez, K. H. Ploog, A. Melnikov and A. D. Wieck, *Appl. Phys. Lett.* **91**, 072514 (2007).

[45] A. Ney, T. Kammermeier, E. Manuel, V. Ney, S. Dhar, K. H. Ploog, F. Wilhelm and A. Rogalev, *Appl. Phys. Lett.* **90**, 252515 (2007).

[46] L. Liu, P. Y. Yu, Z. Ma and S. S. Mao, *Phys. Rev. Lett.* **100**, 127203 (2008).

[47] Y. Gohda and A. Oshiyama, *Phys. Rev. B* **78**, 161201 (2008).

[48] C. Mitra and W. Lambrecht, presented at *Symp. D. Materials Research Society Fall Meeting 2008*, December 2 (2008).

[49] N. Teraguchi, A. Suzuki, Y. Nanishi, Y.-K. Zhou, M. Hashimoto and H. Asahi, *Solid State Commun.* **122**, 651 (2002).

[50] Y. K. Zhou, S. W. Choi, S. Kimura, S. Emura, S. Hasegawa and H. Asahi, *J. Supercond. Novel Magn.* **20**, 429 (2007).

[51] J. K. Hite, R. M. Frazier, R. Davies, G. T. Thaler, C. R. Abernathy and S. J. Pearton, *Appl. Phys. Lett.* **89**, 092119 (2006).

[52] P. Junod, A. Menth and O. Vogt, *Phys. Kondens. Mater.* **8**, 323 (1969).

[53] J. K. Hite, R. M. Frazier, R. P. Davies, G. T. Thaler, C. R. Abernathy, S. J. Pearton, J. M. Zavada, E. Brown and U. Hömmerich, *J. Electron. Mater.* **36**, 391 (2007).

[54] Y. K. Zhou, S. W. Choi, S. Emura, S. Hasegawa and H. Asahi, *Appl. Phys. Lett.* **92**, 062505 (2008).

[55] V. Bougrov, M. E. Levinshtein, S. L. Rumyantsev and A. Zubrilov, in *Properties of Advanced Semiconductor Materials GaN, AlN, InN, BN, SiC, SiGe*, eds. M. E. Levinshtein, S. L. Rumyantsev and M. S. Shur (John Wiley & Sons, Inc., New York, 2001), pp. 1–30.

[56] H. Tang, J. B. Webb, J. A. Bardwell, S. Rolfe and T. MacElwee, *Solid State Electron.* **44**, 2177 (2000).

[57] J. W. Johnson, J. Han, A. G. Baca, R. D. Briggs, R. J. Shul, J. R. Wendt, C. Monier, F. Ren, B. Luo, S. N. G. Chu, D. Tsvetkov, V. Dmitriev and S. J. Pearton, *Solid State Electron.* **46**, 513 (2001).

[58] U. K. Mishra, P. Parikh and Y.-F. Wu, *Proc. IEEE* **90**(6), 102 (2002).

[59] X. L. Wang, C. M. Wang, G. X. Hu, J. X. Wang, T. S. Chen, G. Jiao, J. P. Li, Y. P. Zeng and J. M. Li, *Solid State Electron.* **49**, 1387 (2005).

[60] J. Han, M. H. Crawford, R. J. Shul, J. J. Figiel, M. Banas, L. Zhang, Y. K. Song, H. Zhou and A. V. Nurmikko, *Appl. Phys. Lett.* **73**, 1688 (1998).

[61] J. Shakya, K. Knabe, K. H. Kim, J. Li, J. Y. Lin and H. X. Jiang, *Appl. Phys. Lett.* **86**, 091107 (2005).

[62] E. Monroy, F. Calle, J. L. Pau, E. Muñoz, F. Omnès, B. Beaumont and P. Gibart, *J. Cryst. Growth* **230**, 537 (2001).

[63] S. J. Pearton, C. R. Abernathy, M. E. Overberg, G. T. Thaler, D. P. Norton, N. Theodoropoulou, A. F. Hebard, Y. D. Park, F. Ren, J. Kim and L. A. Boatner, *J. Appl. Phys.* **93**, 1 (2003).

[64] J. K. Hite, Influence of Gadolinium as a Dopant in III-Nitride Dilute Magnetic Semiconductors, http://purl.fcla.edu/fcla/etd/UFE0017405, PhD dissertation, University of Florida, USA (2006).

[65] F.-Y. Lo, A. Melnikov, D. Reuter, Y. Cordier and A. D. Wieck, *Appl. Phys. Lett.* **92**, 112111 (2008).

[66] F.-Y. Lo, A. Melnikov, D. Reuter, Y. Cordier and A. D. Wieck, presented at *Symp. D. Materials Research Society Fall Meeting 2008*, December 2 (2008) [and to be published in the Materials Research Society Symposium Proceedings].

[67] S. Emura, M. Takahashi, H. Tambo, A. Suzuki, T. Nakamura, Y.-K. Zhou, S. Hasegawa and H. Asahi, presented at *Symp. D. Materials Research Society Fall Meeting 2008*, December 2 (2008) [and to be published in the Materials Research Society Symposium Proceedings].

[68] U. Vetter, J. Zenneck and H. Hofass, *Appl. Phys. Lett.* **83**, 2145 (2003).

[69] J. M. Zavada, N. Nepal, J. Lin, K. H. Kim, H. X. Jiang, J. Hite, G. T. Thaler, R. M. Frazier, C. R. Abernathy and S. J. Pearton, Photoluminescence from Gd-implanted AlN and GaN epilayers, in *Advances in III-V Nitride Semiconductor Materials and Devices*, eds. C. R. Abernathy, H. Jiang, J. M. Zavada, Proc. 955E (Materials Research Society, 2007), 0955-I10-0.

Chapter Four

Electronic Structure and Lattice Site Location of Mn in III-Mn-V Ferromagnetic Semiconductors

K. Alberi, K. M. Yu and W. Walukiewicz*

Materials Science Division, Lawrence Berkeley National Laboratory, Berkeley, CA 94720, USA

**w_walukiewicz@lbl.gov*

We present here a review of our experimental and theoretical works leading to our current understanding of the factors affecting ferromagnetic coupling between Mn spins in group III-Mn-V ferromagnetic semiconductors. The significance of the interstitial Mn defects, the Fermi level controlled incorporation of substitutional Mn atoms and the maximum hole concentration were established by a series of experiments with different material structures and processing. Our results indicate that any improvement in the magnetic properties of this material system will require decoupling of the doping process from the introduction of magnetic moments. We have also shown that the electronic structure of the Mn impurity band can be described in terms of a band anticrossing model. The model explains hole transport properties and the effect of alloying on the magnetic properties of group III-Mn-As. Moreover, it can also be used to predict trends in the magnetic properties of other ternary and dilute quaternary III-Mn-V semiconductors.

4.1 INTRODUCTION

Dilute magnetic semiconductors (DMS), in which magnetic impurities are incorporated into standard semiconductors, have been studied for almost four decades.[1] Initial effort in this area mostly focused on group II-VI semiconductors containing Mn atoms, wherein a large variety of II-Mn-VI alloys were synthesized in bulk and epitaxial thin film forms, and investigation of their properties provided a basis for understanding itinerant carrier-mediated magnetic coupling.[2] Subsequent progress in semiconductor epitaxial growth allowed the synthesis of III-Mn-V compounds with significant Mn concentrations.[3,4] Instead of acting as an

Handbook of Spintronic Semiconductors by W M Chen & I A Buyanova

www.panstanford.com
978-981-4267-36-6

isovalent cation, Mn has a deficiency of one valence electron in III-V compounds, allowing it to act as an electrically active acceptor impurity. In fact, as has been demonstrated in the case of $Ga_{1-x}Mn_xAs$, III-V DMS grown under proper conditions can exhibit large p-type conductivity.

The successful synthesis of group III-Mn-V DMS with a Curie temperature T_C exceeding 150 K[5–7] led to the quest for alloys with T_Cs greater than room temperature. It was evident that the T_C could be improved by increasing the concentration of isolated substitutional Mn atoms in the lattice. Additionally, the significant variations in the T_Cs of various III-Mn-V alloys indicated that the structure of the electronic states in the vicinity of the valence band edge plays a key role in determining the strength of the magnetic coupling. Therefore, the identification of the factors limiting Mn incorporation on substitutional sites as well as a comprehensive understanding of the mechanism responsible for the magnetic coupling are necessary for optimizing the T_C in these systems. This chapter addresses these two issues. The first section reviews structural studies of the location of the Mn atoms in III-Mn-V DMS and the role of Fermi energy and Mn doping levels on site location. The second section describes the influence of Mn on the valence band structure of the host that is used to explain the trends in the electrical and optical properties of group III-Mn-V DMS and their alloys.

4.2 LATTICE SITE LOCATION OF Mn IN $Ga_{1-x}Mn_xAs$

4.2.1 *Self-compensation and Mn Location in GaMnAs*

The equilibrium solubility of Mn in III-V semiconductors is known to be at most $10^{19}\,cm^{-3}$ (about 0.002 mole fraction). Thus, to incorporate the higher Mn concentrations into the III-V lattice that are required for achieving ferromagnetism, it is necessary to resort to low temperature molecular beam epitaxy (LT-MBE), in which strong non-equilibrium growth conditions are realized. Typically, such growth is carried out at a substrate temperature in the range of 200–280°C.[8–10] The low growth temperatures necessary to prevent the formation of thermodynamically more stable second phases such as MnAs result, however, in the formation of native defects, giving rise to self-compensation and ultimately to a limitation of T_C.

The Zener model of ferromagnetism in $Ga_{1-x}Mn_xAs$ considers only divalent substitutional Mn atoms, Mn_{Ga}, which act as acceptors.[8,11] Although each Mn atom in the Ga sublattice is in principle expected to contribute a hole to the system, in practice it was found that the hole concentration in this material is significantly lower than the Mn concentration (by a factor of 2–3), especially as the Mn concentration increased.[8,9] Potashnik *et al.* showed that both the T_C and the conductivity in $Ga_{1-x}Mn_xAs$ alloys for $x > 0.05$ saturate in optimally annealed materials.[12] These observations suggest that, as x increases, more of the Mn acceptors are compensated by donors, and a rising fraction of Mn spins ceases to participate in ferromagnetism. Such self-compensation effects in turn evince the existence of a T_C-limiting mechanism in $Ga_{1-x}Mn_xAs$ grown by the standard LT-MBE. Therefore, despite intense research efforts, similar maximum values of T_C only up to 170 K were achieved in thin $Ga_{1-x}Mn_xAs$ films prepared in different laboratories with rather different values of x ranging from $\sim$0.05 to 0.10 and annealed at temperatures in the range between 150 and 280°C.[7,9]

Due to the non-equilibrium LT-MBE growth, it is reasonable to assume that defects related to the Mn site location play an important role in the self-compensation of $Ga_{1-x}Mn_xAs$. In addition to the Mn_{Ga} acceptors, the most important Mn-related defects include: *interstitial* Mn (Mn_I) commensurate with the zinc-blende lattice structure and Mn atoms precipitated out to form different phases such as Mn clusters or MnAs inclusions. In the latter case, Mn resides at random sites (Mn_{rand}) that are incommensurate with the zinc-blende lattice. Similar to all interstitials of metal atoms, Mn interstitials act as double donors, and thus each Mn_I tends to passivate two substitutional Mn acceptors. On the other hand, the Mn atoms which form precipitates, e.g., MnAs inclusions, do not contribute either holes or electrons, and are not included in the Zener picture of magnetization.

4.2.2 Channeling Rutherford Backscattering and Particle Induced X-ray Emission

Several atomic scale studies of the local environment of Mn in $III_{1-x}Mn_xV$ alloys using extended X-ray absorption fine structure (EXAFS) have been reported.[13–15] While accurate determination of the nearest and next-nearest neighbor distances ($< \pm 0.005$ Å), as well as a measure of local disorder, can be achieved using EXAFS, this technique is not well suited for measuring coordination numbers (error $\sim$ 20%) or for chemically distinguishing between neighbors with small atomic number differences (i.e., Ga and As). Consequently, impurity locations in the lattice can only be indirectly inferred from these studies by multi-parameter fitting of the EXAFS spectra to calculated model structures.

Ion channeling techniques have been applied extensively to determine lattice site locations of impurities in semiconductors and metals and are therefore important tools for investigating the incorporation of Mn into III-V compounds.[16,17] The specific location of Mn in $Ga_{1-x}Mn_xAs$ was determined by simultaneously performing channeled particle-induced X-ray emission (c-PIXE) and channeled Rutherford backscattering spectrometry (c-RBS) using a 2.0 MeV $^4He^+$ beam generated by a 2.5 MeV Van de Graaff accelerator. Because Mn atoms have a lower atomic number than the host Ga and As atoms, small concentrations of Mn in GaAs cannot be detected. The PIXE technique is element-specific and well suited for this work since characteristic X-rays of Mn, Ga and As do not interfere with each other. The location of Mn in the GaAs host can be identified by comparing the Mn (PIXE) and GaAs (RBS) angular scans, using the normalized yield as a function of tilt angle around the $\langle 100 \rangle$, $\langle 110 \rangle$ and $\langle 111 \rangle$ axial channels. Here the normalized yield for the RBS (χ_{GaAs}) or the PIXE Mn X-ray signals (χ_{Mn}) is defined as the ratio of the channeled yields to the corresponding unaligned "random"

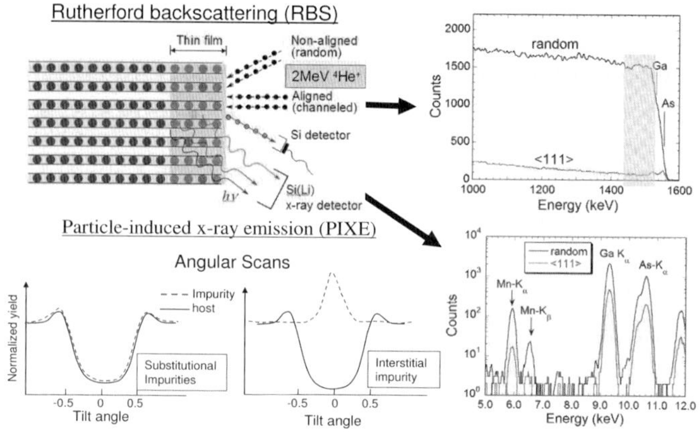

Figure 4.1. A schematic representation of the simultaneous RBS and PIXE measurements of a GaMnAs thin film on GaAs. RBS and PIXE spectra from a 100 nm thick $Ga_{0.92}Mn_{0.08}As$ thin film in the non-aligned (random) and $\langle 111 \rangle$ aligned directions are shown on the right. Schematic diagrams of the angular scans of the impurity (PIXE) and host (RBS) signals for a substitutional and interstitial impurity are also shown.

yields. Furthermore, considering that the ion beam penetrates several microns into the target, most of the PIXE yield pertaining to Ga and As came from the GaAs substrate and were not used in comparison with the Mn signal.

Figure 4.1 shows a schematic representation of the simultaneous RBS and PIXE measurements of a $Ga_{1-x}Mn_xAs$ thin film on GaAs. RBS and PIXE spectra from a 100 nm thick $Ga_{0.92}Mn_{0.08}As$ thin film in the non-aligned (random) and $\langle 111 \rangle$ aligned directions are displayed on the right. Angular scan plots of the impurity PIXE signals and the host RBS signals around the axial channeling direction (typically tilting from $-2°$ to $+2°$) give a quantitative identification of the impurity location in the crystal lattice. Schematic diagrams of the angular scans of the impurity and host signals for a substitutional and interstitial impurity are also shown in Fig. 4.1. Notice that the ion flux in an axial channel is focused by the atomic potential and is higher than the flux in the non-aligned case (flux peaking). This gives rise to a strong peak in the angular scan for an interstitial impurity located in the middle of an axial channel.

4.2.3 Detection of Mn Interstitials in GaMnAs

Atoms in the interstitial positions (hexagonal or tetrahedral) in a diamond lattice are shadowed by the host atoms when viewed along both the $\langle 100 \rangle$ and $\langle 111 \rangle$ axes, but are exposed in the $\langle 110 \rangle$ axial channel.[16,17] Therefore, identification of the interstitial and substitutional impurities by ion channeling analysis can only be achieved through measurement in several axial directions (and sometimes also planar channels). Fig. 4.2 shows the angular scans of Mn (PIXE) and GaAs host (RBS) signals about the $\langle 100 \rangle$, $\langle 110 \rangle$ and $\langle 111 \rangle$ axes for a 110 nm thick

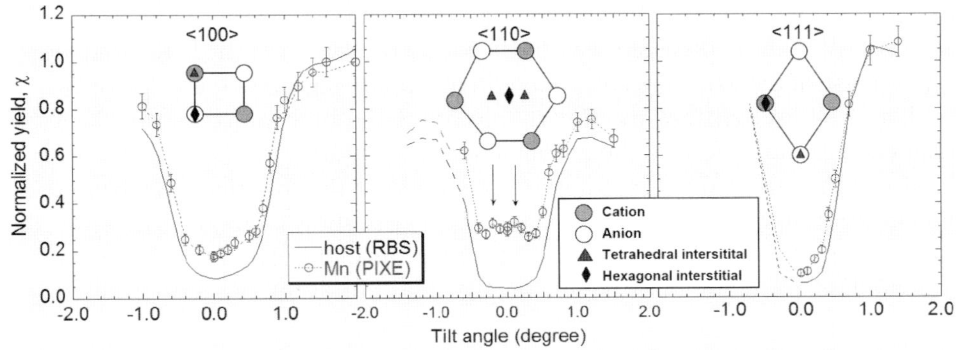

Figure 4.2. Mn and GaAs angular scans about the ⟨100⟩, ⟨110⟩ and ⟨111⟩ channels for a 110 nm thick $Ga_{1-x}Mn_xAs$ film with $x = 0.09$ (top row). The lattice locations of Mn as revealed by the projections for various axial channels are shown in the bottom row.

$Ga_{0.91}Mn_{0.09}As$ film. The ⟨100⟩ and ⟨111⟩ angular scans of the Mn PIXE signals closely follow those of the GaAs RBS signals, suggesting that along these two directions the Mn atoms are shadowed by the host GaAs atoms. The slightly higher χ_{Mn} along these two axes indicates, however, that a small fraction of Mn atoms are present in the form of random clusters, which probably act as precursors of larger precipitates when samples are annealed at high temperature ($> 400°C$).

In contrast to the ⟨100⟩ and ⟨111⟩ scans, the value of χ_{Mn} along the ⟨110⟩ axis is much higher than for χ_{GaAs}. This unambiguous signature suggests that the Mn atoms do not all reside at substitutional sites, and instead a fraction of these non-random Mn atoms must be located at *interstitial* sites. Schematics showing the various lattice locations as revealed by the projections for the axial channels are also included in Fig. 4.2. The tetrahedral interstitials give rise to a double-peak feature in the ⟨110⟩ angular scan due to the flux peaking effect of the ion beam in the channel. A double-peak feature was indeed observable in the ⟨110⟩ scan data. The fraction of these interstitial Mn has been roughly estimated to be ~15% of the total Mn in the specimen.[18]

Ion channeling studies on a series of $Ga_{1-x}Mn_xAs$ samples with x ranging from 0.02 to 0.1 revealed that the Mn_I concentration increases (from 5 to 15%) with the total Mn content. The upper panel of Fig. 4.3 displays the Mn_{Ga} and Mn_I as measured by channeling for a wide range of alloy compositions. A monotonous increase in the Mn_I concentration is observed in the $Ga_{1-x}Mn_xAs$ film with rising x. Interstitial Mn is expected to weaken the ferromagnetism of $Ga_{1-x}Mn_xV$ alloys for several reasons. First, compensation by the double Mn_I donors reduces the hole concentration to an approximate value of $[Mn_{Ga}]-2 \times [Mn_I]$; second, it has been theoretically demonstrated that interstitial Mn does not contribute to the Zener-type Mn-Mn exchange due to its negligible $p-d$ coupling.[19] Finally, because Mn_I are both highly mobile and positively charged, they are expected to drift to interstitial sites *adjacent* to the negatively charged Mn_{Ga} acceptors and form antiferromagnetic Mn_I–Mn_{Ga} pairs, thus canceling the magnetic moment of

Mn_{Ga}.[18–20] The density of Mn atoms that contribute to the ferromagnetism of the $Ga_{1-x}Mn_xV$ alloy is then reduced to the value of $[Mn_{Ga}]-[Mn_I]$. Consequently, any increase of the Mn_I concentration will inevitably suppress the value of T_C.

It is well known that the determination of the hole concentration in ferromagnets is complicated by the so-called *anomalous Hall effect* (AHE), characteristic of conducting ferromagnets.[21,22] To circumvent this problem, we have used the electrochemical capacitance voltage (ECV) profiling method to measure the depth distribution of uncompensated Mn acceptors in the $Ga_{1-x}Mn_xAs$ layers.[23,24] This method provides information on the distribution of the net space charge

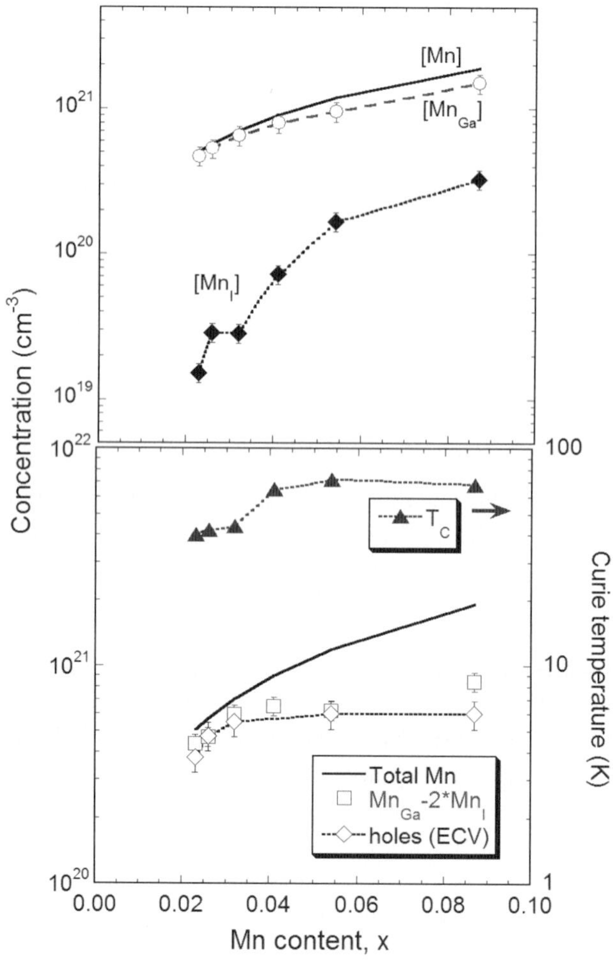

Figure 4.3. Concentrations of interstitial and substitutional Mn (top panel) and the net uncompensated Mn acceptor concentration ($[Mn_{Ga}]-2\times[Mn_I]$) (lower panel) as measured by channeling experiments on as-grown $Ga_{1-x}Mn_xAs$ films. Hole concentrations obtained by electrochemical capacitance–voltage (ECV) profiling for x ranging from 0.02 to 0.09 and corresponding Curie temperatures T_C are shown in the lower panel for the same films.

in the depletion region. A comparison of the net uncompensated Mn acceptors ($[Mn_{Ga}]-2 \times [Mn_I]$) and the hole concentration obtained by ECV profiling, along with the total Mn concentration in the samples, is presented in the lower panel of Fig. 4.3. The hole concentration agrees well with the net Mn acceptors for $Ga_{1-x}Mn_xAs$ samples with the exception of the $x = 0.09$ sample, where compensation by Mn_I alone cannot fully account for the low hole concentration. It is possible that in the latter case As antisite (As_{Ga}) donors are also compensating some of the substitutional Mn acceptors.[25]

4.2.4 Stability of Mn Interstitials

It is well established that point defects such as As antisites (As_{Ga}) and Mn interstitials (Mn_I) play a crucial role in determining the magnetic properties of $Ga_{1-x}Mn_xAs$.[18–20,25] Since these materials were grown well below normal GaAs MBE growth temperatures (~600°C), Mn_I is also expected to be mobile and thermally unstable. It is therefore foreseeable that T_C can be enhanced by eliminating compensating Mn_I through thermal annealing and/or reducing the film thickness so that diffusion of Mn_I can be promoted by surface effects.

4.2.4.1 Low-temperature annealing of $Ga_{1-x}Mn_xAs$

Reports in the literature show that appropriate low-temperature annealing of $Ga_{1-x}Mn_xAs$ can increase the Curie temperature, total magnetic moment, and hole concentration through improvement of material homogeneity and alteration of the magnetization temperature dependence to a more mean-field-like behavior.[22,26,27] For example, Hayashi *et al.* demonstrated an improvement in the Curie temperature and the crystallinity of $Ga_{1-x}Mn_xAs$ by a post-MBE growth thermal annealing.[26] Potashnik *et al.* subsequently carried out a systematic study of the effects of annealing $Ga_{1-x}Mn_xAs$ and established, in particular, that the Curie temperature T_C and magnetization $M(T)$ both strongly depend on the annealing temperature and duration.[22]

Here, we review our systematic investigation into the influence of the annealing temperature on the electronic and magnetic properties of $Ga_{1-x}Mn_xAs$, with special emphasis on the correlation between the location of Mn ions in the lattice and the magnetic properties. Epilayers with a wide range of Mn concentrations were annealed at temperatures $260°C < T_a < 350°C$ in an N_2 atmosphere.[28] The time of annealing was varied from 0.5 hours to 3 hours.

Temperature dependent resistivity measurements indicate that the conductivity and T_C in samples with low Mn concentrations (e.g., $x \sim 0.032$) show no pronounced dependence on annealing. Conversely, highly doped $Ga_{1-x}Mn_xA$ (e.g., $x \sim 0.083$) samples, both as-grown and annealed at 260°C, 280°C, 300°C and 310°C, show a clearly metallic behavior, with the distinct resistivity peak that occurs around T_C. Optimum results were obtained by annealing at 280°C for a time between 1.0 and 1.5 hours. Additionally, the T_C increased from 70 to 110 K as the annealing temperature was raised up to 280°C but dropped with further increase. After 1.5 hours of annealing a slight decline in T_C was also observed, in agreement with the results of Potashnik *et al.*[22]

The $Ga_{1-x}Mn_xAs$ epilayer with $x = 0.083$ annealing at $T_a = 280°C$ for 1.0 hour exhibited the resistivity peak at $T_\rho = 127$ K. SQUID magnetization measurements on this optimally-annealed sample yielded a T_C of 115 K. In fact, resistivity and SQUID measurements also revealed good agreement between the magnetic and transport results for all samples included in the investigation. Fig. 4.4 presents the temperature dependence of the magnetization M for a typical $Ga_{1-x}Mn_xAs$ sample with high Mn concentrations ($x = 0.081$) for various annealing temperatures (1.0 hour annealing time). The magnetization was measured by a SQUID magnetometer in a weak magnetic field of 10 Gauss after the sample had been magnetized at a higher field of about 1,000 Gauss applied parallel to the sample surface. The Curie temperature of the as-grown $Ga_{1-x}Mn_xAs$ sample was determined to be about 67 K. A dramatic increase in both T_C and the magnetization was observed under optimized annealing conditions (1.0 hour annealing at 280°C). Finally, Fig. 4.4 shows that annealing above the optimal temperature (in this case at 350°C) produces an equally dramatic drop in T_C to a value below that of the as-grown sample.

The observed improvement of the magnetic and electrical properties in $Ga_{1-x}Mn_xAs$ upon low-temperature annealing can, most plausibly, be attributed to the redistribution of highly mobile Mn_I. Investigation of the Mn atom distribution at various lattice sites for different annealing conditions was carried out by channeled RBS and PIXE analysis, and the results were correlated with the concentration of the holes and of the uncompensated magnetic moments of Mn^{++}, to provide a better understanding of how the location of Mn within the $Ga_{1-x}Mn_xAs$

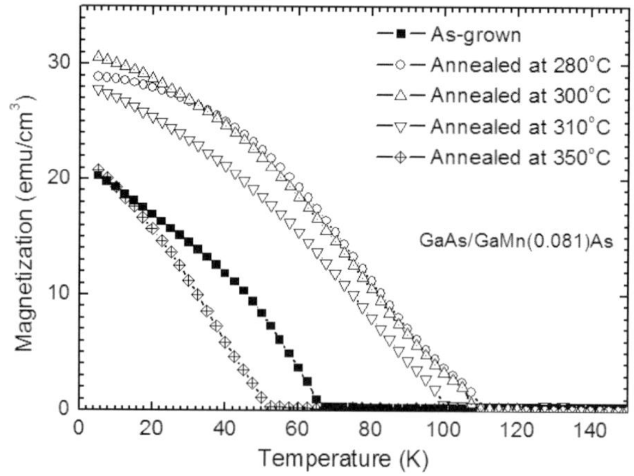

Figure 4.4. Magnetization of a $Ga_{1-x}Mn_xAs$ sample with a high Mn content ($x = 0.081$) as a function of temperature after various annealing procedures. The annealing temperatures are indicated in the figure. Note that the Curie temperature and the saturation magnetization are significantly enhanced by low-temperature annealing. When the annealing is carried out at temperatures higher than 300°C, however, both the Curie temperature and the magnetization are seen to drop monotonically. The highest Curie temperature for this series is 111 K.

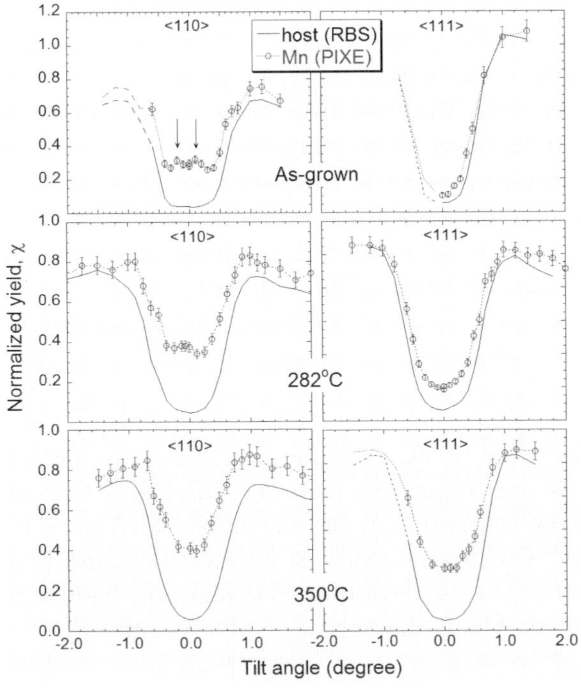

Figure 4.5. Angular scans about the ⟨110⟩ and ⟨111⟩ axes for three $Ga_{1-x}Mn_xAs$ samples originating from the same growth, but subjected to different annealing conditions.

lattice affects the ferromagnetism of this material.

The PIXE (Mn) and RBS (GaAs) angular scans about the ⟨110⟩ and ⟨111⟩ axes are plotted in the first and the second columns of Fig. 4.5 for three $Ga_{1-x}Mn_xAs$ samples: as-grown, annealed at 282°C, and annealed at 350°C. Again, the "double-peak" feature of the Mn scan in the ⟨110⟩ axis indicates the presence of Mn_I in the sample. The fraction of the Mn_I has been roughly estimated to be ∼15% of the total Mn concentration of the sample. This double-peak feature is considerably less prominent for the sample annealed at 282°C, indicating that the concentration of Mn_I has been reduced by the annealing process. This result strongly suggests that the Mn interstitials are unstable and precipitate out upon annealing to form Mn clusters and/or MnAs inclusions. Furthermore, the high and nearly equal channeled Mn signal along all three channeling scan directions in the sample annealed at 350°C implies that, in addition to intersititial Mn, a significant fraction of the Mn_{Ga} have also left their original substitutional positions to form random precipitates (Mn_{rand}).

As discussed earlier, Mn_I act as compensating centers for the substitutional Mn_{Ga} acceptors, thus significantly reducing the concentration of holes. Moreover, Mn_I are highly mobile and can drift preferentially to interstitial sites immediately adjacent to negatively charged Mn_{Ga} acceptors, forming antiferromagnetically aligned Mn_I–Mn_{Ga} pairs that cancel the contribution of Mn_{Ga} to the magnetization of the $Ga_{1-x}Mn_xAs$ system as a whole.[19] The above channeling results clearly indicate that annealing the sample at 282°C (only slightly above the growth temperature of ~250°C) can break these relatively weakly bound Mn_I–Mn_{Ga} pairs and release the highly mobile Mn_I to diffuse away, presumably to form precipitates and/or clusters. Substitutional Mn_{Ga} ions are then left to act as electrically-active acceptors as well as uncompensated magnetic moments. Annealing at 282°C has therefore led both to a higher hole concentration and to a higher saturation magnetization, as confirmed by ECV and SQUID measurements, respectively.

A summary of the channeling, T_C and hole concentration measurements is presented in Fig. 4.6. The upper panel shows fractions of Mn atoms at the various sites — substitutional (Mn_{Ga}), interstitial (Mn_I), and in random-cluster form (Mn_{rand}) — obtained from the angular scans shown in Fig. 4.5. The lower panel shows the values of T_C measured by SQUID magnetization as well as the hole concentrations measured by ECV for the same sample subjected to different annealing procedures. It is clear from these measurements that annealing-induced enhancement of hole concentration and uncompensated spins both contribute to the increase of T_C. The large changes in the electronic and magnetic properties induced by annealing can also be attributed to rearrangement of the sites occupied by the highly unstable Mn_I within the $Ga_{1-x}Mn_xAs$ lattice.

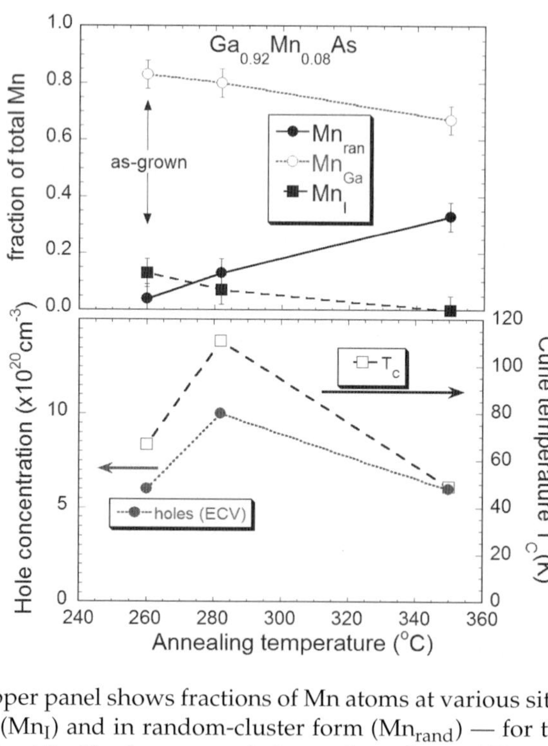

Figure 4.6. The upper panel shows fractions of Mn atoms at various sites — substitutional (Mn_{Ga}), interstitial (Mn_I) and in random-cluster form (Mn_{rand}) — for the samples used in angular scans in Fig. 4.5. The lower panel shows the values of T_C measured by SQUID magnetization, and of the hole concentration measured by ECV for the same samples.

4.2.4.2 Film thickness effect on Mn_I

Recently, quite a few reports on LT-MBE $Ga_{1-x}Mn_xAs$ with T_C over 150 K have been published. These measurements have been made mostly on very thin (< 50 nm) $Ga_{1-x}Mn_xAs$ films after annealing at optimal conditions.[6,27,29–32] However, $T_C > 110$ K in samples thicker than 100 nm has not been clearly demonstrated.[33] As a result, surface and/or interfacial effects may play some role in the improved T_C found in very thin layers that has not yet been addressed in the preceding discussion. This section explores this question by investigating the effects of Mn location in $Ga_{1-x}Mn_xAs$ as a function of film thickness.[34]

A series of as-grown $Ga_{1-x}Mn_xAs$ films with thicknesses ranging from 14 to 200 nm and Mn concentration in the range $0.07 < x < 0.1$ were studied for this purpose. SQUID magnetometry measurements of $Ga_{0.92}Mn_{0.08}As$ reveal T_C values of 110 and 65 K for the thin (14 nm) and the thick (200 nm) samples, respectively. Ion channeling results show that the Mn atoms in the 14 nm $Ga_{0.9}Mn_{0.1}As$ layer are incorporated either at substitutional Ga sites (Mn_{Ga} fraction ~70%) or at random positions that are incommensurate with the lattice (Mn_{rand} fraction ~30%). The fraction of Mn_I in thin samples (< 15 nm) is below the detection limit of the channeling technique, which is less than 2% of the total Mn concentration. These results contrast sharply with the results for $Ga_{1-x}Mn_xAs$ films thicker than 50 nm, where a substantial fraction of Mn is found in interstitial positions. The higher T_C in the thin sample is clearly due to the absence of compensating Mn_I.[18,35]

Edmonds *et al.* reported that Mn_I are relatively mobile and have a tendency to out-diffuse to the surface during post-growth low-temperature annealing.[27] Such Mn_I out-diffusion was found to be governed by an energy barrier of ~0.7 eV. As a result of the Mn_I diffusion, a Mn-rich oxide layer was detected on the surface of $Ga_{1-x}Mn_xAs$ that can be etched off by HCl.[34] We believe that the large fraction of Mn_{rand}(~30%) that we observe in the 14 nm $Ga_{1-x}Mn_xAs$ film, as described above, originates from the Mn in the surface oxide layer due to the outdiffusion of Mn_I. The reduced thickness of the film allows the out-diffusion process to occur during growth and sample cooling, and the Mn-rich layer subsequently oxidizes when the film is exposed to air.

Channeling RBS and PIXE measurements on the 14 nm layer after etching in HCl show that the removal of the ~2 nm thick surface layer leads to a ~25% reduction in the total Mn and a reduction of Mn_{rand} from ~30% to 15%.[34] This suggests that the oxide surface layer is Mn-rich and contains $\sim 8\times 10^{14}/\text{cm}^2$ of Mn_{rand}. The ~15% Mn_{rand} still present in the etched sample most probably exists in the form of small Mn-related clusters in the film. Furthermore, magnetization measurements before and after the HCl etch display essentially identical T_C and saturation magnetization values, suggesting that the Mn_{rand} in the oxide layer do not participate in the carrier-mediated ferromagnetism of the sample.

This interpretation of the RBS/PIXE results is confirmed by X-ray absorption spectroscopy (XAS) measurements in total electron yield mode at room temperature.[34] Figure 4.7 shows the Mn *2p–3d* XAS spectra for the 14 nm thick sample before and after the HCl etch. As other researchers have demonstrated previously, the XAS spectrum for the as-grown sample exhibits a main peak for the L3 level at 640 eV, with a shoulder at ∼0.5 eV lower in energy.[36,37] Ishiwata *et al.* attributed this low-energy shoulder to metastable paramagnetic defects due to coupling with excess As.[37] They argued that after low-temperature annealing these defects transform into the ferromagnetic component responsible for the higher energy peak. Figure 4.7 clearly shows that only the low-energy peak remains after the Mn-rich oxide layer is removed by HCl etching. We have also performed XAS on a $Zn_{0.88}Mn_{0.12}O_{0.02}Te_{0.98}$ alloys and observed the same peak at slightly higher energy than 640 eV, suggesting that it is indeed related to Mn–O bonds. The high-energy peak in the XAS spectrum from the as-grown $Ga_{1-x}Mn_xAs$ film is therefore due to Mn in the oxide layer, while the lower energy peak in the doublet arises from Mn_{Ga}. This is in agreement with the recent report by Edmonds *et al.*, who also identified the presence of an Mn-rich oxide layer and correlated this low-energy peak with strong X-ray magnetic circular dichroism, thus attesting to its ferromagnetic origin.[27,36]

Figure 4.8 shows the distribution of Mn atoms in various lattice sites in $Ga_{1-x}Mn_xAs$ films ($x \sim 0.07$–0.10) with thicknesses ranging from 14 to 120 nm obtained by ion channeling studies. A monotonic increase in the Mn_I fraction and a corresponding decrease in the Mn_{rand} fraction are observed with increasing film thickness. Above a thickness of 60 nm, however, the relative amounts of Mn_I and Mn_{rand} remain rather constant. The distribution of Mn in the 14 and 22 nm $Ga_{1-x}Mn_xAs$ films after HCl etching are also shown in Fig. 4.8. An increase in the fraction of Mn_{Ga} is observed in both these thin samples due to the removal of the Mn-rich oxide surface layer.

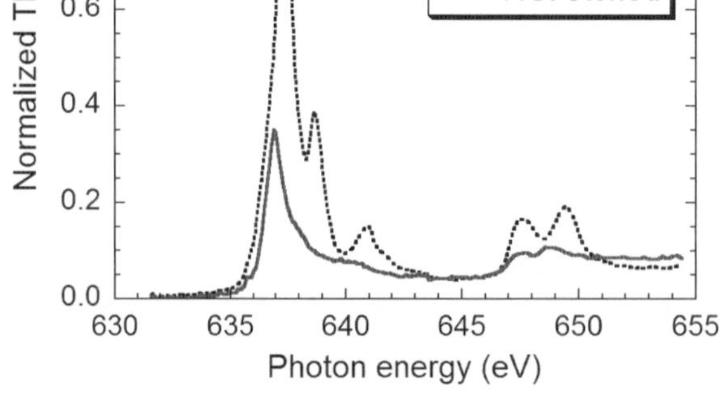

Figure 4.7. Mn 2*p* XAS spectra for a 14 nm thick $Ga_{1-x}Mn_xAs$ sample, as-grown and after HCl etching.

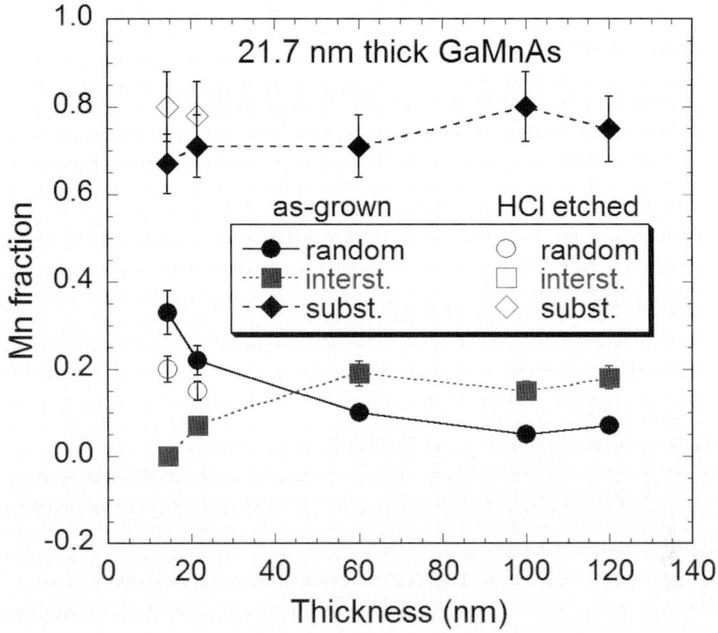

Figure 4.8. Fractions of Mn occupying various lattice sites — substitutional (Mn_{Ga}), interstitial (Mn_I), and in random-cluster form (Mn_{rand}) — measured by channeling techniques for $Ga_{1-x}Mn_xAs$ samples with film thicknesses between 14 and 120 nm. Data for 14 and 21.7 nm $Ga_{1-x}Mn_xAs$ films etched by HCl are also shown as open symbols.

To understand the origin of this thickness dependence, we note that the density of Mn atoms per unit area in the surface oxide layer of the thin samples is relatively independent of film thickness, and is approximately equal to $\sim 8\times10^{14}/cm^2$. This value is close to the density of Ga sites on the (001) surface, indicating that the out-diffusion of Mn_I during growth, or after exposure to air, can be limited by the accumulation of approximately one monolayer of Mn on the surface. Note that the areal density of Mn_I is $\sim 5\times10^{14}\,cm^{-2}$ in a 15 nm thick $Ga_{0.9}Mn_{0.1}As$ film with 15% Mn_I. For films with thicknesses of ~10 nm, the diffusion length of Mn_I in the film is comparable to the film thickness, and it is therefore conceivable that the out-diffusion of Mn_I can account for the elimination of Mn_I in very thin films.[27] This hypothesis is confirmed by Koeder *et al.*, who observed a pronounced increase of holes towards the surface of the $Ga_{1-x}Mn_xAs$ film despite the apparent uniform Mn distribution throughout.[38] Therefore, at higher film thicknesses, Mn_I out-diffusion affects only the outer thin layer of the film and is limited by the accumulation of ~1 monolayer of Mn on the surface, while the remaining Mn_I remains incorporated in the bulk of the layer. The electronic and magnetic properties of these thick films are then determined by a balance between Mn_{Ga}, Mn_I, and Mn_{rand}.

4.2.5 Maximum Carrier Concentration and T_C in III-Mn-V Semiconductors

The ECV measurements shown in Fig. 4.3 suggest that the hole concentration in $Ga_{1-x}Mn_xAs$ has a tendency to saturate at the Mn content of about 3–4%. To discuss this issue, it is useful to invoke the amphoteric native defect model, which has been shown to provide good qualitative predictions of the maximum carrier concentration achievable by doping in a wide variety of semiconductors (including both III–V and II–VI materials).[39–42] The model relates the type and concentration of native defects responsible for dopant compensation to the location of the Fermi level with respect to an internal energy reference, the *Fermi level stabilization energy*, E_{FS}. E_{FS} is located at ~4.9 eV below the vacuum level and is the same for all III–V and II–VI semiconductors. Materials with the conduction band located close to E_{FS} can be easily doped n-type and, similarly, those whose valence bands are close to E_{FS} are easily doped p-type. In III–V semiconductors, the maximum and minimum energy locations of the Fermi level typically do not deviate by more than 1.0 eV in either direction relative to E_{FS}.

In the specific case of GaAs the conduction band is located at $E_{FS} + 0.9$ eV and the valence band is at $E_{FS} - 0.5$ eV. The closer proximity of the valence band to E_{FS} sets higher limits on the maximum hole concentration compared to that of conduction electrons. For p-type GaAs, free hole concentrations as high as mid $10^{20}/cm^3$ have been achieved via Zn diffusion in GaAs, in good agreement with the prediction of the amphoteric defect model.[40]

From the observed dependence of $[Mn_I]$ on the Mn concentration in $Ga_{1-x}Mn_xAs$, it is understood that an increase in x raises the concentration of Mn_{Ga} in $Ga_{1-x}Mn_xAs$ pushing the Fermi energy to its maximum level E_{Fmax} (i.e., to the condition where the hole concentration p approaches p_{max}). This dependence of the defect formation energy on the Fermi level and the resulting free carrier saturation are illustrated in Fig. 4.9. Note that due to the reduced growth temperature of $Ga_{1-x}Mn_xAs$ (190–280°C), the formation energy of Mn_{Ga} is lower than it would be for instances of equilibrium growth, which allows E_{Fmax} to exist at a lower energy with respect to the valence band edge than in GaAs grown under standard conditions. The hole saturation limit, p_{max}, is correlated with the relatively constant hole concentrations of about $6 \times 10^{20}/cm^3$ for as-grown $Ga_{1-x}Mn_xAs$ samples with Mn concentrations higher than 3% (shown in Fig. 4.3) and about $1 \times 10^{21}/cm^3$ for optimally annealed $Ga_{0.92}Mn_{0.08}As$ samples (shown in Fig. 4.6). As this limit is reached, the formation energies of Mn_{Ga} acceptors and compensating Mn_I become comparable. Introduction of additional Mn into the $Ga_{1-x}Mn_xAs$ beyond this point is expected to result in a downward shift of the Fermi energy that would in turn raise the formation energy of negatively charged Mn_{Ga} acceptors. Subsequently, as the formation of Mn_{Ga} acceptors becomes energetically unfavorable, an increasing fraction of Mn will then be incorporated in the form of Mn_I donors and/or electrically inactive MnAs or Mn clusters. Given that the ferromagnetism in this system is related to the uncompensated Mn spins and

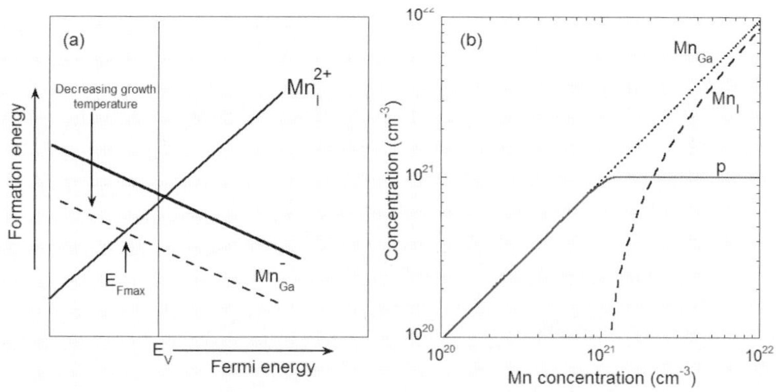

Figure 4.9. Schematics showing (a) the formation energy dependence of Mn_{Ga} and Mn_I on the position of the Fermi level, and (b) the saturation of hole concentration resulting from compensation by Mn_I as a function of Mn concentration.

is mediated by holes, such a Fermi-level-induced hole saturation effect necessarily imposes a limit on the Curie temperature of the system.

4.2.6 Co-doping of GaMnAs with Shallow Dopants

The Zener model of ferromagnetism proposed by Dietl *et al.* predicts that[11]

$$T_C = Cxp^{1/3}, \tag{4.1}$$

where x is the mole fraction of substitutional Mn^{++} ions, p is the hole concentration, and C is a constant specific to the host material. Because the ferromagnetic ordering depends on the concentration of uncompensated holes, the use of an additional acceptor impurity with a shallow acceptor level and high solubility in GaAs, such as Be, to increase the total hole concentration appears to be a promising route to improve T_C.[43] However, as we have discussed in the last section the dependence of the defect formation energy on the Fermi level results in saturation of the hole concentration and sets an upper bounds to T_C. A number of systematic experiments have been undertaken to explore the effect of creating additional carriers on the properties of $Ga_{1-x}Mn_xAs$ by Be doping, which will be described below.[35,44,45]

4.2.6.1 Additional p-type doping with Be

$Ga_{1-x}Mn_xAs$ is a strongly compensated material when it is heavily doped with Mn, so it is instructive to begin by discussing studies of extrinsic Be doping in $Ga_{1-x}Mn_xAs$ involving specimens with Mn concentration below the hole saturation level of about 1×10^{21} cm^{-3}. A series of $Ga_{1-x}Mn_xAs$ samples was fabricated for this purpose with a fixed Mn concentration, with a systematically increasing Be doping level. Yuldashev *et al.* reported a monotonic rise in T_C (from 40 to 50 K) by enhancing the Be doping for these low Mn samples, which is consistent with

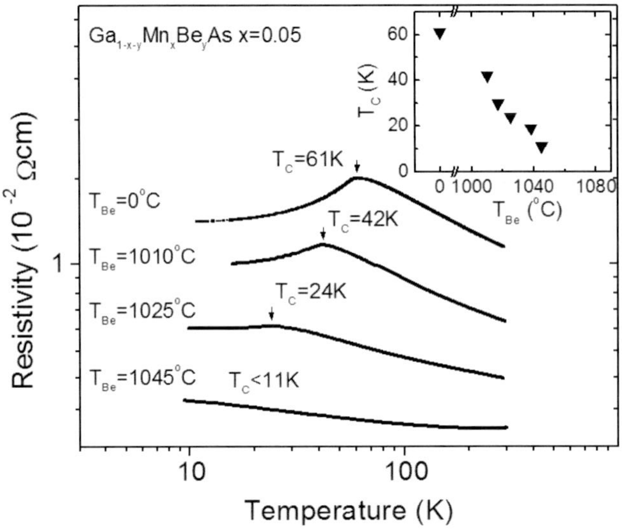

Figure 4.10. Temperature dependence of the resistivity for undoped and Be-doped samples of $Ga_{1-x}Mn_xAs$ with high Mn content ($x = 0.05$) at zero magnetic field. The dependence of T_C on Be doping of these samples is shown in the inset.

the $p^{1/3}$ dependence of the T_C assumed by the Zener model.[44] These authors also reported that the increase of the hole concentration p measured by the Hall effect with higher Be co-doping does indeed correspond to a true increase in p. As the Mn concentration and thus also the hole concentration are below the hole saturation level, the formation energies of compensating defects (Mn_I and antisites) remain higher than the formation energy of Mn_{Ga}. Hence, the introduction of Be acceptors during the growth of $Ga_{1-x}Mn_xAs$ with low x does improve the hole concentration without affecting the concentration of Mn_{Ga} leading to an increase of T_C.

While Be co-doping of $Ga_{1-x}Mn_xAs$ with low Mn concentrations enhances p and T_C, the incorporation of Be into alloys with $x > 0.05$ leads to a strong *decrease* in T_C.[45–47] This effect is clearly visible in Fig. 4.10, where the resistivity peak shifts to lower T with increasing Be doping level, which is controlled by the Be cell temperature T_{Be} during MBE growth.[45] This strong drop of T_C (see the inset in Fig. 4.10) was also confirmed by studies of AHE and by direct magnetization measurements using SQUID. Moreover, despite the reduction in T_C with increased Be co-doping, the hole concentration remained *nearly constant*, as displayed in Fig. 4.11.[46,47] This conclusion was determined by ECV for all samples and was additionally corroborated by Hall measurements for samples with a high Be content [see Fig. 4.11(b)], which were not ferromagnetic and thus not affected by AHE. Here, the Be content shown in the figure was estimated from the lattice constant determined by X-ray diffraction that was calibrated by RHEED intensity oscillations.

To obtain physical insight into the causes of this surprising behavior, the relationship between the location of Mn in the lattice and the ferromagnetism of

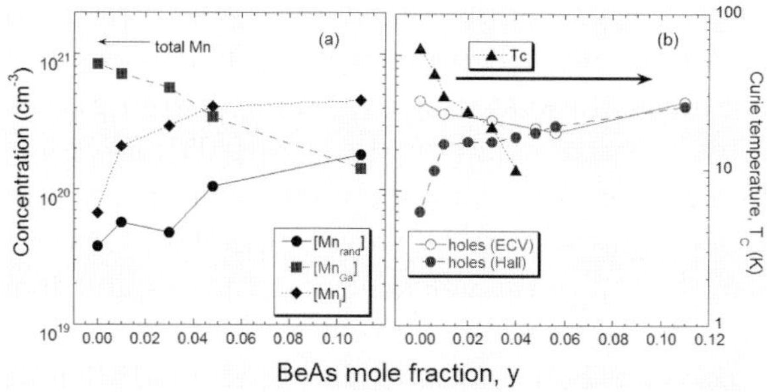

Figure 4.11. (a) Concentration of Mn atoms at various inequivalent sites in the GaMnAs epilayers obtained from the angular scan analysis. (b) T_C together with hole concentrations measured by ECV and by Hall resistivity for $Ga_{1-x-y}Mn_xBe_yAs$ films as a function of Be concentration y.

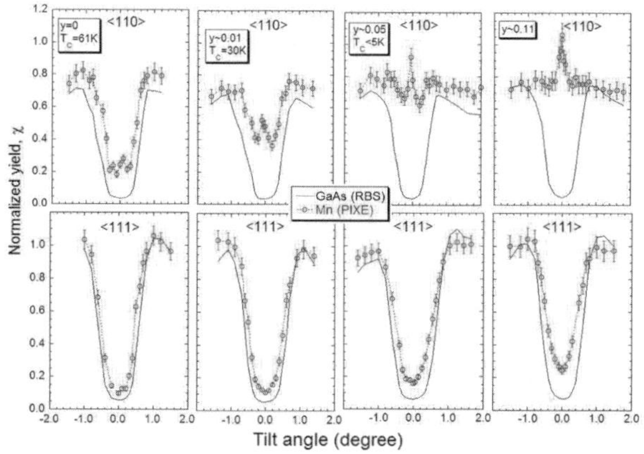

Figure 4.12. Angular scans about the $\langle 110 \rangle$ and $\langle 111 \rangle$ channeling axes for undoped and for Be doped $Ga_{1-x}Mn_xAs$ samples.

the $Ga_{1-x-y}Mn_xBe_yAs$ layers was investigated using c-PIXE/RBS. Figure 4.12 shows the PIXE and RBS angular scans about the $\langle 110 \rangle$ and $\langle 111 \rangle$ axes for the $Ga_{1-x}Mn_xAs$ and $Ga_{1-x-y}Be_yMn_xAs$ films as a function of y.[35] The $\langle 110 \rangle$ Mn angular scans show a definite peak at the center of the [110] channel that intensifies with an increasing level of Be doping — a clear signature that the concentration of Mn interstitials is amplified as more Be is introduced into the alloy lattice. Concurrently, the normalized yields χ_{Mn} in the $\langle 111 \rangle$ scans also rise gradually and deviate from the corresponding host scans. Hence, the formation of Mn-containing random clusters not commensurate with the GaAs lattice is tied to Be incorporation.

The fractions of Mn atoms at the various lattice locations measured from the angular scans are shown in Fig. 4.11(a). Since the fraction of Mn_I as well as random Mn-based precipitates Mn_{rand} in the $Ga_{1-x-y}Mn_xBe_yAs$ increases monotonically with Be content, the net concentration of uncompensated Mn ions that can participate in the ferromagnetism ($[Mn_{Ga}]-[Mn_I]$) decreases strongly with increasing

y, and the reduction of active spins will naturally lead to the disappearance of ferromagnetism. Additionally, when the number of active spins becomes smaller than the number of holes, the system is driven into a spin-glass state. Notice that as the Be fraction y approaches that of Mn ($x = 0.05$), the concentration of Mn_I becomes larger than that of Mn_{Ga}, and the number of uncompensated spins, now that of Mn_I, increases again with y. However, since Mn_I do not hybridize with p states of the holes at the top of the valence band, those spins do not participate in ferromagnetism, and the system remains non-ferromagnetic.

The relatively constant hole concentration at $x \geq 0.05$ [shown in Fig. 4.11(b)] demonstrates that the hole concentration in these $Ga_{1-x-y}Mn_xBe_yAs$ samples is at its saturation limit. We recall that at this limit the formation energies of Mn_{Ga} acceptors and compensating Mn_I become comparable. Therefore, the introduction of additional Be acceptors into $Ga_{1-x-y}Mn_xBe_yAs$ leads to a downward shift of the Fermi energy, which in turn increases the formation energy of negatively charged Mn_{Ga} acceptors and reduces the formation energy of Mn_I donors. As a result, Mn is more likely to be incorporated into the system in the form of Mn_I donors and/or electrically inactive precipitates, leading, as argued earlier, to a drastic drop in T_C with increasing y.

4.2.6.2 Modulation doping of $Ga_{1-x}Mn_xAs$ in heterostructures

Studies of low-temperature annealing and extrinsic Be co-doping of $Ga_{1-x}Mn_xAs$ layers have clearly demonstrated that a strong increase of the Mn_I concentration occurs at the expense of Mn_{Ga} as E_F decreases. However, lowering the position of E_F can affect the creation of Mn_I only when the additional holes are already present during the growth of $Ga_{1-x}Mn_xAs$. Extrapolating from the $Ga_{1-x}Mn_xAs$ behavior discussed above, the magnetic properties of a GaMnAs quantum well (QW) in a modulation-doped $Ga_{1-y}Al_yAs/Ga_{1-x}Mn_xAs/Ga_{1-y}Al_yAs$ heterostructure depend on which of the $Ga_{1-y}Al_yAs$ layers is doped with Be. An increase of T_C is expected when the Be-doped barrier is grown *after* the $Ga_{1-x}Mn_xAs$ QW since the value of p increases *after* the ferromagnetically-active region has already been formed. Conversely, a drop in T_C is expected when the Be-doped barrier is grown *before* the $Ga_{1-x}Mn_xAs$ QW, because the holes are transferred from the barrier to the $Ga_{1-x}Mn_xAs$ during deposition, which results in a downward shift of the Fermi energy and redistribution of the Mn atoms to interstitial sites.

To experimentally demonstrate the crucial role that layer growth sequencing plays in the formation of Mn_I states, and hence in controlling T_C, a series of low-temperature $Ga_{1-y}Al_yAs/Ga_{1-x}Mn_xAs/Ga_{1-y}Al_yAs$ QW structures were deposited in the following geometries: a structure with Be doping in the first barrier ("inverted modulation doped QW," I-MDQW), an undoped structure, and a structure with Be doping in the second barrier ("normal MDQW" — N-MDQW).[48] As predicted, the T_C increases (to 98 K) for the N-MDQW sample compared with the undoped sample ($T_C = 80$ K). In contrast, the T_C of the I-MDQW sample decreases (to 62 K) in agreement with the model of Fermi-energy-dependent creation of Mn_I.

The enhanced incorporation of Mn into interstitial positions in structures in which the Be-doped barrier was grown before the deposition of the $Ga_{1-x}Mn_xAs$ well was unambiguously confirmed by c-RBS/PIXE experiments.[49] Figure 4.13 presents the $\langle 110 \rangle$ and $\langle 111 \rangle$ angular scans for the $Ga_{1-x}Mn_xAs$ layers in both the I-MDQW and the N-MDQW geometries. The difference in the respective Mn locations for I-MDQW and N-MDQW is clearly detected by the Mn PIXE signal in angular scans about the $\langle 110 \rangle$ axial channel shown in the right-hand panel. The much higher χ_{Mn} in the $\langle 110 \rangle$ scan for the I-MDQW indicates that the concentration of Mn_I is higher in the case when the Be-doped barrier layer is grown prior to depositing of the $Ga_{1-x}Mn_xAs$ QW. The fractions of Mn_I are estimated to be ~20% for I-MDQW and ~11% for N-MDQW.

The RBS/PIXE results presented in Fig. 4.13 provide direct experimental evidence that the manner of incorporation of Mn into $Ga_{1-x}Mn_xAs$ is directly controlled by the Fermi energy during the LT-MBE growth itself. These data also rule out the possibility that other effects, such as the competition between Mn and Be atoms to occupy the same substitutional sites, might be responsible for the increase in the number of Mn_I, as the Mn and Be atoms are spatially separated in the modulation-doped structures. Our studies of inverted and normal MDQWs provide further strong support for the model of Fermi-level-induced limitation of T_C in $Ga_{1-x}Mn_xAs$.[35]

4.2.6.3 Donor co-doping of $Ga_{1-x}Mn_xAs$

The last section explores the effects that extrinsic p-type co-doping can have on the Mn incorporation into GaAs and the resulting ferromagnetic properties of the alloy. However, relatively little attention has been given to doping of GaMnAs by donors.[50] A recent report by Cho *et al.* investigated the impact of Si doping

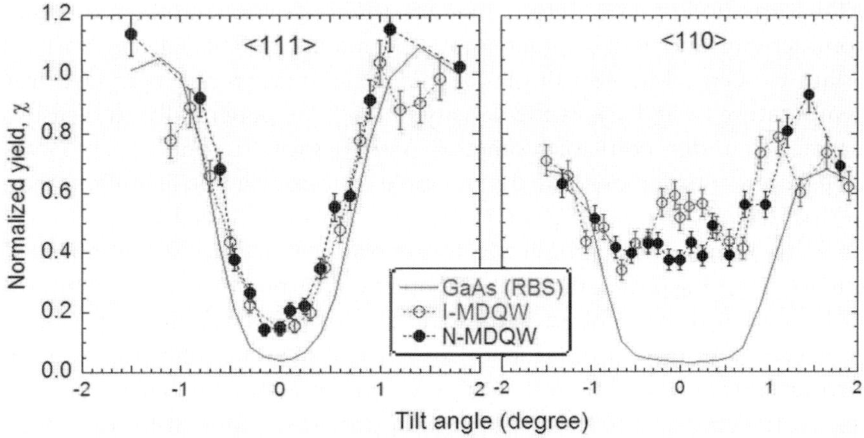

Figure 4.13. Normalized yields of the Mn (PIXE) and GaAs (RBS) signals for a $Ga_{0.938}Mn_{0.062}As$ QW layer in both I-MDQW and N-MDQW geometries as a function of the incident tilt angles for angular scans about the $\langle 110 \rangle$ and $\langle 111 \rangle$ channeling axes.

(with [Si]$\sim$ $1 \times 10^{20}\,\mathrm{cm}^{-3}$) of $Ga_{1-x}Mn_xAs$ films.[51] They found that Si-doping of $Ga_{1-x}Mn_xAs$ with low Mn concentration (e.g., $x < 0.033$) moderates T_C in comparison to undoped $Ga_{1-x}Mn_xAs$ (same x) simply through p reduction via compensation. The p-type nature of $Ga_{1-x}Mn_xAs$ places the Fermi level very close to the valence band edge, and the formation energy of donor defects is low. The Si atoms are therefore expected to occupy Ga sites and form donors that will compensate holes and reduce the T_C of the alloy. ECV measurements confirm that the hole concentration p in the undoped and Si-doped films is 3×10^{20} and $2 \times 10^{20}/\mathrm{cm}^3$, respectively, which is in excellent agreement with the expectation that Si donors (Si_{Ga}) compensate Mn holes in these samples.

Figure 4.14(b) displays the temperature dependence of the resistivity of as-grown $Ga_{0.84}Mn_{0.16}As$ and $Ga_{0.84}Mn_{0.16}As$:Si films.[51] The resistivity of the undoped $Ga_{0.84}Mn_{0.16}As$ shows a strongly insulating behavior that can be attributed to the energetically favorable formation of Mn_I. By contrast, the resistivity of $Ga_{0.84}Mn_{0.16}As$:Si is *lower* than that of the undoped sample. Changes in the T_C of Si-doped and undoped GaMnAs for high x induced by annealing (190°C for 2 hours) is displayed in Fig. 4.14(a). The T_C of $Ga_{0.84}Mn_{0.16}As$:Si is discernibly higher than that of the undoped $Ga_{0.84}Mn_{0.16}As$ for both the as-grown and annealed cases. A comparison of the hysteresis loops of the as-grown and Si-doped $Ga_{0.84}Mn_{0.16}As$, taken at 5 K with external magnetic fields applied along the in-plane $\langle 100 \rangle$ direction, is included in Fig. 4.14(c), where a much more abrupt magnetization reversal is observed for the sample with Si doping. Since c-PIXE measurements determine that the effective Mn concentrations $x_{\mathrm{eff}} = x(Mn_{Ga}) - x(Mn_I)$, representing the number of *active* Mn spins that can interact with the holes, are 0.094 and 0.075 for $Ga_{0.84}Mn_{0.16}As$ and $Ga_{0.84}Mn_{0.16}As$:Si, respectively, the weak ferromagnetism of the $Ga_{0.84}Mn_{0.16}As$ cannot be explained entirely by Mn_I.

It was speculated that Si doping in LT-MBE grown $Ga_{1-x}Mn_xAs$ can suppress the formation of As_{Ga} defects that are double donors and therefore improve the conductivity and ferromagnetism of the material. This idea is particularly important for $Ga_{1-x}Mn_xAs$ samples with $x > 0.1$ that were grown at extremely low temperature (<180°C), as the density of As_{Ga} "exponentially" increases with decreasing T_s under non-stoichiometric As-rich growth condition.[52] However, ECV measurements demonstrate that the hole concentration in as-grown undoped $Ga_{0.84}Mn_{0.16}As$ ($p = 6 \times 10^{20}/\mathrm{cm}^3$) is *larger* than in Si-doped $Ga_{0.84}Mn_{0.16}As$:Si ($p = 4 \times 10^{20}/\mathrm{cm}^3$). Even at high Mn concentrations, Si doping continues to suppress p in $Ga_{0.84}Mn_{0.16}As$ rather than enhancing it through the prevention of As_{Ga} defect formation during the growth. The observed behaviors of transport and ferromagnetic properties of Si-doped $Ga_{1-x}Mn_xAs$ can be interpreted through a detailed understanding of the manner in which Mn alters the valence band structure of the III–V host. The following section provides a look at the role of Mn in GaAs and other Ga-V compounds and predicts trends in their transport and magnetic properties.

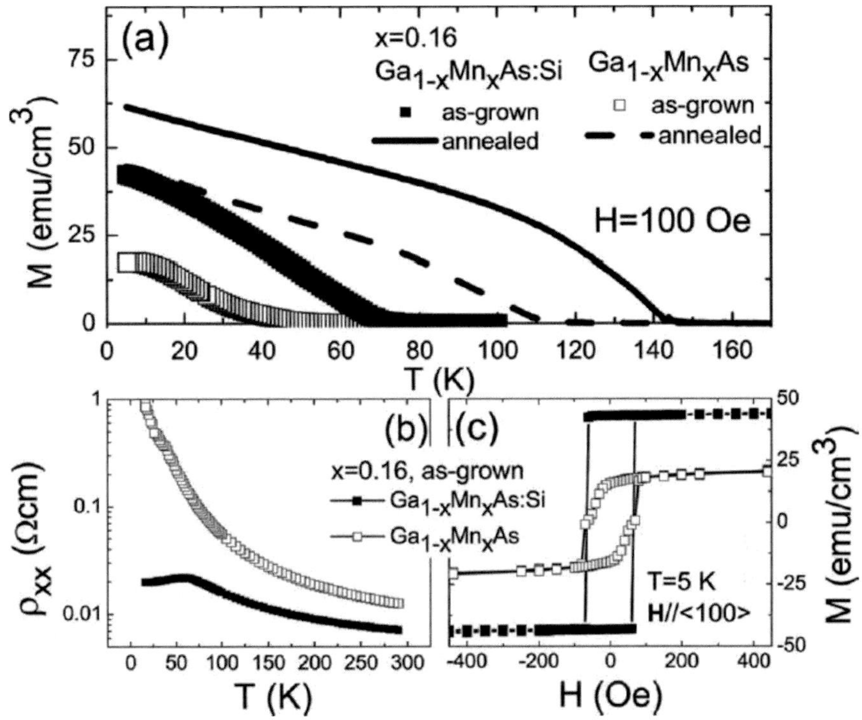

Figure 4.14. (a) Temperature dependence of magnetization of $Ga_{0.84}Mn_{0.16}As$ and $Ga_{0.84}Mn_{0.16}As$:Si, (b) temperature dependence of resistivity, and (c) hysteresis loop at 5 K of as-grown $Ga_{0.84}Mn_{0.16}As$ and $Ga_{0.84}Mn_{0.16}As$:Si. The magnetic field is applied along the in-plane ⟨100⟩ direction.

4.3 ELECTRONIC STRUCTURE OF Mn IN GaMnAs

According to the Zener model of DMS, the T_C depends on the concentration of isolated substitutional Mn atoms, the hole concentration and the valence band structure. The preceding section focused on the limitations of substitutional Mn_{Ga} incorporation and the maximum hole concentration imposed by the location of the host valence band edge with respect to the Fermi level stabilization energy, E_{FS}. This work suggests that the material properties could be engineered simply by selecting a host based on its band alignment to E_{FS}. For example, the valence band edge of GaSb is located only about 0.1 eV below the E_{FS}, so the limitations on the maximum hole concentration should be less severe. This argument holds only if one assumes that the free holes are responsible for the magnetic coupling. Despite the initial success and common acceptance of the Zener model, it was argued quite early on that the holes localized in an Mn-derived impurity band are responsible for the ferromagnetic coupling in $Ga_{1-x}Mn_xAs$ rather than free holes.[53] More recently, a series of infrared absorption experiments on $Ga_{1-x}Mn_xAs$ provided further support for an impurity band model.[54] The disadvantage of such a model, however, was that there was no simple description of the impurity band structure. In this section we discuss the application of the recently developed band anticrossing (BAC) model to describe the electronic structure of the Mn impurity band in group III–V compounds. We show that the model can be used not only to explain the experimental observations of $Ga_{1-x}Mn_xAs$ alloys but also to predict trends in the magnetic properties of other III-Mn-V semiconductors.

The incorporation of Mn into GaAs and other III–V compounds not only changes the ferromagnetic behavior of the material, but as described above, it drastically alters the transport properties of the host compared to the addition of other group II dopants. As with most non-magnetic acceptors in semiconductors, the shallow p states of the Mn atoms have long been thought to merge with the valence band of GaAs at sufficient concentrations $(x > 0.02)$ according to the Mott theory of a metal-to-insulator transition (MIT).[11,55] Under this framework contained within the Zener model, hole transport at these alloy compositions occurs within the extended states of the valence band and mediates ferromagnetism. Unlike the behavior of non-magnetic acceptors, however, the exchange energy produced by p–d hybridization tends to more strongly localize the Mn states. The sufficiently strong exchange interactions between the Mn and GaAs states predicted by dynamic mean field theory and *ab initio* first-principles calculations support the concept of Mn as a deep center with a fixed energy level regardless of the host rather than a shallow acceptor.[56–58] As a result, new support exists for the persistence of a detached Mn-derived impurity band at much higher x.

The deep center nature of the Mn p states suggests that the electronic structure of $Ga_{1-x}Mn_xAs$ may behave in a manner related to "highly mismatched" semiconductor alloys (HMA) containing isoelectronic elements with similarly localized states (i.e., N or Bi in GaAs).[59,60] Formed from Bloch functions across several bands, the wave functions of deep trap states are extended in k-space. Consequently, their energy levels remain constant with respect to the vacuum level and do not track changes in the position of the host band edges, as do most shallow states. Those with A_1, or s-like, symmetry are typically located near the conduction band edge of most semiconductors, while those of T_2, or p-like, symmetry often lie near the valence band edges.[61] The relative position of the localized T_2 state of Mn in Ga-V compounds is displayed in Fig. 4.15 as an illustration of this effect. The real-space short-range potential of Mn ultimately prevents the overlap of its p states and therefore restricts the broadening and merging of an initial impurity band with the host valence band. Dynamic mixing of the localized impurity and delocalized host states in HMA has been demonstrated to lead to a restructuring of the valence band into E_- and E_+ sub-bands through an anticrossing interaction analogous to that found in a degenerate two-level system.[60] In this section, we apply the band anticrossing (BAC) theory to $Ga_{1-x}Mn_xAs$ and related DMS systems, which yields a valence band structure that is consistent with the experimental results pertaining to the hole effective mass, mobility and MIT observed in these alloys.

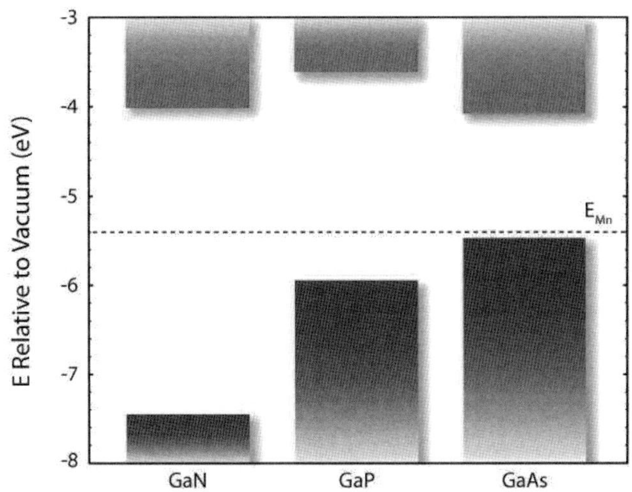

Figure 4.15. Conduction and valence band alignments of Ga-V compounds relative to the Mn acceptor level, E_{Mn}.

4.3.1 Valence Band Anticrossing Model

Application of the valence band anticrossing (VBAC) theory to $\mathrm{Ga}_{1-x}\mathrm{Mn}_x$-V alloys offers quantitative assessment of the valence band structure and provides detailed information on the development of an independent Mn-derived impurity band. The hybridization of Mn and host states is treated in the **kp** formalism and is described by a 12×12 Hamiltonian, which consists of the 6×6 Kohn Luttinger matrix expressing the extended host p states and the six time-reversal symmetry-invariant wave functions of the localized Mn p states[60,62]:

$$H_V = \begin{vmatrix}
H & \alpha & \beta & 0 & \frac{i\alpha}{\sqrt{2}} & -i\sqrt{2}\beta & V(x) & 0 & 0 & 0 & 0 & 0 \\
\alpha^* & L & 0 & \beta & \frac{iD}{\sqrt{2}} & i\sqrt{\frac{3}{2}}\alpha & 0 & V(x) & 0 & 0 & 0 & 0 \\
\beta^* & 0 & L & -\alpha & -i\sqrt{\frac{3}{2}}\alpha^* & \frac{iD}{\sqrt{2}} & 0 & 0 & V(x) & 0 & 0 & 0 \\
0 & \beta^* & -\alpha^* & H & -i\sqrt{2}\beta^* & \frac{-i\alpha^*}{\sqrt{2}} & 0 & 0 & 0 & V(x) & 0 & 0 \\
\frac{-i\alpha^*}{\sqrt{2}} & \frac{-iD}{\sqrt{2}} & i\sqrt{\frac{3}{2}}\alpha & i\sqrt{2}\beta & S & 0 & 0 & 0 & 0 & 0 & V(x) & 0 \\
i\sqrt{2}\beta^* & -i\sqrt{\frac{3}{2}}\alpha^* & \frac{-iD}{\sqrt{2}} & \frac{i\alpha}{\sqrt{2}} & 0 & S & 0 & 0 & 0 & 0 & 0 & V(x) \\
V(x) & 0 & 0 & 0 & 0 & 0 & E_{\mathrm{Mn}} & 0 & 0 & 0 & 0 & 0 \\
0 & V(x) & 0 & 0 & 0 & 0 & 0 & E_{\mathrm{Mn}} & 0 & 0 & 0 & 0 \\
0 & 0 & V(x) & 0 & 0 & 0 & 0 & 0 & E_{\mathrm{Mn}} & 0 & 0 & 0 \\
0 & 0 & 0 & V(x) & 0 & 0 & 0 & 0 & 0 & E_{\mathrm{Mn}} & 0 & 0 \\
0 & 0 & 0 & 0 & V(x) & 0 & 0 & 0 & 0 & 0 & E_{\mathrm{Mn\text{-}SO}} & 0 \\
0 & 0 & 0 & 0 & 0 & V(x) & 0 & 0 & 0 & 0 & 0 & E_{\mathrm{Mn\text{-}SO}}
\end{vmatrix}$$

The parameters that describe the host valence band structure following the formulation of the Kohn–Luttinger matrix are defined as:

$$H = -\frac{\hbar^2}{2m_0}[(k_x^2 + k_y^2)(\gamma_1 + \gamma_2) + k_z^2(\gamma_1 - 2\gamma_2)] + \Delta E_{\mathrm{VBM}} \cdot x,$$

$$L = -\frac{\hbar^2}{2m_0}[(k_x^2 + k_y^2)(\gamma_1 - \gamma_2) + k_z^2(\gamma_1 + 2\gamma_2)] + \Delta E_{\mathrm{VBM}} \cdot x,$$

$$\alpha = \sqrt{3}\frac{\hbar^2}{m_0}[k_z(k_x - ik_y)\gamma_3],$$

$$\beta = \frac{\sqrt{3}}{2}\frac{\hbar^2}{m_0}[(k_x^2 - k_y^2)\gamma_2 - 2ik_xk_y\gamma_3],$$

$$D = L - H,$$

$$S = \frac{1}{2}(L + H) - \Delta_0 - \Delta E_{\mathrm{SO}} \cdot x,$$

where the Luttinger parameters and spin-orbit splitting energy were set to the corresponding values of the host semiconductor. The single energy values of E_{Mn} and $E_{\mathrm{Mn\text{-}SO}}$ represent the energetic positions of the localized heavy/light hole and spin-orbit split-off Mn states, respectively, relative to the host valence band edge. The strength of the dynamic mixing of the host and Mn states is regulated by the hybridization energy, V:

$$V = \langle P_X^A|U|X\rangle = \langle P_Y^A|U|Y\rangle = \langle P_Z^A|U|Z\rangle = C_{\mathrm{Mn}}\sqrt{x},$$

where configurational over the random distribution of Mn states simplifies the interaction of the GaAs wave functions, P^A, and Mn Bloch states, X, Y, or Z, with localized potential, U, to the product of an empirically determined coupling parameter, C_{Mn}, and the square-root of the fraction of $\mathrm{Mn_{Ga}}$, x. The coupling parameter depends largely on the degree to which the Mn states are localized within the host and thus largely captures the effects of p–d exchange. Therefore, this simplified model based on p state interaction provides a valid representation of the impurity and valence band edges.

In the application of the VBAC model to $\mathrm{Ga_{1-x}Mn_xAs}$, E_{Mn} was set to the Mn acceptor level, while C_{Mn} was experimentally established by photomodulated reflectance (PR) spectroscopy measurements of MBE-grown Be-doped $\mathrm{Ga_{1-x}Mn_xAs}$ films.[24,35] The anticrossing interaction not only sustains the existence of an independent Mn-derived E_+ sub-band, but it also drives the two bands apart with increased state mixing. This splitting therefore can be detected experimentally in the shift of the alloy band gap as the valence band edge (E_- sub-band) is pushed downward in energy. The PR spectra from several alloys, displayed in Fig. 4.16, show an apparent increase in the band-gap energy from 1.42 eV ($x = 0$)

to 1.46 eV ($x = 0.041$) that is associated only with an increase in the Mn concentration and is consistent with independent measurements reported in the literature.[63] The shift in the valence band edge extracted from the change in E_g was then fit by the VBAC model with a coupling parameter value of $C_{Mn} = 0.39$ eV, which is accordant with other alloys containing impurities with p states of comparable localization.[64]

4.3.2 Mobility and the Metal-to-Insulator Transition

4.3.2.1 GaMnAs

The detailed picture of the $Ga_{1-x}Mn_xAs$ valence band structure provides an excellent foundation from which to understand the transport properties of this alloy. At the outset of this discussion it is important to note that the transport properties of this and other DMS alloys are quite sensitive to the quantity and location of Mn atoms in the lattice as well as the concentration of defects, and significant variation in the material processing and quality is expected to produce some deviation in experimentally measured quantities. As a general trend, however, the hole mobilities of GaAs doped with non-magnetic, shallow acceptors (μ =10–50 cm^2/Vs) are at least an order of magnitude greater than those of $Ga_{1-x}Mn_xAs$ (μ = 1–5 cm^2/Vs) with commensurate hole concentrations.[64–66] Spin disorder scattering is negligible at room temperature, leaving ionized impurity scattering as the principle mobility-limiting mechanism in these materials and is expressed as[69]

$$\mu_{II} = \frac{\varepsilon^2 k_F}{2\pi e^3 \hbar N_i \left(\ln(1+\xi) - \dfrac{\xi}{1+\xi}\right)} \left(\frac{dE}{dk}\right)^2,$$

where ε is the dielectric constant, k_F is the Fermi wave vector, N_i is the concentration of ionized impurities, and

$$\xi = (2k_F R)^2.$$

The screening length, R, as a function of the carrier concentration, p, is given as

$$\frac{1}{R^2} = \left(\frac{3p}{\pi}\right)^{1/3} \frac{4m_{\text{eff}}e^2}{\varepsilon\hbar^2}.$$

Both the magnetic and non-magnetic films are assumed to contain corresponding defect densities based on the similarity of their growth conditions, and therefore a disparity in the hole effective mass between the two materials may account for experimentally observed trends in the mobility. The heavy hole effective mass of $Ga_{1-x}Mn_xAs$ may be obtained from the dispersion curves of the Mn-derived impurity band, depicted in Fig. 4.17, according to the relation:

$$m_{\text{eff}} = \frac{\hbar k}{\partial E/\partial k} \quad \text{at } k = k_F.$$

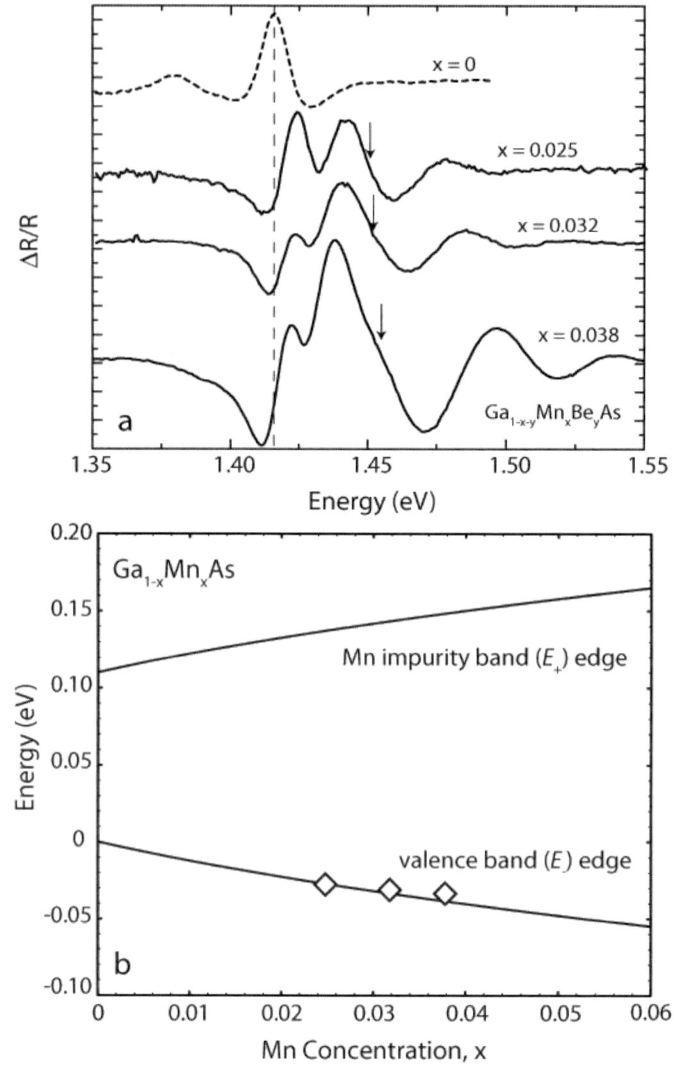

Figure 4.16. (a) PR spectra of $Ga_{1-x-y}Mn_xBe_yAs$ grown on GaAs. The transition corresponding to the band-gap of the film, denoted by an arrow, shifts to higher energy with an increase in Mn concentration but shows no movement with a change in the Be concentration. The dashed line marks the GaAs transition. (b) E_+ and E_- heavy hole band edge positions as a function of Mn concentration determined by the VBAC model. The shift in the $Ga_{1-x-y}Mn_xBe_yAs$ valence band edge ascertained from the movement in band-gap relative to that of GaAs is also displayed.

A representative picture of the hole mass in $Ga_{1-x}Mn_xAs$ was obtained assuming near complete substitution of Mn and moderate compensation by charged defects, whereby the VBAC model predicts effective masses upwards of $m_{\rm eff} = 30\, m_{\rm e}$ for $0.01 < x < 0.05$. These calculations are in agreement with experimentally determined values reported by Burch *et al.*[54] The trend in the hole mobility of $Ga_{1-x}Mn_xAs$, also displayed in Fig. 4.17, is reproduced quite well with such a large effective mass, while that of the GaAs films doped with non-magnetic acceptors (Be, C, Zn) is fit with the standard GaAs heavy hole mass of $m_{\rm eff} = 0.47\,m_{\rm e}$. The analysis was carried out with a sizeable ionized impurity concentration of $N_i = 2.5p$, but is considered reasonable in light of the low growth temperatures required to incorporate dopants on the order of an atomic percent. It is interesting to note that if the mobility of $Ga_{1-x}Mn_xAs$ were to be calculated with the effective mass of holes in the valence band, the ionized impurity concentration would need to be nearly 40 times the hole concentration to fit the experimental trend.

The metal-to-insulator transition in $Ga_{1-x}Mn_xAs$, observed at $x = 0.02$, is also verified by the VBAC model despite the picture of hole transport in a detached

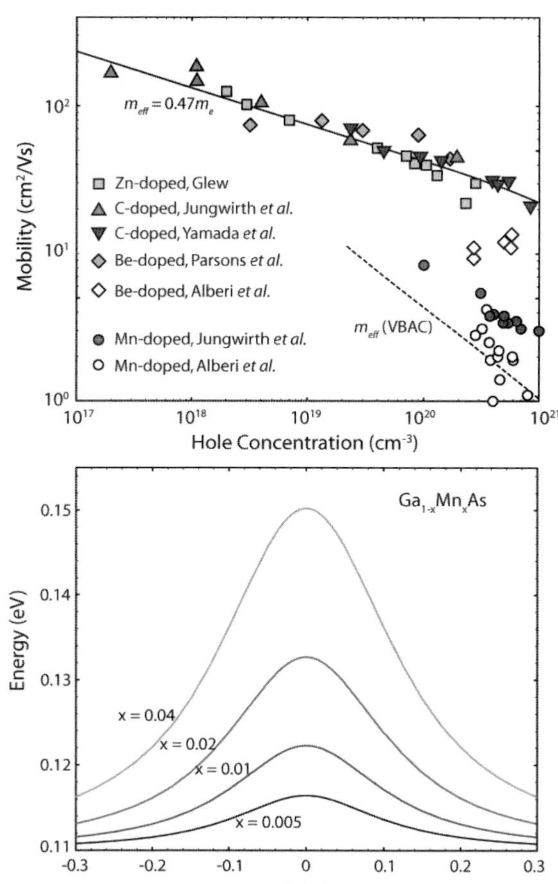

Figure 4.17. (a) Trends in the hole mobility in $Ga_{1-x}Mn_xAs$ versus Ga(Be, Zn, C)As as a function of hole concentration. (b) Heavy hole impurity band dispersion curves at several Mn concentrations. As x increases, the band widens and the states become more extended in character.

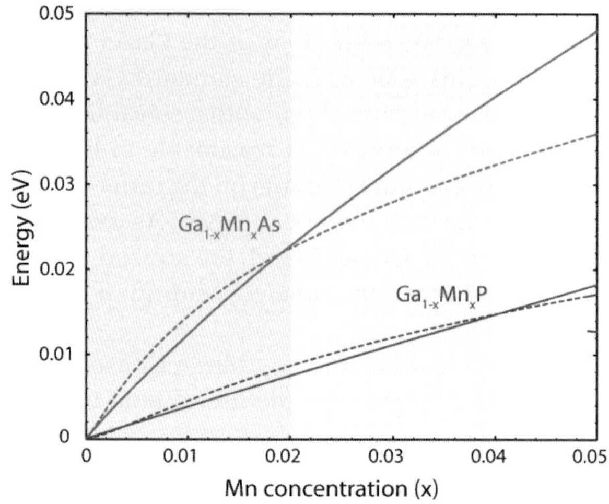

Figure 4.18. Impurity band width of $Ga_{1-x}Mn_xAs$ and $Ga_{1-x}Mn_xP$ (solid lines) versus the lifetime broadening of the heavy hole impurity band states as a function of Mn concentration.

impurity band. Instead of a transfer of conduction from the impurity band to the valence band as the states of the two overlap, the MIT may occur solely within the impurity band as its width becomes great enough to support carrier scattering. Quantitative assessment of this transition is made in terms of the lifetime broadening of the hole states in the impurity band:

$$\partial E = \frac{\hbar e}{\mu m_{\text{eff}}}.$$

At low Mn concentrations, the impurity band is much narrower than the energy uncertainty of the hole states, and transport is maintained by carriers hopping from one state to the next. At higher x, however, anticrossing-induced hybridization widens the band, and once in excess of dE, the carriers may travel freely from state to state. The VBAC model predicts this transition to occur roughly at $x = 0.02$, as shown in Fig. 4.18, assuming $N_i = 1.5_p$ and a 40% compensation ratio.

The concept of extended transport within the impurity band is further confirmed in $Ga_{1-x}Mn_xAs$ alloys with high Mn concentrations ($x > 0.1$). Without compensating donors, the Fermi level lies low in the impurity band, and transport is dominated by states of more localized character compared to those at the band edge. Indeed, hole mobilities derived from resistivity and ECV measurements are as low as 0.5 cm^2/Vs in $Ga_{0.84}Mn_{0.16}As$. However, doping with Si moves the Fermi level upward in the band to states of more extended character and improves the mobility to 2.3 cm^2/Vs.

4.3.3 $Ga_{1-x}Mn_x$-V Alloys

The concept of an anticrossing interaction applies to other $Ga_{1-x}Mn_x$-V alloy systems as well, where their behavior may be extrapolated from the trend in the valence band alignment to E_{Mn} and the relative strength of the p–d exchange. The lower absolute energy position of the GaP valence band places the fixed Mn p-level well within the gap at $E_{Mn} = 0.4\,\text{eV}$. Subsequent reduction in the extent of p state hybridization allows the impurity band states to remain fairly localized. Consequently, the VBAC model predicts that the band width will be narrow and the MIT is forced to a higher Mn concentration of $x = 0.04$. Due to the lack of experimental data providing information on the hybridization strength in this system, C_{Mn} was assumed similar to that of $Ga_{1-x}Mn_xAs$. Although the p–d exchange in $Ga_{1-x}Mn_xP$ is strengthened by the close proximity of the Mn-3d states to the GaP valence band edge, any enhancement of C_{Mn} will most likely be slight in comparison to the effect of the energy separation of the p states in this case. Very little experimental evidence exists for the MIT of $Ga_{1-x}Mn_xN$, but based on the projected energy separation between the GaN valence band edge and E_{Mn} the MIT is assumed to extend to very high x.

Finally, carrier transport within an independent impurity band has adverse consequences for the optimization of DMS properties through the engineering of quaternary alloys.[70] The introduction of a second element on the group V sublattice produces additional carrier scattering through alloy disorder that lowers the hole mobility and enhances the lifetime broadening of the impurity band states. Alloy disorder-limited mobility is expressed as[71]

$$\mu_{AD} = \frac{h^3 e}{8\pi^2 m_{\text{eff}}^2 k_F |V_{AD}|^2 \Omega y},$$

where Ω is the unit cell volume, y is the concentration of the group V alloyed element, and V_{AD} is the scattering potential. Typically, V_{AD} is assigned as the band offset energy between the two alloyed compounds. The strong dependence of carrier scattering on the heavy hole effective mass ensures that alloying on the order of only a few atomic percent is needed to disrupt the balance between impurity band width and state broadening and transform a metallic alloy into an insulator. Indeed, experiments demonstrate that under fixed Mn concentrations at which metallic transport occurs in $Ga_{1-x}Mn_xAs$, the addition of P or N induces an MIT at $0.016 < y < 0.024$ and $y < 0.003$, respectively. The VBAC model substantiates these findings, as shown in Fig. 4.19.

Trends in T_C may be interpreted on the basis of the transport properties discussed above. According to the Ruderman–Kittel–Kasuya–Yosida (RKKY)-Zener model, T_C is dependent upon itinerant hole movement throughout the lattice to couple Mn spin-states and is therefore tied to their mean free path[72]

$$l = v_F \tau = \frac{h k_F \mu}{e}.$$

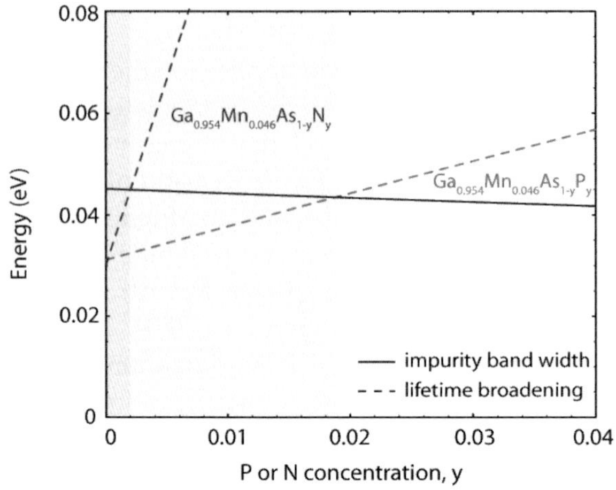

Figure 4.19. Impurity band width versus lifetime broadening energy of the heavy hole impurity band states of $Ga_{1-x-y}Mn_xAs_{1-y}P_y$ and $Ga_{1-x-y}Mn_xAs_{1-y}N_y$ as a function of P or N concentration.

The theoretically expected drop in the hole mobility of $Ga_{1-x}Mn_x$-V alloys with increased localized character of the impurity band states confirms the experimentally observed trend that the T_C drops as the group V element moves up the period from As to N. Likewise, Si-doped GaMnAs alloys see an increase in T_C due to the filling of the impurity band to states of more delocalized nature. Finally, the alloy disorder scattering in $Ga_{1-x}Mn_xAs_{1-y}P_y$ and $Ga_{1-x}Mn_xAs_{1-y}N_y$ also accounts for the drastic reduction in T_C upon a slight increase in y as well.

4.4 CONCLUSIONS

In this chapter we have presented our current understanding of the factors affecting ferromagnetic coupling between Mn spins in group III-Mn-V ferromagnetic semiconductors. We have shown that in the case of $Ga_{1-x}Mn_xAs$ there is overwhelming evidence for the Fermi level controlled incorporation of substitutional Mn atoms and the maximum hole concentration that can be achieved in this material. The results indicate that any improvement in the magnetic properties of this material system will require decoupling of the doping process from the introduction of magnetic moments. This objective, however, cannot be realized when the source of the magnetic moment is also an electrically active acceptor. We have also shown that the electronic structure of the Mn impurity band can be described in terms of the band anticrossing model. The model explains hole transport properties and the effect of alloying on magnetic properties of group III-Mn-As. It also points to the unique situation of $Ga_{1-x}Mn_xAs$ in which the Mn level is sufficiently localized so that the impurity band does not merge with the valence band but is located close enough to the valence band to form a broad impurity band through the band anticrossing interaction.

ACKNOWLEDGMENTS

The authors would like to thank close collaborators (X. Liu, Y. Cho, T. Wojtowicz and J. K. Furdyna) at the University of Notre Dame for providing the samples and performing many resistivity and magnetic measurements. Our co-workers at LBNL (M. Scarpulla, P. Stone and O. Dubon) also provided significant contributions to many aspects of III-Mn-V reviewed in this article. Support for the preparation of this manuscript is provided by the Director, Office of Science, Office of Basic Energy Sciences, Division of Materials Sciences and Engineering, of the U.S. Department of Energy under Contract No. DE-AC02-05CH11231.

References

[1] T. Kasuya and A. Yanase, *Rev. Mod. Phys.* **40**, 684 (1968).
[2] J. K. Furdyna and J. Kossut (eds.), *Semiconductor and Semimetals*, Vol. 25 (Academic, New York, 1988).
[3] H. Ohno, H. Munekata, T. Penney, S. von Molnar and L. L. Chang, *Phys. Rev. Lett.* **68**, 2664 (1992).
[4] H. Ohno, A. Shen, F. Matsukura, A. Oiwa, A. Endo, S. Katsumoto and Y. Iye, *Appl. Phys. Lett.* **69**, 363 (1996).
[5] K. W. Edmonds, K. Y. Wang, R. P. Campion, A. C. Neumann, N. R. S. Farley, B. L. Gallagher and C. T. Foxon, *Appl. Phys. Lett.* **81**, 4991 (2002).
[6] K. C. Ku, S. J. Potashnik, R. F. Wang, S. H. Chun, P. Schiffer, N. Samarth, M. J. Seong, A. Mascarenhas, E. Johnston-Halperin, R. C. Myers, A. C. Gossard and D. D. Awschalom, *Appl. Phys. Lett.* **82**, 2302 (2003).
[7] A. H. MacDonald, P. Schiffer and N. Samarth, *Nature Mater.* **4**, 195 (2005).
[8] H. Ohno, *Science* **281**, 951 (1998).
[9] H. Ohno, *J. Magn. Mater.* **200**, 110 (1999).
[10] J. K. Furdyna, P. Schiffer, Y. Sasaki, S. J. Potashnik and X. Y. Liu, in *Optical Properties of Semiconductor Nanostructures*, NATO Science Series, Vol. 81, eds. M. L. Sadowski, M. Potemski and M. Grynberg (Kluwer, Dordrecht, 2000), p. 211.
[11] T. Dietl, H. Ohno, F. Matsukura, J. Cibert and D. Ferrand, *Science* **287**, 1019 (2000).
[12] S. J. Potashnik, C. K. Ku, R. Mahendiran, S. H. Chun, R. F. Wang, N. Samarth and P. Schiffer, *Phys. Rev. B* **66**, 012408 (2002).
[13] Y. L. Soo, S. W. Huang, Z. H. Ming, Y. H. Kao, H. Munekata and L. L. Chang, *Phys. Rev. B* **53**, 4905 (1996).
[14] R. Shioda, K. Ando, T. Hayashi and M. Tanaka, *Phys. Rev. B* **58**, 1100 (1998).
[15] Y. L. Soo, G. Kioseoglou, S. Kim, X. Chen, H. Luo, Y. H. Kao, L.-H. J.in, H. H. Hsieh, T. Y. Hou, C. T. Chen, Y. Sasaki, X. Liu and J. K. Furdyna, *Phys. Rev. B* **67**, 214401 (2003).
[16] L. C. Feldman, J. W. Mayer and S. T. Picraux, *Materials Analysis by Ion Channeling* (Academic, New York, 1982).
[17] J. R. Tesmer and M. A. Nastasi (eds.), *Handbook of Modern Ion Beam Materials Analysis* (Materials Research Society, Pittsburgh, 1995).
[18] K. M. Yu, W. Walukiewicz, T. Wojtowicz, I. Kuryliszyn, X. Liu, Y. Sasaki and J. K. Furdyna, *Phys. Rev. B* **65**, 201303(R) (2002).
[19] J. Blinowski and P. Kacman, *Phys. Rev. B* **67**, 121204(R) (2003).
[20] F. Máca and J. Mašek, *Phys. Rev. B* **65**, 235209 (2002).

[21] C. M. Hurd, in *The Hall Effect and Its Applications*, eds. C. L. Chienand and C. W. Westgate (Plenum, New York, 1980), p. 43.
[22] S. J. Potashnik, K. C. Ku, S. H. Chun, J. J. Berry, N. Samarth and P. Schiffer, *Appl. Phys. Lett.* **79**, 1495 (2001).
[23] P. Blood, *Semicond. Sci. Technol.* **1**, 7 (1986).
[24] K. M. Yu, W. Walukiewicz, T. Wojtowicz, W. L. Lim, X. Liu, Y. Sasaki, M. Dobrowolska and J. K. Furdyna, *Appl. Phys. Lett.* **81**, 844 (2002).
[25] S. Sanvito and N. A. Hill, *Appl. Phys. Lett.* **78**, 3493 (2001).
[26] T. Hayashi, Y. Hashimoto, S. Katsumoto and Y. Iye, *Appl. Phys. Lett.* **78**, 1691 (2001).
[27] K. W. Edmonds, P. Boguslawski, K. Y. Wang, R. P. Campion, S. N. Novikov, N. R. S. Farley, B. L. Gallagher, C. T. Foxon, M. Sawicki, T. Dietl, M. B. Nardelli and J. Bernholc, *Phys. Rev. Lett.* **92**, 037201 (2004).
[28] I. Kuryliszyn, T. Wojtowicz, X. Liu, J. K. Furdyna, W. Dobrowolski J.-M. Broto, O. Portugall, H. Rakoto and B. Raquet, *J. Supercond.* **16**, 63 (2003).
[29] D. Chiba, K. Takamura, F. Matsukura and H. Ohno, *Appl. Phys. Lett.* **82**, 3020 (2003).
[30] D. Chiba, K. M. Yu, W. Walukiewicz, Y. Nishitani, F. Matsukura and H. Ohno, *J. Appl. Phys.* **103**, 07D136 (2008).
[31] M. Adell, L. Ilver, J. Kanski, V. Stanciu, P. Svedlindh, J. Sadowski, J. Z. Domagala, F. Terki, C. Hernandez and S. Charar, *Appl. Phys. Lett.* **86**, 112501 (2005).
[32] K. Y. Wang, R. P. Campion, K. W. Edmonds, M. Sawicki, T. Dietl, C. T. Foxon, and B. L. Gallagher, *AIP Conf. Proc.* **772**, 333 (2005).
[33] S. Mack, R. C. Myers, J. T. Heron, A. C. Gossard and D. D. Awschalom, *Appl. Phys. Lett.* **92**, 192502 (2008).
[34] K. M. Yu, W. Walukiewicz, T. Wojtowicz, J. Denlinger, M. A. Scarpulla, X. Liu and J. K. Furdyna, *Appl. Phys. Lett.* **86**, 042102 (2005).
[35] K. M. Yu, W. Walukiewicz, T. Wojtowicz, W. L. Lim, X. Liu, U. Bindley, M. Dobrowolska and J. K. Furdyna, *Phys. Rev. B* **68**, 041308(R) (2003).
[36] K. W. Edmonds, N. R. S. Farley, R. P. Campion, C. T. Foxon, B. L. Gallagher, T. K. Johal, G. van der Laan, M. MacKenzie, J. N. Chapman and E. Arenholz, *Appl. Phys. Lett.* **84**, 4065 (2004).
[37] Y. Ishiwata, M. Watanabe, R. Eguchi, T. Takeuchi, Y. Harada, A. Chainani, S. Shin, T. Hayashi, Y. Hashimoto, S. Katsumoto and Y. Iye, *Phys. Rev. B* **65**, 233201 (2002).
[38] A. Koeder, S. Frank, W. Schoch, V. Avrutin, W. Limmer, K. Thonke, R. Sauer, A. Waag, M. Krieger, K. Zuern, P. Ziemann, S. Brotzmann and H. Bracht, *Appl. Phys. Lett.* **82**, 3278 (2003).
[39] W. Walukiewicz, *Appl. Phys. Lett.* **54**, 2094 (1989).
[40] W. Walukiewicz, *Mat. Res. Soc. Symp. Proc.* **300**, 421 (1993).
[41] W. Walukiewicz, *Physica B* **302–303**, 123 (2001).
[42] K. M. Yu, W. Walukiewicz, L. Y. Chan, R. Leon, E. E. Haller, J. M. Jaklevic and C. M. Hanson, *J. Appl. Phys.* **74**, 86 (1993).
[43] P. Specht, R. C. Lutz, R. Zhao, E. R. Weber, W. K. Liu, K. Bacher, F. J. Towner, T. R. Stewart and M. Luysberg, *J. Vac. Sci. Technol. B* **17**, 1200 (1999).
[44] S. U. Yuldashev, H. Im, V. Sh. Yalishev, C. S. Park, T. W. Kang, S. Lee, Y. Sasaki, X. Liu and J. K. Furduna, *Jpn. J. Appl. Phys.* **42**, 6256 (2003).
[45] S. Lee, S. J. Chung, I. S. Choi, Sh. U.Yuldashev, H. Im, T. W. Kang, W. L. Lim, Y. Sasaki, X. Liu, T. Wojtowicz and J. K. Furdyna, *J. Appl. Phys.* **93**, 8307 (2003).
[46] J. K. Furdyna, X. Liu, W. L. Lim, Y. Sasaki, T. Wojtowicz, I. Kuryliszyn, S. Lee, K. M. Yu and W. Walukiewicz, *J. Korean Phys. Soc.* **42**, S579 (2003).
[47] T. Wojtowicz, J. K. Furdyna, X. Liu, K. M. Yu and W. Walukiewicz, *Physica E* **25**, 171 (2004).

[48] T. Wojtowicz, W. L. Lim, X. Liu, M. Dobrowolska, J. K. Furdyna, K. M. Yu, W. Walukiewicz, I. Urgaftman and J. R. Meyer, *Appl. Phys. Lett.* **83**, 4220 (2003).

[49] K. M. Yu, W. Walukiewicz, T. Wojtowicz, W. L. Lim, X. Liu, M. Dobrowolska, and J. K. Furdyna, *Appl. Phys. Lett.* **84**, 4325 (2004a).

[50] M. A. Scarpulla, K. M. Yu, W. Walukiewicz and O.D. Dubon, *Proc. 27th Int. Conf. Physics of Semiconductors*, Flagstaff, AZ, July 26-30, 2004; *AIP Conf. Proc.* **772**, 1367 (2005).

[51] Y. J. Cho, K. M. Yu, X. Liu, W. Walukiewicz and J. K. Furdyna, *Appl. Phys. Lett.* **93**, 262505 (2008).

[52] M. Missous and S. O'Hagan, *J. Appl. Phys.* **75**, 3396 (1994).

[53] A. Van Esch, L. Van Bockstal, J. De Boeck, G. Verbanck, A. S. van Steenbergen, P. J. Wellmann, B. Grietens, R. Bogaerts, F. Herlach and G. Borghs, *Phys. Rev. B* **56**, 13103 (1997).

[54] K. S. Burch, D. B. Shrekenhamer, E. J. Singley, J. Stephens, B. L. Sheu, R. K. Kawakami, P. Schiffer, N. Samarth, D. D. Awschalom and D. N. Basov, *Phys. Rev. Lett.* **97**, 087208 (2006).

[55] N. F. Mott, *Rev. Modern Phys.* **40**, 677 (1968).

[56] Y. Zhang and S. Das Sarma, *Phys. Rev. B* **72**, 125303 (2005).

[57] E. H. Hwang and S. Das Sarma, *Phys. Rev. B* **72**, 035201 (2005).

[58] S. Sanvito, P. Ordejon and N. A. Hill, *Phys. Rev. B* **63**, 165206 (2001).

[59] J. Wu, W. Walukiewicz, K. M. Yu, J. D. Denlinger, W. Shan, J. W. Ager III, A. Kimura, H. F. Tang and T. F. Kuech, *Phys. Rev. B* **70**, 115214 (2004).

[60] K. Alberi, J. Wu, W. Walukiewicz, K. M. Yu, O. D. Dubon, S. P. Watkins, C. X. Wang, X. Liu, Y.-J. Cho and J. Furdyna, *Phys. Rev. B* **75**, 045203 (2007).

[61] H. P. Hjalmarson, P. Vogl, D. J. Wolford and J. D. Dow, *Phys. Rev. Lett.* **44**, 810 (1980).

[62] R. People and S. K. Sputz, *Phys. Rev. B* **41**, 8431 (1990).

[63] B. Beschoten, P. A. Crowell, I. Malajovich, D. D. Awschalom, F. Matsukura, A. Shen and H. Ohno, *Phys. Rev. Lett.* **83**, 3073 (1999).

[64] K. Alberi, K. M. Yu, P. R. Stone, O. D. Dubon, W. Walukiewicz, T. Wojtowicz, X. Liu and J. Furdyna *Phys. Rev. B* **78**, 075201 (2008).

[65] R. W. Glew, *J. Cryst. Growth* **68**, 44 (1984).

[66] T. Jungwirth, J. Sinova, A. H. MacDonald, B. L. Gallagher, V. Novak, K. W. Edmonds, A. W. Rushforth, R. P. Campion, C. T. Foxon, L. Eaves, E. Olejnik, J. Masek, S.-R. E. Yang, J. Wunderlich, C. Gould, L. W. Molenkamp, T. Dietl and H. Ohno, *Phys. Rev. B* **76**, 125206 (2007).

[67] T. Yamada, E. Tokumitsu, K. Saito, T. Akatsuka, M. Miyauchi, M. Konagai and K. Takahashi, *J. Cryst. Growth* **95**, 145 (1989).

[68] J. D. Parsons and F. G. Krajenbrink, *J. Electrochem. Soc.* **130**, 1782 (1983).

[69] W. Zawadzki and W. Szymanska, *Phys. Stat. Sol. B* **45**, 415 (1971).

[70] P. R. Stone, K. Alberi, S. K. Z. Tardif, J. W. Beeman, K. M. Yu, W. Walukiewicz and O. D. Dubon, *Phys. Rev. Lett.* **101**, 087203 (2008).

[71] J. W. Harrison and J. R. Hauser, *Phys. Rev. B* **13**, 5347 (1976).

[72] S. Das Sarma, E. H. Hwang and D. J. Priour Jr., *Phys. Rev. B* **70**, 161203(R) (2004).

Chapter Five

$Ga_{1-x}Mn_xP$ Synthesized by Ion Implantation and Pulsed-Laser Melting

Peter R. Stone and Oscar D. Dubon*

Department of Materials Science and Engineering, University of California, 210 HMMB #1760, Berkeley, CA 94720, USA
Materials Science Division, Lawrence Berkeley National Laboratory, CA 94720, USA
**oddubon@berkeley.edu*

Michael A. Scarpulla

Departments of Materials Science & Engineering and Electrical & Computer Engineering, University of Utah, UT 84112, USA

Kin Man Yu

Materials Science Division, Lawrence Berkeley National Laboratory, CA 94720, USA

The synthesis of single-crystalline epitaxial thin films of the carrier-mediated ferromagnetic phase of $Ga_{1-x}Mn_xP$ and $Ga_{1-x}Mn_xP$-based quaternary alloys using ion implantation and pulsed-laser melting (II-PLM) has allowed for the exploration of the effect of anion substitution on ferromagnetism in $Ga_{1-x}Mn_x$-pnictide systems. Despite displaying significantly greater hole localization than the canonical $Ga_{1-x}Mn_xAs$ system many of the properties, including the dependence of the Curie temperature on x and hole concentration, X-ray absorption spectroscopy and magnetic circular dichroism lineshapes, and manipulation of the magnetic anisotropy by carrier concentration and epitaxial strain, are substantially similar in the two materials. Furthermore, the combination of far infrared photoconductivity and THz spectroscopy indicates that the carriers responsible for ferromagnetic exchange are localized within an impurity band that remains unmerged with the GaP valence band for at least $x \leq 0.042$. These remarkable findings suggest that ferromagnetism in $III_{1-x}Mn_xV$ materials exists on a continuum in terms of carrier localization and that localized carriers primarily of cation d character are capable of effectively mediating ferromagnetic exchange.

Handbook of Spintronic Semiconductors by W M Chen & I A Buyanova

www.panstanford.com
978-981-4267-36-6

5.1 INTRODUCTION

The ability to simultaneously manipulate the electronic (charge) and magnetic (spin) functionalities in a single device is essential to the development of spin-based electronics or *spintronics*.[1,2] One strategy for obtaining a material with such dual functionality is to replace a few percent of the host atoms in a semiconductor with a magnetically active impurity, resulting in a so-called diluted magnetic semiconductor. If a long-range exchange interaction spontaneously couples the impurity spins, then the material is said to be a ferromagnetic semiconductor. One particularly well-studied subset of ferromagnetic semiconductors is that in which a few percent of cations of a III-V semiconductor is replaced by Mn. When Mn substitutes for the group III element, its localized 3*d* electrons provide a local magnetic moment while its acceptor state is a source of holes. The kinetic *p*–*d* exchange interaction between the holes and Mn moments gives rise to ferromagnetism. Due to this interaction the ferromagnetic Curie temperature (T_C) is proportional to both the concentration of substitutional Mn moments, x, and the hole concentration, p. The carrier-mediated nature of exchange gives rise to a number of interesting phenomena including the large magnetoresistance,[3,4] the strong anomalous Hall effect,[5] the so-called giant planar Hall effect[6] and related anisotropic magnetoresistance,[7,8] the control of ferromagnetism by electrical gating,[9] large magnetoptical effects,[10] and the control of ferromagnetism using circularly polarized light.[11,12]

Since its first synthesis in 1996,[13] $Ga_{1-x}Mn_xAs$ has become the most well understood and extensively studied $III_{1-x}Mn_xV$ ferromagnetic semiconductor. Yet the detailed nature of ferromagnetic exchange in $Ga_{1-x}Mn_xAs$ and other Mn-doped III-V materials remains unclear. While many experimental results in $Ga_{1-x}Mn_xAs$ are well explained by the Zener kinetic exchange model, which assumes that ferromagnetism is mediated by itinerant or weakly localized holes that are of valence band character,[14–16] there is a growing body of literature which suggests that holes residing in an impurity band can be responsible for ferromagnetic exchange.[17–20] Hence, even in this canonical system the character of the mediating holes — impurity band-like, valence band-like, or mixed — is still not clear. Furthermore, despite extensive experimental and theoretical efforts there has been little progress in understanding ferromagnetic exchange in other Ga-Mn-pnictide systems. Both deciphering the character of the mediating holes and understanding how anion substitution affects exchange are crucial to establishing accurate models of ferromagnetism and transport in III-Mn-V materials.

Changing the semiconductor host from GaAs to GaP has allowed us to explore these two important problems. With an acceptor level of $\sim$400 meV, the Mn acceptor level in GaP[21] is nearly four times deeper than it is in GaAs (110 meV)[22] leading to significantly more localized hole states. The shorter bond length in GaP is also predicted to enhance the *p*–*d* hybridization between Mn *d* and anion *p* states.[23,24] Mn-doped GaP thus represents an important model system in which to explore the interplay of carrier localization, hybridization, and ferromagnetic exchange.

Early reports of ferromagnetism in Mn-doped GaP focused largely on polycrystalline materials grown either by low-temperature molecular beam epitaxy (LT-MBE) and/or ion implantation followed by rapid thermal annealing.[25–27] While these materials showed unusual ferromagnetic behavior and rather high Curie temperatures ($\sim$300 K), the origin of ferromagnetism was unclear, and the presence of a carrier-mediated phase was never clearly established. Later, hole-mediated ferromagnetism in $Ga_{1-x}Mn_xP$ was demonstrated by us in samples synthesized by the combination of ion implantation and pulsed-laser melting (II-PLM).[28–30] Ion-channeling analysis and transmission electron microscopy (TEM) of II-PLM formed $Ga_{1-x}Mn_xP$ revealed that the films are single crystalline and epitaxial and have the majority of Mn atoms incorporated substitutionally on Ga sites, or Mn_{Ga}.[30] Carrier-mediated ferromagnetism was established in these materials through the combination of magnetotransport and magnetometry. Additionally, photoconductive spectroscopy provided evidence that the carriers responsible for ferromagnetic exchange are localized in an Mn-derived impurity band that is detached from the valence band at doping levels up to 4.2%.[29,31] Nonetheless, carrier-mediated exchange is remarkably similar in $Ga_{1-x}Mn_xP$ and $Ga_{1-x}Mn_xAs$ despite the former showing non-metallic transport and strongly insulating behavior at low temperatures, suggesting that carrier-mediated ferromagnetism in $III_{1-x}Mn_xV$ systems exists on a continuum in terms of localization of the hole states.

In this chapter we will discuss the synthesis and properties of $Ga_{1-x}Mn_xP$ grown by II-PLM. The layout of this chapter is as follows. Section 5.2 details the II-PLM growth procedure. The collective evidence for carrier-mediated ferromagnetism is presented in Sec. 5.3 while confirmation that the carriers responsible for exchange are localized in an impurity band is presented in Sec. 5.4. Selected properties of $Ga_{1-x}Mn_xP$ films are discussed in Secs. 5.5 through 5.7. Finally, we offer some concluding remarks and future perspectives on $Ga_{1-x}Mn_xP$ and other $III_{1-x}Mn_xV$ materials in Secs. 5.8 and 5.9.

5.2 ION IMPLANTATION AND PULSED-LASER MELTING (II-PLM)

The Mn-doping levels of 10^{20}–10^{21} cm^{-3} required for ferromagnetism in GaP are well in excess of the equilibrium solubility of Mn in GaP, which is on the order of 10^{19} cm^{-3}.[32] Hence, the synthesis of $Ga_{1-x}Mn_xP$ necessitates the use of non-equilibrium processing. For the most part LT-MBE has been the method of choice for the incorporation of Mn in excess of its solubility limit in the canonical $Ga_{1-x}Mn_xAs$ system. However, as was mentioned in Sec. 5.1, the substitutional incorporation of Mn into a single crystalline, epitaxial $Ga_{1-x}Mn_xP$ thin film has yet to be demonstrated by LT-MBE. The combination of ion implantation and pulsed-laser melting is an alternative process that can be used to incorporate dopants well in excess of their equilibrium solubility. Laser irradiation can deposit large amounts of heat within micron length scales of a solid surface. With pulsed lasers with duration on the order 10^{-8} sec, it is possible to melt materials and subsequently achieve 10^9–10^{10} K/s cooling rates.[33] Such fast quench rates lead to kinetically controlled crystal solidification mechanisms which allow the incorporation of dopants in excess of their equilibrium solubility without the precipitation of second phases. The II-PLM process has previously been used to produce epitaxial, single-crystalline $Ga_{1-x}Mn_xAs$ thin films,[30,34] which show identical static magnetic and electrical properties to those of their LT-MBE grown counterparts having the same Mn_{Ga} concentration.[35]

The details of the II-PLM process for the growth of $Ga_{1-x}Mn_xP$ are now discussed. 50 keV $^{55}Mn^+$ ions are implanted into (001)-oriented, n-type (background doping of 10^{16}–10^{17} cm^{-3}) GaP substrates to doses ranging from 2.5×10^{15}–2.0×10^{16} cm^{-2}. This leads the near-surface region of the substrate to be supersaturated with Mn and amorphized from damage caused by implantation of high-energy Mn ions. In order to repair this damage, as well as incorporate the Mn^+ atoms substitutionally on Ga sites, the sample is irradiated with a single 0.44 ± 0.04 J/cm^2 pulse from a KrF ($\lambda = 248$ nm) excimer laser with duration ~32 ns, FWHM 23 ns and peak intensity at 16 ns. The energy from the laser pulse melts the amorphous region into the undamaged crystalline substrate. Solidification then occurs in crystallographic registry with the underlying substrate, resulting in a single-crystalline, epitaxial thin film. The large cooling rates resulting from the short temporal duration of the laser pulse causes the solid–liquid interface to move at velocities on the order of several meters per second. The rapid interface velocity does not allow sufficient time for the diffusion of atomic species to and from the interface, which is necessary for solidification under local thermodynamic equilibrium. Thus, the equilibrium solubility limits can be overcome via this kinetically-controlled "solute trapping" process.[36] TEM analysis has shown that II-PLM $Ga_{1-x}Mn_xP$ films are single crystalline, though the first ~25 nm from the surface are highly defective.[30] This poorly regrown layer is readily removed by etching in concentrated HCl for ~24 hours[29,30] resulting in epitaxial $Ga_{1-x}Mn_xP$ grown on GaP.

Secondary ion mass spectrometry (SIMS) analysis for the total Mn concentration as a function of depth for a representative $Ga_{1-x}Mn_xP$ thin film is shown in Fig. 5.1(a). The film is characterized by a Mn distribution that is approximately 100 nm in width and peaks ~20–30 nm from the film's surface, typical for most of the $Ga_{1-x}Mn_xP$ films that will be discussed herein. We note, however, that films with lower Mn implant doses tend to have a more uniform distribution of Mn atoms throughout the film thickness. An inhomogeneous distribution of Mn is also present in $Ga_{1-x}Mn_xAs$ formed by II-PLM; nonetheless, good quantitative agreement is found between the magnetic and transport properties of II-PLM grown $Ga_{1-x}Mn_xAs$ and LT-MBE formed materials.[35] This important finding establishes that reliable results can be obtained in II-PLM formed materials, including the $Ga_{1-x}Mn_xP$ films presented herein despite non-uniformity in the Mn concentration. We define x as the peak Mn_{Ga} concentration as studies of II-PLM synthesized $Ga_{1-x}Mn_xAs$ have shown that its magnetic and transport properties are dominated by the region of the film with maximum Mn_{Ga} concentration and that properties such as T_C and conductivity for II-PLM formed films with peak concentration x are quantitatively similar to those from LT-MBE films having the same x.[28,35]

To evaluate x for a particular $Ga_{1-x}Mn_xP$ film the peak in the total Mn concentration determined by SIMS is multiplied by the fraction of Mn atoms residing substitutionally on Ga sites (f_{sub}). The lattice location of the Mn atoms in the film is determined by simultaneous Rutherford backscattering (RBS) and particle-induced X-ray emission (PIXE) measurements in the channeling mode, in particular along the ⟨110⟩ and ⟨111⟩ axial channels. This method has been used previously in $Ga_{1-x}Mn_xAs$ to determine the fractions of Mn atoms residing on either substitutional or interstitial (f_I) sites.[37] Figure 5.1(b) compares PIXE spectra (Mn K_α and K_β signals) from a $Ga_{1-x}Mn_xP$ film prepared identically to that presented in Fig. 5.1(a) when the $^4He^+$ beam is intentionally non-aligned (random) and aligned along the ⟨110⟩ and ⟨111⟩ directions of the crystal. The reduced Mn X-ray signals for the ⟨110⟩ and ⟨111⟩ aligned spectra in comparison to the random

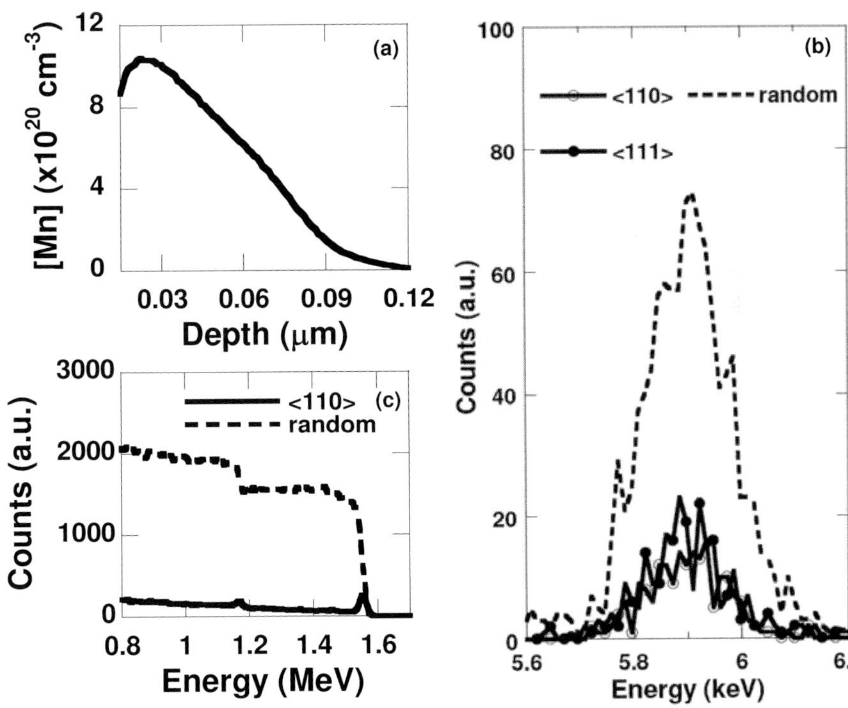

Figure 5.1. (a) Total Mn concentration as a function of depth for a $Ga_{1-x}Mn_xP$ prepared by implanting 1.5×10^{16} 50 kV Mn^+ ions into GaP followed by PLM with a single $0.44\,J/cm^2$ laser pulse. (b) PIXE spectra of an identically prepared film with the ion beam in ⟨110⟩, and ⟨111⟩ channeling alignment and random alignment. (c) Channeling RBS spectra for ⟨110⟩ channeling alignment and random alignment for an identically prepared sample. Based on these data $x = 0.034$ for this sample (see text for further details).

spectrum indicate that a large fraction of the Mn atoms occupies commensurate lattice sites in the zinc-blende structure. The $\langle 110 \rangle$ and $\langle 111 \rangle$ scans are nearly identical. Any atoms occupying the tetrahedral interstitial site in the zinc-blende lattice would be exposed in the $\langle 110 \rangle$ channels but not the $\langle 111 \rangle$ channels. Therefore, if an appreciable fraction of Mn were present at this interstitial site (as is the case for as-grown LT-MBE $Ga_{1-x}Mn_xAs$), one would expect a larger Mn PIXE signal for the $\langle 110 \rangle$ spectrum than the $\langle 111 \rangle$ spectrum.[37] As this is *not* the case, the presence of Mn occupying interstitial positions can be excluded in II-PLM $Ga_{1-x}Mn_xP$. The fraction of substitutional Mn can be evaluated by the following approximate formula

$$f_{\text{sub}} = \frac{1 - \chi_{\text{min}}^{\text{impurity}}}{1 - \chi_{\text{min}}^{\text{lattice}}}, \tag{5.1}$$

where χ_{min} denotes the normalized yields of either the lattice or the impurity, i.e., the ratio of the channeled signal and the random signal.[38] For the Mn impurities χ_{min} comes from the ratio of the integrated intensities of the channeled and non-aligned PIXE signals. For the $Ga_{1-x}Mn_xP$ sample in Fig. 5.1, $\chi_{\text{min}}^{\text{impurity}}$ is $\sim$0.23. For the GaP host, χ_{min} is evaluated from channeling RBS, which is shown in Fig. 5.1(c). Integrating the two spectra in Fig. 5.1(c) over the interval 1.479–1.535 MeV, which corresponds to particles backscattered from Ga atoms in the $Ga_{1-x}Mn_xP$ thin film, $\chi_{\text{min}}^{\text{lattice}}$ is found to be 0.04. Substituting these values into Eq. (5.1) we find $f_{\text{sub}} \approx 80\%$. The maximum total Mn concentration is $1.04 \times 10^{22}\,\text{cm}^{-3}$ and the Ga sublattice concentration in GaP is $2.47 \times 10^{22}\,\text{cm}^{-3}$ resulting in $x = 0.034$ for this sample.

Electrochemical capacitance-voltage (ECV) measurements were performed to measure the depth-resolved net acceptor concentration,[39] and are shown in Fig. 5.2 alongside SIMS data for an identical sample. The proportionality between the ECV and SIMS is approximately 70% throughout the depth of the film. The error bars

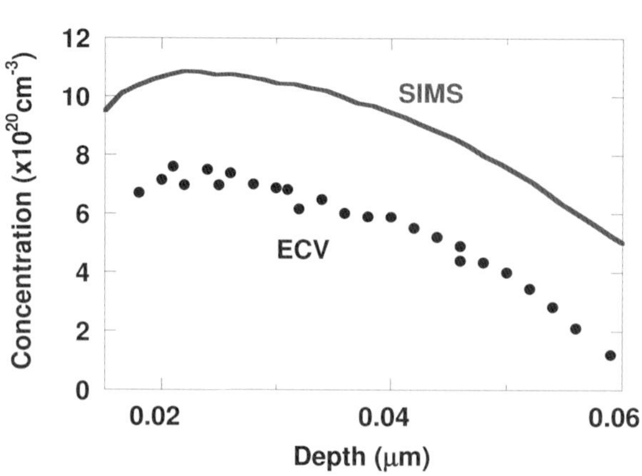

Figure 5.2. SIMS and ECV data from identically prepared $Ga_{1-x}Mn_xP$ films. A constant factor scales the SIMS and ECV curves suggesting a constant substitutional Mn fraction throughout the film thickness.

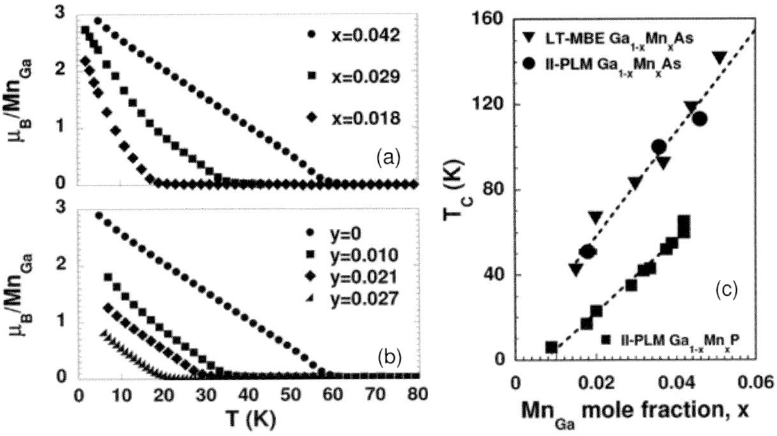

Figure 5.3. Temperature dependence of the magnetization as a function of (a) x (after Ref. 31) and (b) p (after Ref. 40). Data were taken with the applied field parallel to an in-plane $\langle 110 \rangle$ direction. Samples in panel (b) have $x = 0.041$ except for $y = 0$, which has $x = 0.042$. (c) T_C as a function of x for II-PLM grown $Ga_{1-x}Mn_xP$ and $Ga_{1-x}Mn_xAs$ as well as LT-MBE grown $Ga_{1-x}Mn_xAs$. The dashed lines are linear fits to the data. II-PLM data are from Refs. 31 and 83. LT-MBE data are from Ref. 74.

for ECV, f_{sub} determined by PIXE, and SIMS are all on the order of 5–10%. Therefore, within the overall experimental error, the ECV profile and SIMS profile scaled by the Mn_{Ga} fraction obtained from RBS/PIXE analysis are identical. These results suggest that the Mn substitutional fraction is virtually constant throughout the depth of the film assuming that nearly all of the Mn_{Ga} acceptors are electrically active. Such information cannot be directly inferred from ion beam analysis which averages over the entire volume of the thin film because PIXE lacks depth resolution. These data help to justify the determination of peak substitutional x discussed above.

5.3 EVIDENCE OF THE CARRIER-MEDIATED PHASE OF $Ga_{1-x}Mn_xP$

$Ga_{1-x}Mn_xP$ synthesized by II-PLM exhibits all of the major characteristics differentiating carrier-mediated ferromagnetism from ferromagnetism arising from second phase precipitates. Figure 5.3(a) shows thermomagnetic curves for several $Ga_{1-x}Mn_xP$ films with different values of x.[31] As is expected for a carrier-mediated ferromagnetic phase, T_C is proportional to the concentration of Mn_{Ga} moments. Analogous data is presented in Fig. 5.3(b) for the $Ga_{1-x}Mn_xP_{1-y}S_y$ system where ferromagnetism is deliberately suppressed through the incorporation of non-magnetic donors.[40,41] T_C decreases monotonically with y (at constant x) in $Ga_{1-x}Mn_xP_{1-y}S_y$ providing further evidence of carrier-mediated ferromagnetism. A similar decrease in T_C was demonstrated by co-doping $Ga_{1-x}Mn_xP$ with Te.[29]

Figure 5.3(c) compares the dependence of T_C with x for $Ga_{1-x}Mn_xAs$ and $Ga_{1-x}Mn_xP$. While T_C is lower in $Ga_{1-x}Mn_xP$ than $Ga_{1-x}Mn_xAs$ for a given value of x because of stronger hole localization, the dependence of T_C on x is substantially similar in these two materials. Indeed, both sets of data are well described by linear trends, though we note that other fits (e.g., $x^{4/3}$) also reasonably describe the $Ga_{1-x}Mn_xP$ data over this dilute range of x. Linear extrapolation of the $Ga_{1-x}Mn_xP$ data to room temperature implies that $x \sim 0.18$ is required for room temperature ferromagnetism in this system. It is important to note that the data for both II-PLM-formed and LT-MBE-formed $Ga_{1-x}Mn_xAs$ follow the same linear trend indicating that II-PLM processing of materials has little influence on T_C as was noted in Sec. 5.2. Hence, the properties of $Ga_{1-x}Mn_xP$ discussed herein are intrinsic to the material and should not be associated with the alternative processing method.

Figure 5.4 presents the field dependence of the Hall resistance for $Ga_{0.958}Mn_{0.042}P$ at different temperatures.[29] At $T = 50$ K (below the film's T_C of 60 K) the Hall resistance clearly reflects the sample's out-of-plane magnetization demonstrating the dominance of the anomalous component to the Hall resistance. The magnitude of the anomalous component increases nearly linearly with temperature for $T < T_C$, in parallel to the linear rise in the magnetization with decreasing temperature observed by SQUID magnetometry when the magnetic field is applied normal to the film plane. Upon moving through T_C the ferromagnetic to paramagnetic phase transition is reflected by the changes in the Brillouin-like lineshape of the Hall resistance at $T = 100$ and 300 K. The slope of the $T = 300$ K line implies a hole concentration on the order of 2×10^{20} cm^{-3}, though such a value can only be viewed as a rough estimate in light of the strong anomalous Hall term that may still be present even at $T = 5T_C$.[42] The large value of the anomalous Hall signal and its dominance at low temperatures is a manifestation of the spin polarization of holes, which provides further evidence of the strong coupling of transport and magnetism in the carrier-mediated phase of $Ga_{1-x}Mn_xP$.[43]

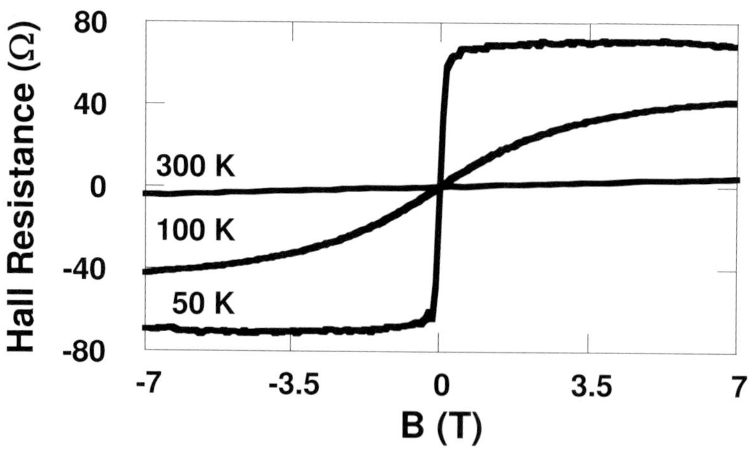

Figure 5.4. Field dependence of the Hall resistance for $Ga_{0.958}Mn_{0.042}P$. After Ref. 29.

The origins of ferromagnetism in $Ga_{1-x}Mn_xP$ have been directly probed by X-ray magnetic circular dichroism (XMCD) spectroscopy performed at the Mn $L_{3,2}$ absorption edges. Briefly, XMCD measures the difference in absorption of right and left circularly polarized light at the L_3 $(2p_{3/2} \rightarrow 3d)$ and L_2 $(2p_{1/2} \rightarrow 3d)$ edges. As these excitations have spin-orbit split initial states, the photoelectrons are preferentially excited according to spin, which provides element-specific information about the spin-polarized $3d$ band structure responsible for ferromagnetism.[44] Figure 5.5 shows X-ray absorption spectra for the magnetization vector parallel (I^+) and anti-parallel (I^-) to the helicity vector of circularly polarized light for $Ga_{0.966}Mn_{0.034}P$. The XMCD $(I^+ - I^-)$ spectrum is also shown. A strong XMCD signal is observed at both the Mn L_3 and L_2 absorption edges indicating a strong spin polarization of the Mn $3d$-derived density of states (DOS) at the Fermi energy (E_F). Furthermore, the element-specific magnetic moments calculated by the so-called "sum rule" analysis[45,46] of the XMCD data are in reasonable agreement with those calculated from the combination of SQUID magnetometry and ion beam analysis, which provides unambiguous evidence that the atomistic origin of ferromagnetism in these materials is Mn_{Ga} moments.[47]

Remarkably, the Mn $L_{3,2}$ XAS and XMCD lineshape in $Ga_{1-x}Mn_xP$ is almost identical to that observed in annealed LT-MBE and II-PLM $Ga_{1-x}Mn_xAs$.[48,49] As the XAS and XMCD lineshapes are strongly influenced by the hybridization between the Mn d and anion p orbitals, the similarity of the XMCD spectra indicates that the local environment and electronic structure of Mn_{Ga} atoms are similar in $Ga_{1-x}Mn_xP$ and $Ga_{1-x}Mn_xAs$ despite considerable differences in acceptor binding energy and carrier localization. This result, along with the similar dependence of T_C on x [Fig. 5.3(c)], magnetotransport characteristics (Fig. 5.4), and control

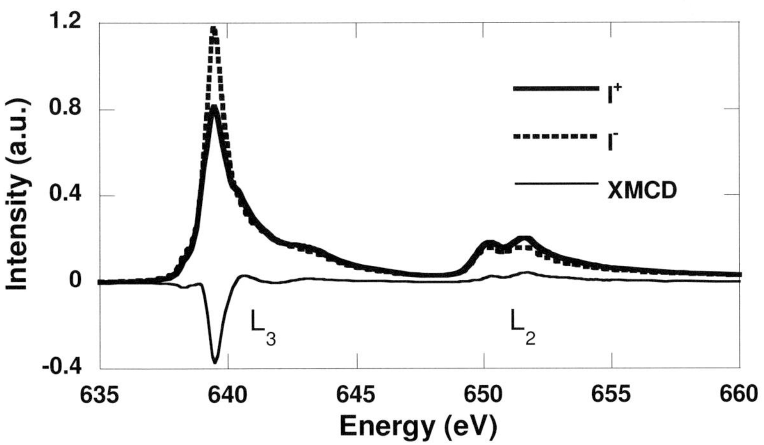

Figure 5.5. X-ray absorption spectra for circular photon helicity parallel (I^+) and anti-parallel (I^-) to the sample magnetization for $Ga_{0.966}Mn_{0.034}P$. The XMCD spectrum $(I^+ - I^-)$ is also shown. Data were measured 30° grazing to the sample plane in a field of 0.54 T at $T = 17$ K and collected in total electron yield mode. After Ref. 47.

of the magnetic anisotropy by epitaxial strain and carrier-concentration (Sec. 5), indicates that similar carrier-mediated ferromagnetic exchange mechanisms are operative in the two materials systems.

5.4 THE DETACHED Mn-DERIVED IMPURITY BAND IN $Ga_{1-x}Mn_xP$

While there is still considerable controversy over whether the holes responsible for ferromagnetic exchange in $Ga_{1-x}Mn_xAs$ are of valence band.[15,50,51] or impurity band character,[17–20] the holes which mediate ferromagnetic exchange in $Ga_{1-x}Mn_xP$ have been shown to reside in a Mn-derived impurity band which remains unmerged with the valence band for $x \leq 0.042$.[29,31] The presence of this gap has been indicated by three experimental techniques: variable temperature resistivity, low-temperature far-infrared photoconductivity spectroscopy, and THz spectroscopy. Figure 5.6 shows the temperature dependent sheet resistivity for $Ga_{0.958}Mn_{0.042}P$. The high-temperature behavior indicates a thermally-activated process with activation energy near 30 meV. This activation energy differs from the 400 meV acceptor binding energy; however at the $\sim 10^{21}$ cm^{-3} doping levels realized in $Ga_{0.958}Mn_{0.042}P$ even very compact wave functions will have significant overlap, leading to the formation of an impurity band. Thus, it is reasonable to hypothesize that the observed activation energy may be due to thermal excitation from a broadened impurity band to the valence band. Furthermore, samples with lower Mn_{Ga} concentration have larger values for this activation energy, which is consistent with a picture of a broadening of the impurity band with increasing x.[29,31] A decrease in the activation energy is also observed by compensating the Mn acceptors by Te donors.[29] Further details of variable temperature electrical transport can be found in Sec. 5.6.

Figure 5.7(a) shows the results of far-infrared photoconductivity experiments on $Ga_{0.958}Mn_{0.042}P$. In these measurements a bias is applied across a highly resistive sample, which is illuminated. If photon absorption results in the transition of carriers from bound states to extended, current-carrying states, an increase in conductivity is detected. The dashed line in Fig. 5.7(a) shows the spectrum incident on the sample as measured by a Golay cell (i.e., the instrument response). The solid line in Fig. 5.7(a) reproduces this spectrum except in two regions. The dip at $\sim$45 meV is due to absorption by an optical phonon mode. The region below 26 meV, where the sample shows no photoconductivity yet where there is significant spectral weight in the incident spectrum, is clear evidence of an excitation gap for transitions between bound impurity states and delocalized valence band states for holes. The magnitude of the excitation gap determined by the onset of the photoconductive response is in good agreement with that determined from variable temperature transport. Figure 5.7(b) compares the photoconductive response for $Ga_{1-x}Mn_xP$ samples with $x = 0.042$ and $x = 0.032$. The decrease in the Mn_{Ga} concentration causes the photoexcitation edge to increase to $\sim$65 meV. This observation is consistent with the hypothesis of an unmerged impurity band,

and variable temperature transport measurements (Sec. 5.6). Additionally, in Te-compensated samples the photoconductive response begins to shift to lower energies as the temperature is increased, which is consistent with increased thermal population of states closer to the edge of the partially-emptied impurity band.[29]

Finally, THz spectroscopy measurements on $Ga_{0.958}Mn_{0.042}P$ ($T_C = 60$ K) show that the Drude free carrier response of valence band holes is strongly

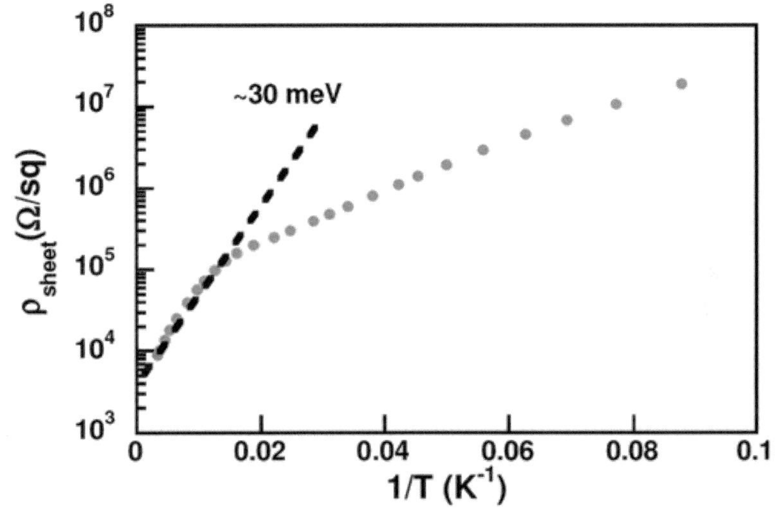

Figure 5.6. Sheet resistivity versus $1/T$ for $Ga_{0.958}Mn_{0.042}P$. The dashed line emphasizes the ~30 meV slope of the high temperature data.

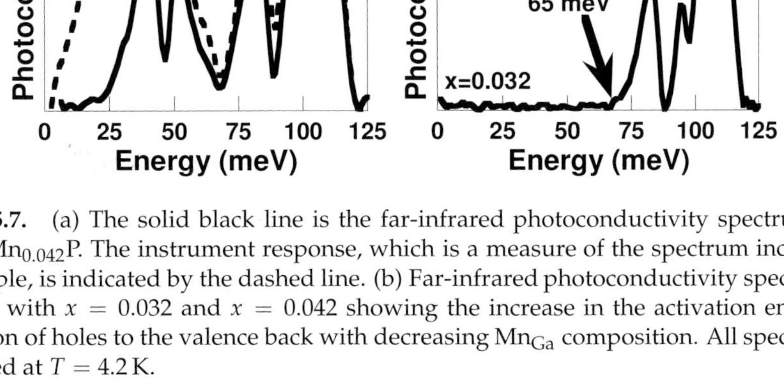

Figure 5.7. (a) The solid black line is the far-infrared photoconductivity spectrum from $Ga_{0.958}Mn_{0.042}P$. The instrument response, which is a measure of the spectrum incident on the sample, is indicated by the dashed line. (b) Far-infrared photoconductivity spectra from samples with $x = 0.032$ and $x = 0.042$ showing the increase in the activation energy for ionization of holes to the valence back with decreasing Mn_{Ga} composition. All spectra were measured at $T = 4.2$ K.

suppressed with decreasing temperature.[52] The activation energy of this carrier freeze-out process is ∼44 meV, which is in reasonable agreement with the gap between the impurity band and valence band extracted from variable temperature transport and photoconductivity results. Furthermore, the Drude free carrier response decreased to undetectable levels near $T = 80\,K$, which is *greater* than T_C. The negligible concentration of valence band holes at the onset of ferromagnetism provides strong evidence for ferromagnetism mediated by impurity band carriers.

5.5 MAGNETIC ANISOTROPY

The strong coupling between holes and Mn moments in $III_{1-x}Mn_xV$ ferromagnetic semiconductors gives rise to a rich and complex parameter space which governs the magnetic anisotropy. Experimental studies of $III_{1-x}Mn_xV$ materials, and in particular $Ga_{1-x}Mn_xAs$, have indicated that the orientation of the magnetic easy axis can be changed by varying the temperature,[53] carrier concentration,[54] and epitaxial strain[55] of the film. These contributions to the magnetic anisotropy in $Ga_{1-x}Mn_xAs$ have been shown to be consistent with mean-field models of ferromagnetism in which the exchange is mediated by itinerant holes of valence band character.[54,56] In this section, we show that many of these anisotropic effects are also exhibited by $Ga_{1-x}Mn_xP$ thin films. Hence, theories pertaining to the origin of the various magnetic anisotropies in $III_{1-x}Mn_xV$ must take into account the fact that localized, impurity-band holes are capable of mediating the same anisotropic exchange interactions as the holes in metallic $Ga_{1-x}Mn_xAs$.

As in the case of $Ga_{1-x}Mn_xAs$, one can rotate the magnetic easy axis of $Ga_{1-x}Mn_xP$ from in-plane to out-of-plane by manipulating the epitaxial strain

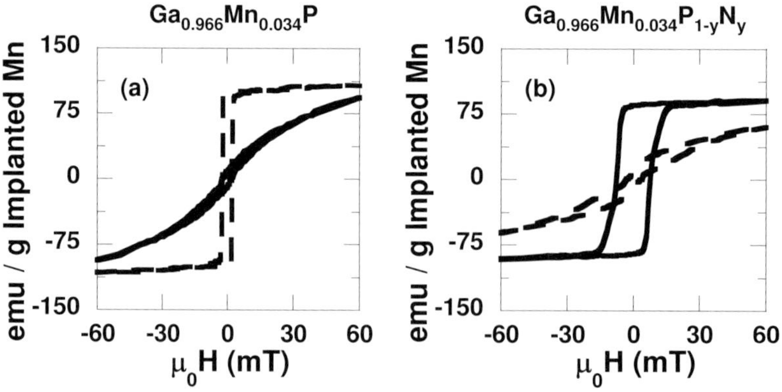

Figure 5.8. Field dependence of the magnetization for (a) $Ga_{0.966}Mn_{0.034}P$ and (b) $Ga_{0.966}Mn_{0.034}P_{1-y}N_y$. Curves with dashed lines were measured with the field parallel to an in-plane ⟨110⟩ direction and those denoted by solid lines were measured with the field oriented parallel to the out-of-plane [001] direction. The value of y for the sample in (b) was estimated to be 0.01.

state of the film. Figure 5.8 compares the field dependence of the magnetization for $Ga_{0.966}Mn_{0.034}P$ films with and without N incorporated on the anion sublattice. The N atoms were introduced into the P sublattice by implantation of 5×10^{15} N^+/cm^2 at 33 kV into Mn-implanted GaP prior to PLM. The implantation energy was chosen to produce a N concentration profile that best overlapped the Mn_{Ga} distribution. Previous studies in $GaAs_{1-y}N_y$ have demonstrated that these implant parameters result in $y \sim 0.01$, and we take that as an estimate of the N content of the film presented in Fig. 5.8(b). The $Ga_{0.966}Mn_{0.034}P$ thin film grown on GaP exhibits an in-plane magnetic easy axis while the film with $y \sim 0.01$ has an easy axis perpendicular to the film plane. The change in magnetic easy axis corresponds to a change in the strain state of the ferromagnetic semiconductor thin film. The $Ga_{0.966}Mn_{0.034}P$/GaP film is in compressive strain due to the larger Mn cations substituting for a few percent Ga. The introduction of ~1% of the significantly smaller N atoms in place of P on the anion sublattice puts the film $Ga_{0.966}Mn_{0.034}P_{1-y}N_y$/GaP film into tensile strain, resulting in a rotation of the easy axis to the out-of-plane [001] direction.

The contributions to the magnetic anisotropy for $Ga_{1-x}Mn_xP$ films grown in compressive strain have been studied in more detail by Bihler *et al.*[57] Figure 5.9 shows the dependence of the ferromagnetic resonance field for rotations about the [110] (out-of-plane rotation) and [001] (in-plane rotation) directions for $Ga_{0.958}Mn_{0.042}P$ grown on GaP. As is expected for films grown in compressive strain, a strong increase in the resonance field is observed as the field is rotated towards the film plane normal. Simulation of the FMR data has indicated that the demagnetization field accounts for only about one fourth of the effective out-of-plane uniaxial anisotropy field, implying the presence of a strong tetragonal distortion in the $Ga_{0.958}Mn_{0.042}P$ film, which gives rise to a large magnetocrystalline term.[55,58,59] Cubic and uniaxial contributions to the in-plane magnetic anisotropy, which are much smaller in magnitude, are apparent from the in-plane rotations. Interestingly, the sign of the in-plane cubic anisotropy is negative, which gives rise to $\langle 110 \rangle$ easy axes in the plane of the film, which has previously only been observed in $In_{1-x}Mn_xAs$.[60] A uniaxial contribution to the in-plane anisotropy lifts the symmetry between the $\langle 110 \rangle$ directions: the $[1\bar{1}0]$ direction is magnetically preferred over the [110] direction. This in-plane uniaxial component to the in-plane magnetic anisotropy seems general to $III_{1-x}Mn_xV$ systems as it is also observed in $In_{1-x}Mn_xAs$[61] and $Ga_{1-x}Mn_xAs$.[53,54,62] We note that the presence of this uniaxial component to the magnetic anisotropy in films grown by both II-PLM and LT-MBE indicates that the symmetry breaking mechanism responsible for the appearance of the uniaxial component to the in-plane magnetic anisotropy is not related to solid/vacuum surface reconstructions[55,63] unless similar reconstructions take place along the advancing liquid–solid interface during PLM. Thin finding is in agreement with recently published work comparing the magnetic anisotropy of II-PLM and LT-MBE formed $Ga_{1-x}Mn_xAs$ thin films.[64]

The in-plane uniaxial anisotropy causes the magnetization reversal process to differ when the magnetic field is applied parallel to $[1\bar{1}0]$ and [110] as is illustrated

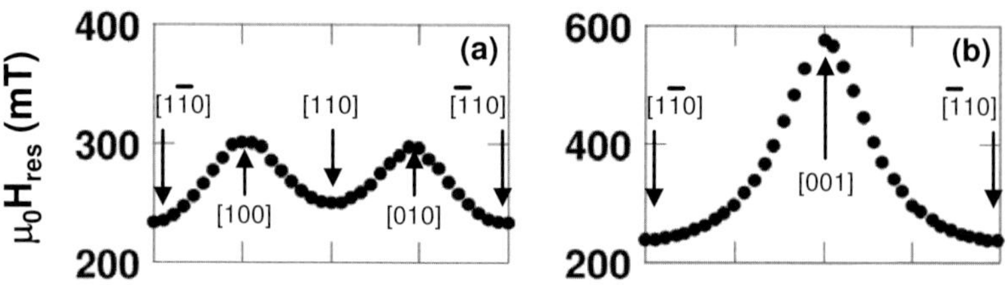

Figure 5.9. Dependence of the ferromagnetic resonance field for rotations about the (a) [001] and (b) [110] axes. Data collected at $T = 5\,\text{K}$ with $\omega/2\pi = 9.3\,\text{GHz}$. The [001] direction is normal to the film plane. After Ref. 57.

in Fig. 5.10(a). For H||[1$\bar{1}$0] magnetization reversal occurs by one spin switch that occurs at the coercive field. However, for H||[110] magnetization reversal occurs by a multistep spin switching process which has been successfully modeled as occurring due to a combination of non-coherent spin switching and coherent spin rotation.[65] The interplay of the in-plane uniaxial and cubic anisotropy energies leads to a complex free energy landscape in which the magnetization flips to an energy minimum near the easy axis, rotates about the easy axis, and then switches again an alignment parallel to the magnetic field.[57]

The in-plane uniaxial anisotropy field can be tuned through co-doping the $Ga_{1-x}Mn_xP$ film with a few percent of sulfur.[66] The magnitude of the in-plane uniaxial anisotropy field along [110] increases with increasing S concentration. This, in parallel with a decrease in the energy required to nucleate and grow domain walls, causes a change in the lineshape of the M(H) curves for H||[110] from "kinked" to "wasp-waisted" as illustrated in Fig. 5.10(b). This behavior is reminiscent of that observed for perpendicular magnetization reversal in lightly doped $Ga_{1-x}Mn_xAs$[67]; in both cases the "wasp-waisted" hysteresis loops are observed when the cubic and uniaxial anisotropy fields are of similar magnitude. We attribute the changes in the magnetic anisotropy in $Ga_{1-x}Mn_xP_{1-y}S_y$ to a decrease

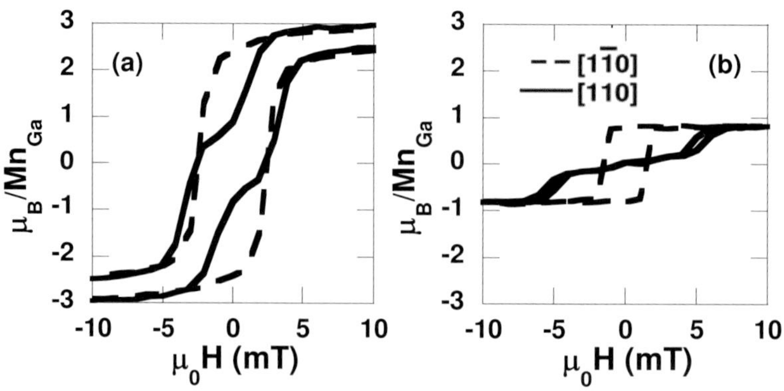

Figure 5.10. Magnetization as a function of applied magnetic field for (a) $Ga_{0.958}Mn_{0.042}P$ and (b) $Ga_{0.958}Mn_{0.042}P_{0.973}S_{0.027}$. After Ref. 66.

in the carrier concentration due to compensation of holes by S donors. Such an explanation is consistent with results in $Ga_{1-x}Mn_xAs$ where a reorientation of the magnetic easy axis from $[1\bar{1}0]$ to [110] is caused by annealing, which increases p. In $Ga_{0.958}Mn_{0.042}P$ $[1\bar{1}0]$ starts as the easy axis and becomes increasingly magnetically preferred over [110] as p is lowered (y is increased).

5.6 ELECTRICAL TRANSPORT

All $Ga_{1-x}Mn_xP$ samples synthesized to date ($x \leq 0.042$) are electrically insulating. For the most heavily doped $Ga_{1-x}Mn_xP$ samples (e.g., $x = 0.042$) there are two distinct thermally-activated processes by which conduction occurs between $T =$ 300 and 10 K, as can be readily seen from the two linear regions of Fig. 5.11(a). This behavior is described well by the following relation:

$$\rho = \left(\frac{1}{\rho_1}\exp\left(\frac{-\varepsilon_1}{k_B T}\right) + \frac{1}{\rho_3}\exp\left(\frac{-\varepsilon_3}{k_B T}\right)\right)^{-1}, \quad (5.2)$$

where the activation energies ε_1 and ε_3 and pre-exponential factors ρ_1 and ρ_3 are fitting parameters.[68] The first term in Eq. (5.2) is attributed to band conduction of extrinsic carriers excited across an energy gap, ε_1, which is the dominant transport mechanism at high temperatures. The second term results from nearest neighbor hopping between adjacent impurity sites with activation energy ε_3. For $Ga_{0.958}Mn_{0.042}P$ the best fit to Eq. (5.2) occurs for $\varepsilon_1 = 31$ meV and $\varepsilon_3 = 6$ meV. The extracted value of ε_1 is much less than that of the isolated Mn acceptor as was noted in Sec. 5.4, which is attributed to the formation and broadening of an Mn impurity band at these high doping concentrations. This value is in good agreement with the photoconductivity edge of 26 meV in $Ga_{0.958}Mn_{0.042}P$ (Sec. 5.4).

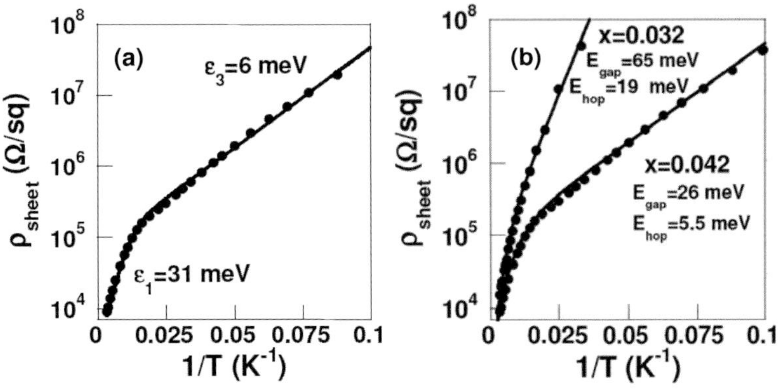

Figure 5.11. (a) ρ_{sheet} as a function of inverse temperature for $Ga_{0.958}Mn_{0.042}P$ (symbols) along with the best fit to Eq. (5.2) (solid line). (b) ρ_{sheet} as a function of inverse temperature for $Ga_{1-x}Mn_xP$ with $x = 0.032$ and 0.042. Solid lines in (b) are the results of numerical modeling described in more detail in the text.

While this simple phenomenological model provides reasonable agreement with experiment, it contains several implicit assumptions, which are not entirely consistent with impurity band transport. It is assumed that the activation energy ε_1 for exciting holes from impurity states to the valence band remains constant over the entire temperature range. This is not valid for a broadened impurity band since the energy barrier will be altered as changes in the population of the impurity band shift E_F. This, and other, drawbacks of Eq. (5.2) can be addressed by more rigorous numerical modeling of the carrier statistics. In this model the impurity band is assumed to be symmetric about the isolated Mn acceptor energy of 400 meV above a parabolic valence band. The energy gap (E_{gap}) between the valence band maximum and impurity band minimum is set to the photoconductivity edge (Sec. 5.4). This energy gap is the most important parameter of the model; changing the energy dependence of the impurity band DOS accounts for only slight differences in the predicted resistivity. A truncated Gaussian density of states gives good agreement with experiment while DOS $\propto 1 - \text{const.} \cdot E^2$ gave the best agreement and no ambiguity as to the impurity band width. The low temperature behavior is accounted for by the inclusion of a nearest-neighbor hopping term with energy barrier (E_{hop}) in analogy to the second term in Eq. (5.2). The modeling is illustrated in Fig. 5.11 for $x = 0.042$ and $x = 0.032$. The model is in good agreement with the experimental data in both cases, which illustrates two important points. First, the numerical modeling supports the interpretation of the empirical activation energies ε_1 and ε_3 as originating due to impurity band to valence band transitions and nearest-neighbor hopping, respectively. Secondly, the unmerged Mn-derived impurity band model of hole transport occurring by parallel interband and nearest-neighbor hopping mechanisms is consistent with the temperature dependence of the resistivity for $300\,\text{K} \geq T \geq 10\,\text{K}$.

Figure 5.12 shows the temperature dependence of the sheet resistivity as a function of Mn_{Ga} concentration, illustrating that the sheet resistivity increases as x

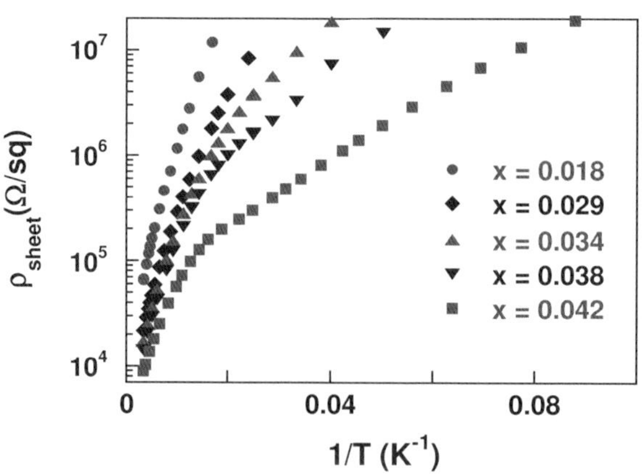

Figure 5.12. Temperature dependence of the sheet resistivity for $Ga_{1-x}Mn_xP$ of varying x. After Ref. 31.

is decreased. This behavior is to be expected from the previously discussed model. The reduced concentration of Mn_{Ga} results in a tighter impurity band, which is separated by a larger energy gap from the valence band. Therefore, in terms of the empirical relation (Eq. 5.2), ε_1 increases with decreasing x. For $x > 0.03$ the transition to hopping transport is apparent as the temperature is lowered. Clearly, ε_3 increases with decreasing x since nearest-neighbor hopping is hindered by the larger mean distance between Mn_{Ga} impurities. When $x < 0.03$, reliable data could only be collected for a rather narrow temperature range due to the rapidly increasing impedances and the development of non-Ohmic contacts at low temperature. In these materials a change to hopping-dominated transport (ε_3) is not visible. The nonlinearity of the Arrhenius plot is accounted for by curvature caused by the detailed changes in E_F, which is important for $E_{gap} > k_BT$ and related to the detailed impurity band edge shape. It is expected that the transition to hopping transport would only become apparent at lower temperatures but may never be resolved because of the extreme impedances of thin films of $Ga_{1-x}Mn_xP$.

5.7 STABILITY OF Mn_{Ga} DEFECT IN $Ga_{1-x}Mn_xP$: BEHAVIOR UPON EXTENDED ANNEALING

The Mn_{Ga} defect in $Ga_{1-x}Mn_xP$ is unstable with respect to thermal annealing above 300°C similar to LT-MBE $Ga_{1-x}Mn_xAs$.[37] Annealing of $Ga_{0.966}Mn_{0.034}P$ at 325°C for 15 min in flowing N_2 gas leads to a reduction of T_C from 40 to 30 K while annealing at 400°C under similar conditions completely destroys ferromagnetism in the film. Ion beam analysis demonstrates that annealing at successively higher temperatures results in the relocation of a larger fraction of Mn_{Ga} moments to sites incommensurate with the lattice, which is consistent with the lowering of T_C. Interestingly, the drop in T_C cannot be attributed solely to the reduction of x. Figure 5.13 compares T_C as a function of x for as-grown and annealed $Ga_{1-x}Mn_xP$. The variation in x for as-grown films was achieved by changing the dose of implanted Mn^+ ions [cf. Fig. 5.3(c)]. Annealing at 375°C for 15 min reduced x from 0.034 to 0.028 and T_C from 43 to 8 K. From the trend in the as-grown material, one would expect a film with $x = 0.028$ to exhibit a T_C of ~33 K. Therefore, even though the total Mn_{Ga} concentration is equivalent in these films, annealing leads to a thermally-driven redistribution of the Mn_{Ga} impurities that is detrimental to ferromagnetism. This behavior suggests that the relocation of Mn_{Ga} to incommensurate sites leads to partial compensation or to the formation of defect complexes which render a large portion of Mn moments ferromagnetically inactive. The instability of the Mn_{Ga} defect at temperatures greater than 300°C indicates that ferromagnetism in Mn-doped GaP in materials exposed to extended thermal treatment above this temperature (i.e., GaP ion implanted with Mn and subsequently annealed) is unlikely to originate from the dilute alloy, hole-mediated ferromagnetic phase.[31]

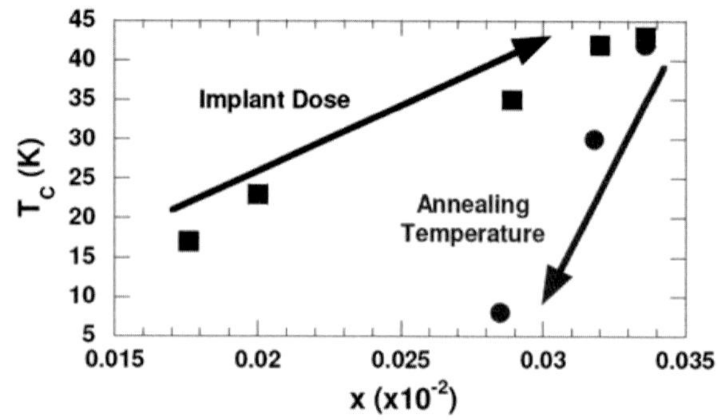

Figure 5.13. Dependence of T_C on x in $Ga_{1-x}Mn_xP$ where the variation in x has been achieved either by changing the implant dose (squares) or annealing (circles). After Ref. 31.

5.8 COMPARISON OF $Ga_{1-x}Mn_xP$ TO OTHER $III_{1-x}Mn_xV$ MATERIALS

Perhaps the most important conclusion that can be drawn from our work in $Ga_{1-x}Mn_xAs$ and $Ga_{1-x}Mn_xP$ and comparing it to other $III_{1-x}Mn_xV$ systems is the importance of localization in limiting the T_C of homogeneous, dilute, hole mediated ferromagnetic phases in $III_{1-x}Mn_xV$ semiconductors. It is expected that smaller lattice constants lead to greater wave function overlap between holes and Mn ions which will strengthen the *p–d* exchange interaction ($J_{p-d} = |N_o\beta|$).[69] This has been expected to increase the effective coupling between different Mn ions (J_{Mn-Mn}) through their interactions with a single hole leading to higher T_C. However, localization of holes will lessen the interaction between a hole associated with a specific Mn ion and other Mn ions in the crystal. Thus, in systems with more strongly localized holes as a result of strong Coloumb interactions or disorder, the effective coupling J_{Mn-Mn} through a mediating hole will be smaller. The balance between J_{p-d} and spatial localization determines J_{Mn-Mn} and therefore, T_C. The issue of localization especially as it applies to $Ga_{1-x}Mn_xN$ has also been discussed in Refs. 70–72. The importance of localization was also pointed out qualitatively in Ref. 14.

The scaling of T_C with lattice constant implied from mean-field theories stems from the assumption that holes are completely delocalized, i.e., each hole interacts with each Mn_{Ga} moment in the crystal. According to the mean field theory of hole-mediated ferromagnetism

$$T_C \propto x N_0 \beta^2, \tag{5.3}$$

where x is the Mn composition, N_o is the site concentration, which is inversely proportional to the lattice constant cubed a_o^3, and $|N_o\beta| = J_{p-d}$ is the *p–d* exchange constant. Thus, a plot of $(T_C/x)^{1/2}$ versus $a_o^{3/2}$ will be proportional to J_{p-d} for mean-field systems.[14] This simple analysis ignores the effect of compensation which is always present in real samples. Here, we consider only the highest T_C samples reported for different materials with $0 < x < 0.1$, which should represent the highest

x, lowest compensation samples achieved over this range of compositions, allowing for some general comparison to be made between the different materials in the dilute limit.

Figure 5.14 illustrates the dependence of $(T_C/x)^{1/2}$ on nearest-neighbor distance for the record highest T_C samples of a variety of $III_{1-x}Mn_xV$ ferromagnetic semiconductors.[12,29,35,73–78] To allow for the inclusion of wurtzite $Ga_{1-x}Mn_xN$ samples, the cation nearest-neighbor distance is used as the length parameter rather than a_0. Again, we note that the data point for II-PLM $Ga_{1-x}Mn_xAs$ is nearly identical to that of LT-MBE $Ga_{1-x}Mn_xAs$, demonstrating that II-PLM produces comparable results.[35] The expected increasing trend in the J_{p-d} parameter $(T_C/x)^{1/2}$ with decreasing nearest-neighbor distance is seen for InSb, GaSb, InAs, and GaAs hosts. The striking break in this trend is seen for $Ga_{1-x}Mn_xP$ and the low-T_C $Ga_{1-x}Mn_xN$ data, where $(T_C/x)^{1/2}$ *decreases* (with decreasing nearest-neighbor distance) even as the scaling for larger lattice constant materials would suggest that it should continue to increase. We attribute this behavior to localization effects on ferromagnetic exchange. In strongly localized systems, using Eq. (5.3) to extract J_{p-d} is not valid. As a result mean-field models incorrectly predict high T_Cs for compounds such as $Ga_{1-x}Mn_xN$, which have small lattice constants but have highly localized carriers associated with the mid-gap, Mn-related

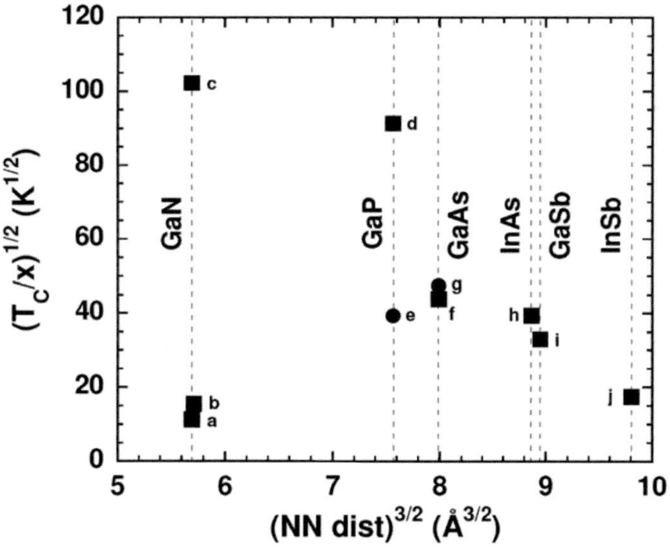

Figure 5.14. Scaling plot for the T_C of multiple $III_{1-x}Mn_xV$ ferromagnetic semiconductors with nearest-neighbor cation distance according to Eq. (5.3). The ordinate value would be strictly proportional to J_{p-d} for systems with no hole compensation and negligible hole localization. Circles refer to II-PLM samples prepared by the authors, while squares are taken from the literature. References; a = Ref. 76 (wurtzite), b = Ref. 73 (zinc-blende), c = Ref. 77, d = Ref. 27, e = Ref. 29, f = Ref. 74, g = Ref. 35, h = Ref. 12, i = Ref. 75 using $x = 0.023$ and $T_C = 25$ K to exclude phase-separated samples, j = Ref. 78.

levels.[79] The Mn acceptor is approximately four times deeper in GaP than in GaAs (~0.4 eV versus 0.11 eV) and the Mn^{2+}/Mn^{3+} charge transfer level lies in mid gap for GaN. The relative positions of the band edge and the impurity charge transfer levels as well as the energetics of the acceptor/donor must be considered in detail.

For illustrative comparison, we also include data for two notable reports of high-T_C Mn-doped GaP (data point d[27]) and GaN (data point c[77]) in which the ferromagnetism was not conclusively proven to originate from the same dilute, random, hole-mediated ferromagnetic phase found in $Ga_{1-x}Mn_xAs$ and related materials. The data from these reports are clearly above the trend of increasing J_{p-d} established in the larger lattice constant materials. The results on $Ga_{1-x}Mn_xP$ presented herein are consistent with well-substantiated data from both zinc-blende and wurtzite $Ga_{1-x}Mn_xN$ demonstrating T_Cs in the 5–15 K range,[73,76] indicating that the low T_C results in $Ga_{1-x}Mn_xN$ are more consistent with the hole-mediated dilute ferromagnetic phase akin to that observed in $In_{1-x}Mn_xSb$, $Ga_{1-x}Mn_xSb$, $In_{1-x}Mn_xAs$, $Ga_{1-x}Mn_xAs$, and $Ga_{1-x}Mn_xP$.

Polaronic models in which localized holes polarize a cloud of Mn spins are more consistent with the results of $Ga_{1-x}Mn_xP$; however, their predictive power is only qualitative, and applicability may not be appropriate for highly-doped, metallic $Ga_{1-x}Mn_xAs$.[80] Recent progress in the synthesis of Mn-doped $GaAs_{1-y}P_y$[81–83] and $Ga_{1-y}Al_yAs$[84] provides a promising route for exploring the effects of partial anion and/or cation substitution on inter-Mn exchange. Initial results indicate that localization plays a crucial role in determining T_C in these quaternary systems. Thus, the results on both $Ga_{1-x}Mn_xP$ presented herein as well as these alloys elucidate the critical need to integrate localization parameters into models if they are to accurately describe the fundamental nature of hole-mediated ferromagnetic exchange in these novel semiconductors.

5.9 SUMMARY

$Ga_{1-x}Mn_xP$ is a novel ferromagnetic semiconductor alloy in which ferromagnetism is mediated by holes that are localized in a detached Mn-derived impurity band for $x \leq 0.042$. Despite this, $Ga_{1-x}Mn_xP$ thin films exhibit all of the hallmarks of carrier-mediated exchange that are present in the canonical $Ga_{1-x}Mn_xAs$ system. The ferromagnetic Curie temperature in $Ga_{1-x}Mn_xP$ is linear in x with a slope close to that measured for $Ga_{1-x}Mn_xAs$. $L_{3,2}$ XMCD spectra provide unambiguous evidence that the origins of ferromagnetism in $Ga_{1-x}Mn_xP$ are due to Mn_{Ga} acceptors which have a local electronic environment identical (within experimental resolution) to that of Mn_{Ga} $Ga_{1-x}Mn_xAs$. Moreover, the spin-polarized holes in $Ga_{1-x}Mn_xP$ mediate the same anisotropic exchange interactions as $Ga_{1-x}Mn_xAs$, which allows for the tuning of the magnetic easy axis by temperature, epitaxial strain, and carrier concentration. The combined, optical, electrical, and magnetic properties of this materials system make it an important medium for exploring the interplay among carrier localization, band structure and ferromagnetic exchange.

ACKNOWLEDGMENTS

The authors are indebted to numerous collaborators and colleagues who have contributed to the collective body of experimental work in $Ga_{1-x}Mn_xP$, particularly R. Farshchi, R. Chopdekar, S. Tardif, C. Bihler, M. S. Brandt, J. W. Beeman, and Y. Suzuki. The authors thank W. Walukiewicz and T. Dietl for fruitful discussions regarding the theory of ferromagnetic semiconductors. Work on this chapter was supported by the Director, Office of Science, Office of Basic Energy Sciences, Division of Materials Sciences and Engineering, of the U.S. Department of Energy under Contract No. DE-AC02-05CH11231. P. R. Stone acknowledges support from NSF and NDSEG.

References

[1] S. A. Wolf, D. D. Awschalom, R. A. Buhrman, J. M. Daughton, S. von Molnar, M. L. Roukes, A. Y. Chtchelkanova and D. M. Treger, *Science* **294**, 1488 (2001).
[2] I. Zutic, J. Fabian and S. Das Sarma, *Rev. Mod. Phys.* **76**, 323 (2004).
[3] K. W. Edmonds, R. P. Campion, K. Y. Wang, A. C. Neumann, B. L. Gallagher, C. T. Foxon and P. C. Main, *J. Appl. Phys.* **93**, 6787 (2003).
[4] T. Omiya, F. Matsukura, T. Dietl, Y. Ohno, T. Sakon, M. Motokawa and H. Ohno, *Physica E* **7**, 976 (2000).
[5] H. Ohno, *Science* **281**, 951 (1998).
[6] H. X. Tang, R. K. Kawakami, D. D. Awschalom and M. L. Roukes, *Phys. Rev. Lett.* **90**, 107201 (2003).
[7] S. T. B. Goennenwein, S. Russo, A. F. Morpurgo, T. M. Klapwijk, W. Van Roy and J. De Boeck, *Phys. Rev. B* **71**, 193306 (2005).
[8] F. Matsukura, M. Sawicki, T. Dietl, D. Chiba and H. Ohno, *Physica E* **21**, 1032 (2004).
[9] H. Ohno, D. Chiba, F. Matsukura, T. Omiya, E. Abe, T. Dietl, Y. Ohno and K. Ohtani, *Nature* **408**, 944 (2000).
[10] M. K. Jain, *Diluted Magnetic Semiconductors* (World Scientific, Singapore, River Edge, 1991).
[11] S. Koshihara, A. Oiwa, M. Hirasawa, S. Katsumoto, Y. Iye, C. Urano, H. Takagi and H. Munekata, *Phys. Rev. Lett.* **78**, 4617 (1997).
[12] T. Slupinski, H. Munekata and A. Oiwa, *Appl. Phys. Lett.* **80**, 1592 (2002).
[13] H. Ohno, A. Shen, F. Matsukura, A. Oiwa, A. Endo, S. Katsumoto and Y. Iye, *Appl. Phys. Lett.* **69**, 363 (1996).
[14] T. Dietl, H. Ohno and F. Matsukura, *Phys. Rev. B* **63**, 195205 (2001).
[15] T. Dietl, H. Ohno, F. Matsukura, J. Cibert and D. Ferrand, *Science* **287**, 1019 (2000).
[16] H. Ohno, *Science* **281**, 951 (1998).
[17] K. Alberi, K. M. Yu, P. R. Stone, O. D. Dubon, W. Walukiewicz, T. Wojtowicz, X. Liu and J. Furdyna, *Phys. Rev. B* **78**, 075201 (2008).
[18] K. Ando, H. Saito, K. C. Agarwal, M. C. Debnath and V. Zayets, *Phys. Rev. Lett.* **100** (2008).
[19] K. S. Burch, D. B. Shrekenhamer, E. J. Singley, J. Stephens, B. L. Sheu, R. K. Kawakami, P. Schiffer, N. Samarth, D. D. Awschalom and D. N. Basov, *Phys. Rev. Lett.* **97**, 087208 (2006).

[20] J. Okabayashi, A. Kimura, O. Rader, T. Mizokawa, A. Fujimori, T. Hayashi, and M. Tanaka, *Phys. Rev. B* **64**, 125304 (2001).

[21] B. Clerjaud, *J. Phys. C Solid State* **18**, 3615 (1985).

[22] M. Linnarsson, E. Janzen, B. Monemar, M. Kleverman and A. Thilderkvist, *Phys. Rev. B* **55**, 6938 (1997).

[23] A. H. Macdonald, P. Schiffer and N. Samarth, *Nature Mater.* **4**, 195 (2005).

[24] J. Masek, J. Kudrnovsky, F. Maca, J. Sinova, A. H. MacDonald, R. P. Campion, B. L. Gallagher and T. Jungwirth, *Phys. Rev. B* **75**, 045202 (2007).

[25] M. E. Overberg, B. P. Gila, C. R. Abernathy, S. J. Pearton, N. A. Theodoropoulou, K. T. McCarthy, S. B. Arnason and A. F. Hebard, *Appl. Phys. Lett.* **79**, 3128 (2001).

[26] M. E. Overberg, B. P. Gila, G. T. Thaler, C. R. Abernathy, S. J. Pearton, N. A. Theodoropoulou, K. T. McCarthy, S. B. Arnason, A. F. Hebard, S. N. G. Chu, R. G. Wilson, J. M. Zavada and Y. D. Park, *J. Vac. Sci. Technol.* **20**, 969 (2002).

[27] N. Theodoropoulou, A. F. Hebard, M. E. Overberg, C. R. Abernathy, S. J. Pearton, S. N. G. Chu and R. G. Wilson, *Phys. Rev. Lett.* **89**, 107203 (2002).

[28] O. D. Dubon, M. A. Scarpulla, R. Farshchi and K. M. Yu, *Physica B* **376**, 630 (2006).

[29] M. A. Scarpulla, B. L. Cardozo, R. Farshchi, W. M. Hlaing Oo, M. D. McCluskey, K. M. Yu and O. D. Dubon, *Phys. Rev. Lett.* **95**, 207204 (2005).

[30] M. A. Scarpulla, U. Daud, K. M. Yu, O. Monteiro, Z. Liliental-Weber, D. Zakharov, W. Walukiewicz and O. D. Dubon, *Physica B* **340**, 908 (2003).

[31] R. Farshchi, M. A. Scarpulla, P. R. Stone, K. M. Yu, I. D. Sharp, J. W. Beeman, H. H. Silvestri, L. A. Reichert, E. E. Haller and O. D. Dubon, *Solid State Commun.* **140**, 443 (2006).

[32] S. A. Abagyan, G. A. Ivanov, G. A. Koroleva, Y. N. Kuznetsov and Y. A. Okunev, *Sov. Phys. Semicond.* **9**, 243 (1975).

[33] P. Baeri, *Mater. Sci. Eng. A* **178**, 179 (1994).

[34] M. A. Scarpulla, O. D. Dubon, O. Montiero, M. R. Pillai, M. J. Aziz and M. C. Ridgway, *Appl. Phys. Lett.* **82**, 1251 (2003).

[35] M. A. Scarpulla, R. Farshchi, P. R. Stone, R. V. Chopdekar, K. M. Yu, Y. Suzuki and O. D. Dubon, *J. Appl. Phys.* **103**, 073913 (2008).

[36] C. W. White, B. R. Appleton and S. R. Wilson, in *Laser Annealing of Semiconductors*, eds. J. M. Poate and J. W. Mayer (Academic Press, New York, 1982).

[37] K. M. Yu, W. Walukiewicz, T. Wojtowicz, I. Kuryliszyn, X. Liu, Y. Sasaki and J. K. Furdyna, *Phys. Rev. B* **65**, 201303 (2002).

[38] L. C. Feldman, J. W. Mayer and S. T. Picraux, *Materials Analysis by Ion Channeling: Submicron Crystallography* (Academic Press, New York, 1982).

[39] K. M. Yu, W. Walukiewicz, T. Wojtowicz, W. L. Lim, X. Liu, Y. Sasaki, M. Dobrowolska and J. K. Furdyna, *Appl. Phys. Lett.* **81**, 844 (2002).

[40] M. A. Scarpulla, P. R. Stone, I. D. Sharp, E. E. Haller, O. D. Dubon, J. W. Beeman and K. M. Yu, *J. Appl. Phys.* **103**, 123906 (2008).

[41] P. R. Stone, M. A. Scarpulla, R. Farshchi, I. D. Sharp, J. W. Beeman, K. M. Yu, E. Arenholz, J. Denlinger, E. E. Haller and O. D. Dubon, in *Proc. 28th Intl. Conf. on the Physics of Semiconductors*, eds. by W. Jantsch and F. Schäffler (Springer, New York, 2007), p. 1177.

[42] D. Ruzmetov, J. Scherschligt, D. V. Baxter, T. Wojtowicz, X. Liu, Y. Sasaki, J. K. Furdyna, K. M. Yu and W. Walukiewicz, *Phys. Rev. B* **69**, 155207 (2004).

[43] T. Jungwirth, J. Sinova, J. Masek, J. Kucera and A. H. MacDonald, *Rev. Mod. Phys.* **78**, 809 (2006).

[44] J. Stohr and Y. Wu, in *New Directions in Research with Third-Generation Soft X-ray Synchrotron Radiation Sources*, Vol. 254, eds. A. S. Schlachter and F. J. Wuilleumier (Kluwer, London, 1994).

[45] P. Carra, B. T. Thole, M. Altarelli and X. D. Wang, *Phys. Rev. Lett.* **70**, 694 (1993).

[46] B. T. Thole, P. Carra, F. Sette and G. Vanderlaan, *Phys. Rev. Lett.* **68**, 1943 (1992).

[47] P. R. Stone, M. A. Scarpulla, R. Farshchi, I. D. Sharp, E. E. Haller, O. D. Dubon, K. M. Yu, J. W. Beeman, E. Arenholz, J. D. Denlinger and H. Ohldag, *Appl. Phys. Lett.* **89**, 012504 (2006).

[48] K. W. Edmonds, N. R. S. Farley, R. P. Campion, C. T. Foxon, B. L. Gallagher, T. K. Johal, G. van der Laan, M. MacKenzie, J. N. Chapman and E. Arenholz, *Appl. Phys. Lett.* **84**, 4065 (2004).

[49] K. W. Edmonds, N. R. S. Farley, T. K. Johal, G. van der Laan, R. P. Campion, B. L. Gallagher and C. T. Foxon, *Phys. Rev. B* **71**, 064418 (2005).

[50] T. Dietl, J. Konig and A. H. MacDonald, *Phys. Rev. B* **64**, 241201(R) (2001).

[51] T. Jungwirth, J. Sinova, A. H. MacDonald, B. L. Gallagher, V. Novak, K. W. Edmonds, A. W. Rushforth, R. P. Campion, C. T. Foxon, L. Eaves, E. Olejnik, J. Masek, S. R. E. Yang, J. Wunderlich, C. Gould, L. W. Molenkamp, T. Dietl and H. Ohno, *Phys. Rev. B* **76**, 125206 (2007).

[52] B. Schmid, R. Huber and R. A. Kaindl, personal communication.

[53] M. Sawicki, F. Matsukura, T. Dietl, G. M. Schott, C. Ruester, G. Schmidt, L. W. Molenkamp and G. Karczewski, *J. Supercond.* **16**, 7 (2003).

[54] M. Sawicki, K.-Y. Wang, K. W. Edmonds, R. P. Campion, C. R. Staddon, N. R. S. Farley, C. T. Foxon, E. Papis, E. Kaminska, A. Piotrowska, T. Dietl and B. L. Gallagher, *Phys. Rev. B* **71**, 121302(R) (2005).

[55] X. Liu, Y. Sasaki and J. K. Furdyna, *Phys. Rev. B* **67**, 205204 (2003).

[56] M. Abolfath, T. Jungwirth, J. Brum and A. H. MacDonald, *Phys. Rev. B* **63**, 054418 (2001).

[57] C. Bihler, M. Kraus, H. Huebl, M. S. Brandt, S. T. B. Goennenwein, M. Opel, M. A. Scarpulla, P. R. Stone, R. Farshchi and O. D. Dubon, *Phys. Rev. B* **75**, 214419 (2007).

[58] C. Bihler, H. Huebl, M. S. Brandt, S. T. B. Goennenwein, M. Reinwald, U. Wurstbauer, M. Doppe, D. Weiss and W. Wegscheider, *Appl. Phys. Lett.* **89**, 012507 (2006).

[59] W. Limmer, M. Glunk, J. Daeubler, T. Hummel, W. Schoch, R. Sauer, C. Bihler, H. Huebl, M. S. Brandt and S. T. B. Goennenwein, *Phys. Rev. B* **74**, 205205 (2006).

[60] X. Liu, W. L. Lim, Z. Ge, S. Shen, M. Dobrowolska, J. K. Furdyna, T. Wojtowicz, K. M. Yu and W. Walukiewicz, *Appl. Phys. Lett.* **86**, 112512 (2005).

[61] P. T. Chiu, S. J. May and B. W. Wessels, *J. Appl. Phys.* **99**, 083907 (2006).

[62] D. Hrabovsky, E. Vanelle, A. R. Fert, D. S. Yee, J. P. Redoules, J. Sadowski, J. Kanski and L. Ilver, *Appl. Phys. Lett.* **81**, 2806 (2002).

[63] U. Welp, V. K. Vlasko-Vlasov, A. Menzel, H. D. You, X. Liu, J. K. Furdyna and T. Wojtowicz, *Appl. Phys. Lett.* **85**, 260 (2004).

[64] Y. J. Cho, M. A. Scarpulla, Y. Y. Zhou, Z. Ge, X. Liu, M. Dobrowolska, K. M. Yu, O. D. Dubon and J. K. Furdyna, *J. Appl. Phys.* **104** (2008).

[65] X. Liu, W. L. Lim, L. V. Titova, M. Dobrowolska, J. K. Furdyna, M. Kutrowski and T. Wojtowicz, *J. Appl. Phys.* **98**, 063904 (2005).

[66] P. R. Stone, C. Bihler, M. Kraus, M. A. Scarpulla, J. W. Beeman, K. M. Yu, M. S. Brandt and O. D. Dubon, *Phys. Rev. B* **78**, 214421 (2008).

[67] L. V. Titova, M. Kutrowski, X. Liu, R. Chakarvorty, W. L. Lim, T. Wojtowicz, J. K. Furdyna and M. Dobrowolska, *Phys. Rev. B* **72**, 165205 (2005).

[68] B. I. Shklovskii and A. L. Efros, *Electronic Properties of Doped Semiconductors* (Springer, Berlin, 1984).

[69] W. A. Harrison and G. K. Straub, *Phys. Rev. B* **36**, 2695 (1987).

[70] G. Bouzerar, T. Ziman and J. Kudrnovsky, *Europhys. Lett.* **69**, 812 (2005).

[71] P. Mahadevan and A. Zunger, *Appl. Phys. Lett.* **85**, 2860 (2004).

[72] K. Sato, W. Schweika, P. H. Dederichs and H. Katayama-Yoshida, *Phys. Rev. B* **70**, 201202 (2004).

[73] K. W. Edmonds, S. V. Novikov, M. Sawicki, R. P. Campion, C. R. Staddon, A. D. Giddings, L. X. Zhao, K. Y. Wang, T. Dietl, C. T. Foxon and B. L. Gallagher, *Appl. Phys. Lett.* **86**, 152114 (2005).

[74] T. Jungwirth, K. Y. Wang, J. Masek, K. W. Edmonds, J. Konig, J. Sinova, M. Polini, N. A. Goncharuk, A. H. MacDonald, M. Sawicki, A. W. Rushforth, R. P. Campion, L. X. Zhao, C. T. Foxon and B. L. Gallagher, *Phys. Rev. B* **72**, 165204 (2005).

[75] F. Matsukura, E. Abe and H. Ohno, *J. Appl. Phys.* **87**, 6442 (2000).

[76] E. Sarigiannidou, F. Wilhelm, E. Monroy, R. M. Galera, E. Bellet-Amalric, A. Rogalev, J. Goulon, J. Cibert and H. Mariette, *Phys. Rev. B* **74** (2006).

[77] S. Sonoda, S. Shimizu, T. Sasaki, Y. Yamamoto and H. Hori, *J. Cryst. Growth* **237–239**, 1358 (2002).

[78] T. Wojtowicz, G. Cywinski, W. L. Lim, X. Liu, M. Dobrowolska, J. K. Furdyna, K. M. Yu, W. Walukiewicz, G. B. Kim, M. Cheon, X. Chen, S. M. Wang and H. Luo, *Appl. Phys. Lett.* **82**, 4310 (2003).

[79] T. Graf, S. T. B. Goennenwein and M. S. Brandt, *Phys. Status Solidi B* **239**, 277 (2003).

[80] A. Kaminski and S. Das Sarma, *Phys. Rev. Lett.* **88** (2002).

[81] A. Lemaitre, A. Miard, L. Travers, O. Mauguin, L. Largeau, C. Gourdon, V. Jeudy, M. Tran and J.-M. George, *Appl. Phys. Lett.* **93**, 021123 (2008).

[82] A. W. Rushforth, M. Wang, N. R. S. Farley, R. P. Campion, K. W. Edmonds, C. R. Staddon, C. T. Foxon and B. L. Gallagher, *J. Appl. Phys.* **104**, 073908 (2008).

[83] P. R. Stone, K. Alberi, S. K. Z. Tardif, J. W. Beeman, K. M. Yu, W. Walukiewicz and O. D. Dubon, *Phys. Rev. Lett.* **101**, 087203 (2008).

[84] A. W. Rushforth, N. R. S. Farley, R. P. Campion, K. W. Edmonds, C. R. Staddon, C. T. Foxon, B. L. Gallagher and K. M. Yu, *Phys. Rev. B* **78**, 085209 (2008).

Chapter Six

InMnAs Thin Films and Heterostructures

B. W. Wessels

Materials Research Center, Northwestern University,
Evanston, IL 60208, USA
b-wessels@northwestern.edu

We review the growth and basic magnetic, magnetoresistive and magneto-optical properties of the magnetic semiconductor InMnAs system. Furthermore, the application of this material to spin transport in heterojunctions is described.

6.1 INTRODUCTION

Magnetic compound semiconductors have been widely investigated for their unique semiconducting and magnetic properties. While much of our knowledge about these materials results from the extensive prior work on II–VI compounds, their technological importance is limited since their ground state is antiferromagnetic.[1] With the discovery of ferromagnetism in transition metal-doped III–V compounds, research has now focused on their development. The first ferromagnetic III–V compound to be synthesized was InMnAs.[2] In contrast to doped II–VI compounds where the solubility of transition metals is appreciable, this is not the case for III–V compounds. The limited solubility necessitates the use of thin film techniques where non-equilibrium concentrations of magnetic impurities can be incorporated so that ferromagnetism is realized. The first ferromagnetic semiconductor alloy $In_{1-x}Mn_xAs$, was prepared by molecular beam epitaxy.[2] In order to obtain a single phase magnetic compound that shows ferromagnetism, Mn concentration is limited to $x = 0.02$–0.10. Since this concentration is beyond the equilibrium solubility limit, low temperature molecular beam epitaxy LT-MBE is required. Initial studies indicated that the InMnAs alloys are ferromagnetic with a Curie temperature (T_C) on the order of 10 K. Subsequent studies indicated that Curie temperature was closer to 50 K.[3] The interest in these materi-

Handbook of Spintronic Semiconductors by W M Chen & I A Buyanova

www.panstanford.com
978-981-4267-36-6

als increased dramatically with the discovery of $Ga_{1-x}Mn_xAs$ with a T_C of greater than 100 K.[4] As a result of this discovery, much of the subsequent research focused on this system as well as the exploration of other potential magnetic semiconductors including nitrides, oxides, silicon and germanium.

Nevertheless, significant work continues on the (In,Mn)As alloy system to improve its magnetic properties especially T_C. One approach was to use digital doping techniques during LT-MBE.[5] Another approach was to use buffer layers of GaSb to minimize epitaxial strain effects.[6] While researchers used LT-MBE nearly exclusively to prepare InMnAs thin films, we decided to explore metalorganic vapor phase epitaxy (MOVPE) as a technique to deposit this compound. This was based on our early work on transition metal-doped III–V thin films, which demonstrated that MOVPE enabled preparation of a wide variety of transition metal-doped III–V compounds. The compounds exhibited excellent electronic properties for dilute concentrations.[7] Using MOVPE we subsequently demonstrated that single phase magnetic InMnAs could be deposited at 500°C, some 200°C higher than that used in MBE.[8] Furthermore, the films are ferromagnetic with a Curie temperature of 330 K.[9, 10]

In this chapter we review the growth and basic magnetic, magnetoresistive and magneto-optical properties of the InMnAs ferromagnetic semiconductor system. Furthermore, the application of this material to magnetic bipolar heterojunctions is described.

6.2 EPITAXIAL FILM DEPOSITION

The low solubility of Mn in InAs limits the crystal growth techniques to those that enable non-equilibrium incorporation of the dopants. The predominant technique is molecular beam epitaxy, where it has been shown that manganese can be incorporated at a concentration of $x = 0.07$. Deposition temperatures of less than 300°C, however, are needed to obtain single phase, epitaxial thin films by MBE.[11] An important issue for many of the early studies using MBE was that the films were n-type. This resulted despite the fact that Mn in InAs is a shallow acceptor. Presumably, the high donor concentration was due to background donor impurities. Addition of Mn leads to compensated n-type films with electron concentrations of the order of 10^{18}–10^{19} cm^{-3}. Subsequent studies showed that p-type films could be prepared with hole concentrations of 10^{19} cm^{-3}. That many of the initial magnetic property measurements were performed on n-type materials had a profound influence on conclusions about the suitability of InMnAs as a spintronic material.

An alternative method we introduced for the deposition of InMnAs was MOVPE. Our early studies on transition metal-doping of III–V compounds showed that there were a number of suitable precursors available. In contrast, however, the deposition temperature is 200 to 300°C higher than that used for MBE of InMnAs. When grown at 500°C, the MOVPE films are single phase for concentrations up to $x = 0.1$ as shown in Fig. 6.1. Above this concentration the

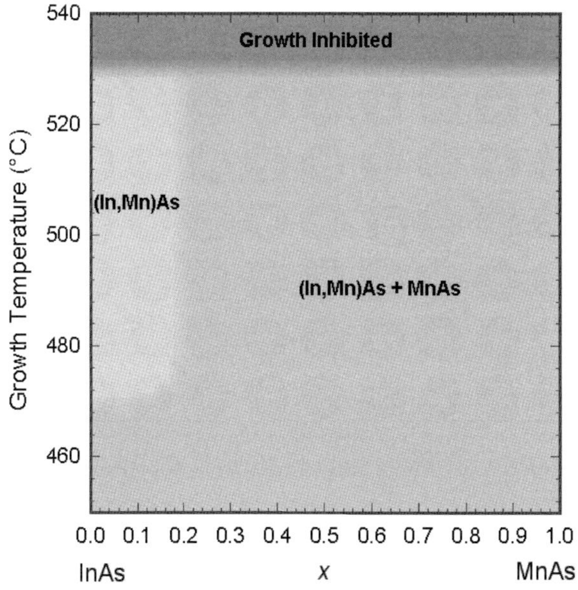

Figure 6.1. Regions of phase stability for MOVPE InMnAs.

films show formation of secondary phases, in this case MnAs. MnAs has a NiAs crystal structure and can be readily detected by X-ray diffraction and transmission electron microscopy. Single phase films for a substrate temperature of $T_{sub} = 475$–520°C have been achieved by MOVPE as determined by X-ray diffraction. Analytical electron microscopy has also been performed on the films and has shown that there is a homogeneous Mn distribution.[12] Mn substitutes for In in the zinc-blende lattice as determined by extended X-ray absorption fine structure (EXAFS) analysis.[13] Four-fold co-ordination of the Mn is observed for single phase films. The presence of MnAs is observed as a change in co-ordination of Mn in the EXAFS spectra. EXAFS spectra for both MBE and MOVPE InMnAs have been compared. Differences are observed mainly in the outer co-ordination spheres. Since the concentration of Mn in the alloys is several percent formation of atomic scale, clusters would be expected even for random alloys. At these concentrations, some of the Mn atoms would be expected to substitute at second nearest-neighbor sites. Mn in the form of dimers, trimers and tetramers could potentially form,[14] as shown in Fig. 6.2. Theoretical calculations indicate that formation of these clusters is favored thermodynamically.

EXAFS was measured to determine whether clusters form in the InMnAs; initial analysis indicates that dimers and trimers form. Formation of clusters of Mn acceptors can have potentially large effects on electronic and magnetic properties of the films. Theory indicates that cluster formation can enhance the Curie temperature.[15] Furthermore, theory shows that dopant disorder can influence the Ruderman–Kittel–Kasuya–Yosida (RKKY) interactions.[16]

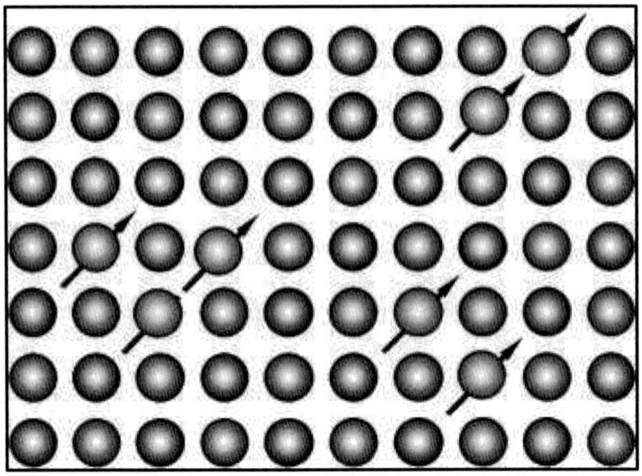

Figure 6.2. Formation of dimers and trimers from substitutional magnetic atoms.

6.3 ELECTRONIC PROPERTIES OF InMnAs

When Mn substitutes for In in InMnAs, it acts as an effective mass acceptor. Thus, InMnAs films are nominally p-type provided the background donor concentration is low. Nevertheless, early studies indicate InMnAs grown by MBE were n-type when deposited at temperatures below 270°C. p-type films are obtained for temperatures in the range of 270–370°C. In the case of MOVPE films, the films are always p-type. For single phase films, hole concentrations are typically 2×10^{17} to $2 \times 10^{18}\,\text{cm}^{-3}$. Two phase films where MnAs precipitates are present have hole concentration of $3 \times 10^{18}\,\text{cm}^{-3}$.[17]

The ionization energy for Mn in InMnAs has been measured. The ionization energy is 23 meV up to an acceptor concentration $1.0 \times 10^{18}\,\text{cm}^{-3}$ and decreases to 17 meV for $1.5 \times 10^{18}\,\text{cm}^{-3}$. Degenerate behavior is observed for two phase films with $p = 2.7 \times 10^{18}\,\text{cm}^{-3}$. The doping behavior suggests that the hole concentration is limited by the formation of a second phase. Nevertheless, the magnitude of the hole concentration is somewhat surprising in that the concentration of Mn is nearly a factor of one hundred higher than the measured hole concentration. A question arises: why is Mn not more electrically active in nominally single phase films? One possibility is that most Mn is tied up in precipitates or atomic-scale clusters where the Mn is electrically inactive. Raebiger *et al.* have addressed this possibility for GaMnAs using density functional theory (DFT) calculations.[18] They found that when Mn is in a cluster, denoted as Mn_i, where i is the number of substitutional Mn atoms forming the cluster, it forms one shallow hole per cluster. Thus, for $i = 5$ only one in five manganese atoms would be electrically active. Presumably, all the Mn in the cluster would still be magnetic.

The temperature dependence of the resistivity and hole concentration have been measured for MOVPE and MBE InMnAs. For hole concentrations of less than $2 \times 10^{18}\,\text{cm}^{-3}$ the MOVPE films are semiconducting and metallic above that concentration. For comparison, the concentration using the Mott criterion assuming a hydrogenic impurity state $a_B N_A^{1/3} = 0.25$ is $N_A = 2 \times 10^{18}\,\text{cm}^{-3}$.

6.4 MAGNETIC PROPERTIES

The basic mechanism for the ferromagnetism in magnetic semiconductors is currently debated. The presence of free carriers in InMnAs along with ferromagnetism have led to a model based on carrier mediated ferromagnetism. An RKKY mechanism has been proposed where free carriers mediate the ferromagnetism.[19] Mean-field theory predicts that the Curie temperature should scale with Mn concentration and hole concentration as $xp^{1/3}$. Good agreement between theory and experiment has been noted for GaMnAs. In contrast, the agreement for the InMnAs system has not been as good. The observation of T_C of 10 K for InMnAs films with carrier concentration of $10^{18}\,\text{cm}^{-3}$ is inconsistent with calculations. Based on calculations InMnAs with a hole concentration of 10^{18} should have a T_C of 8 K instead of 40 K observed experimentally. Furthermore, for recent studies of MBE grown InMnAs using post-growth annealing T_C as high as 90 K have been achieved for a hole concentration of $1 \times 10^{20}\,\text{cm}^{-3}$.[20] This value is well above that predicted by mean-field theory.[19]

The case for ferromagnetism for InMnAs grown by MOVPE is even more complex. Films with a carrier concentration of $10^{18}\,\text{cm}^{-3}$ have Curie temperatures of 330 K as determined by SQUID magnetometry. Furthermore, the Curie temperature is nearly independent of carrier concentration. The measured Curie temperature of 330 K is nearly the same as that of MnAs, which suggests the high temperature is of the same or similar origin. However, as indicated, X-ray diffraction and TEM analysis of the layers did not indicate the presence of a second phase with a different crystal structure. One possibility is that the MnAs is present as coherent precipitates with a zinc-blende structure. Such precipitates have been observed in GaMnAs, but would be unexpected in InMnAs since the lattice constant of zinc-blende MnAs is predicted to be much smaller than InAs.[21, 22] Thus, incoherent precipitates should form due to energetics and should be observable by X-ray diffraction or TEM. Another possibility is that the MnAs is present in InAs as clusters consisting of several atoms as previously described. These clusters presumably would be of a size that is well below the detection limit for both X-ray diffraction and high resolution TEM.

6.5 MAGNETORESISTANCE AND MAGNETOTRANSPORT

Magnetotransport studies can give insight into the magnetic properties of thin films. The longitudinal resistivity of InMnAs shows a complex temperature and magnetic field dependence. The magnetoresistance of the MOVPE InMnAs grown films has been measured at low temperatures.[23] A weak negative magnetoresistance with a value of less than one percent was observed, as shown in Fig. 6.3.

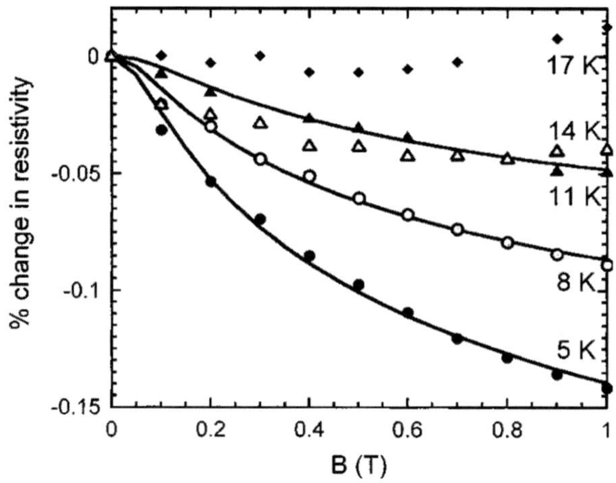

Figure 6.3. Magnetoresistance versus field for InMnAs films (after Ref. 17).

For temperatures of 11 K and below, the data were described by a Khosla–Fischer model for magnetoresistance where *sp–d* interactions are considered.[24] The magnetoresistance followed the semi-empirical equation: $\Delta\rho/\rho = -B_1^2 \ln(1 + B_2^2 H^2)$, where B_1 and B_2 are constants and H is the magnetic field. The basis for the equation is Toyazawa's localized magnetic moment model, where free carriers are scattered by the localized spin of the magnetic impurity atoms. As the magnetic field increases, the amount of scattering decreases due to freezing out of magnetic fluctuations, resulting in a negative magnetoresistance.

For temperatures above 11 K the simplified Khosla–Fischer equation does not hold for InMnAs and higher-order terms must be considered. Both negative and positive contributions to the magnetoresistance are needed. As seen in Fig. 6.3 at 17 K and 1 T the magnetoresistance becomes slightly positive, indicating that a second scattering mechanism prevails. The second mechanism was unidentified. It was speculated that the positive magnetoresistance was due to a decrease in the density of states with magnetic field. Recently, theoretical models have been proposed for semiconductors where the presence of magnetic impurities can lead to enhanced scattering and positive magnetoresistance.[25] One proposed mechanism is that carrier scattering results from the built-in spatial fluctuation of the localized spins. This scattering mechanism would increase with increasing magnetic field due to the increase in the local Zeeman band splitting. The enhanced spin-dependent scattering by the external magnetic field is predicted to be especially large for semiconductors with electrically active magnetic impurities. It should be noted that the same magnetic impurities that give rise to the negative magnetoresistance at low magnetic fields and temperature give rise to the positive magnetoresistance at high fields. This presumably results from contributions from higher-order terms in H dominating at high fields.

Khosla and Fischer proposed a semi-empirical equation to describe the complex dependence of magnetoresistance on magnetic field for doped semiconductors at higher fields.[24] The magnetoresistance is given by $\Delta\rho/\rho = -aH + bH^2$, where a and b are constants independent of field but dependent on temperature and magnetic ion concentration. In this case the quadratic H^2 term could be due to scattering from spatial fluctuations of the Mn concentration. Since fluctuations also increase with alloy composition, the parameter b should increase with x. Consequently, the positive magnetoresistance should be dominant in alloys with large magnetic ion concentrations. The positive magnetoresistance should also dominate at high temperatures.

6.6 BIPOLAR MAGNETIC SEMICONDUCTOR DEVICE STRUCTURES

With the availability of MOVPE epitaxial p-type layers of InMnAs, magnetic pn heterojunctions have been fabricated and their magnetotransport characterized. The heterojunctions consist of p-type InMnAs layers deposited on n-type semiconductor InAs (100) oriented substrates.[26] The heterojunctions show a positive giant magnetoresistance (GMR). The heterojunctions exhibited excellent rectifying characteristics. The $J - V$ characteristics are given by $J = J_{\text{diff}} + J_{\text{gr}} + J_{\text{tun}} + J_{\text{leakage}}$, where J_{diff} is the diffusion current, J_{gr} is the generation current, J_{tun} is the tunneling current, and J_{leakage} is the leakage current.

The magnetotransport characteristics of the junctions were initially measured at fields up to 0.25 T. The junctions exhibited a positive magnetoresistance at 295 K. The J–V characteristics in the presence of a magnetic field are described by

$$J = J_o \exp(V - IR_o - IR(H))/nkT,$$

where R_o is the junction series resistance and n is the ideality factor. The term $R(H)$ is the magnetoresistance of the junction. Figure 6.4 shows the magnetoresistance characteristics for measurements up to 9 T.[27] At fields below 1 T junction magnetoresistance is sublinear in field. At a higher field the magnetoresistance, however, becomes linear in field and shows no sign of saturation. The observed positive magnetoresistance is counter to what is predicted for a magnetic semiconductor using heterojunction theory[28]; in that case, a negative magnetoresistance was predicted for the junctions.

The origin of the positive giant magnetoresistance observed in the InMnAs/InAs heterojunctions is not understood and is currently under study. One possibility is that spin transport determines its magnetoresistance. Presumably, the bands of the InMnAs in the heterojunction are spin split in the presence of a magnetic field, and enhanced scattering of the spin split carriers is observed.[29] The sign of the magnetoresistance will depend on the specific spin scattering mechanism. For low fields and low temperatures a negative magnetoresistance for junctions would be expected from the low temperature magnetoresistance measurements on epitaxial InMnAs and the theory of scattering in magnetic semiconductors. At high fields a positive magnetoresistance should be observed.[24] As is evident in Fig. 6.4 a positive giant magnetoresistance of 700% is observed at room

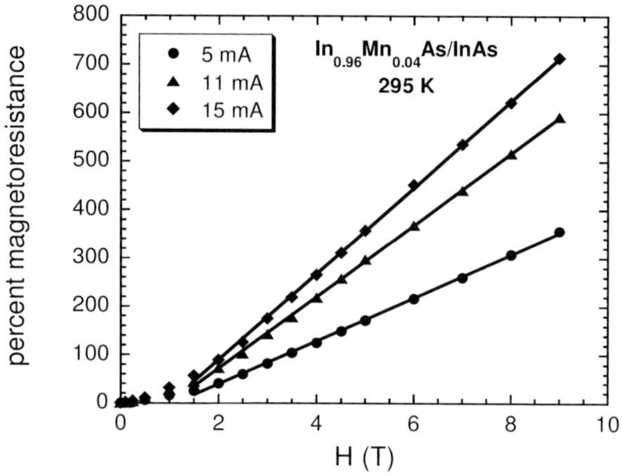

Figure 6.4. Magnetoresistance for an InMnAs/InAs *pn* junction (after Ref. 27).

temperature. Measurements on other heterojunctions indicated that a GMR of 1300% could be obtained. In order to determine whether the GMR saturates with field, high field measurements were performed up to 18 T; the I–V characteristics are shown in Fig. 6.5. The magnetoresistance increases linearly with field and no saturation is observed.

To explain the junction magnetotransport, a model is proposed whereby the valence band of the magnetic semiconductor is spin split. This splitting leads to polarization of carriers having either spin up or spin down. Carriers with different spin states will have different scattering cross-sections and mobilities. Recent theoretical calculations indicate that majority spins will have higher mobilities that increase with field whereas minority spins will have lower mobilities that

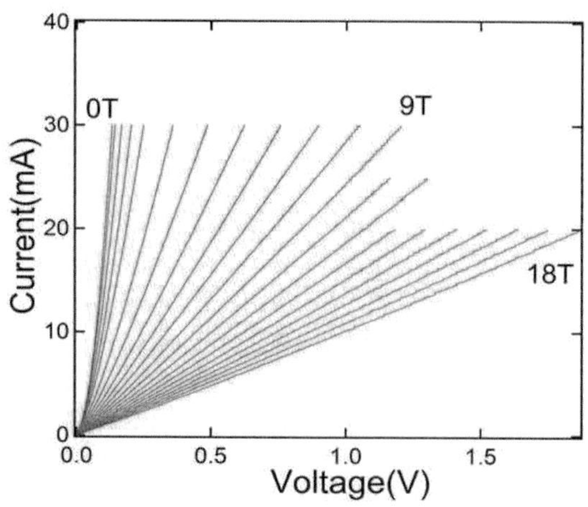

Figure 6.5. Current–voltage characteristics for an InMnAs/InAs *pn* junction measured at 300 K and fields up to 18 T (after Ref. 30).

decrease with field.[25] The sign of the junction magnetoresistance will depend on the spin scattering mechanism, which in turn depends on temperature, magnetic field strength, alloy composition and its spatial distribution.

6.7 MAGNETO-OPTICAL PROPERTIES

As to the nature of the magnetic species leading to the high Curie temperature and spin-polarized transport of the MOVPE InMnAs material, magnetic circular dichroism (MCD) can give the required insight. MCD of InMnAs has been recently reviewed.[30] As indicated by both theory and experiments there is a tendency for InMnAs to form magnetic complexes even above that expected for the random alloy. This tendency can be observed in the magneto-optical properties. Epitaxial InMnAs has been shown to exhibit both visible and mid-infrared magneto-optical Kerr effect (MOKE) at room temperature. Visible MCD spectra have indicated the presence of two transitions, both of which depend on Mn composition.[31] The calculated dependence of the MCD on x for different cluster sizes is given by the following equations:

$$\theta_1 = A_1[N_0 x(1-x)^{12}], \tag{6.1}$$
$$\theta_2 = A_2[12N_0 x^2(1-x)^{18}], \tag{6.2}$$
$$\theta_3 = A_3[24N_0 x^3(1-x)^{22}], \tag{6.3}$$
$$\theta_4 = A_4[48N_0 x^4(1-x)^{24}], \tag{6.4}$$

where θ_i is the MCD angle and the subscript i is the number of substitutional Mn atoms that form the cluster. The term A_i is a constant that depends on the exchange interaction and wavelength and N_o is the number of cation sites. For low Mn concentration the $(1-x)$ term can be ignored but for higher concentrations it needs to be considered. As can be seen from the equations, a super-linear dependence of θ_i on x is predicted for $i > 1$. This dependence can be seen in Fig. 6.6, which shows the concentration dependence of the visible MCD spectra for two transitions. The calculated dependence of the measured MCD is shown in Fig. 6.7. The best fit is for $i = 4$, that is, tetramers form whereby the Mn forms a cluster with three other Mn which substitute on the second nearest-neighbor sites.

Alternatively, this can be described as a cluster consisting of As surrounded by four nearest-neighbor Mn. The clusters maintain their tetrahedral symmetry. Thus, the cluster can be identified as Mn_iAs, where $i \leq 4$.

The clusters are presumably ferrimagnetic and their magnetic moments can be estimated from theory.[32] Its moment can be expressed as $(5i-3)\mu_B$, where i is the number of Mn atoms in the cluster. For the tetramer $i = 4$ the magnetic moment of the cluster is $17\mu_B$ or $4.1\mu_B$/Mn atom. The Mn clusters with large magnetic moments could be responsible for the high T_C observed in the MOVPE InMnAs. It also could be the reason why T_C is independent of Mn concentration if the Curie temperature is determined by the clusters and their interactions. It is interesting

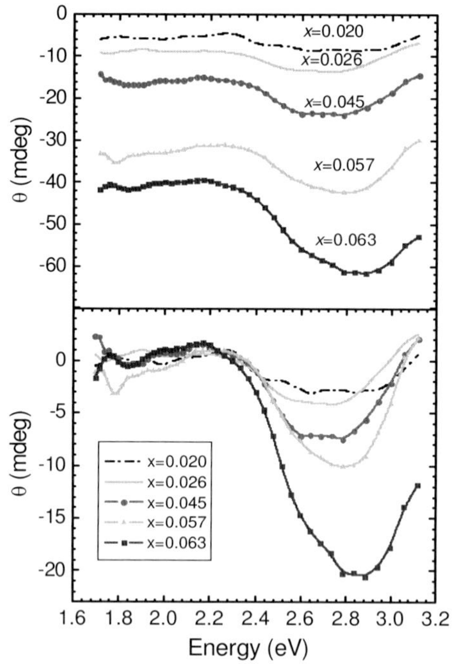

Figure 6.6. MCD of MOVPE InMnAs at 295 K (a) with broad transition and (b) subtracting broad transition (after Ref. 32).

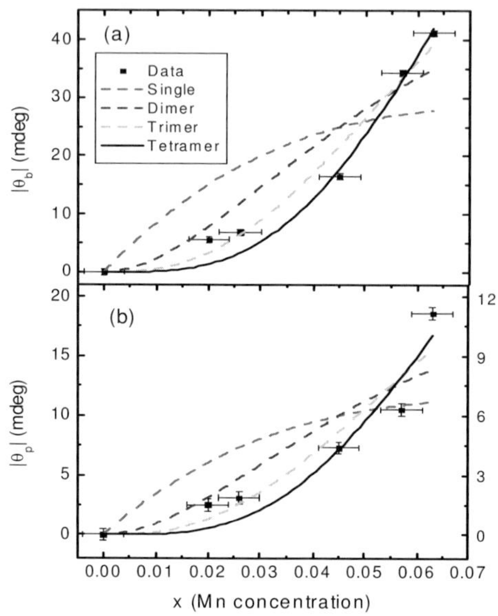

Figure 6.7. Concentration dependence of MCD for two different transitions: (a) broad transition and (b) peak (after Ref. 32).

to point out that the building block of MnAs with a zinc-blende structure is the tetramer Mn_4As. The calculated T_C of this structure is on the order of 320°C.[33]

6.8 CONCLUSIONS

The ferromagnetic semiconductor InMnAs shows considerable promise as a spintronic material. Since the equilibrium solubility of Mn in InAs is low, thin film deposition techniques are needed in order to stabilize the ferromagnetic alloy. In contrast to MBE using the MOVPE thin film technique, nominally phase pure InMnAs alloy films with a Curie transition above room temperature have been realized. It has been proposed that alloy disorder plays an important role in stabilizing the high T_C ferromagnetism. The detailed nature of the ferromagnetic interaction is not known and remains controversial. A mean-field approach, however, may be too simplistic in light of the presence of disorder for the alloy compositions of interest. MOVPE InMnAs has been used in the fabrication of magnetic semiconductor *pn* junction devices. The devices show a giant magnetoresistance at room temperature that indicates the potential importance of these narrow-gap semiconductor magnetic alloys for practical magnetoelectronic and spintronic devices.

ACKNOWLEDGMENTS

The author acknowledges the extensive contributions of members of the Electronic and Photonic Materials Group at Northwestern University. This research was supported by the NSF under grants DMR-0511523 and DMR-0804479 and the AFOSR under grant FA9550-07-1-0381.

References

[1] J. K. Furdyna, *J. Appl. Phys.* **64**, R29 (1988).
[2] H. Munekata, H. Ohno, S. von Molnar, A. Segmuller, L. L. Chang and L. Esaki, *Phys. Rev. Lett.* **63**, 1849 (1989).
[3] T. Slupinski, A. Oiwa, S. Yanagi and H. Munekata, *J. Cryst. Growth* **237–239**, 1326 (2002).
[4] H. Ohno, *Science* **281**, 951 (1998).
[5] G. Acbas *et al.*, *Physica E* **20**, 382 (2004).
[6] S. Koshihara *et al.*, *Phys. Rev. Lett.* **78**, 4617 (1997).
[7] K. Huang and B. W. Wessels, *Appl. Phys. Lett.* **52**, 1155 (1988).
[8] A. J. Blattner, J. Lensch and B. W. Wessels, *J. Electron. Mater.* **30**, 1408 (2001).
[9] A. J. Blattner and B. W. Wessels, *J. Vac. Sci. Technol. B* **20**, 1582 (2002).
[10] A. J. Blattner and B. W. Wessels, *Appl. Surf. Sci.* **221**, 155 (2004).
[11] Y. Iye *et al.*, *Mat. Sci. Eng. B* **63**, 88 (1999).
[12] A. J. Blattner, P. L. Prabhumirahi, V. Dravid and B. W. Wessels, *J. Cryst. Growth* **259**, 8 (2003).
[13] Y. Soo *et al.*, *Appl. Phys. Lett.* **84**, 481 (2004).
[14] M. van Schlifgaarde and O. N. Myrasov, *Phys. Rev. B* **63**, 233205 (2001).

[15] G. Bouzerar, T. Ziman and J. Kudrnovsky, *Appl. Phys. Lett.* **85** 4941 (2004).
[16] D. Wang and S. Xiong, *Chin. Phys. Lett.* **25**, 1102 (2008).
[17] S. J. May, A. J. Blattner and B. W. Wessels, *Physica B*, **340–342**, 870 (2003).
[18] H. Raebiger, A. Ayuela and J. von Boehm, *Phys. Rev. B* **72**, 014465-7 (2005).
[19] T. Dietl, H. Ohno and F. Matsukura, *Phys. Rev. B* **63**, 195205 (2001).
[20] T. Schallenberg and H. Munekata, *Appl. Phys. Lett.* **89**, 042507 (2006).
[21] M. Moreno *et al.*, *Phys. Rev. B* **67**, 235206 (2003).
[22] A. Janotti, S.-H. Wei and L. Belliaiche, *Appl. Phys. Lett.* **82**, 766 (2003).
[23] S. J. May, A. J. Blattner and B. W. Wessels, *Phys. Rev. B* **70** 073303 (2004).
[24] R. P. Khosla and J. R. Fischer, *Phys. Rev. B* **10**, 4084 (1970).
[25] M. Foygel and A. G Petukhov, *Phys. Rev. B* **76**, 205202 (2007).
[26] S. May and B. W. Wessels, in *Conf. on Physics and Chemistry of Semiconductor Interfaces* (2005); *J. Vac. Sci. Technol. B* **23**, 1769 (2005).
[27] S. J. May and B. W. Wessels, *Appl. Phys. Lett.* **88**, 072105-07 (2006).
[28] J. Fabian, I. Zutic and S. Das. Sarma, *Phys. Rev. B* **66**, 165301 (2002).
[29] N. Rangaraju and B. W. Wessels, *J. Vac. Sci. Technol. B* **26**, 1526 (2008).
[30] B. W. Wessels, *New J. Phys* **10**, 055008 (2008).
[31] P. T. Chiu and B. W. Wessels, *Phys. Rev. B* **76**, 165201 (2007).
[32] B. K. Rao and P. Jena, *Phys. Rev. Lett.* **89**, 185504 (2002).
[33] J. Osirio-Guillen, Y. J. Zhao, S.V. Barabash and A. Zunger, *Phys. Rev. B* **74** 035305 (2006).

Chapter Seven

Magnetic Doping of Group IV Semiconductors

Mustafa M. Özer

Materials Science and Technology Division,
Oak Ridge National Laboratory, Oak Ridge, Tennessee 37831, USA

Changgan Zeng

Hefei National Laboratory for Physical Sciences at Microscale,
University of Science and Technology of China,
Hefei, Anhui 230026, China

Hanno H. Weitering

Department of Physics and Astronomy, The University of Tennessee,
Knoxville, Tennessee 37996, USA, and
Materials Science and Technology Division, Oak Ridge National Laboratory,
Oak Ridge, Tennessee 37831, USA
hanno@utk.edu

This review addresses recent developments in the general area of group IV dilute magnetic semiconductors (DMS), focusing on fundamental physics issues as well as the promises and limitations of the actual materials from an experimentalist's perspective. This class of materials is of strategic significance in developing a spin-based information technology that will be compatible with the existing silicon platform. The fundamental physics issues surrounding the mechanisms of electrical transport and magnetism in group IV DMS revolve around the weak-coupling versus strong-coupling pictures of ferromagnetic exchange, the presence of delocalized valence-band holes versus impurity-band carriers, and the nature of the ferromagnetic transition. Materials issues revolve around the solid solubility of Mn dopants, spinodal decomposition, substitutional versus interstitial incorporation of the magnetic dopants, formation of superparamagnetic precipitates, magneto-resistance, transport percolation and the true nature of the observed magnetic order. This review specifically focuses on Mn-doped germanium grown via conventional molecular beam epitaxy. New developments such as the use of surfactants and n-type co-dopants will be highlighted as promising avenues for future research.

Handbook of Spintronic Semiconductors by W M Chen & I A Buyanova

www.panstanford.com
978-981-4267-36-6

7.1 GENERAL INTRODUCTION

The concept of the field effect transistor or FET, the control of electron charge in a solid, was patented by J. E. Lilienfeld in January 1930.[1] After twenty-five years of research the correct materials system and configuration was identified, which resulted in the silicon revolution. Analogously, the control of electron spin is now at a conceptual stage, awaiting the science and engineering breakthroughs to create new technologies.[2,3] As opposed to the FET, it only took ten years from discovery to commercialization of the so-called giant magneto-resistance (GMR) effect[4,5] or "spin valve," which is now widely used in read-heads for magnetic hard disk drives. Companies like Honeywell, Motorola, and IBM have already been developing random access memory (RAM) based on all-metal spin valves and magnetic tunnel junctions which will likely replace the traditional semiconductor RAM because of demonstrated high speed, high density, non-volatility, and radiation hardness. An even more imaginative scenario with potentially much greater impact, would be the integration of spintronics and semiconductors.[2,6–8] Semiconductor spintronic devices may one day allow "qubit" (quantum bit) operations for quantum computing, which would be most easily implemented within a silicon-based materials platform. Implementation of spin control is not just an improvement over existing technology, however. It foretells entirely new technological concepts.

The control of the electron spin in semiconductor devices, specifically "spin injection" across interfaces, is presently at a conceptual stage. There are two ways to realize spin injection[9]: (i) fabricate a ferromagnetic-metal/semiconductor heterostructure, or (ii) use a dilute magnetic semiconductor (DMS) as the spin source.[10,11] The first method is hampered by chemical intermixing and/or lattice mismatch at the metal/semiconductor interface, which usually results in significant loss of spin polarization. The large "conductivity mismatch" between the metal and semiconductor further limits the spin injection efficiency for "transparent" metal/semiconductor contacts,[12] though significant injection efficiency has been realized by spin-dependent tunneling through a Schottky barrier.[13] Low Curie temperatures (T_C) generally limit the second method involving a DMS.

From a technological perspective, spin injection in group IV semiconductors such as Si and Ge would represent a milestone development[14–17] because this would allow integration of spintronics with the current industrial standard which is heavily dominated by Si and SiO_2. Furthermore, Si is highly appealing because of its long spin decoherence time, an essential ingredient for spin-based quantum bit operations and error correction.[17] Spin injection in Si is difficult to measure and quantify.[16] Standard optical techniques applied to III–V semiconductors are not easily implemented for Si because of its indirect band-gap. On the materials side, it is both difficult to realize a high quality ferromagnetic metal/silicon interface for spin injection, or to magnetically dope Si without formation of spurious intermetallic precipitates.

The discovery of "high-temperature ferromagnetism" in Mn-doped Ge at the Naval Research Laboratory[14] seemed to present a major step in the right direction toward solving these obstacles. Ge is fully compatible with the Si platform while SiGe CMOS technology has already been commercialized. This milestone paper triggered a large flurry of activity that was mostly aimed at characterizing and possibly improving the quality and functionality of doped Mn:Ge material. Some papers have even reported on the magnetic doping of Si.[18,19] In the following sections, we will discuss first the fundamentals of magnetic doping in Ge (or Si) and then present an overview of the current status in this field.

7.2 MAGNETIC DOPING OF CONVENTIONAL SEMICONDUCTORS: THEORETICAL CONSIDERATIONS

The magnetic properties of a semiconductor containing magnetic ions depend on the energies of states derived from the magnetic *d* or *f* shells and on their interactions with the valence and conduction bands of the host material. For this reason, the magnetic dopant is usually chosen to be Mn.[6–8] For many years, researchers have tried to make II–VI compound semiconductors (e.g., CdTe) magnetic by substitutional Mn-doping.[20] The Mn solubility in II–VI is high as Mn replaces the divalent cations. Mn neither introduces nor binds carriers but does give rise to a localized spin state of $S = 5/2$ ($3d^5$). Hybridization between the Mn 3*d* states and cation *p* states leads to a short-range "super-exchange" coupling between the magnetic moments that is antiferromagnetic.[6–8]

The situation is different in III–V semiconductors.[6–8] Substitutional Mn dopants in GaAs replace the trivalent Ga atoms. Yet, in the simplest picture the Mn dopants still prefer to be divalent so each substitutional Mn must steal an electron from the valence band, thus introducing a valence-band hole. The short-range antiferromagnetic coupling between the Mn^{2+} ions is readily overwhelmed by ferromagnetic interactions that are mediated by valence-band holes: the semiconductor becomes ferromagnetic at temperatures usually below room temperature. In the following, we will slightly modify this picture in our discussion of substitutional doping in Ge.

Figure 7.1 shows a simple molecular orbital diagram for a substitutional Mn dopant in Ge.[22] The diagram includes the effects of crystal field splitting and exchange. The Mn 3*d*-orbitals are split into the two-fold degenerate e_g and three-fold degenerate t_{2g} levels due to crystal field effects in a tetrahedral coordination.[21] Note that the level ordering is opposite to that of octahedral coordination. Tetrahedral crystal field effects are usually small, meaning that the electronic configuration should be a "high spin" configuration. This is illustrated in Fig. 7.1, showing large exchange splitting on top of the crystal field splitting. The spin-up and spin-down t_{2g} and e_g level hybridize with the sp^3-like dangling bond orbitals of the neighboring Ge atoms, resulting in hybrid orbitals that are predominantly dangling-bond-like (dangling bond hybrid or DBH)[23] or 3*d*-like (crystal field resonance or CFR).[23] Specifically, the e_g orbitals remain non-bonding whereas the t_{2g} orbitals form bonding and antibonding hybrids. With the four Ge atoms contributing one

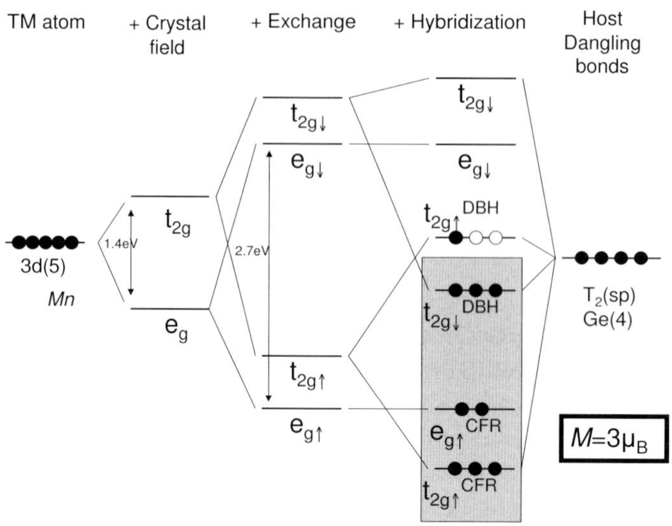

Figure 7.1. Schematic molecular orbital diagram of an Mn atom with four Ge ligands in tetrahedral coordination, showing the combined effects of crystal field splitting, magnetic exchange, and *p*–*d* hybridization.[22] The notation DBH and CFR was taken from Ref. 23.

electron each and five 3*d*-electrons supplied by the Mn atom (both 4*s* electrons go into the deeper lying a_{1g} orbitals), we must accommodate a total of nine electrons. This completely fills the $t_{2g}\uparrow$, $e_g\uparrow$ and $t_{2g}\downarrow$ levels, leaving one electron (or two holes) in the antibonding $t^*_{2g}\uparrow$ orbital. This adds up to a total of five up-spins in the CFRs and two hole spins in the DBHs, producing a net moment of $3\mu_B$ per Mn dopant. In the simplest ionic picture, this could be interpreted as a Mn^{2+} $3d^5$ configuration with two anti-parallel holes on the Ge ligands. In the case of GaAs, the As ligands would provide a total of five electrons instead of four, producing a configuration that could be described as Mn $3d^5$+ hole.

Density functional theory (DFT) calculations by Stroppa *et al.*,[24] employing the all-electron full potential linearized augmented plane wave (FLAPW) method within the local spin density approximation (LSDA), of a 32-atom supercell containing one Mn atom (i.e., $Mn_{0.03125}Ge_{0.96875}$) are consistent with this simple picture. In these calculations, the spin-up density of states correspond to the $t_{2g}\uparrow$, $e_g\uparrow$ orbitals whereas the states near the Fermi level have an antibonding $t^*_{2g}\uparrow$ character. The $t_{2g}\downarrow$ levels are centered at about 1 eV below E_F whereas the $e_g\downarrow$ levels are located at about 0.5 eV above E_F. The crystal field splittings and exchange splitting of the e_g states amount to about 1.4 and 2.7 eV, respectively; values that are similar to those in Mn-doped GaAs. The calculations suggest a total magnetic moment of $3.1\mu_B$ per Mn atom of which $3\mu_B$ is localized on the Mn atom. Note that in the ionic picture, the local moment on the Mn atom would be $5\mu_B$ as expected from Hund's rule for intra-atomic exchange. The smaller moment indicates a strong hybridization between the Mn 3*d* states of t_{2g} symmetry and the Ge dangling bond orbitals. The integer net moment suggests that Mn_xGe_{1-x} could be a half-metallic ferromagnet near the 3% doping level. Indeed, the density of states of the minority spin channel reduces to zero at the Fermi level in these calculations.[24]

The magnetic coupling mechanism was further explored by calculating the LSDA total energy differences for ferromagnetic and antiferromagnetic spin alignments of two Mn atoms inside a 32-atom[14] or 64-atom[25] supercell. In these DFT calculations, Mn atoms were placed at various locations inside the supercell so as to capture the effects of Mn–Mn separation and site distribution on the magnetic coupling. The calculations in Ref. 14 indicate that the Mn–Mn interaction is antiferromagnetic at very short distances and ferromagnetic at longer distances, consistent with the existence of competing short-range super-exchange and long-range kinetic exchange interactions. In the paper by Zhao *et al.*,[25] the interactions were even found to oscillate, conforming to the Ruderman–Kittel–Kasuya–Yosida (RKKY) formalism and weak coupling picture.

Experimentally, Mn-doped Ge turns out to be strongly insulating[14,26,27] and thus it is natural to challenge the RKKY or Stoner band pictures as the starting point in our discussion of the ferromagnetic coupling mechanism. Indeed, it has been pointed out that the strongly localized nature of the moments and strong Coulomb interactions should be included in even the most minimal model, effects that cannot be captured adequately within an effectively one-electron (i.e., mean field) LSDA band structure calculation.[6–8] Nonetheless, it seems almost generally accepted that the ferromagnetism in "conventional" DMS is mediated by holes.[6–8]

Although the quantum mechanical underpinnings of magnetism are well understood, there is no generic "theory of magnetism" that universally applies to the many different materials systems exhibiting ferromagnetism. DFT can be very powerful for systems exhibiting itinerant exchange but may not always be very useful in capturing the physics of strongly correlated systems. Model Hamiltonians seem to be better suited to capture the physics of correlated systems although their usefulness is often compromised by the simplicity of the model and, quite often, the inability to extract exact solutions from the model. Nonetheless, the general qualitative picture of the physics is as follows.[6,8,26,27]

At low doping concentrations, isolated Mn impurities in Ge produce deep acceptor levels at approximately 0.16 eV above the valence band maximum.[28] Accordingly, the effective Bohr radius of the bound hydrogenic hole states should only be of the order of one or at most a few interatomic distances. Furthermore, the exchange interaction J_{pd} between the magnetic impurity and the hole spin is believed to be antiferromagnetic[6–8,14] (as is also indicated by the simple molecular orbital diagram in Fig. 7.1). As the doping level increases, the acceptor levels merge into a narrow impurity band whose width is determined by the hopping integral t. Most states in the impurity band will be localized, however, due to the introduced substitutional disorder, meaning that the Fermi level is located below the mobility edge of the impurity band.[29]

In the strong coupling regime ($J_{pd} \gg t$) or highly dilute doping limit, ferromagnetic exchange between the Mn atoms can be mediated by carrier hopping in the impurity band. This hopping is sometimes described as a variant of the

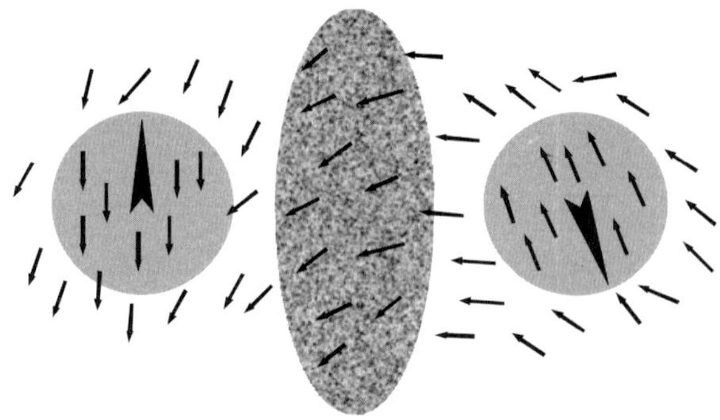

Figure 7.2. Formation and near overlap of two bound magnetic polarons. In this picture, the Mn concentration greatly exceeds the hole density. Within the radius of the polaron, all Mn spins (small arrows) are antiferromagnetically aligned to the hole spin (large arrow). The polaron radius increases with decreasing temperature. Ferromagnetism arises when polarons overlap (after Refs. 31 and 32).

double-exchange mechanism known from the manganite literature[30] or, in cases where the carrier density is only a small fraction of the total Mn density, the percolation of bound magnetic polarons.[31–33] In this picture, a hole binds magnetically to an impurity spin at a temperature scale that is defined by the exchange integral J_{pd}, thus forming a polaronic object (Fig. 7.2).[30] As the temperature is lowered, individual polarons increase their size and start to overlap until percolation takes place at the ferromagnetic T_C.

At the other end of the spectrum or weak coupling limit ($J_{pd} \ll t$), impurity bands are quite broad and overlap with the valence band continuum. The width of the band increases with doping concentration, which finally loses its independent nature. Ferromagnetism is then more appropriately described in terms of itinerant valence band holes that mediate the coupling between local Mn spins via an exchange interaction that can be described by the mean-field Zener model or RKKY exchange.[6–8]

We note that the doping dependence of T_C is qualitatively different for the strong and weak coupling regimes.[30] When $J \to \infty$, each carrier is tightly bound and the impurity band width ultimately vanishes because of the strong contraction of the Bohr radius. If the spin polarized impurity band is completely full (i.e., one carrier per Mn), hopping processes will be forbidden when the Mn spins are ferromagnetically aligned. Since T_C in this regime is controlled by the delocalization energy in the impurity band, T_C maxes out at half-filling (as in the double-exchange model). On the other hand, in the weak coupling limit, the mean-field Zener model predicts that T_C increases monotonically with the doping level or hole density.[34]

Summarizing, the key question is whether ferromagnetism in Mn_xGe_{1-x} should be interpreted in terms of impurity band physics or in terms of itinerant exchange. As we will show in this review, the situation seems to have settled in favor of the impurity band picture, although materials issues strongly blur these fundamental physics issues.

7.3 MAGNETIC DOPING OF CONVENTIONAL SEMICONDUCTORS: PRACTICAL CONSIDERATIONS

Applications of DMS require that they be magnetic at room temperature. All the literature so far indicates that for experimentally accessible materials systems, the key toward higher T_C is to increase the hole density as much as possible. Creating high hole densities is a well-recognized problem. The hole density is coupled to the Mn density and, consequently, T_C is severely limited by the insolubility of Mn in conventional DMS. For instance, the thermodynamic solubility of Mn in Ge or GaAs is well below one percent while ferromagnetism near 100 K generally requires doping levels that are of the order of five percent or more.[6–8] The key to circumvent the thermodynamic solubility limit has been the use of non-equilibrium growth techniques. Molecular beam epitaxy (MBE) has produced amazing results: $Ga_{1-x}Mn_xAs$ thin films displayed a spontaneous ferromagnetic ordering with critical temperatures well above 100 K for Mn concentrations on the order of 5% ($x \approx 0.05$).[6–8] Further increases of Mn content, however, do not lead to larger hole concentrations or higher T_C due to carrier-compensation, clustering of Mn, or formation of intermetallic precipitates. On the other hand, post-annealing generally improves the magnetic properties.[35]

The initial report of T_C in Mn-doped Ge of about 114 K rivals those on Mn-doped GaAs.[14] Meanwhile, theoretical calculations based on the Zener model predicted that the Curie temperature of Mn-doped Si could even be higher.[6–8] It has become increasingly clear, however, that the kinetic solubility of Mn under low temperature growth conditions remains extremely small and that many of the magnetic properties should be attributed to intermetallic precipitates. An overview of those works is presented in the following sections.

7.4 Mn_xGe_{1-x}: A SILICON COMPATIBLE DMS

A thorough understanding of the ferromagnetic exchange mechanism at dilute doping levels is obscured by a number of obstacles. As we already pointed out in the previous sections, precise knowledge of the location and distribution of the magnetic impurities in the semiconductor host matrix is essential for understanding the origin of the observed ferromagnetism in DMS structures. This is particularly true because the exchange interaction strongly depends on the distance between — and local arrangements of — the Mn spins.[36] The RKKY interaction has always been an appealing concept in dilute magnetic systems, including the DMS, but strictly speaking it only applies to impurity spins in metallic systems. In the case of a lightly doped DMS, the cut-off length of the RKKY interaction is simply too short for observing the characteristic oscillatory switching between ferromagnetic and antiferromagnetic behavior.[37] Theories on DMS must take into account many other complicating factors such as carrier compensation and the imbalance between Mn impurity concentration and hole density, strong Coulomb interactions, strong disorder and localization, magnetic anisotropy and spin-orbit coupling, geometrical spin frustration, etc. RKKY theory and first-principles methods would thus seem poorly equipped to tackle these complex materials systems.

In our experimental reports during the past several years, we have attempted to address many of these issues. Empirically, T_C tends to increase with increasing doping concentration.[14,38] This observation instigated many studies in the high doping limit, motivated by the potential technological implications. However, one might argue that *fundamental* studies should instead begin in the dilute limit. The striking correlation between electrical transport and magnetic behavior underscores the importance of investigating vastly different concentration regimes. Furthermore, while both Ge and GaAs are fairly simple conventional semiconductors, their magnetic and transport properties reveal intriguing differences. This has motivated us and many other groups to shift attention to the group IV DMS. Studies of group IV DMS not only complement those of the much more extensively studied III–V systems but also suggest interesting analogies with other doped materials systems such as complex oxides.

Mn is notoriously insoluble in Ge, meaning that doping requires a non-equilibrium synthesis approach. Molecular beam epitaxy (MBE)[40–46] and ion implantation[47–50] have so far been the most widely used techniques. Molecular beam epitaxy involves co-deposition of Mn and Ge onto an atomically clean Ge (100) substrate under ultrahigh vacuum conditions. While MBE growth of high mobility devices is typically carried out under conditions that are much closer to thermodynamic equilibrium, i.e., higher temperature and thus producing superior crystalline quality, MBE growth of DMS is performed at much lower temperature to prevent the formation of precipitates. The art of a good MBE experiment for DMS growth is to find the optimum compromise that minimizes precipitate formation and maximizes crystalline quality by properly adjusting the temperature and growth rate. Ion implantation combined with solid-phase epitaxial growth (SPEG) is a low-temperature, kinetically limited process aimed at overcoming the thermodynamic limit imposed by solid solubility. In SPEG, layer-by-layer recrystallization of an implant-amorphized and -doped layer occurs at relatively low temperature such that local crystalline order is restored without the long-range diffusion that permits clustering or precipitation. Under such kinetically constrained conditions, SPEG can lead to "solute trapping" of impurities at concentrations that are well above their solubility limit — up to a few percent.

Recently, subsurfactant growth[38] has been demonstrated as a new method to achieve homogeneous doping profiles in the dilute limit. The general idea is to convert the menacing insolubility of Mn into a virtue by putting it to use as a surfactant during MBE growth. In these experiments, a submonolayer of Mn is deposited first at low temperature while multilayers of Ge are deposited next at a higher temperature. During the first step, Mn atoms quickly find their way to a favorable interstitial location right underneath the Ge dimers of the Ge $(100)2 \times 1$ surface reconstruction (hence the terminology "subsurfactant"). Because of its insolubility, Mn will stay near the growth front during the Ge epi-layer growth.

Substitutional doping is then accomplished when some of the subsurfactant atoms are inevitably trapped at substitutional locations as they try to find their way to the growth front. The main difference with traditional MBE is that the latter involves co-deposition of Mn and Ge at some elevated temperature whereas subsurfactant epitaxy involves sequential deposition steps at two different temperatures. For details, see Ref. 38. In the following section, we will mostly concentrate on the transport and magnetic properties of Mn-doped Ge (100) films, grown by MBE and by subsurfactant epitaxy. During the past several years, the general understanding and appreciation of the complex morphology issues related to doping inhomogeneity and intermetallic precipitates has advanced tremendously. The reader who is willing to make an extensive literature search in this subject should be warned that the conclusions regarding the magnetic and transport properties of these materials drawn in many earlier studies later turned out to be related to their complex nanoscale morphology.

7.4.1 Magnetic Properties

It is well known that solid phase epitaxy of Mn on a Ge (111) substrate can produce epitaxial metallic ferromagnets such as Mn_5Ge_3 ($T_C = 296$ K)[15] and $Mn_{11}Ge_8$ ($T_C = 270$ K).[15] Similarly, a number of studies have shown that MBE co-deposition of Mn and Ge on Ge (100) substrates often results in Mn_5Ge_3, $Mn_{11}Ge_8$, Mn_5Ge_2 or several other unknown precipitates when growth experiments are conducted near ~100°C or above.[27,51–53] The Mn_5Ge_3 precipitates can easily be identified from the shape of the magnetization curve, $M(T)$; their fingerprint is a T_C at or close to room temperature (Fig. 7.3). Given the intrinsic complexity of DMS materials with regard to their transport and magnetism, and specifically the deeper meaning of $\partial^2 M/\partial T^2$ in DMS theory (see, e.g., Refs. 6–8, 31, 32), it is highly desirable to avoid or at least minimize the formation of these precipitates. Carefully controlled studies have shown that these precipitates may be greatly reduced by co-depositing Mn and Ge at around 70°C substrate temperature for nominal Mn concentrations in the range from 1–10%. Further reduction of temperature will result in severe loss in the overall single-crystal quality.[26] As we will see, there is very little room for obtaining precipitate-"free" DMS, even when doping levels are reduced by an order of magnitude.[54] The temperature window for obtaining

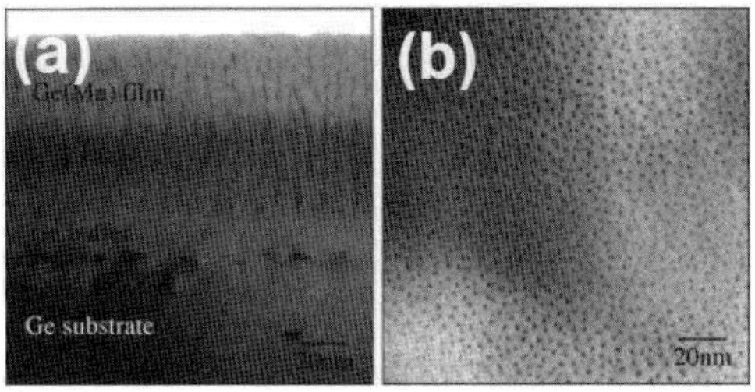

Figure 7.3. Temperature dependent magnetization of as-grown $Mn_{0.05}Ge_{0.95}$ films for various growth temperatures T_S. The applied magnetic field is 0.1 T for all measurements.[27]

precipitate-free DMS is also extremely small, maybe of the order of 30°C or even less.[27] In fact, many of the studies claiming the formation of a precipitate-free material (including our own),[26] later turned out to exhibit significant phase separation at the nanoscale.[55]

The low-temperature growth recipe greatly reduces the formation of intermetallic precipitates, but it does not necessarily (or generally) produce DMS samples that are free of dopant clusters. Figure 7.4 shows the transmission electron microscopy (TEM) image of a 70 nm thick DMS sample with a 5% Mn concentration grown at 80°C.[55] The vertical, dark, thin features are Mn rich "nanocolumns" that extend from the DMS/Ge buffer layer interface to the film surface. The average column diameter is 2 nm and their pair correlation length is 7 nm. According to the combined electron energy loss spectroscopy and ion channeling study, most of the Mn is incorporated into the nanocolumns while the surrounding Ge matrix is largely (though not completely) depleted of Mn. The apparent stoichiometry of the columns is $Mn_{17}Ge_{83}$ and their structure appears very disordered.[55] A very similar morphology was obtained in the work of Jamet *et al.*, who performed the growth at 130°C.[56] It is striking to observe that, although the nanoscale morphologies in both studies seem to be quite similar, the 50°C difference in the growth temperature leads to qualitative differences in the structure, composition and spatial distribution of the columns, as well as their magnetic (Fig. 7.3) and magneto-transport properties (which will be addressed below). Specifically, the average column diameter and spacing in the work of Jamet *et al.* are 3 nm and 10 nm, respectively, and the nanocolumn's composition seems to be close to that of an unknown more-or-less crystalline $MnGe_2$ phase.

In the following, we discuss the magnetic properties of a 70 nm thick DMS sample grown at 70°C with a 5% Mn concentration.[27] This sample should be very similar to the one by Li described above, which was grown at 80°C. Figure 7.5 shows the magnetization of the 70°C sample as a function of temperature in 0.1 and 1 T applied magnetic field normal to the film. The remanent magnetization is also shown. The sample shows no measurable magnetic response above 200 K, meaning that the sample does not contain Mn_5Ge_3 or $Mn_{11}Ge_8$ precipitates in amounts that could be detectable by SQUID magnetometry (see also Fig. 7.3). This in turn also suggests that the Mn-rich nanocolumns in the work of Li *et al.*[55] probably do not consist of a *known* intermetallic ferromagnetic phase. The saturation moment was determined from field-dependent magnetization isotherms and amounts to about 1 μ_B per Mn atom at 5 K. Comparing this number to the theoretical moment of 3 μ_B,[14,24] it would appear that only one-third of the Mn atoms are magnetically active. While the magnetization at 5 K is clearly hysteric, the system loses its ferromagnetic hysteresis above 12 K. So above this temperature, the magnetization is only field induced. Note that the concavity of $M(T)$ like the one shown in Figs. 7.3 and 7.5 is typically associated with strong coupling behavior.[57] However, the reader should be cautioned that this argument only applies to the spontaneous magnetization below T_C[57] and generally not to the field induced magnetization above T_C.

The high temperature tail of the d.c. susceptibility, which is defined as M/H, follows Curie–Weiss behavior. This becomes evident when plotting the inverse susceptibility or H/M versus temperature for different field values (Fig. 7.5, inset). This produces a set of straight lines crossing the temperature axis at 112 ± 5 K, which indicates a divergence of the *extrapolated* susceptibility. We denote this temperature as T_C^* to distinguish this extrapolated temperature from a macroscopically defined Curie temperature T_C.

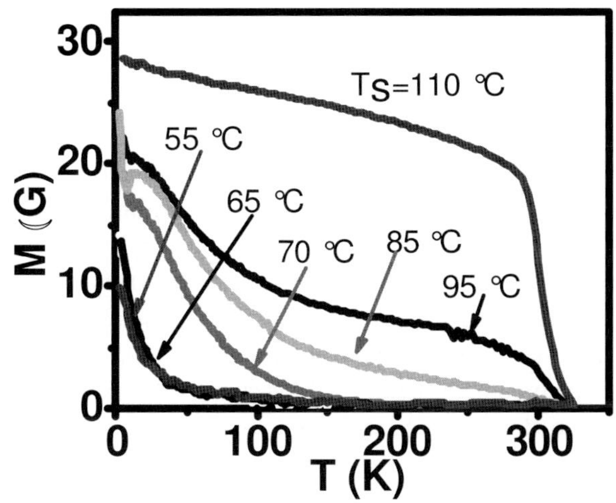

Figure 7.4. TEM images of an as-grown sample showing the formation of Mn-rich nanocolumns as a result of nanoscale phase separation. (a) Cross-sectional TEM image. (b) Plan-view TEM image.[55]

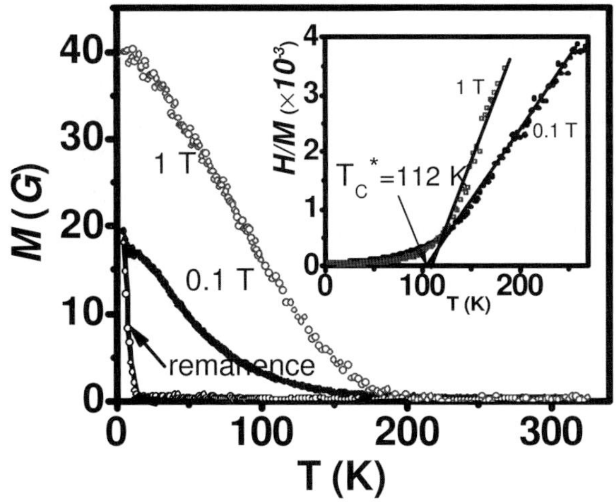

Figure 7.5. Remanence and field-dependent magnetization of a 5% Mn sample at 0.1 and 1 T magnetic field. The inset shows the Curie–Weiss extrapolation for this sample at the corresponding fields.[27]

The disappearance of remanence above 12 K suggests that global ferromagnetism vanishes. This is confirmed by a.c. susceptibility measurements.[27] A sharp cusp in the real part of the a.c. magnetization at 12 K coincides perfectly with the onset of the remanent d.c. magnetization. These observations indicate that global ferromagnetic order sets in at $T_C \ll T_C^*$, 12 K, for the 5% sample. Notice that the cusp temperature increases almost linearly with the nominal Mn concentration up to the 9% level. In contrast, T_C^* saturates at about 5%.[26]

The absence of remanent magnetization for $T_C < T < T_C^*$ begs the question whether T_C could be a super-paramagnetic blocking temperature, possibly related to the nanocolumnar structure shown in Fig. 7.4. According to Li *et al.*,[27] the field-dependent magnetization data above 40 K collapse onto the Langevin function of a super-paramagnet; however, the magnetization data below 40 K significantly deviate from this Langevin function. Notice that T_C is even much lower, 12 K, meaning that, strictly speaking, the system is neither ferromagnetic nor super-paramagnetic for $12\,\mathrm{K} < T < 40\,\mathrm{K}$. Furthermore, the super-paramagnetic blocking temperature T_B of *non-interacting* spin clusters should depend logarithmically on the measurement time, which is inversely proportional to probe frequency in the a.c. magnetic measurements. The magnitude of the factor $\Delta T_B / T_B \Delta(\ln f)$ in a.c. susceptibility measurements is typically of order 0.1 in super-paramagnets.[58] However, frequency-dependent measurements by Li *et al.*[27] did not reveal any measurable shift of T_C over three decades of frequency within the resolution of the experiment (0.2 K). This would be indicative of true ferromagnetism as opposed to super-paramagnetism, or it could indicate the presence of *strongly-interacting* magnetic clusters. On the other hand, Jaeger *et al.* did in fact report a shifting peak position with driving frequency with $\Delta T_B / T_B \Delta(\ln f) \approx 0.06$ at the 4% doping level.[59] Finally, considering the relatively low saturation fields of only a few Tesla, it is also unlikely that Mn_xGe_{1-x} should be categorized as a conventional spin-glass system because the saturation fields in typical spin-glasses are usually much higher.[60]

Samples similar to the one shown in Fig. 7.4 acquire significantly different magnetic properties upon post-annealing at 200°C for about 2 hours. First, the remanence survives much longer, up to about 125 K instead of the 12 K of the as-grown sample and the low-temperature saturation moment increases from 1.0 μ_B to 1.5 μ_B. Secondly, a.c. measurements exhibit additional peaks at higher temperatures, most notably a pronounced maximum at 125 K and a shoulder near 60 K which are shifting to higher and lower temperatures, respectively, under an applied d.c. field. These radical changes have been attributed to a changing of the local Mn content or compositional changes inside the nanocolumns. The increased Mn content in the nanocolumns was indeed indicated by ion-channeling experiments, suggesting a stoichiometry close to $Mn_{0.3}Ge_{0.7}$.[55]

A more complete picture of the phase diagram was provided by Devillers *et al.*,[61] who explored a wider range of temperatures and compositions. The formation of the nanocolumns is a manifestation of two-dimensional spinodal decomposition. At low temperature ($T < 120$°C), the nanocolumns have relatively small radii and more or less behave as weakly-coupled super-paramagnetic objects, generally producing a very weak ferromagnetic response. Mn_5Ge_3 nanoclusters typically form above 180°C according to Devillers,[61] or 80°C according to Li *et al.*[27] At intermediate growth temperatures, nanocolumns and nanocrystals can appear in a variety of forms, with some samples producing an amazingly strong ferromagnetic response with T_Cs in excess of 400 K.[61] Generally speaking, the overall magnetic response of these MBE grown Mn_xGe_{1-x} systems contains signatures of a super-paramagnetic blocking transition somewhere between 200 K and 300 K, which is associated with the presence of Mn_5Ge_3 or similar intermetallic nanoclusters, and a ferromagnetic or spin-glass freezing transition near 12 K, which should be attributed to the presence of the nanoclumns.[59] This is qualitatively illustrated in Fig. 7.6. Devillers *et al.*[61] classified the magnetic phase diagram roughly into four different regimes, namely (i) a dilute paramagnet; (ii) mixed phase with super-paramagnetic intermetallic precipitates and weakly ferromagnetic nanocolumns; (iii) high T_C nanocolumns; and (iv) Mn_5Ge_3. The surprisingly strong ferromagnetism ($T_C > 400$ K) of nanocolumns grown in a narrow temperature window around 130°C remains to be explained.[61]

7.4.2 Transport Properties

7.4.2.1 Zero field conductivity

The electronic configuration of an individual Mn atom greatly depends on whether it occupies a substitutional or interstitial location inside the host matrix. A substitutional Mn atom produces a deep acceptor level at 160 meV above the Ge valence band[28] while interstitial Mn probably behaves as a double donor.[62] This has great consequences for the carrier type and density and its possible role in mediating long-range ferromagnetic interactions. In principle, this can be studied systematically by varying the relative ratio of the substitutional and interstitial Mn atoms, for instance, via a post-annealing experiment.[55] Here, we discuss the magneto-transport properties of the as-grown versus the post-annealed samples.

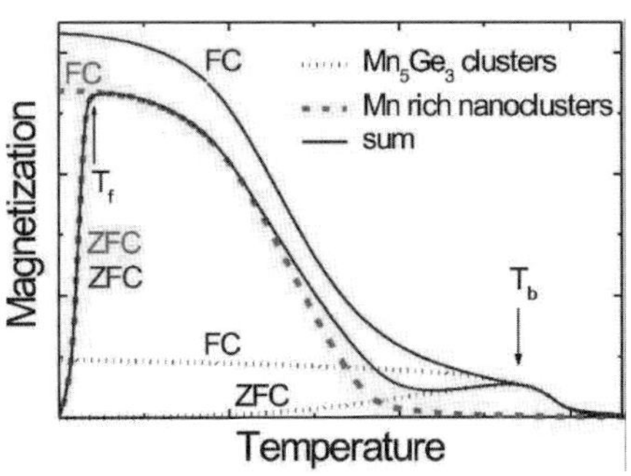

Figure 7.6. Qualitative illustration of the combined ZFC and FC magnetization of two kinds of nanoscale clusters. Mn_5Ge_3 clusters (black dotted curves) yield the blocking temperature T_B; Mn-rich clusters (red dashed curves) dictate a freezing temperature T_f.[59]

The zero-field resistivity $\rho(T)$ of as-grown Mn_xGe_{1-x} thin films shows an overall insulating behavior up to the 9% doping level, except perhaps for the presence of a shallow dip just below T_C (Fig. 7.7). The local maximum in the resistivity at T_C does not represent a metal–insulator transition, however, because the resistivity still diverges as $T \rightarrow 0\,\mathrm{K}$ [Fig. 7.7(b)]. Even though a similar feature has been observed in the $Ga_{1-x}MnAs$ system, there the resistivity generally does not diverge except in the very dilute limit.[63,64] Moreover, the Mn_xGe_{1-x} sample with highest Mn concentration (9%) no longer exhibits this dip, showing only a shoulder around T_C. If this feature were associated with a metal–insulator transition, then one would have expected the 9% sample to be metallic because, generally speaking, DMS systems should become metallic at such degenerate doping levels.

The insulating behavior at low temperature is usually interpreted in terms of a variable range hopping mechanism or a variant thereof. If the Fermi level is located below the mobility edge of an Mn-induced impurity band, then the transport mechanism is thermally activated (or phonon assisted). Generally, for doped semiconductors, the temperature dependent resistivity is given by[65]

$$\rho(T) \propto T^{\alpha} \exp[(T_0/T)^{\beta}], \tag{7.1}$$

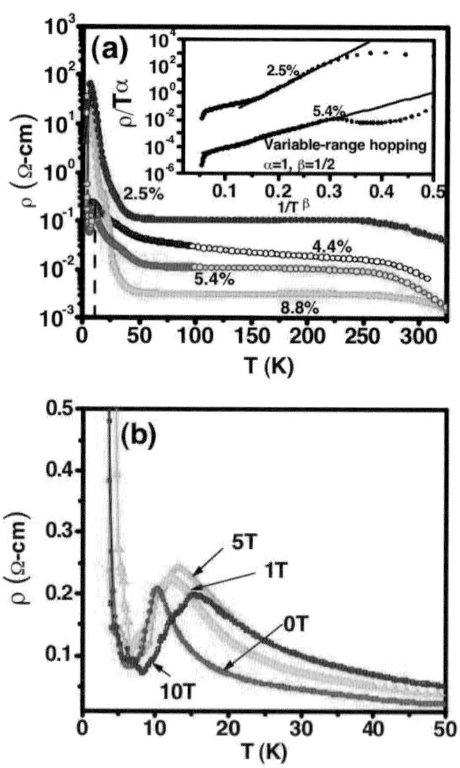

Figure 7.7. (a) Temperature-dependent resistivity of as-grown (70°C) $Mn_{0.05}Ge_{0.95}$ with various doping concentrations. Inset shows the fitting to the Efros–Shklovskii hopping formula. (b) Resistivity in various magnetic fields, showing a progressive change in the peak position.[27]

where α and β are a non-universal power exponent and a hoping exponent, respectively. The best fit to our data above T_C [inset Fig. 7.7(a)] indicates that $\alpha = 1$ and $\beta = 1/2$, suggesting Efros–Shklovskii hopping over approximately 1–2 decades of resistivity. Finally, note the progressive change of the peak position in an applied magnetic field [Fig. 7.7(b)], indicating the close connection between the resistivity and magnetization.

7.4.2.2 Giant magneto-resistance

The magneto-resistance, defined as $\Delta\rho/\rho_0 = [\rho(H) - \rho(0)]/\rho(0)$, shows a very striking contrast between as-grown and post-annealed samples.[55] As-grown samples exhibit a giant positive magneto-resistance between 50 and 200 K [Fig. 7.8(a)]. Post-annealed samples, however, exhibit both positive and negative contributions to the magneto-resistance, the negative one dominating at high magnetic field [Fig. 7.8(b)]. The absolute magnitude of the magneto-resistance effect is also dramatically lower. The magneto-resistance of the post-annealed samples is typical of a metallic DMS system and can be attributed to a reduction of spin disorder scattering in the presence of a magnetic field.[66]

The dramatic contrast between the positive giant magneto-resistance of the as-grown samples and the weak negative magneto-resistance of the post-annealed samples has been attributed to the conversion of interstitial Mn into substitutional Mn. Ion-channeling experiments on the as-grown sample (with nominal Mn concentration of 5%) indicated that about 24% of the Mn atoms are substitutional, 12% are interstitial, and the remaining 64% are incorporated into the nanocolumns.[55] Because substitutional Mn is a single acceptor and interstitial Mn presumably a double donor,[62] the n- and p-type carriers in the as-grown samples should be perfectly compensated and the system should be strongly insulating. On the other hand, ion-channeling results from the post-annealed samples indicate that post-annealing converts nearly all of the interstitial Mn into substitutional Mn (the stoi-

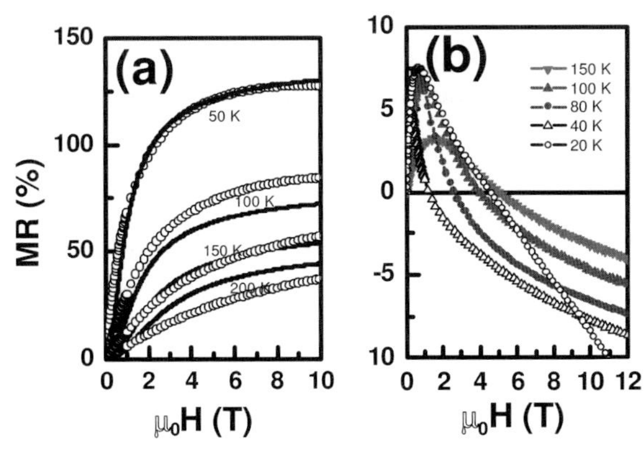

Figure 7.8. Magneto-resistance of an (a) as-grown and (b) post-annealed $Mn_{0.05}Ge_{0.95}$ sample.[55]

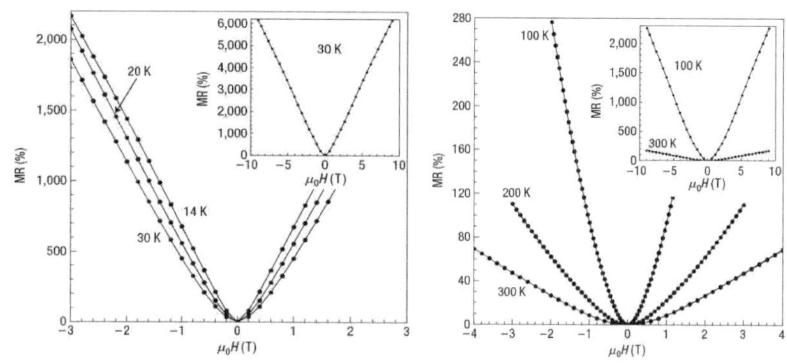

Figure 7.9. Magneto-resistance for perpendicular magnetic fields. (a) Low temperature data with inset showing high-field behavior at 30 K. (b) High temperature data with inset showing high-field behavior at 100 and 300 K.[56] (Reprinted by permission from Macmillan Publishers Ltd: Nature Materials.[56])

chiometry and size of the nanocolumns also changes). The net result is an increase in the substitutional fraction to about 1.8%, thus producing a hole density of about $8 \times 10^{20}\,\text{cm}^{-3}$. The most likely explanation of the negative magneto-resistance after post-annealing is that the system undergoes an insulator-to-metal transition in the Mn-induced impurity band, as a result of the substitutional incorporation of Mn. This notion would also be consistent with the metallic behavior of samples grown at 130°C.[56] Notice, however, the 130°C material of Jamet *et al.* exhibits an unexplained *giant orbital magneto-resistance* (Fig. 7.9), which unlike the results of Li *et al.* (Fig. 7.8) appears to be totally unrelated to the magnetization. This of course reinforces the notion that the magneto-resistance is extremely sensitive to the dopant distribution.

The giant positive magneto-resistance of the as-grown material in Figs. 7.8(a) and 7.9 is intriguing and it would be highly desirable to acquire deeper understanding of the underlying mechanism. In the following, we will show that the giant positive magneto-resistance of Fig. 7.8(a) may be the result of magnetic anisotropy and antiferromagnetic interactions that are associated with the columnar nanophase morphology (the giant orbital magneto-resistance in Fig. 7.9 still remains unexplained).

The resistivity of as-grown Mn_xGe_{1-x} samples does not exhibit a significant exponential temperature dependence over a wide range of temperatures, $50 < T < 200\,\text{K}$. An obvious explanation is that at these temperatures Mn acceptors are fully ionized and band-to-band excitations are still rare so that electrical transport must be due to free holes.[55,67] In this temperature range, the hole density should be more or less constant and approximately equal to the density of uncompensated (i.e., substitutional) Mn acceptors. In this case the magneto-resistance can be attributed to a field-dependent hole mobility, which in turn is determined by their scattering off randomly distributed ionic spins S having a volume concentration xN, where N is the number of Ge atoms per unit volume. The scattering arises from an exchange interaction of strength $\beta N \approx 1$ eV. Following Refs. 68 and 69 and neglecting magneto-crystalline anisotropy of the system one can find the scattering probability[55,67]:

$$\frac{1}{\tau_k^{\pm}} = \frac{mk}{\pi\hbar^3}\left(\frac{\beta}{2g\mu_B}\right)^2 \left\{ \begin{array}{l} k_B T \left[\chi_{\parallel} + 2\chi_{\perp} F^{\pm} \left(1 \pm \dfrac{2m\Delta}{\hbar^2 k^2}\right)^{1/2}\right] \\ + (g\mu_B)^2 N x (1-x) \langle S_z \rangle^2 \end{array} \right\}, \qquad (7.2)$$

where $\tau_k^{\pm}$ is the scattering time calculated within the first Born approximation for a free carrier with wave vector k and spin up (+) or down (−), assuming a simple parabolic dispersion with effective mass m. Here g is the Landé factor of the ionic spin, and μ_B is the Bohr magneton. Also, $\langle S_z \rangle = SB_S(S\alpha) = M(H)/g\mu_B x N$ is the thermodynamically averaged projection of the ionic spin onto the direction of magnetic field with $M(H)$ being the magnetization and $B_S(S\alpha)$ the Brillouin function, where $\alpha \approx g\mu_B H/T$ in the paramagnetic phase; $\chi_{\parallel} = \partial M/\partial H$ and $\chi_{\perp} = M/H$ are the longitudinal and transverse magnetic susceptibilities, respectively. In Eq. (2), $F^{\pm} = 1$ if the Zeeman splitting of the valence band $\Delta \approx xN\beta SB_S(S\alpha) \leq \hbar^2k^2/2m$, otherwise $F^{+} = 0$ and $F^{-} = 0$.

The first two terms in Eq. (7.2) describe the scattering by thermal fluctuations of the ionic spins and the last one represents scattering by spatial fluctuations of the local concentrations of magnetic ions. For ordered magnetic semiconductors, such as EuSe or ErAs, where magnetic ions form a regular lattice ($x = 1$), Eq. (7.2) coincides with the expression obtained by Haas.[69] In the absence of thermal spin fluctuations when $\chi_{\parallel,\perp} = 0$ we recover the expression for the scattering probability in disordered nonmagnetic alloys.[70]

Given the scattering probability one can calculate the mobilities $\mu^{\pm}$ of the free carriers in the spin-split sub-bands.[69] Then for the sufficiently large magnetic fields, such that the Zeeman splitting $\Delta > T$, taking into account the majority-spin carriers only one can estimate the magneto-resistance of the non-degenerate magnetic semiconductor in the impurity-depletion temperature region as[55,67]

$$\frac{\Delta\rho}{\rho(0)} \sim \frac{\mu(0)}{\mu^{+}(H)} - 1 = \frac{k_B T[\chi_{\parallel}(H) + 2\gamma\chi_{\perp}(H)] + (g\mu_B)^2 N x(1-x)\langle S_z \rangle^2}{k_B T(1+2\gamma)\chi_{\parallel}(0)} - 1. \qquad (7.3)$$

In deriving Eq. (7.3) from Eq. (7.2) we have introduced a phenomenological magneto-crystalline anisotropy factor $\gamma < 1$. It can be seen that the terms proportional to $\chi(H)$ decrease with the magnetic field H, thus increasing the mobility due to the suppression of the thermal spin fluctuations, and giving rise to a negative contribution to the magneto-resistance. Simultaneously, scattering by the spatial fluctuations of the magnetic ions (last term) increases with H since the magnitude of the random Zeeman splittings increases with H, thus decreasing the mobility and giving rise to a positive term in the magneto-resistance. Our analysis shows that, in the absence of magnetic anisotropy ($\gamma = 1$), contrary to conclusions made in Ref. 71, these competing spin-scattering mechanisms must lead to a small but *negative* magneto-resistance.

In the temperature range 50 K to 200 K, the magnetization data of the as-grown samples collapse onto the Langevin function $L(y)$ (see Ref. 27). This is a signature of a superparamagnetic system with the effective spin of the magnetic clusters $S \gg 1$. For such a system, $B_S(y) \rightarrow L(y)$ with $y = Sg\mu_B H/T$ and, according to Eq. (7.3) magneto-resistance $\Delta\rho/\rho(0) = 0$ if $\gamma = 1$. However, an account of (dipole–dipole) antiferromagnetic coupling between magnetic clusters would lead to positive magneto-resistance even if $\gamma = 1$. Indeed, if we replace the argument of $L(y)$ with $y = Sg\mu_B H/(T + T_{AF})$, where $T_{AF} > 0$ is the antiferromagnetic temperature,[20] we obtain a positive magneto-resistance, $\Delta\rho/\rho(0) = L^2(y)T_{AF}/T$. The introduction of the magneto-crystalline anisotropy factor $\gamma < 1$, which suppresses the transverse thermodynamic fluctuations of the ionic spins, has a similar effect on the magneto-resistance. Figure 7.8(a) shows a fit to the high-temperature magneto-resistance data, using Eq. (7.3) with $T_{AF} = 15\,\text{K}$ and $\gamma = 0.35$. While the agreement is not perfect, the theoretical expression Eq. (7.3) works amazingly well, considering the fact that it simultaneously fits all four magneto-resistance isotherms over a very wide temperature range, reproducing both the magnitude and general shape of the isotherms with only two parameters: T_{AF} and γ.

The following picture thus emerges.[55] The nanocolumns are locally ferromagnetic. The magnetic dipolar coupling between the columns competes with the carrier mediated exchange interactions. Due to the near perfect compensation of the host matrix, the carrier mediated exchange is very weak and the dominant dipolar coupling most likely favors an approximate antiferromagnetic alignment between the nanocolumns. However, magnetic remanence remains finite near 0 K due to the geometrical magnetic frustration in the antiferromagnetic array of nanocolums. Post-annealing increases the substitutional Mn fraction and hole concentration and the dipolar coupling is easily overwhelmed by the carrier mediated exchange. Consequently, the positive magneto-resistance is strongly suppressed, leading to a weak negative magneto-resistance, characteristic of carrier scattering by thermodynamic spin fluctuations.

7.4.2.3 Anomalous Hall effect

Studies of the anomalous Hall effect generally provide valuable information on the transport mechanisms. For a metallic ferromagnet, the Hall resistivity is given by

$$\rho_{\text{Hall}} = R_0 B + \mu_0 R_s M, \tag{7.4}$$

where B is the magnetic induction and R_0 and R_s are the normal and anomalous Hall coefficients, respectively. Here, we will apply a similar phenomenological expression to the Mn_xGe_{1-x} DMS system and identify the roles of R_0 and R_s in the nonmetallic regime.

For small induced magnetization values and thin film geometry, the applied field H_a will be simply denoted by $H = B/\mu_0$ and will be equal to the field inside the sample. So, for a generally field-dependent R_0 and R_s we write

$$\rho_{\text{Hall}}(H) = R_0(H)\mu_0 H + \mu_0 R_s(H) M. \tag{7.5}$$

As expected from the second term, the Hall resistivity of the as-grown DMS sample rises quickly with the applied field at temperatures below T_C^*. This is due to the large field-induced magnetization overwhelming the normal Hall effect (Fig. 7.10).[27] $R_s(H)$ was obtained by dividing the Hall resistivity by the magnetization, thus ignoring $R_0(H)$ for the time being. For our nonmetallic 5.4% sample grown at 70°C, it was found that $R_s(H)$ varies linearly with $\rho(H)$ between 20 and 200 K (Fig. 7.11).[27]

Because the slope of $\rho_{\text{Hall}}(H)$ is more or less constant above ~5 T, it is possible to extract the normal Hall coefficient R_0. Still, one must take into account the field dependence of both $M(H)$ and $R_s(H)$. Experimentally, $M(H)$ appears to saturate at high fields but $R_s(H)$ is strongly field-dependent because of its dependence

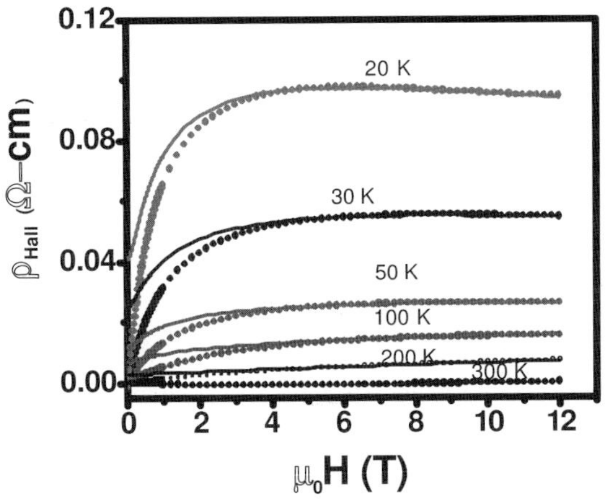

Figure 7.10. Field-dependent Hall resistivity of the as-grown $Mn_{0.05}Ge_{0.95}$ sample at various temperatures. Solid lines show fittings to Eq. (7.5) (see text).[27]

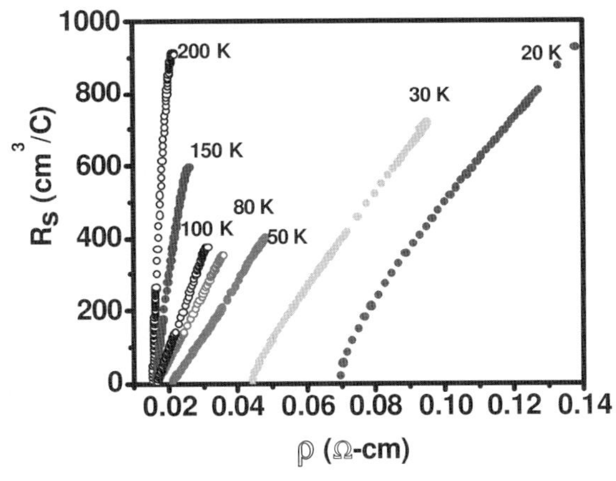

Figure 7.11. Linear relation between the anomalous Hall coefficient $R_s(H)$ and longitudinal resistance $\rho(H)$ at various temperatures for the as grown $Mn_{0.05}Ge_{0.95}$ sample.[27]

on $\rho(H)$ or the magneto-resistance, which is indeed very large for the as-grown material, as we discussed before. Based on the observation that $(\partial\rho_{\mathrm{Hall}}/\partial H)_T$ is constant at *high field*, we infer that up to first order, R_0 should be independent of the field and that $R_s(H)$ should be linear in $\rho(H)$ and H:

$$\left(\frac{\partial\rho_{\mathrm{Hall}}}{\partial H}\right)_T \cong \mu_0 R_0 + \alpha'\mu_0 \left(\frac{\partial\rho}{\partial H}\right)_T M_T, \tag{7.6}$$

where M_T is the saturation magnetization at temperature T. $R_s(H = 0)|_T$ can be obtained by extrapolating the linear segments of $\rho_{\mathrm{Hall}}(H)$ to zero field for a given temperature. The validity of this extrapolation can be cross checked by fitting $\rho_{\mathrm{Hall}}(H)$ using the experimental values of $M(H)$, $\rho(H)$ and R_0. As expected, fitting works pretty well at high fields (Fig. 7.10) while the deviation of the low field data points from the fit must be attributed to the complicated field dependence of the magnetization and transport properties. The outcome of this exercise is a plot of the normal and anomalous Hall coefficients as a function of temperature (Fig. 7.12). The positive sign of R_0 indicates hole conduction at all temperatures. Another important conclusion is that the sign reversal of $(\partial\rho_{\mathrm{Hall}}/\partial H)_T$ at high fields (i.e., the slope going from positive at high temperature to negative at low temperature), could easily have been misinterpreted as sign reversal of the normal Hall coefficient if the large magneto-resistance had not been taken into account.

The temperature dependence of R_0 and $R_s(H = 0)$ supports the variable-range hole hopping picture because both diverge as $T \to 0$ K.[65,72] Furthermore, R_0 and $R_s(H = 0)$ roughly follow the leading hopping term $\exp(T_0/T)^{1/2}$ between T_C and 50 K (Fig. 7.12).

7.4.3 Generalized Polaron Percolation Picture

It is evident that the electrical transport properties are closely related to the magnetic properties while both are strongly connected to the intriguing nanophase morphology of the Mn_xGe_{1-x} material. On the magnetic side, it is clear that T_C^* ought to be interpreted as some "cluster temperature" or paramagnetic blocking

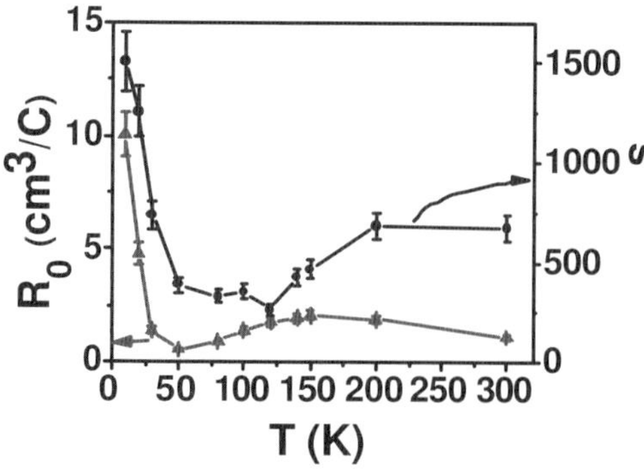

Figure 7.12. Normal (R_0) and anomalous Hall coefficient $R_s(H)$ versus temperature.[27]

phenomenon related to the presence of intermetallic precipitates,[59] while $T_C \ll T_C^*$ clearly implies the existence of long-range ferromagnetic order, which in turn seems to be related to the coupling between nanocolumns.[55] The p-type transport is thermally activated and also shows a clear connection with the magnetic properties via the magneto-resistance, anomalous Hall effect, and the existence of a local resistivity maximum at T_C [Fig. 7.7(b)]. The magnetic properties and magneto-resistance are moreover strongly dependent on the Mn substitutional fraction in the crystalline host matrix. Clearly, magnetism and transport must be intimately related but the fundamental issues regarding the detailed nature of the exchange coupling mechanism have become blurred due to the effect of the precipitates. Nonetheless, the key ingredients describing the physics of these materials probably should include the impurity band concept and likely the possibility of magnetic cross talk between intermetallic and or nanocolumnar clusters, which is somehow mediated by p-type carriers.

In this context, one could think of generalizing the magnetic polaron picture of Das Sarma and co-workers[57,31] by redefining the concept of a polaronic cluster. Strictly speaking, a polaronic seed contains a single hole that is magnetically bound to an impurity spin at a temperature scale that is defined by the exchange integral J_{pd}, thus forming a polaronic object. Alternatively, if the dopant distribution were inhomogeneous, T_C^* would be the effective ferromagnetic transition temperature of a "physical" cluster containing several holes. The physical clusters may be viewed as regions with excessive acceptor concentrations containing "puddles" of bound holes, which would have to tunnel through regions with depleted acceptor concentrations in order to contribute to the electrical transport. Intermetallic precipitates and nanocolumns are of course also good examples of such physical clusters. As the temperature drops below T_C^*, these polaronic or physical clusters increase their "magnetic interaction radius" and eventually percolate to form an infinite ferromagnetic "cluster" at $T_C = 12\,\mathrm{K}$ ($x = 0.05$). This ferromagnetic percolation would establish a transport path with a strongly diminished exchange contribution to the hopping barriers.[32]

For strongly localized carriers and low doping concentrations, the size of a bound magnetic polaron grows logarithmically slowly as the temperature decreases.[32] The resistivity will very much depend on the actual percolation scenario and hopping trajectories of the hole, which would probably be sample dependent, but it is expected to increase monotonically upon cooling as the vanishing exchange barrier is offset by the decreased hopping probability at lower temperature. Alternatively, at higher doping levels, ferromagnetic percolation and transport percolation do coincide if dopants are grouped into clusters, leading to an abrupt reduction of the transport activation barrier (or vanishing exchange barrier) just below T_C.[32] The latter scenario is perfectly consistent with the observed dip in the resistivity below T_C. The field-induced shift of the resistivity maximum in Fig. 7.7(b) further corroborates the close connection between magnetic and transport percolation, as in other doped polaronic systems where inhomogeneities play an important role.[73]

7.4.4 Ultra Dilute Regime of Mn_xGe_{1-x} DMS

So far, most of our discussion has been based on DMS samples with Mn concentrations around 5%, which in terms of semiconductor physics and devices seems quite excessive. Only very few studies have focused on the extremely dilute limit. However, studies in the extreme dilute limit may provide better understanding of the magnetic and transport properties because there are far fewer precipitates. Here, we will discuss two methods for obtaining Mn concentrations of about 0.25%, namely, subsurfactant epitaxy (as discussed in Sec. 7.4.4.1) and conventional MBE using extremely low rates of Mn effusion (Sec. 7.4.4.2).

7.4.4.1 Subsurfactant epitaxy

The motivation behind this work is two-fold. The general idea is based on the so-called "surfactant" effect known from semiconductor hetero-epitaxy,[74] and in the particular case of Mn in Ge, the approach was inspired from first-principles total energy calculations.[75] The calculations predict that up to 0.5 monolayer of Mn can easily be trapped in the interstitial locations right underneath the Ge–Ge dimers on the Ge (100)2×1 surface reconstruction. If the Mn deposition is done at low temperature (e.g., 150 K), then the trapped Mn atoms cannot hop back to the surface because this process requires a rather high activation energy.[38] Indeed, STM images clearly show that while room temperature deposition of Mn results in the formation of Mn clusters on the surface, low-temperature deposition yields a fairly smooth surface, showing clear indications that the Mn atoms are trapped under the surface. Once the atoms are trapped, the sample can be safely warmed up to 85°C. As already mentioned, this temperature is high enough for epitaxial Ge growth. As new epitaxial Ge layers are grown, the subsurface interstitials must float towards the vacuum interface so as to maintain their identity and preferred status of subsurface interstitial. This floating phenomenon is also natural because Mn does not dissolve in bulk Ge. However, during this process a small fraction of the Mn atoms will be trapped inside the Ge matrix and most likely occupy the substitutional sites.[75] The upward floating process continues until the subsurface layer is depleted of Mn. According to secondary ion mass spectrometry (SIMS) depth profile analyses, an initial deposit of 0.5 monolayer at 150 K results in an amazingly flat vertical doping profile (Fig. 7.13). This resulting bulk concentration of Mn is $\sim$0.25%, showing no trace of Mn clusters or precipitates in scanning transmission electron microscopy imaging (Fig. 7.14). Using this method, it is not possible to increase the Mn content beyond 0.25% but one can reduce the nominal doping content by depositing less than 0.5 monolayer during the initial step.

The most striking property of this 0.5 monolayer subsurfactant sample is that it has a very high Curie temperature ($>$ 400 K). The saturation magnetization of the 0.25% sample at 2 K is about 2.8 μ_B per Mn, which is quite close to the theoretical value of 3 μ_B for a substitutional Mn atom.[14,25] The subsurfactant samples also exhibit a strong anomalous Hall effect that increases with the nominal Mn content, as expected for an intrinsic DMS (Fig. 7.15). The anomalous Hall effect can

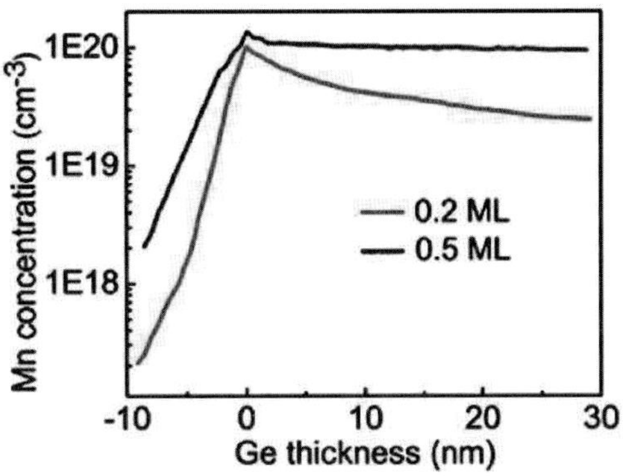

Figure 7.13. SIMS concentration profile of a subsurfactant specimen with initial Mn deposits of 0.5 and 0.2 monolayer made at 150 K. 0 nm indicates the location of the interface between the Ge buffer layer and epilayer.[38]

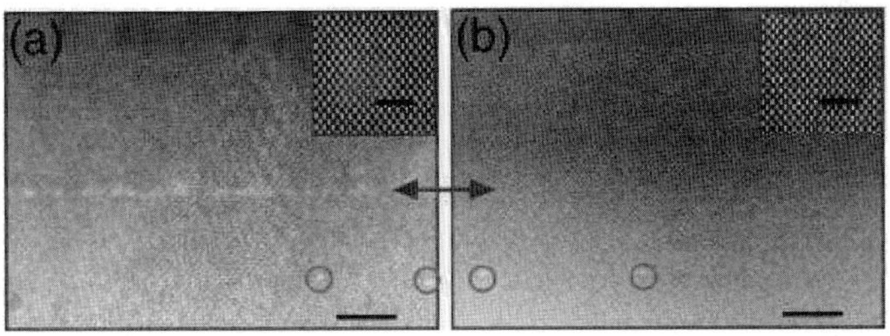

Figure 7.14. TEM images of a DMS thin film with (a) an initial Mn deposit done at room temperature, and (b) are initial Mn deposit done at 150 K. Insets show higher magnification images. The double pointed arrow indicates the interface between the buffer layer and Mn-doped epilayer. Circles track the line defects at the substrate/buffer layer interface. The scale bars correspond to 20 nm and 2 nm in the main images and insets, respectively. Both images show a thin capping layer on top.[38]

be observed up to at least 300 K [Fig. 7.15(b)], which is consistent with the magnetic measurements. In contrast, when the initial Mn deposit was done at room temperature, the resulting samples did not exhibit an anomalous Hall effect.

Figure 7.15(c) also shows the temperature-dependent resistivity of the 0.5 monolayer subsurfactant sample. The zero-field resistivity decreases with increasing temperature up to at least 60 K, indicating semiconducting behavior. Interestingly, the data below 2.5 K reveal a magnetic-field induced insulator–metal–insulator (I–M–I) transition. The temperature coefficient of the resistivity changes from negative at 0 and 0.2 T to positive at 1 and 5 T, and then back to negative at 10 and 14 T. The I–M–I transition is also reflected in the magneto-resistivity isotherm at 2 K, shown in the insert of Fig. 7.15(c). These data can be rationalized using the Shapiro phase diagram of an impurity band system,[77] assuming that the doping concentration is just below the critical value for the metal-insulator transition, as sketched at the bottom of Fig. 7.15(c). The I–M transition takes place when the

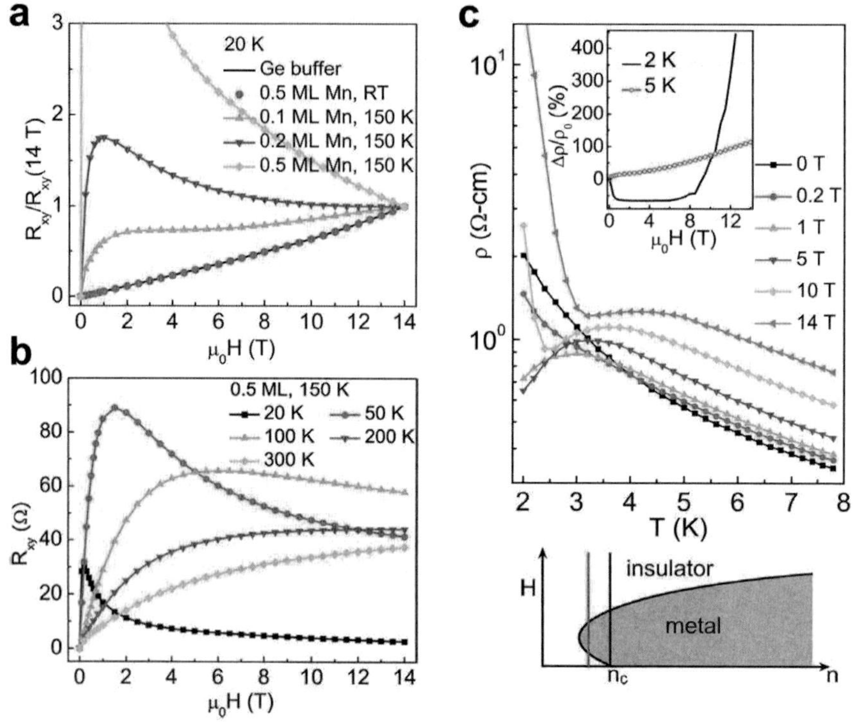

Figure 7.15. Magneto-transport. (a) Hall effect measured at 20 K for different initial Mn coverages and different growth temperatures. The Ge epi-layer thickness is 35 nm. The Hall effect of the Ge buffer layer is also shown for reference. (b) Hall effect of the low temperature sample with 0.5 monolayer initial coverage, measured at various temperatures. (c) Temperature-dependent resistivity of the low temperature sample in various magnetic fields. The inset shows the magneto-resistivity $\Delta\rho/\rho_0 = [\rho(H) - \rho(0)]/\rho(0)$, where $\rho(0)$ and $\rho(H)$ are the thin film resistivities in the absence and presence of the applied magnetic field H, respectively. The bottom panel shows the Shapiro phase diagram of an impurity band system near the metal–insulator transition in a magnetic field. The red line indicates the approximate carrier density of the low temperature (i.e., subsurfactant) sample.[76]

mobility edge drops below the Fermi level as the magnetic field increases from 0 T, while the subsequent M–I transition presumably originates from the shrinkage of the wave function at high field.[77] Using the Mott criterion $a_B n_C^{1/3} \approx 0.25$,[29] with estimated Bohr radius $a_B \approx 3\,\text{Å}$,[28] we infer the critical carrier density to be $n_C \approx 5.8 \times 10^{20}\,\text{cm}^{-3}$, higher than, but on the same order as the doping level of the 0.5 monolayer film ($\sim 1.0 \times 10^{20}\,\text{cm}^{-3}$) determined earlier. The I–M–I transition is absent for samples with an initial deposit of 0.1 and 0.2 monolayer (i.e., lower doping level), consistent with the impurity band picture and the Shapiro diagram.

7.4.4.2 MBE co-deposition in the extreme dilute limit

In the previous examples, we have seen that epitaxial growth of Ge (100) can be accomplished at temperatures as low as ~70°C. For Mn concentrations on the order of a few percent, MBE co-deposition undoubtedly results in the formation of various clusters and precipitates. On the other hand, the subsurfactant method described above has been successful for obtaining cluster-free DMS, at least at the level of our STEM detection capability. However, the *integrated* amount of Mn that can be incorporated into these samples is very small, at most 0.5 monolayer, which is at the borderline of the detection limit of various measurements. For instance, detailed studies of the location and distribution of Mn dopants in the subsurfactant samples are highly desirable, especially in light of the extraordinarily high T_C, which demands explanation. To this end, ion-channeling experiments could provide invaluable information but these measurements require at least an order of magnitude higher doping content.

One alterative possibility for exploring the dilute regime is to *slow down* the MBE growth of Ge while co-depositing Mn at a very slow rate. Because the temperature must be kept reasonably low to avoid the formation of MnGe precipitates, we have limited the Ge deposition rate to 1.3 Å/min. We adjusted the Mn deposition rate so as to obtain a 0.25% atomic ratio. This way, we produced one micron thick DMS films at 200°C and 140°C substrate temperatures and a ~600 nm thick film at 95°C. Reflection high energy electron diffraction (RHEED) patterns of the 200°C and 140°C samples indicated reasonable epitaxial quality but the 95°C sample was not as good. Interestingly, during the initial stages of growth (up to ~150 nm), the RHEED patterns were actually better at the lower growth temperature. This observation seems counterintuitive because higher substrate temperatures normally lead to better epitaxial quality. However, higher growth temperatures also imply a faster upward diffusion of the co-deposited or buried Mn atoms, which in turn could explain the deterioration of the RHEED patterns (due to roughening or clustering).

The magnetic data of these films reveal some systematic variations. The saturation magnetization of the 200°C sample is about 3 μ_B per Mn but it decreases with lower growth temperatures. The remanent magnetization is characterized by two regimes (Fig. 7.16): the magnetization initially reveals a steep drop as the temperature increases from 2 K to about 10 K. This drop is followed by a long tail up to room temperature, suggesting the formation of Mn_5Ge_3 precipitates. Hence, the magnetization at ~10 K (i.e., immediately following the initial drop off) is expected to correlate with the volume of the ferromagnetic precipitates. This precipitate signal drops by about a factor of three for the 140°C sample and is almost nonexistent in the 95°C sample. The magnetization $M(T)$ of the 200°C sample, measured at 1 T, is clearly convex ($\partial^2 M/\partial T^2 < 0$). It becomes concave for the 95°C sample and it is quite linear over a wide temperature range for the 140°C sample.

The zero field cooled (ZFC) data of the 200°C and 140°C samples taken at 100 G field show two clear peaks (Fig. 7.17): one at low temperature and one at high temperature. These peaks are again attributed to the formation of a glassy spin state at low temperature and the superparamagnetic blocking of Mn_5Ge_3 clusters at high temperature.[59] As opposed to the 200°C and 140°C samples, the 95°C sample does not exhibit clear contrast between FC-ZFC magnetization, nor does

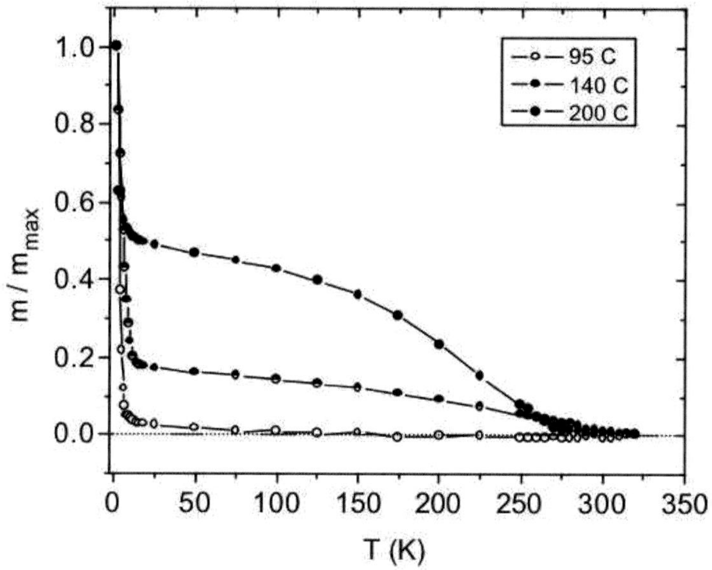

Figure 7.16. Temperature dependent remanent magnetization of DMS films grown at 200°C, 140°C, and 95°C. The 200°C and 140°C samples are ~1000 nm thick and the 95°C sample is ~650 nm thick.[78]

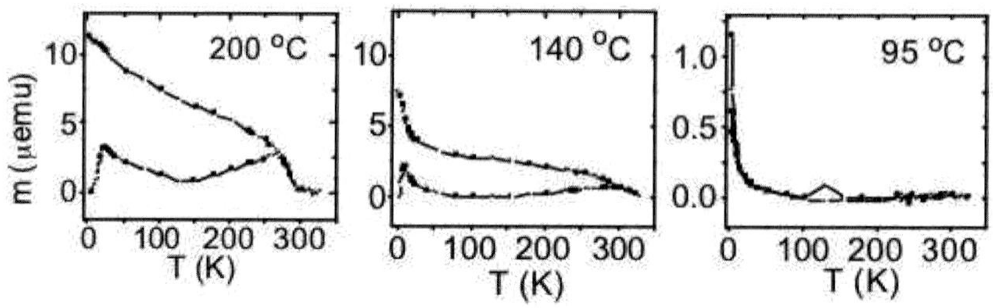

Figure 7.17. FC and ZFC magnetic moments at 100 G applied field parallel to the samples. The growth temperatures are (a) 200°C, (b) 140°C, and (c) 95°C. The 200°C and 140°C samples are ~1000 nm thick. The 95°C sample is ~650 nm thick.[78]

it show the high temperature blocking phenomenon of the ZFC state. All of the above observations indicate a reduction of precipitate formation at lower growth temperatures with the 95°C sample so far being the most promising candidate for future ion-channeling studies. Furthermore, the observations imply that clustering seems almost unavoidable above 95°C, even at very dilute Mn concentrations.

7.5 CONCLUDING REMARKS

In this review, we have addressed many of the fundamental physics issues surrounding the mechanisms of electrical transport and exchange coupling in group IV DMS, as well as the many complicating materials issues arising from the extremely low kinetic and thermodynamic solubility of Mn in Ge. Non-equilibrium growth techniques such as MBE co-deposition and ion implantation do increase the substitutional incorporation of Mn beyond the solid solubility limit but also produce unwanted precipitates. Some of these spinodal precipitates, namely, the columnar nanostructures, are very intriguing in their own right. Their spatial self-organization, magnetic properties and orbital magneto-resistance[56] are quite amazing and still require explanation.

Since the initial reports of ferromagnetism in Ge by Park *et al.*,[14] this field has grown very rapidly and we regret that we had to make a rather narrow selection in choosing our references. Even the number of most closely related references that appeared after 2002, i.e., the ones dealing specifically with the Mn doping of group IV materials, is already approaching 200 at the time this review was written. Furthermore, the reader probably noticed that little attention has been given to Mn doping in silicon. It is the authors' opinion that magnetic doping of Si will be much harder to accomplish than doping of germanium. First of all, whereas Mn prefers to be at a substitutional location in Ge, it prefers to be an interstitial atom in Si.[79,80] In fact, Mn is known to be a rapid interstitial diffuser in Si which facilitates the formation of silicide precipitates. MBE growth of silicon also requires higher growth temperatures, which would make it all the more difficult to avoid precipitates. In this context, it is interesting to note that the first observation of nanocolumn formation in low-temperature MBE growth was made for Mn in Si.[81] This paper, however, did not contain information about their magnetic properties. Overall, it would seem that for Mn_xSi_{1-x} it will be much harder to strike the right balance between maximizing substitutional incorporation, minimizing formation of precipitates, and maximizing the epitaxial quality.

Several ion implantation studies have indicated above room temperature ferromagnetism in Mn_xSi_{1-x}. However, these striking conclusions only seemed to be based on SQUID measurements with little or no information on the transport properties.[18,82,83] These results should thus be interpreted with care. Clearly, the correlation between magnetism and transport in Mn_xSi_{1-x} requires more detailed studies, such as the ones that have been highlighted in this review. In closing, we wish to highlight three recent developments in the general area of group IV DMS. First of all, subsurfactant epitaxy (Sec. 7.4.4.1) seems to be highly promising for doping of Ge.[36] These samples seemed to be free of precipitates, yet their Curie temperatures easily reach room temperature. It is still a mystery how such a diluted system can exhibit high T_C. This finding certainly differs from the itinerant exchange in III–V DMS, which requires metallic doping levels, but it may be more in line with reports on high-T_C ferromagnetism in insulating DMS, including dilute oxide and nitride systems.[84–86] This suggests an interesting commonality between these seemingly different DMS systems. We caution, however, that subsurfactant epitaxy may not work for Si because substitutional trapping of Mn in bulk Si is thermodynamically unfavorable.[79]

This automatically brings us to the second development which is still mostly at the theoretical stage. DFT calculations recently indicated that the site preference of an Mn dopant in Si can be reversed from interstitial to substitutional in the presence of a neighboring n-type dopant.[79] In Ge, an Mn atom is more readily incorporated into the lattice when there is an n-type dopant in its immediate neighborhood, thus forming a stable Mn dopant pair with both impurities located at the substitutional sites. For both systems, it appears energetically favorable and kinetically accessible for an interstitial Mn to occupy a substitutional site next to an n-type dopant, thereby increasing the probability for substitutional incorporation. This effectively enhances the solid solubility of Mn. Intentional co-doping of n- and p-type impurities may thus be a highly promising avenue for future experimental studies.

Finally, we mention that the kinetic solubility of magnetic dopants will most likely be altered at the nanoscale. Recent works indicated the intriguing possibility of realizing above room temperature ferromagnetism and record-breaking hole mobilities *simultaneously* in Mn-doped Ge nanowires.[87] No doubt much remains to be explored in this highly fruitful area at this intersection of condensed matter physics, materials science, and nanotechnology.

ACKNOWLEDGMENT

The experimental work was funded by NSF Grant No. DMR 0606485 and by Oak Ridge National Laboratory, which is supported by the Office of Science of the US Department of Energy under Contract No. DE-AC05-00OR22725.

References

[1] J. E. Lilienfield, *Method and Apparatus for Controlling Electric Currents*, US Patent 1,745,175 (1930).
[2] S. A. Wolf, D. D. Awschalom, R. A. Buhrman, J. M. Daughton, S. von Molnar, M. L. Roukes, A. Y. Chtchelkanova and D. M. Treger, *Science* **294**, 1488 (2001).
[3] I. Zutic, J. Fabian and S. Das Sarma, *Rev. Mod. Phys.* **76**, 323 (2004).
[4] G. Binasch, P. Grunberg, F. Saurenbach and W. Zinn, *Phys. Rev. B* **39**, 4828 (1989).
[5] M. N. Baibich, J. M. Broto, A. Fert, F. N. Vandau, F. Petroff, P. Eitenne, G. Creuzet, A. Friederich and J. Chazelas, *Phys. Rev. Lett.* **61**, 2472 (1988).
[6] C. Timm, *J. Phys. Cond. Matt.* **15**, R1865 (2003).
[7] T. Dietl, *Semiconduct. Sci. Technol.* **17**, 377 (2002).
[8] T. Jungwirth, J. Sinova, J. Masek, J. Kucera and A. H. MacDonald, *Rev. Mod. Phys.* **78**, 809 (2006).
[9] G. Prinz and K. Hathaway, *Phys. Today* **48**, 24 (1995).
[10] R. Fiederling, M. Keim, G. Reuscher, W. Ossau, G. Schmidt, A. Waag and L. W. Molenkamp, *Nature* **402**, 787 (1999).
[11] Y. Ohno, D. K. Young, B. Beschoten, F. Matsukura, H. Ohno and D. D. Awschalom, *Nature* **402**, 790 (1999).
[12] G. Schmidt, D. Ferrand, L. W. Molenkamp, A. T. Filip and B. J. van Wees, *Phys. Rev. B* **62**, R4790 (2000).

[13] A. T. Hanbicki, B. T. Jonker, G. Itskos, G. Kioseoglou and A. Petrou, *Appl. Phys. Lett.* **80**, 1240 (2002).

[14] Y. D. Park, A. T. Hanbicki, S. C. Erwin, C. S. Hellberg, J. M. Sullivan, J. E. Mattson, T. F. Ambrose, A. Wilson, G. Spanos and B. T. Jonker, *Science* **295**, 651 (2002).

[15] C. G. Zeng, S. C. Erwin, L. C. Feldman, A. P. Li, R. Jin, Y. Song, J. R. Thompson and H. H. Weitering, *Appl. Phys. Lett.* **83**, 5002 (2003).

[16] I. Zutic, J. Fabian and S. C. Erwin, *Phys. Rev. Lett.* **97**, 026602 (2006).

[17] A. M. Tyryshkin, S. A. Lyon, A. V. Astashkin and A. M. Raitsimring, *Phys. Rev. B* **68**, 193207 (2003).

[18] M. Bolduc, C. Awo-Affouda, A. Stollenwerk, M. B. Huang, F. G. Ramos, G. Agnello and V. P. LaBella, *Phys. Rev. B* **71**, 033302 (2005).

[19] Y. H. Kwon, T. W. Kang, H. Y. Cho and T. W. Kim, *Solid State Commun.* **136**, 257 (2005).

[20] T. Dietl and J. Spalek, *Phys. Rev. B* **28**, 1548 (1983).

[21] J. E. Huheey, *Inorganic Chemistry, Principles of Structure and Reactivity*, 2nd Edn. (Harper & Row, New York, 1978).

[22] S. C. Erwin, private communication.

[23] A. Zunger and U. Lindefelt, *Phys. Rev. B* **27**, 1191, (1983).

[24] A. Stroppa, S. Picozzi, A. Continenza and A. J. Freeman, *Phys. Rev. B* **68**, 155203 (2003).

[25] Y. J. Zhao, T. Shishidou and A. J. Freeman, *Phys. Rev. Lett.* **90**, 047204 (2003).

[26] A. P. Li, J. Shen, J. R. Thompson and H. H. Weitering, *Appl. Phys. Lett.* **86**, 152507 (2005).

[27] A. P. Li, J. F. Wendelken, J. Shen, L. C. Feldman, J. R. Thompson and H. H. Weitering, *Phys. Rev. B* **72**, 195205 (2005).

[28] H. H. Woodbury and W. W. Tyler, *Phys. Rev.* **100**, 659 (1955).

[29] N. F. Mott, *Metal Insulator Transitions*, (Taylor and Francis, London, 1974).

[30] A. Chattopadhyay, S. Das Sarma and A. J. Millis, *Phys. Rev. Lett.* **87**, 227202 (2001).

[31] A. Kaminski and S. Das Sarma, *Phys. Rev. Lett.* **88**, 247202 (2002).

[32] A. Kaminski and S. Das Sarma, *Phys. Rev. B* **68**, 235210 (2003).

[33] V. M. Galitski, A. Kaminski and S. Das Sarma, *Phys. Rev. Lett.* **92**, 177203 (2004).

[34] T. Dietl, H. Ohno, F. Matsukura, J. Cibert and D. Ferrand, *Science* **287**, 1019 (2000).

[35] K. W. Edmonds, P. Boguslawski, B. L. Gallagher, R. P. Campion, K. Y. Wang, N. R. S. Farley, C. T. Foxon, M. Sawicki, T. Dietl, M. Buongiorno Nardelli and J. Bernholc, *Phys. Rev. Lett.* **92**, 037201 (2004).

[36] G. Kioseoglou, A. T. Hanbicki, C. H. Li, S. C. Erwin, R. Goswami and B. T. Jonker, *Appl. Phys. Lett.* **84**, 1725 (2004).

[37] H. Ohno, *Science* **281**, 951 (1998).

[38] C. Zeng, Z. Zhang, K. Von Benthem, M. F. Chisholm and H. H. Weitering, *Phys. Rev. Lett.* **100**, 066101 (2008).

[39] G. Alvarez, M. Mayr and E. Dagotto, *Phys. Rev. Lett.* **89**, 277202 (2002).

[40] S. Ahlers, D. Bougeard, N. Sircar, G. Abstreiter, A. Trampert, M. Opel and R. Gross, *Phys. Rev. B* **74**, 214411 (2006).

[41] C. Bihler, C. Jaeger, T. Vallaitis, M. Gjukic, M. S. Brandt, E. Pippel, J. Woltersdorf and U. Gosele, *Appl. Phys. Lett.* **88**, 112506 (2006).

[42] H. K. Choi, Y. S. Kim, S. S. A. Seo, I. T. Jeong, W. O. Lee, Y. S. Oh, K. H. Kim, J. C. Woo, T. W. Noh, Z. G. Khim, Y. D. Park and S. H. Chun, *Appl. Phys. Lett.* **89**, 102503 (2006).

[43] S. H. Chun, Y. S. Kim, H. K. Choi, I. T. Jeong, W. O. Lee, K. S. Suh, Y. S. Oh, K. H. Kim, Z. G. Khim, J. C. Woo and Y. D. Park, *Phys. Rev. Lett.* **98**, 026601 (2007).

[44] P. De Padova, J. P. Ayoub, I. Berbezier, J. M. Mariot, A. Taleb-Ibrahimi, M. C. Richter, O. Heckmann, A. M. Testa, D. Fiorani, B. Olivieri, S. Picozzi, and K. Hricovini, *Surf. Sci.* **601**, 2628 (2007).

[45] N. Pinto, L. Morresi, M. Ficcadenti, R. Murri, F. D'Orazio, F. Lucari, L. Boarino and G. Amato, *Phys. Rev. B* **72**, 165203 (2005).
[46] S. Sugahara, K. L. Lee, S. Yada and M. Tanaka, *Jpn. J. Appl. Phys. (Part 2)* **44**(46–49), L1426 (2005).
[47] V. Ko, K. L. Teo, T. Liew, T. C. Chong, T. Liu, A. T. S. Wee, A. Y. Du, M. Stoffel and O. G. Schmidt, *J. Appl. Phys.* **103**, 053912 (2008).
[48] R. Morgunov, M. Farle, M. Passacantando, L. Ottaviano and O. Kazakova, *Phys. Rev. B* **78**, 045206 (2008).
[49] L. Ottaviano, A. Verna, V. Grossi, P. Parisse, S. Piperno, M. Passacantando, G. Impellizzeri and F. Priolo, *Surf. Sci.* **601**, 2623 (2007).
[50] A. Verna, F. D'Orazio, L. Ottaviano, M. Passacantando, F. Lucari, G. Impellizzeri and F. Priolo, *Phys. Status Solidi A* **204**, 145 (2007).
[51] D. Bougeard, S. Ahlers, A. Trampert, N. Sircar and G. Abstreiter, *Phys. Rev. Lett.* **97**, 237202 (2006).
[52] J. S. Kang, G. Kim, S. C. Wi, S. S. Lee, S. Choi, S. Cho, S. W. Han, K. H. Kim, H. J. Song, H. J. Shin, A. Sekiyama, S. Kasai, S. Suga and B. I. Min, *Phys. Rev. Lett.* **94**, 147202 (2005).
[53] Y. Wang, J. Zou, Z. M. Zhao, X. H. Han, X. Y. Zhou and K. L. Wang, *Appl. Phys. Lett.* **92**, 101913 (2008).
[54] M. M. Özer, unpublished.
[55] A. P. Li, C. Zeng, K. van Benthem, M. F. Chisholm, J. Shen, S. V. S. Nageswara Rao, S. K. Dixit, L. C. Feldman, A. G. Petukhov, M. Foygel and H. H. Weitering, *Phys. Rev. B* **75**, 201201(R) (2007).
[56] M. Jamet, A. Barski, T. Devillers, V. Poydenot, R. Dujardin, P. Bayle-Guillemaud, J. Rothman, E. Bellet-Amalric, A. Marty, J. Cibert, R. Mattana and S. Tatarenko, *Nature Mater.* **5**, 653 (2006).
[57] S. Das Sarma, E. H. Hwang and A. Kaminski, *Phys. Rev. B* **67**, 155201 (2003).
[58] S. I. S. Jacobs and C. P. Bean, Fine particles, thin films and exchange anisotropy, in *Magnetism*, Vol. 3, eds. G. T. Rado and H. Suhl (Academic, New York, 1963).
[59] C. Jaeger, C. Bihler, T. Vallaitis, S. T. B. Goennenwein, M. Opel, R. Gross and M. S. Brandt, *Phys. Rev. B* **74**, 045330 (2006).
[60] J. A. Mydosh, *Spin Glasses: An Experimental Introduction* (Taylor and Francis, London, 1993).
[61] T. Devillers, M. Jamet, A. Barski, V. Poydenot, P. Bayle-Guillemaud, E. Bellet-Amalric, S. Cherifi and J. Cibert, *Phys. Rev. B* **76**, 205306 (2007).
[62] S. C. Erwin and A. G. Petukhov, *Phys. Rev. Lett.* **89**, 227201 (2002).
[63] A. H. MacDonald, P. Schiffer and N. Samarth, *Nature Mater.* **4**, 195 (2005).
[64] A. Van Esch, L. Van Bockstal, J. De Boeck, G. Verbanck, A. S. van Steenbergen, P. J. Wellmann, B. Grietens, R. Bogaerts, F. Herlach and G. Borghs, *Phys. Rev. B* **56**, 13103 (1997).
[65] W. Allen, E. G. Gwinn, T. C. Kreutz and A. C. Gossard, *Phys. Rev. B* **70**, 125320 (2004).
[66] F. Matsukura, H. Ohno, A. Shen and Y. Sugawara, *Phys. Rev. B* **57**, R2037 (1998).
[67] M. Foygel and A. G. Petukhov, *Phys. Rev. B* **76**, 205202 (2007).
[68] P. G. De Gennes and J. Friedel, *J. Phys. Chem. Sol.* **4**, 71 (1958).
[69] C. Haas, *Phys. Rev.* **68**, 531 (1968).
[70] A. L. Efros and M. E. Raikh, in *Optical Properties of Mixed Crystals*, eds. R. J. Elliot and I. P. Ipatova (Elsevier, New York, 1988), p. 133.
[71] C. Michel, P. J. Klar, S. D. Baranovskii and P. Thomas, *Phys. Rev. B* **69**, 165211 (2004).
[72] A. A. Burkov and L. Balents, *Phys. Rev. Lett.* **91**, 057202 (2003).
[73] M. Jaime, H. T. Hardner, M. B. Salamon, M. Rubinstein, P. Dorsey and D. Emin, *Phys. Rev. Lett.* **78**, 951 (1997).
[74] M. Copel, M. C. Reuter, E. Kaxiras and R. M. Tromp, *Phys. Rev. Lett.* **63**, 632 (1989).

[75] W. G. Zhu, H. H. Weitering and E. Kaxiras, *Phys. Rev. Lett.* **93**, 126102 (2004).
[76] C. G. Zeng and H. H. Weitering, unpublished.
[77] B. Shapiro, *Philos. Mag. B* **50**, 241 (1984).
[78] M. M. Özer and H. H. Weitering, unpublished.
[79] W. G. Zhu, Z. Y. Zhang and E. Kaxiras, *Phys. Rev. Lett.* **100**, 027205 (2008).
[80] Z. Z. Zhang, B. Partoens, K. Chang and F. M. Peeters, *Phys. Rev. B* **77**, 155201 (2008).
[81] Y. Zhang, Q. Jiang, D. J. Smith and J. Drucker, *J. Appl. Phys.* **98**, 033512 (2005).
[82] P. R. Bandaru, J. Park, J. S. Lee, Y. J. Tang, L.-H. Chen, S. Jin, S. A. Song and J. R. O'Brien, *Appl. Phys. Lett.* **89**, 112502 (2006).
[83] F. M. Zhang, X. C. Liu, J. Gao, X. S. Wu, Y. W. Du, H. Zhu, J. Q. Xiao and P. Chen, *Appl. Phys. Lett.* **85**, 786 (2004).
[84] Y. Matsumoto, *Science* **291**, 854 (2001).
[85] K. A. Griffin, A. B. Pakhomov, C. M. Wang, S. M. Heald and K. M. Krishnan, *Phys. Rev. Lett.* **94**, 157204 (2005).
[86] J. M. D. Coey, M. Venkatesan and C. B. Fitzgerald, *Nature Mater.* **4**, 173 (2005).
[87] M. I. van der Meulen, N. Petkov, M. A. Morris, O. Kazakova, X. Han, K. L. Wang, A. P. Jacob and J. D. Holmes, *Nano Lett.* **9**, 50 (2009).

Chapter Eight

Dynamics of Localized Spins in Non-Magnetic Semiconductors

T. A. Kennedy

Naval Research Laboratory, Code 6877, Washington DC 20375, USA
tavdkennedy@comcast.net

Electron spin is envisioned as the useful property for new applications in electronics, optical communications and quantum information technology. Recent work on ensembles of spins in normal (non-magnetic) semiconductors motivated by this vision is reviewed in this paper. The states of the spins are initialized, controlled and read out using light of suitable energy and duration. The recent progress in theory, materials and experimental techniques is substantial. In the area of spin lifetimes, a microsecond T_2 and a millisecond T_1 have been reported in III-V semiconductors. Demonstrations of coherent population trapping and the conversion of light-polarization to electronic spin point the way toward future applications.

8.1 INTRODUCTION

The topic of this review originates from the discovery of long spin lifetimes in conventional (non-magnetic) semiconductors. Lifetimes of several nanoseconds were observed at room temperature in *n*-type ZnSe[1] and lifetimes of hundreds of nanoseconds were observed at low temperatures for *n*-type GaAs.[2] In both cases the doping levels were high enough that the electrons were delocalized. This implied the possibility of spin transport, which was then explored. But the results also raised a question about spin lifetimes for samples with less doping where the spins are localized and ultimately non-interacting. In this limit, the neutral donors become like a gas of atoms in a semiconductor environment.

A second type of artificial atom in a semiconductor is the quantum dot. In a very stimulating paper, optically controlling spin in a quantum dot was proposed for quantum information processing.[3] All this background led to two distinct but related areas of research. In the first area, techniques and samples were developed to study single atoms and single quantum dots. The primary driving force for

Handbook of Spintronic Semiconductors by W M Chen & I A Buyanova

www.panstanford.com
978-981-4267-36-6

this research is quantum computing and other quantum information processing. A review of this work has been given.[4] In the second area, ensembles of donors and of quantum dots with concentrations low enough that the "atoms" are isolated were studied. While this work has some bearing on quantum information, it is directly suited for the development of materials for optical applications. The effects that are valuable for applications are electromagnetically induced transparency and slow light[5] and coherence-induced negative refractive index.[6] The spin dynamics of ensembles of localized spins in non-magnetic semiconductors is the subject of this review.

A number of previous reviews are useful to this subject. First, there is the classic book by Meier and Zakharchenya.[7] More recent and closer to the present topic are the book by Awschalom, Loss and Samarth[8] and the article by Zutic, Fabian and Das Sarma.[9] Very recently, the coherent spin dynamics of carriers was reviewed by Yakovlev and Bayer.[10] The corresponding work in atomic physics is described clearly in the book by Suter.[11] Finally, the review by Geschwind and Romestain[12] is useful in understanding spin-flip Raman scattering.

This chapter is organized as follows. Section 8.2 gives the theory that shows how light and spin connect in a semiconductor. The basic physics of spin dynamics and control provides the next sections: Sec. 8.3 is on spin initialization, Sec. 8.4 on spin lifetimes, and Sec. 8.5 on spin control. There are a few remarks on optical detection of spin at the end of Sec. 8.5. Next, there are two sections of effects important to applications. Section 8.6 is on coherent population trapping, and Sec. 8.7 is on the transfer of light polarization to spin coherence. Section 8 concludes the review.

8.2 THEORY

Spin states are long-lived because the spin degree of freedom is often decoupled from the orbital degree of freedom. However, the spin states need to be initialized, manipulated and read out in order for them to be useful. Addressing the long-lifetime states is possible through other states for which there is spin-orbit interaction. There are two general approaches that make use of spin-orbit to address spin: electrical and optical. In this review we will be concerned with the optical approach.

It might seem that the connection between spin and light is weak and accidental. The opposite is true. Spin states in matter are strongly connected to polarization states of light when there is sufficiently strong spin-orbit interaction for one of the optically active states. Thus, there is a great advantage to using resonant or near-resonant light to address spin. This section will describe how this connection works for the case of a semiconductor such as GaAs.

Gallium arsenide is a direct band-gap semiconductor with the minimal band-gap occurring at the center of the Brillouin zone. In order to consider its optical properties, it is sufficient to use $\mathbf{k}\cdot\mathbf{p}$ theory.[9,13] The valence band is p-like and can be represented by states with symmetry X, Y and Z. Since we want to treat the effect of the spin-orbit interaction, we form eigenstates of the orbital angular momentum operator. These are of the form $|l, m_l\rangle$ and there are three:

$$|1,\,1\rangle = |(1/2)^{-1/2}(X + iY)\rangle,$$
$$|1,\,0\rangle = |Z\rangle,$$
$$|1,\,-1\rangle = |(1/2)^{-1/2}(X - iY)\rangle.$$

The spin orbit operator, $\lambda\mathbf{l}\cdot\mathbf{s}$, will then produce eigenstates of the total angular momentum: j, j_z for the hole states. This leads to three sets of states. First, for the conduction band there are two states: spin up and spin down (see Table 8.1). Two sets come from the valence band. The first has $j = 3/2$ and four states as shown in the table. The second has $j = 1/2$ and is split off from the band edge by 0.34 eV in GaAs. These last states are sufficiently far away in energy that they will not be considered here.

Light will induce transitions between the conduction and valence band states generating an electron and a hole. Linearly polarized light will induce a linear oscillating dipole and circularly polarized light a rotating dipole. To illustrate how this works we consider left circularly polarized light ($\sigma+$) propagating in the z-direction. The dipole operator for this light is written as $(X + iY)$. If we consider the matrix element for this operator between the heavy hole state $|3/2,\,3/2\rangle$ and the electron state $|1/2, 1/2\rangle$, we find that absorption will produce a spin polarized hole and electron along the direction of the light. This spin polarization arises from the spin-orbit interaction coupling the orbital and spin degrees of freedom for the hole states.

Applying polarized light to selected electron and hole states allows a great variety of access to the resulting spin states. Earlier applications used non-resonant light to induce spin polarization proportional to the intensity of the light. These then involved the square of the matrix elements and described either absorption processes to polarize spin or emission processes to detect spin. These effects are still useful. However, the development of tunable lasers, especially the Ti:sapphire laser, has made possible resonant, coherent interactions. These involve the matrix element itself and open up more possibilities.

Table 8.1 Band edge states for GaAs.

Band/symmetry	j, m_j	Wave function
Conduction	$\|1/2, 1/2\rangle$	$\|S, \uparrow\rangle$
Γ_6	$\|1/2, -1/2\rangle$	$\|S, \downarrow\rangle$
Valence	$\|3/2, 3/2\rangle$	$\|(1/2)^{1/2}(X + iY)\uparrow\rangle$
Γ_8	$\|3/2, 1/2\rangle$	$\|(1/6)^{1/2}[(X + iY)\downarrow +2Z\uparrow]\rangle$
	$\|3/2, -1/2\rangle$	$\|-(1/6)^{1/2}[(X - iY)\uparrow -2Z\downarrow]\rangle$
	$\|3/2, -3/2\rangle$	$\|(1/2)^{1/2}(X - iY)\downarrow\rangle$

In the paradigm of quantum information technology, the spin process is divided into initializing a particular spin state, allowing that state to evolve or applying a control pulse to modify the state, and detecting the state at a subsequent time.[14] These processes require that the spin lifetimes (for phase and energy) are long compared to the optical pulses that are used. Clearly, there is an advantage in using ultrafast (picosecond) pulses since many operations are possible during a spin lifetime.

As semiconductor materials have developed and pulse techniques refined, exciting results have been published describing initialization with greater fidelity, control with greater precision and detection with greater sensitivity. The materials have been grown and selected to give longer lifetimes and more isolated optical transitions. The overall progress in these different areas has led to demonstrations of effects important to quantum optics applications such as slow light and quantum information applications such as the conversion of the polarization state of light to the polarization state of a spin. These new capabilities are described in the remainder of this review.

8.3 SPIN INITIALIZATION

We consider a low density of electrons localized in a wide quantum well. The electrons transfer to the well from remotely placed donors in the surrounding barrier material. This system presents three advantages for optical interactions with spin. First, the inhomogeneous broadening in a wide quantum well is small. The electrons become localized by fluctuations in the well-width, but these are small and produce little inhomogeneity in either the energy of the optical transition or the g-factor of the localized electrons. Second, the changing potential in the growth direction splits the light and heavy-hole states of the valence band. The heavy-hole states are lower in energy and this produces a level scheme that is simpler than that for bulk GaAs. Third, the localized electrons are present in the dark — they have an infinite lifetime. This contrasts with the excitonic state for this system, which has a lifetime of about 60 ps.[15] Spin lifetimes without the doping would be limited to this value. However, there is an optical transition and excited state associated with the doped, localized electron called the trion.

The spin states for the electron–trion system are shown in Fig. 8.1. The left-hand side shows the states in a single-particle picture and the right for a multi-particle or exciton picture. The ground state consists of a single electron in the confined conduction band state. Optically adding an electron–hole pair produces the excited state with two electrons in a singlet state and a hole. Because the electron spins are paired off, the spin properties of this trion state are determined entirely by the hole.

The selection rules for this system are particularly simple and powerful (see the right-hand side of Fig. 8.1). We take as the basis the spin states along the direction of the light propagation. Then σ^+ light will excite only the $+1/2$ to $+3/2$

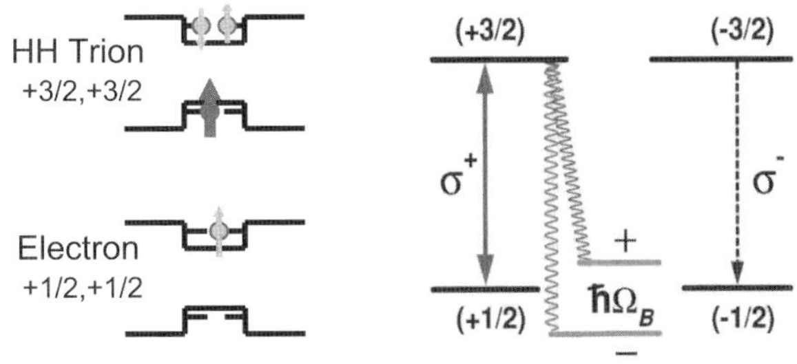

Figure 8.1. Energy diagrams for the electron and its optically excited state, the trion. Left: Single-particle diagrams for the electron and for the trion. Right: Exciton picture in the light basis showing the simple selection rules for circularly polarized light. The center of the figure shows the splitting of the electron states in the field basis.

states and σ^- light will excite only the $-1/2$ to $-3/2$ states. Applying a π-pulse of σ^+ light will populate the $+3/2$ state and empty the $+1/2$ state. Spontaneous emission will then empty the ensemble of $+3/2$ states with an exponential time dependence characterized by the spontaneous emission time. Hole relaxation, which would mix the $+3/2$ and $-3/2$ states, is slower in this system than the spontaneous emission rate.

The addition of a transverse magnetic field leads to a very effective method to initialize the spin of the ground-state electrons.[16,17] This field leads to Zeeman spitting of the electron states as described in the basis along its direction. These states are sketched in the right-hand side of Fig. 8.1. We consider a strong field such that the Larmor precession is slow compared to the light pulses (~1 ps) but fast compared to the recombination time for the trion (60 ps). A fast π-pulse of light will move spins that are along the field direction to the trion state. This leaves spins opposite to the light direction and these will begin to precess. As the trions recombine, they return an electron along the light direction, but this phase is now random with respect to the other precessing electrons. Unlike at zero magnetic field, the ground-state spin excited (actually left behind) by the light pulse persists. Half of the population of electrons has been initiated by a single light pulse. The following experiments validate this model.

The experiments were done with a picosecond laser with a pump pulse to initialize the dynamics and a weaker, time-delayed probe pulse to sample the spin polarization.[18] The pump pulse is circularly polarized to address the selection rules described above. The probe pulse is linearly polarized to detect the difference in the response to σ^+ and σ^- polarizations. Since the work is performed in reflection, it is properly described as Kerr rotation but it is formally equivalent to the transmission variation, Faraday rotation. For selectivity, the wavelength of the light is tuned to the heavy-hole trion transition.

The temporal decays for different magnetic field strengths are shown in Fig. 8.2. For $B = 0$, there is a rapid decay to nearly zero signal with a characteristic time of 30 ps. This is the recombination of the trion producing a nearly complete cancellation of the spin orientation produced by the pump pulse. The

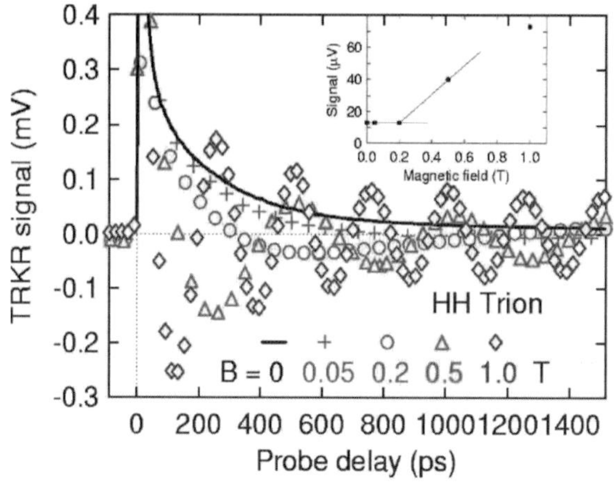

Figure 8.2. Pump-probe response for electrons in a wide quantum well for different magnetic fields. The response for long times increases with increasing field as predicted by the model in the text. This response is plotted in the inset. [Reprinted figure with permission from T. A. Kennedy *et al.*, *Phys. Rev. B* **73**, 045307 (2006), Copyright 2006 by the American Physical Society.]

cancellation shows that hole relaxation is slow in this sample. This in turn indicates that the electrons and holes (trions) are well localized. As magnetic field is increased, precession occurs with a period that becomes shorter. The value of the period is consistent with the g-factor for electrons in this quantum well. The long lifetime of the oscillatory signal also demonstrates that the ground-state electron is responsible. It can be seen in the figure that the long-lived signal becomes significant as the Larmor period approaches the trion lifetime.

All the features expected from the theory are confirmed for these experiments in a wide quantum well. Other semiconductor structures with localized spins show this mechanism as well.[19,20]

8.4 SPIN LIFETIMES

Long spin lifetimes are required for any of the applications envisioned for this area and are desirable also for the experimental research itself. Therefore, we consider what constitute long lifetimes for III-V semiconductors and recount the recent progress in ensemble experiments. The fundamental distinction here is between relaxation between different energy eigenstates of the spin with a characteristic time T_1 and relaxation of the phase for superpositions of eigenstates characterized by a time T_2. The phase relaxation can be further divided between dephasing within an ensemble, an inhomogeneous process described by T_2^*, and the ultimate loss of phase by single spin described by T_2. For localized spins in a III-V semiconductor, the dephasing is controlled by hyperfine interactions with the 100% non-zero nuclear spins of the lattice. This interaction leads to T_2^* of 1–10 ns.[21] For a single spin, the phase memory is ultimately limited by T_1. Thus, it depends on temperature and magnetic field but it may reach the millisecond range.

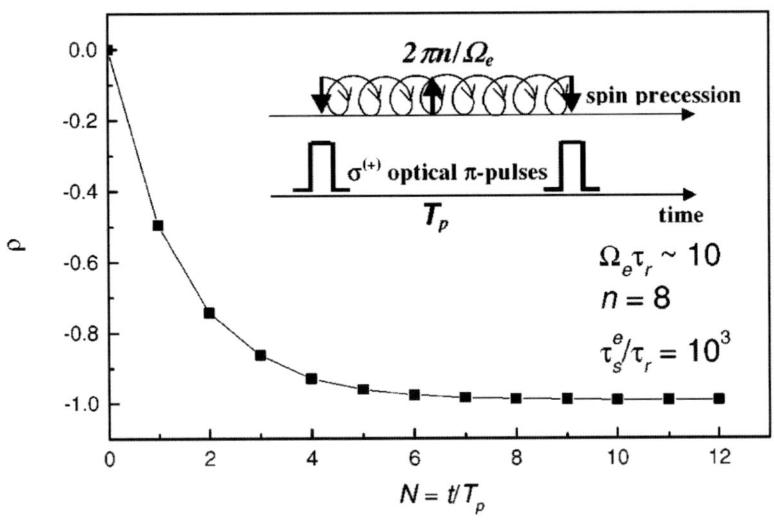

Figure 8.3. Degree of spin polarization versus number of pulses. After eight pulses the spin is fully synchronized to the optical pulses. The insets show the precessing spin and pulses from the optical train. [Reprinted figure with permission from A. Shabaev *et al.*, *Phys. Rev. B* **68**, 201305(R) (2003), Copyright 2003 by the American Physical Society.]

In this section we review results on lifetimes for two types of ensembles in III-Vs. The first is InGaAs self-assembled quantum dots where there is a lot of inhomogeneity in the size and shape that leads to inhomogeneity in the optical transition energy and in the electron g-factor. The second is a very low density of donors in an otherwise very pure GaAs layer. Here, there is a very homogeneous set of centers arising from isolated Si donor impurities.

In the previous section, initialization of electron spin coherence was discussed for the electron–trion system in a quantum structure. Here, we consider strongly confined dots in particular since their widely spaced energy levels lead to the possibility of very long phase memory. Instead of restricting the discussion to the effect of a single pulse, we consider the typical experimental situation involving a train of pulses with a period of 12 ns. This is typical for a mode-locked Ti:sapphire laser. With a pulse train and a long T_2, new possibilities emerge.

If we consider first spins with a single g-factor, it is possible to tune the Larmor frequency of their precession such that there is synchronization between the precession and the repetition period of the laser (see Fig. 8.3).* For σ^+ pulses of area π, the first pulse will leave half of the spins in the spin-down state, as described previously. As long as they precess coherently at a multiple of the laser repetition rate, these spins will return to the spin-down state just in time for the next σ^+ pulse, to which they will not respond. Of the spins whose phases were randomized by absorption of the first pulse and the subsequent spontaneous emission, half will be excited by the second pulse, leaving half in the spin-down state. After six to eight

* The argument here is from Ref. 16 but is extended from a single spin to an ensemble of spins.

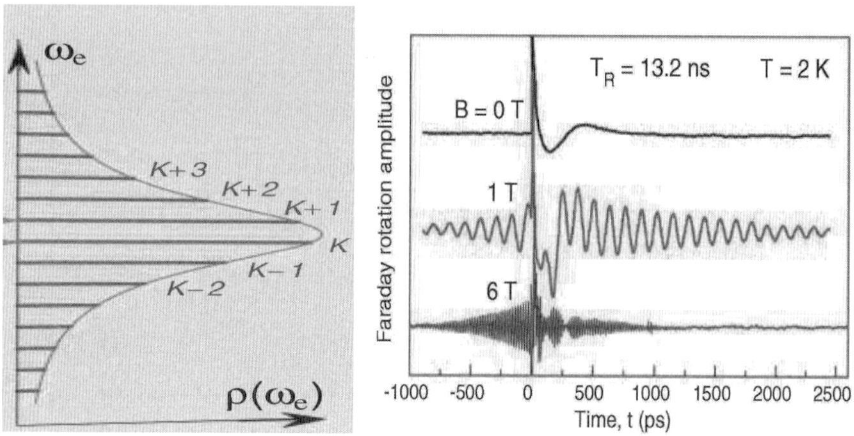

Figure 8.4. Spin modes versus Larmor frequency (left) and pump-probe data showing electron spin precession (right). The inhomogeneous broadening of the dots produces a range of Larmor frequencies within which about 10 modes occur (left). The spins in these modes come back into phase as time approaches zero, the time of another pulse (right). From A. Greilich *et al.*, *Science* **313**, 341 (2006). [Reprinted with permission from AAAS.]

pulses, nearly all the spins have been synchronized and are insensitive to the light pulses. This synchronization will persist until T_2.

In reality, InGaAs and InAs self-assembled quantum dots have a wide range of g-factors and resulting Larmor frequencies.[22,23] For this case there is a set of frequencies each separated by one oscillation that will satisfy the phase-synchronization condition (see Fig. 8.4). The situation resembles the modes in the laser and results in mode-locking of these phase-matched spins.

In the experiments on InGaAs dots,[22] the coherence decays for positive delays in a time characteristic of the range of Larmor frequencies. That is, it is set by the inhomogeneity in the g-factors for the ensemble. However, the spins that are in the phase-matching modes come back into phase for negative delays as the time for the next pulse approaches. This is an obvious sign of a T_2 that far exceeds the T_2^*. In fact, the repetition period can now be changed to reveal the time for loss of coherence. A value of 3.0 μs was reported. This may correspond to the time for a temporal fluctuation in the hyperfine field.[21]

Theoretical work has predicted microsecond to millisecond times for T_2 and millisecond times for T_1 in quantum dots.[24,25] Work in single quantum dots[26] has revealed a coherence time exceeding 1 μs; and the above work shows that microsecond times for coherence can be seen in an ensemble. An important advance in the T_1 times has been obtained for donors in GaAs[27] and this is described next.

The experiment was performed on an MBE-grown GaAs sample using a time-resolved photoluminescence technique. The layer was 10 μm thick and had a donor concentration of 5×10^{13} cm^{-3}. This denotes a collection of isolated donors forming a very homogeneous ensemble. In fact, the photoluminescence from a

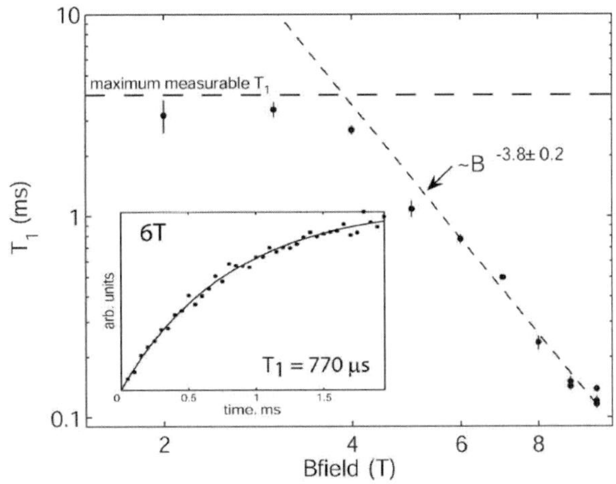

Figure 8.5. Spin-lattice relaxation time (T_1) as a function of magnetic field for donors in GaAs. The inset shows the temporal recovery of spin polarization at 6 T. For low fields, the time is limited instrumentally to 3.5 ms. For high fields, the time decreases as B to the -3.8 power. [Reprinted figure with permission from K.-M. Fu *et al.*, *Phys. Rev. B* **74**, 121304(R) (2006), Copyright 2006 by the American Physical Society.]

common excited state to the two spin-split electron states is resolved for moderate and high magnetic fields. There are three parts to the time-resolved technique. First, the electron spin is polarized by a long optical pulse applied to the spin $-1/2$ to excited state transition. This places a large population in the $+1/2$ state. Second, there is a variable delay (dark) period to allow the spin states to approach thermal equilibrium. Third, there is a short pulse resonant with the $-1/2$ state to measure its population. The data shows an exponential form as the spins states equilibrate (see the inset of Fig. 8.5).

Results were obtained for different magnetic fields at a temperature of 1.5 K (Fig. 8.5). Below 4 T there is a plateau with T_1 equal to 3.5 ms due to the limitations of the equipment. At higher fields the relaxation $(1/T_1)$ increases as B to the fourth power. This is consistent with theory[28] and shows that the relaxation involves a one-phonon process. The result is an important demonstration of the existence of a very long lifetime in GaAs.

8.5 SPIN ROTATION

With the ability to initialize a spin state that is long-lived, one is able to begin the real business of quantum information processing — the rotations of the spin. These are the operations or quantum gates of a quantum information process. Spin rotations are possible with microwave fields but these are limited in speed to around 5 ns. Early on it was realized that optical gates using ultra-fast light pulses would allow many more quantum operations within a spin lifetime.[3] Since then, there have been more suggestions on how to implement optical gates.[29,30]

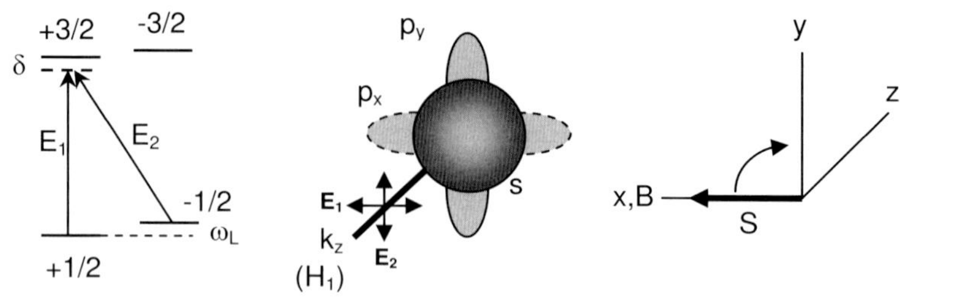

Figure 8.6. Energy levels, wave functions and laboratory axes for spin-flip Raman scattering. Left: Two CW laser fields connect spin-split ground states to one excited state. A detuning of δ is indicated. Center: The dipole coupled electron and hole states oscillate with the polarizations and frequencies of the light fields. There is a magnetic dipole at the difference frequency (H_1). Right: This dipole acts to rotate a spin initially along the x-axis.

An early experiment was explained in terms of an optical Stark effect.[31] In the last couple of years, more experiments have also been performed.[32–35] These generally use some form of spin-flip Raman transition to effect the rotation. In this section, we will describe highlights of this work.

First, it is reasonable to ask how light can produce a pure spin rotation. The answer comes through the spin-flip Raman process[11,12] and is outlined as follows. Consider again the electron–trion energy levels but now in the magnetic field basis (Fig. 8.6, left). With an in-plane magnetic field, the electron spin states are split by $\hbar\omega_L$. The trion states are not split since the in-plane g-factor for holes is small. Circularly polarized light will couple both of the eigenstates to a particular trion state. This produces the classic three-state Λ (lambda) system ideal for Raman scattering.

Next, consider two separate CW light fields E_1 and E_2 that differ in frequency by the Larmor frequency, ω_L. These propagate along the z-direction with linear polarizations in the x- and y-directions (see Fig. 8.6, center). The electron and trion states have wave functions with s and p parts as described in the theory section. Thus, the usual matrix elements for the electric dipole operator will produce oscillating dipoles whose frequencies differ by ω_L. The combined effect, a second order process, will produce a rotating charge in the xy-plane, again at the Larmor frequency. This is an effective H_1 that acts on the ground state spin, which is initially in the x-direction, producing a rotation (Fig. 8.6, right). The amount of rotation depends on the strength of the fields and the duration of the interaction. This is stimulated Raman spin rotation. The two CW fields can be replaced by a circularly polarized light pulse whose spectral bandwidth covers the Larmor frequency. A pico-second pulse has a terahertz bandwidth that easily covers the Larmor frequency for moderate (1–10 T) fields. The spin rotation then depends on the electron g-factor and the strength and duration of the pulse. The process is an inelastic light scattering and can be performed at or near optical resonance. A light pulse produces a pure spin rotation.

Work on electron spins in modulation-doped GaAs quantum wells has demonstrated the importance of performing Raman rotations with the light off-resonance.[34] When the light is on-resonance, there are first-order coherent processes or absorption processes that compete with the Raman rotation. These include the spin-initialization processes discussed in an earlier section. With detuning, the absorption-related processes fall off faster than the Raman process. This favors the selective action of producing a spin rotation. These processes were analyzed carefully by doing two-color experiments in which the Raman pulse detuning was varied.

With the optically initialized spin properties separated, the results for the spin rotations were elucidated.[34] The modulation doping produces an electron–trion system as discussed previously. The initial polarization along the magnetic field direction was thermal and amounted to 4% for $B = 4\,\mathrm{T}$ and $T = 4\,\mathrm{K}$. The detuned control pulses produced Raman rotations of up to $\pi/6$ radians. This value was limited by changes in the system with increasing light intensity that were attributed to heating.

A second work on donor-bound electrons in GaAs confirms and extends the above results.[36] This work makes use of the same system and sample as described in the work demonstrating a long T_1. The experiments were performed at 7 T and 1.5 K. In an initial time period, the donor spin is polarized using a long, narrow-band light pulse. This results in a polarization of 94%. Following some delay, an off-resonant ultrafast pulse is applied to rotate the spin through the Raman process. In the third period, the spin is read out by detecting photoluminescence excited by another long, narrow-band light pulse.

Using a single pulse produced spin rotations of $\pi/4$. Here again, there are limitations that arise as the intensity of the control pulse is increased. Free excitons are created for high light intensity that produce dephasing of the optical process and thus limit the degree of rotation. Applying two separate pulses in succession produced rotations of $\pi/2$. This suggested that multiple pulses could be used to provide an arbitrary rotation.

Future work on optical rotations is needed to obtain larger angles of rotation and better fidelity for the particular operation. New suggestions from theory give direction to these improvements.[37]

For completeness, a few remarks about the optical detection of a spin state are given. Faraday rotation[8] continues to be very valuable for detection. Like in the control operations, it is often advantageous to use off-resonant Faraday rotation.[38] Furthermore, recent work has clarified the need to explore both Faraday rotation and Faraday ellipticity in order to optimize the detection of a spin state.[39]

8.6 APPLICATIONS I: COHERENT POPULATION TRAPPING

Optically active centers with more than two energy levels open up the possibility of controlling one light field with another. This in turn opens up possible applications in the fields of optical communications and quantum information processing. One remarkable example of three-level physics is electromagnetically induced transparency (EIT).[40] In this section, we start with a simple explanation of EIT and its close cousin, coherent population trapping (CPT). Then, we review an important experimental demonstration of CPT using donors in GaAs.

Consider again the three-level system shown in the left-hand side of Fig. 8.7. States 1 and 2 are spin states split by an applied magnetic field. State 3 has an optical electric dipole coupling to both 1 and 2. For CPT and EIT a strong CW laser field is applied to the transition from $2 \rightarrow 3$. The spin dephasing time T_2^* for states 1 and 2 is long with respect to the radiative decay times $3 \rightarrow 1$ and $3 \rightarrow 2$. For this condition, if a probe laser field is scanned through the energy for the $1 \rightarrow 3$ transition, there will be coherence induced in the $1 \rightarrow 2$ levels. There are then two pathways for absorption of the probe light with the possibility of constructive and destructive interference. Destructive interference corresponds to no absorption in the pathway of the resonant probe light. From the optical standpoint, population is trapped in the ground states (1 and 2) leading to the term "coherent population trapping."

CPT is a spectroscopic effect that can be observed for optically thin samples. When the sample is optically thick, the destructive interference constitutes transparency — hence the name "electromagnetically induced transparency."

Demonstrating these effects experimentally requires sharp energy levels and long spin dephasing times. An important realization in semiconductors was made using donors in GaAs.[41] Levels 1 and 2 in the diagram are the split spin states of the neutral donor D^0. These states are optically coupled to a distinguishable level associated with the donor-bound exciton D^0X.

The experiments were performed at 1.5 K with a 7 T magnetic field in the Voigt geometry. A strong coupling field is applied to the transition $2 \rightarrow 3$. A second weaker laser field is scanned across the energy range for the transition $1 \rightarrow 3$.

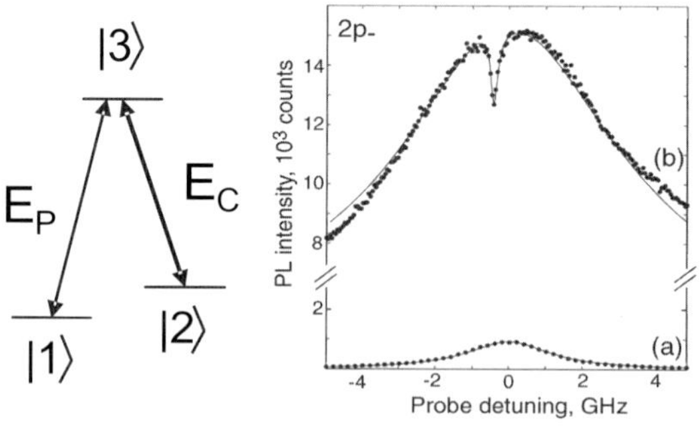

Figure 8.7. Energy levels and photoluminescence versus detuning for coherent population trapping. Left: A strong coupling field (E_C) is applied between levels 2 and 3 and a probe field (E_P) is scanned through the energy difference between levels 1 and 3. Right: With only a probe field, the luminescence denotes a simple absorption (Curve a). With both fields, there is a dip at zero detuning revealing coherent population trapping (Curve b). [Reprinted figure with permission from K.-M. Fu *et al.*, *Phys. Rev. Lett.* **95**, 187405 (2005). Copyright 2005 by the American Physical Society.]

Photoluminescence is monitored that shows the population of state 3. Thus, the experiment is actually an example of photoluminescence excitation (PLE) spectroscopy.

The results are shown in Fig. 8.7 (right). Without the coupling laser, a simple absorption-like curve is recorded that indicates absorption by the $1 \rightarrow 3$ transition. With the coupling laser, stronger photoluminescence is observed with a pronounced dip for zero detuning. This decrease in absorption signifies the coherent population trapping. For the coupling strength used, the dip is about 14%. Separate data show that this value is optimal with smaller dips for lower strengths for the coupling field. The size of the dip is limited for this system because the T_2^* (1–2 ns) is close to the radiative decay time.

This work was the first demonstration of a Λ-system in a semiconductor that utilized true electron ground states.

8.7 APPLICATIONS II: TRANSFER OF LIGHT POLARIZATION TO SPIN COHERENCE

A full quantum information technology requires both qubits that are fixed in space (stationary qubits) and qubits that can be transmitted from one point to another (flying qubits). For this review we have focused on electron spins in a semiconductor host as the stationary qubits. Photons are the natural flying qubits. Interconversion between these two types of qubits is important to developing devices such as the quantum repeater.[42] In this section we explain further the concept of inter-conversion of qubits and then review an important experimental demonstration of the process.

Figure 8.8 shows schematically the inter-conversion process. The polarization state of light can be represented by a point on the Poincaré sphere. Circular

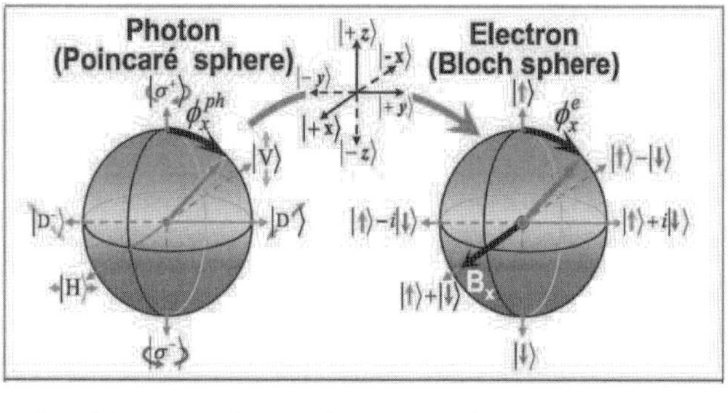

Figure 8.8. Transfer of the state of the polarization of light to the state of the spins of electrons. On the left is the Poincaré sphere with circularly polarized light along the poles. On the right is the Bloch sphere with spin-up and spin-down states along the poles. Optical selection rules generate a correspondence allowing the transfer of the state. [Reprinted figure with permission from H. Kosaka *et al.*, *Phys. Rev. Lett.* **100**, 096602 (2008). Copyright 2008 by the American Physical Society.]

polarization is taken as the basis with states at the poles of the sphere. Linear polarizations occur on the equator. Similarly, electron spin states can be represented by points on the Bloch sphere. The up and down states are taken as the basis with states along the poles. Equal combinations of up- and down-spin occur along the equator with different phases. Faithful inter-conversion is the transfer of light with a particular polarization to electron spin with the corresponding state. Schematically, this process is indicated by the arrow from the Poincaré sphere to the Bloch sphere.

Qubit inter-conversion has been demonstrated for an 11 nm undoped GaAs quantum well.[43] The experiment was performed with the light propagating along z and a 7 T magnetic field along x [see Fig. 8.9(a)]. For this transverse geometry, the electron g-factor is small (−0.21) while the light-hole g-factor is large (−3.5). Thus, there is a large Zeeman splitting for the holes in the 7 T magnetic field. This leads to a V-configuration for the energy levels [see Fig. 8.9(b)]. For pulses of light whose bandwidth is small compared to the light-hole splitting but large compared to the electron splitting, the polarization state of the light can be converted to the electron spin state.

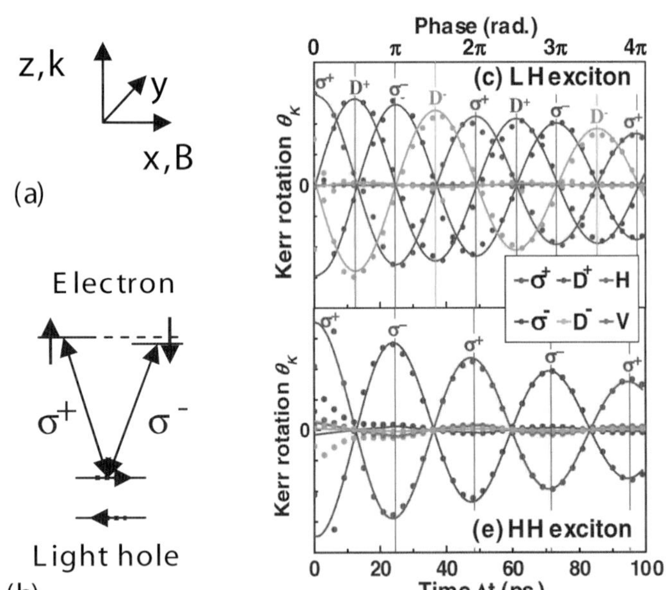

Figure 8.9. Geometry, energy levels and pump-probe results for the conversion of light polarization to spin state. (a) The geometry with light along z and magnetic field along x. (b) The energy levels with the light selecting a V of three levels. The electron spin states are the optically excited states in this case. (c) The change in phase as light polarization is changed with light resonant with the light hole exciton. (d) The reversal of phase that occurs for light resonant with the heavy hole exciton. [Reprinted figure with permission from H. Kosaka *et al.*, *Phys. Rev. Lett.* **100**, 096602 (2008). Copyright 2008 by the American Physical Society.]

Pump-probe experiments with the energies and bandwidths of the light pulses set to match the energies of the V-system verify the qubit conversion. The probe pulse detects the spin along the propagation direction via Kerr rotation. The polarization of the pump pulse is varied in steps from circular to linear and then to the opposite circular polarization. While the probe only detects the projection of the spin along z, the initial phase of the signal reveals the spin direction. As seen in Fig. 8.9(c), the initial phase varies continuously indicating that the electron spin state corresponds to the light polarization state. In contrast, when the light is tuned to the heavy-hole exciton, only spin parallel or anti-parallel to the light direction is generated [Fig. 8.9(d)].

Unlike in the spin initialization and spin control schemes described earlier, there is no requirement in the spin conversion scheme that the pump pulse be strong. In addition, spontaneous emission does not play a role in the conversion. Thus, the concept can be scaled down to a single photon and a single spin. The work is an important demonstration of the coherent transfer of light polarization to electron spin in a semiconductor.

8.8 CONCLUSION

Recent progress in the spin dynamics of ensembles of localized spins in non-magnetic semiconductors has been reviewed. The focus has been on optical methods to couple to the spins, and the physical basis for the coupling of light to electrons and holes in direct-gap semiconductors was described. Electron spin states have been initialized to a high fidelity by pulses of light and pulse trains. Materials and structures have been prepared for which the spin lifetimes approach limits expected from theory. In particular, spin lattice relaxation times of milliseconds and spin coherence times of microseconds have been reported. Recent efforts to control and modify the spin state show moderate success and more progress can be expected in this area soon. Spin states continue to be read out by photoluminescence and Faraday rotation.

The advances in theory, materials and techniques have led to important demonstrations. Coherent population trapping has been shown for donors in GaAs. This is an important precursor to electromagnetically induced transparency and slow light. In addition, the polarization state of light has been faithfully transferred to the spin state of the electrons. This is an important precursor to an optical repeater for quantum information technology.

The area is rich in interesting physics and potential applications, and continuing progress is expected.

ACKNOWLEDGMENTS

The author thanks A. Bracker, S. G. Carter, Al. L. Efros, Sophia Economou, D. Gammon, T. L. Reinecke, A. Shabaev and B. V. Shanabrook for helpful discussions and reading of the manuscript. This work was supported by ONR.

References

[1] J. M. Kikkawa, I. P. Smorchkova, N. Samarath and D. D. Awschalom, *Science* **277**, 1284 (1997).
[2] J. M. Kikkawa and D. D. Awschalom, *Phys. Rev. Lett.* **80**, 4313 (1998).
[3] A. Imamoglu, D. D. Awschalom, G. Burkard, D. P. DiVincenzo, D. Loss, M. Sherwin and A. Small, *Phys. Rev. Lett.* **83**, 4204 (1999).
[4] A. S. Bracker, D. Gammon and V. L. Korenev, *Semicond. Sci. Technol.* **23** 114004 (2008).
[5] C. J. Chang-Hasnain, P. C. Ku, J. Kim and S. L. Chuang, *Proc. IEEE* **9**, 1884 (2003).
[6] J. Kaestel, M. Fleischhauer, S. F. Yelin and R. L. Walsworth, *Phys. Rev. Lett.* **99**, 073602 (2007).
[7] F. Meier and B. P. Zakharchenya (eds.), *Optical Orientation* (North Holland, Amsterdam, 1984).
[8] D. D. Awschalom, D. Loss and N. Samarth (eds.), *Semiconductor Spintronics and Quantum Computation* (Springer, Berlin, 2002).
[9] I. Zutic, J. Fabian and S. Das Sarma, *Rev. Mod. Phys.* **76**, 323 (2004).
[10] D. R. Yakovlev and M. Bayer, in *Spin Physics in Semiconductors*, ed. M. I. Dyakonov (Springer, Berlin, 2008).
[11] D. Suter, *The Physics of Laser-Atom Interactions* (Cambridge University Press, Cambridge, 1997).
[12] S. Geschwind and R. Romestain, in *Light Scattering in Solids IV*, eds. M. Cardona and G. Guentherodt (Springer-Verlag, Berlin, 1984), p. 151.
[13] P. Y. Yu and M. Cardona, *Fundamentals of Semiconductors*, 3rd Edn. (Springer, Berlin, 2001).
[14] D. P. DiVincenzo, in *Semiconductor Spintronics and Quantum Computation*, eds. D. D. Awschalom, D. Loss and N. Samarth (Springer, Berlin, 2002), p. 221.
[15] G. Finkelestein, V. Umansky, I. Bar-Joseph, V. Ciulin, S. Haacke, J.-D. Ganiere and B. Deveaud, *Phys. Rev. B* **58**, 12637 (1998).
[16] A. Shabaev, Al. L. Efros, D. Gammon and I. A. Merkulov, *Phys. Rev. B* **68**, 201305(R) (2003).
[17] S. E. Economou, R.-B. Liu, L. J. Sham and D. G. Steel, *Phys. Rev. B* **71**, 195327 (2005).
[18] T. A. Kennedy, A. Shabaev, M. Scheibner, Al. L. Efros, A. S. Bracker and D. Gammon, *Phys. Rev. B* **73**, 045307 (2006).
[19] J. Tribollet, F. Bernadot, M. Menant, G. Karczewski, C. Testelin and M. Chamarro, *Phys. Rev. B* **68**, 235316 (2003).
[20] M. V. Gurudev Dutt, J. Cheng, B. Li, X. Xu, X. Li, P. R. Berman, D. G. Steel, A. S. Bracker, D. Gammon, S. E. Economou, R.-B. Liu and L. J. Sham, *Phys. Rev. Lett.* **94**, 227403 (2005).
[21] I. A. Merkulov, Al. L. Efros and M. Rosen, *Phys. Rev. B* **65**, 205309 (2002).
[22] A. Greilich, D. R. Yakovlev, A. Shabaev, Al. L. Efros, I. A. Yugova, R. Oulton, V. Stavarache, D. Reuter, A. D. Wieck and M. Bayer, *Science* **313**, 341 (2006).
[23] T. A. Kennedy, A. S. Bracker, S. G. Carter, S. E. Economou, D. Gammon and J. Whitaker, *Proc. SPIE* **6903**, 69030Q (2008).
[24] L. M. Woods, T. L. Reinecke and Y. Lyanda-Geller, *Phys. Rev. B* **66**, 161318(R) (2002).
[25] V. N. Golovach, A. Khaetskii and D. Loss, *Phys. Rev. Lett.* **93**, 16601 (2004).
[26] J. R. Petta, A. C. Johnson, J. M. Taylor, E. A. Laird, A. Yacoby, M. D. Lukin, C. M. Marcus, M. P. Hanson and A. C. Gossard, *Science* **309**, 2180 (2005).
[27] K.-M. C. Fu, W. Yeo, S. Clark, C. Santori, C. Stanley, M. C. Holland and Y. Yamamoto, *Phys. Rev. B* **74**, 121304(R) (2006).
[28] D. Pines, J. Bardeen and C. P. Slichter, *Phys. Rev.* **106**, 489 (1957).

[29] S. E. Economou, L. J. Sham, Y. Wu and D. G. Steel, *Phys. Rev. B* **74**, 205415 (2006).
[30] S. M. Clark, K.-M. Fu, T. D. Ladd and Y. Yamamoto, *Phys. Rev. Lett.* **99**, 040501 (2007).
[31] J. A. Gupta, R. Knobel, N. Samarth and D. D. Awschalom, *Science* **292**, 2458 (2001).
[32] Y. Wu, E. D. Kim, X. Xu, J. Cheng, D. G. Steel, A. S. Bracker, D. Gammon, S. E. Economou and L. J. Sham, *Phys. Rev. Lett.* **99**, 097402 (2007).
[33] Y. Shen, A. M. Goebel and H. Wang, *Phys. Rev. B* **75**, 045341 (2007).
[34] S. G. Carter, Z. Chen and S. T. Cundiff, *Phys. Rev. B* **76**, 201308(R) (2007).
[35] J. Berezovsky, M. H. Mikkelsen, N. G. Stoltz, L. A. Coldren and D. D. Awschalom, *Science* **320**, 349 (2008).
[36] K.-M. Fu, S. M. Clark, C. Santori, C. R. Stanley, M. C. Holland and Y. Yamamoto, *Nature Phys.* **4**, 780 (2008).
[37] S. E. Economou and T. L. Reinecke, *Phys. Rev. Lett.* **99**, 217401 (2007).
[38] T. A. Kennedy, J. Whitaker, A. Shabaev, A. S. Bracker and D. Gammon, *Phys. Rev. B* **74**, 161201(R) (2006).
[39] S. G. Carter, A. Shabaev, S. E. Economou, T. A. Kennedy, A. S. Bracker and T. L. Reinecke, *Phys. Rev. Lett.* **102**, 167403 (2009).
[40] S. E. Harris, *Phys. Today* **50**(7), 36 (1997).
[41] K.-M. C. Fu, C. Santori, C. Stanley, M. C. Holland and Y. Yamamoto, *Phys. Rev. Lett.* **95**, 187405 (2005).
[42] R. Vrijen and E. Yablonovitch, *Physica E* (Amsterdam) **10**, 569 (2001).
[43] H. Kosaka, H. Shigyou, Y. Mitsumori, Y. Rikitake, H. Imamura, T. Kutsuwa, K. Arai and K. Edamatsu, *Phys. Rev. Lett.* **100**, 096602 (2008).

Chapter Nine

Zero-Bias Spin Separation

V. V. Bel'kov[*,†] and S. D. Ganichev[*,‡]

[*]*Terahertz Center, University of Regensburg, 93040 Regensburg, Germany*
[†]*A. F. Ioffe Physico-Technical Institute, Russian Academy of Sciences, 194021 St. Petersburg, Russia*
[‡]*sergey.ganichev@physik.uni-regensburg.de*

Zero-bias spin separation generated by homogeneous optical excitation with terahertz radiation in quantum wells is reviewed. In gyrotropic semiconductor structures spin-dependent asymmetry of electron scattering induces a pure spin current which results in a spin separation. We consider the relaxation mechanism yielding the spin current due to the energy relaxation of a heated electron gas and the excitation mechanism caused by the scattering assisted free carrier absorption. An experimental access to these phenomena provides the application of an external magnetic field converting the spin current into a measurable net electric current. We discuss the microscopic and phenomenological theory of these effects, give an overview of experimental data, and address several applications.

9.1 INTRODUCTION

The spin-orbit coupling provides a versatile tool to generate and to manipulate the spin degree of freedom in low-dimensional semiconductor structures. Band spin splitting, Dyakonov–Perel' spin relaxation and spin manipulation by means of an electric field (for reviews see Refs. 1–11), the spin-galvanic effect where a non equilibrium spin polarization drives an electrical current (for reviews see Refs. 4, 8–14) or the reverse process in which an electrical current generates a nonequilibrium spin polarization[14–22] are all consequences of the spin-orbit interaction. One of the manifestations of the spin-orbit coupling is a generation of pure spin currents in low-dimensional semiconductor structures resulting in a spin separation. Pure spin currents represent a nonequilibrium state, where free carriers with spin up propagate mainly in one direction and an equal number of spin-down carriers propagate in the opposite direction. The state is characterized by zero electric current because electron charge flows of opposite spins cancel each other. However,

Handbook of Spintronic Semiconductors by W M Chen & I A Buyanova

www.panstanford.com
978-981-4267-36-6

this leads to a spatial separation and accumulation of oppositely oriented spins at the edges of the sample. Pure spin currents in semiconductors can be induced by an electric field, like the spin Hall effect (for reviews see Refs. 9–11) or driven by optical means under inter-band optical transitions in non-centrosymmetric bulk and low-dimensional semiconductors (see, for example Refs. 11, 23–25). In gyrotropic semiconductor structures spin currents can also be obtained without driving a current (zero-bias spin separation).[26–28] Zero-bias spin separation is caused by spin-dependent scattering and can be achieved in various ways but all of them must drive the electron gas into a nonequilibrium state. One straightforward method to obtain the zero-bias spin separation is based on electron gas heating followed by a spin-dependent energy relaxation of hot carriers. Electron gas heating can be achieved by, for example, Drude absorption of terahertz (THz) or microwave radiation. Moreover, due to the fact that Drude absorption is accompanied by electron scattering by phonons, static defects, etc. it, itself, becomes in gyrotropic media spin dependent and yields the spin separation. In this chapter we consider microscopic processes resulting in zero-bias spin separation and give an overview of experimental investigations and applications of this phenomenon. The chapter is organized as follows. In Sec. 9.2 an overview of mechanisms yielding the zero-bias spin separation in quantum wells (QWs) is given. We focus the attention here on the generation of pure spin currents due to absorption of terahertz radiation. In Sec. 9.3 we discuss the conversion of pure spin currents into a net electric current by application of an external field which provides an experimental access to the zero-bias spin separation. In Sec. 9.4 the experimental results are presented and discussed in view of the theoretical background. Finally, in Sec. 9.5 we discuss applications of this phenomenon. In particular, we focus on the analysis of bulk inversion asymmetry (BIA) and structural inversion asymmetry (SIA) causing the zero magnetic field band spin splitting in QWs.

9.2 MICROSCOPIC MODEL

Pure spin currents resulting in the zero-bias spin separation come from asymmetry of the electron scattering in gyrotropic semiconductor structures.[26] Scattering of electrons involves a transition from a state with wave vector **k** to a state with wave vector **k**′ which is usually considered to be spin-independent. However, in gyrotropic media, e.g., GaAs/AlGaAs quantum wells or heterojunctions, the spin-orbit interaction adds an asymmetric spin-dependent term to the scattering matrix element[29]:

$$V_{\mathbf{k}\mathbf{k}'} = V_0 + \sum_{\alpha\beta} V_{\alpha\beta}\sigma_\alpha(k_\beta + k'_\beta)\,, \tag{9.1}$$

where V_0 describes the conventional spin-independent scattering and the term proportional to the second rank pseudo-tensor $V_{\alpha\beta}$ yields the asymmetric spin-dependent contribution responsible for the effects described here. The asymmetric spin-dependent scattering matrix element is linear in wave vector **k** and the Pauli spin matrices σ.[a] Microscopically, this term is caused by structural inversion symmetry and/or bulk inversion asymmetry (for recent reviews see Refs. 9–11).[a]

Two mechanisms of the zero-bias spin separation, based on the asymmetric spin-dependent scattering given by Eq. (9.1), have been considered so far.[26] One of them is due to the asymmetry of optical transitions (photoexcitation mechanism) and the other is caused by the energy relaxation of photoexcited carriers (relaxation mechanism). Both mechanisms can be invoked by absorption of terahertz radiation.

9.2.1 Photoexcitation Mechanism of a Pure Spin Current

Absorption of THz radiation with the photon energy not high enough to excite direct inter-band or inter-subband transitions is usually achieved by indirect intra-subband transitions and, therefore, is accompanied by scattering. Due to momentum and energy conservation, intra-subband transitions include a momentum transfer from phonons and impurities to electrons. This type of absorption is also frequently called Drude or free carrier absorption. A process actuating spin separation due to indirect optical transitions is illustrated in Fig. 9.1(a). The figure sketches the Drude absorption via virtual states for a spin-up subband ($s = +1/2$, left panel) and a spin-down subband ($s = -1/2$, right panel) of a gyrotropic quantum well containing a two-dimensional electron gas (2DEG). Vertical arrows indicate optical transitions at the initial state $k_x = 0$ while the horizontal arrows describe an elastic scattering event to a final state with either positive or negative electron wave vector k'_x. While, for simplicity, we have only drawn transitions starting from $k_x = 0$, the argument holds for arbitrary k_x. Due to the spin dependence of scattering described by Eq. (9.1), transitions to positive and negative k'_x-states occur with unequal probabilities. This is indicated by horizontal arrows of different thicknesses. Since the asymmetric part of electron scattering is proportional to components of $[\sigma \times \mathbf{k}']$, probabilities for scattering to positive or negative k'_x are inverted for spin-down and spin-up subbands.[b]

The asymmetry in absorption probabilities causes an imbalance in the distribution of photoexcited carriers for both subbands ($s = \pm 1/2$) between positive and negative k'_x-states. Therefore, electrons with spin-up and spin-down move in opposite directions yielding electron flows $\mathbf{i}_{\pm 1/2}$ within each spin subband. As a result a spin current $\mathbf{j}_s = \frac{1}{2}(\mathbf{i}_{+1/2} - \mathbf{i}_{-1/2})$ is generated causing a spatial spin separation and spin accumulation at the sample edges.

[a]Note that in fact all terms odd in $\mathbf{k}$, including $\mathbf{k}$-cubic terms, may also contribute to spin-dependent asymmetric scattering.

[b]Here we assumed $k_x = 0$, and the presence of SIA only. We also do not consider the effect of a $\mathbf{k}$-linear spin splitting of the electron subband since it does not lead to a significant contribution to the pure spin current.[30]

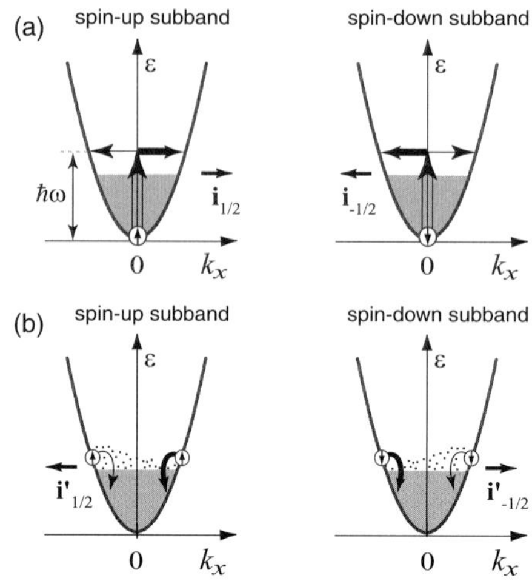

Figure 9.1. Microscopic origin of the zero-bias spin separation. To illustrate the model we draw spin-up and spin-down subbands of the conduction band separately. Scattering matrix elements linear in $\mathbf{k}$ and σ cause asymmetric scattering for both (a) excitation via Drude absorption and (b) energy relaxation. Here, scattering for the spin-up subband is assumed to have a larger probability for positive k_x than that for negative k_x and *vice versa* for the spin-down subband (after Ref. 26).

9.2.2 *Relaxation Mechanism of a Pure Spin Current*

Another root of the pure spin current stems from the energy relaxation of the radiation heated electron gas. Figure 9.1(b) sketches the process of energy relaxation of hot electrons for the spin subbands $s = \pm 1/2$ in a QW containing a two-dimensional electron gas. Energy relaxation processes are shown by curved arrows. Again we take into account the spin-dependent term in the scattering matrix element [see Eq. (9.1)] and obtain that transitions to positive and negative k'_x states occur with unequal probabilities. This fact is indicated in Fig. 9.1(b) by curved arrows of different thickness. The asymmetry causes an imbalance in the distribution of photoexcited carriers between positive and negative k_x-states in both subbands. This in turn yields electron flows $\mathbf{i}'_{\pm 1/2}$, within each spin subband. Similar to the excitation mechanism described above, probabilities of scattering to positive or negative k_x are inverted for spin-down and spin-up subbands, and therefore a pure spin current is generated and spin separation takes place.

9.3 CONVERSION OF PURE SPIN CURRENT INTO SPIN PHOTOCURRENT

A number of experimental methods aimed at the detection of the spin currents and spin separation have been developed. They comprise space resolved Faraday and Kerr rotation spectroscopy,[31] circularly polarized luminescence,[32] and transport using ferromagnetic contacts (see, for example, Refs. 33 and 34). Spin separation also manifests itself by generation of an electric current in the spin-polarized system, as in the well-known anomalous Hall effect (for a recent review see Ref. 11). Experimental investigations of the zero-bias spin separation are based on such a conversion of the spin current into a spin-polarized charge current. This is achieved by applying an external magnetic field, which polarizes spins due to the Zeeman effect. As a result, an imbalance between the fluxes $\mathbf{i}_{\pm 1/2}$ is introduced, giving rise to a net electric current $\mathbf{j} = e\,(\mathbf{i}_{+1/2} + \mathbf{i}_{-1/2})$, where e is the electron charge.[26] The generation of a spin photocurrent caused by the zero-bias spin separation in the presence of a magnetic field can be attributed to the class of magneto-gyrotropic photogalvanic effects.[27, 35]

9.3.1 Spin Photocurrent Due to the Photoexcitation Mechanism

First we consider the generation of the photocurrent caused by the zero-bias spin separation as a result of the photoexcitation mechanism (see Sec. 9.2.1). An external magnetic field $\mathbf{B}$ results in different equilibrium populations of the two spin subbands due to the Zeeman effect. This is shown in Fig. 9.2, where the Zeeman splitting, shifting the two parabolas vertically by $\pm g\mu_B B/2$, is largely exaggerated to simplify the visualization. Obviously, Drude absorption is proportional to the free carrier concentration. Therefore, magnitudes of the fluxes $\mathbf{i}_{\pm 1/2}$ shown in Fig. 9.1(a) are proportional to the free carrier densities in spin-up and spin-down subbands, $n_{\pm 1/2}$. Thus, in a Zeeman spin-polarized system (Fig. 9.2) the oppositely directed fluxes no longer compensate each other and hence yield a net electric current. This current is given by

$$\mathbf{j} = e(\mathbf{i}_{1/2} + \mathbf{i}_{-1/2}) = 4eS\mathbf{j}_s\,, \tag{9.2}$$

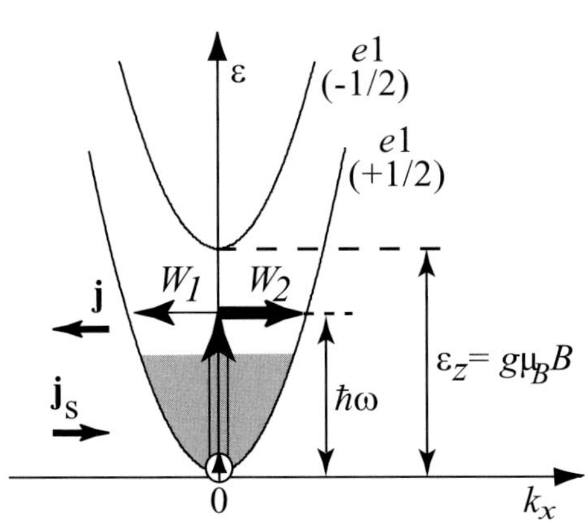

Figure 9.2. Microscopic origin of the spin photocurrent caused by the photoexcitation mechanism in the presence of an in-plane magnetic field. The spin subband $(+1/2)$ is preferably occupied due to Zeeman splitting. The rates of optical transitions for negative and positive k_x are different, $W_1 < W_2$.

where $S = (1/2)(n_{1/2} - n_{-1/2})/(n_{1/2} + n_{-1/2})$ is the average spin per particle and $\mathbf{j}_s$ the pure spin current in the absence of magnetic field. We recall that in equilibrium the average spin is given by

$$S = \varepsilon_Z / 4\bar{\varepsilon}\,, \tag{9.3}$$

where $\varepsilon_Z = -g\mu_B B$ is the Zeeman spin splitting, $\bar{\varepsilon}$ is the characteristic electron energy being equal to the Fermi energy ε_F, or to the thermal energy $k_B T$, for a degenerated or a non-degenerated 2DEG, respectively. Therefore, the photocurrent given by Eq. (9.2) depends linearly on the magnetic field and, in particular, reverses its direction upon the inversion of magnetic field orientation.

In the limiting case of a fully spin-polarized electron gas sketched in Fig. 9.2, electron flow in one of the spin subbands vanishes. Therefore, the electric current becomes independent of the magnetic field strength and carrier statistics and is given by $\mathbf{j} = \mp 2e\mathbf{j}_s$, where $\mp$ corresponds to $\pm$ sign of the Zeeman splitting. Experimentally, full spin polarization in reasonable magnetic fields can be realized in diluted magnetic semiconductors (see Sec. 9.3.3).

A microscopic theory of the spin photocurrent due to the photoexcitation mechanism is developed in Ref. 26. The photocurrent is analyzed within the framework of the spin-density matrix for impurity scattering. The scattering asymmetry induced contribution to the photocurrents is given by

$$\mathbf{j} = \sum_{s\mathbf{k}} e\mathbf{v}_{\mathbf{k}} \delta f_{s\mathbf{k}} = e\frac{2\pi}{\hbar} \sum_{s\mathbf{k}\mathbf{k}'} \tau_p (\mathbf{v}_{\mathbf{k}} - \mathbf{v}'_{\mathbf{k}}) \\ \times \left| M^{\text{ex}}_{s\mathbf{k},s\mathbf{k}'} \right|^2 (f_{s\mathbf{k}'} - f_{s\mathbf{k}}) \delta(\varepsilon_{\mathbf{k}} - \varepsilon_{\mathbf{k}'} - \hbar\omega)\,. \tag{9.4}$$

Here, $\mathbf{v}_{\mathbf{k}} = \hbar\mathbf{k}/m^*$ is the electron velocity, m^* the effective electron mass, $\delta f_{s\mathbf{k}}$ the fraction of the carrier distribution function stemming from optical transitions in the spin subband s, $M^{\text{ex}}_{s\mathbf{k},s\mathbf{k}'}$ the matrix element of the indirect optical transition, $f_{s\mathbf{k}}$ the equilibrium distribution function of carriers in the spin subband s, $\varepsilon_{\mathbf{k}} = \hbar^2 k^2 / 2m^*$ the electron kinetic energy for in-plane motion, and s an index enumerating subbands with spin states $\pm 1/2$ along the direction of the external magnetic field. To the first order in spin-orbit interaction the compound matrix element for the indirect optical transitions via impurity scattering has the form[36]

$$M^{\text{ex}}_{\mathbf{k},\mathbf{k}'} = \frac{eA}{c\omega m^*}\mathbf{e}\cdot(\mathbf{k} - \mathbf{k}')V_{\mathbf{k}\mathbf{k}'} - 2\frac{eA}{c\hbar}\sum_{\alpha\beta} V_{\alpha\beta}\sigma_\alpha e_\beta\,. \tag{9.5}$$

Here, $\mathbf{A} = A\mathbf{e}$ is the vector potential of the electromagnetic wave, c the light velocity and $V_{\mathbf{k}\mathbf{k}'}$ the scattering matrix element given by Eq. (9.1). The first term on the right-hand side of Eq. (9.5) describes transitions involving virtual intermediate states in the conduction band while the second term corresponds to transitions via virtual intermediate states in the valence band.

By using Eqs. (9.1), (9.4) and (9.5) an expression for the electric current $\mathbf{j}$ can be derived. We consider the free carrier absorption to be accompanied by electron scattering from short-range static defects and assume therefore that the matrix element V_0 and the coefficients $V_{\alpha\beta}$ are wave vector independent. For (001)-grown asymmetric QW structures belonging to the C_{2v} point group there are only two nonzero components of the tensor $\mathbf{V}$: V_{xy} and V_{yx}, where

$$x \| [1\bar{1}0]\,, \quad y \| [110]\,. \tag{9.6}$$

For linearly polarized light at normal incidence and an in-plane magnetic field B_y the photocurrent contributions parallel and perpendicular to the magnetic field can be written as

$$j_x = -2(e_x^2 - e_y^2) V_{yx} S_y \frac{e\tau_p}{\hbar V_0} I\eta(\omega)\,, \tag{9.7}$$

$$j_y = -4 e_x e_y V_{xy} S_y \frac{e\tau_p}{\hbar V_0} I\eta(\omega)\,, \tag{9.8}$$

where the photon energy $\hbar\omega$ is assumed to be smaller than the characteristic energy $\bar{\varepsilon}$; I, $\mathbf{e}$ and $\eta(\omega)$ are radiation intensity, polarization vector and absorbance, respectively.

From Eqs. (9.7) and (9.8) it follows that the photoexcitation mechanism results in a polarization-dependent photocurrent. The polarization dependences for transverse and longitudinal photocurrents with respect to the magnetic field orientation are given by

$$j_x \propto (e_x^2 - e_y^2) = \cos 2\alpha\,, \quad j_y \propto 2 e_x e_y = \sin 2\alpha\,, \tag{9.9}$$

where α is the angle between the polarization plane of the light and the x-axis. We note that the polarization behavior of j_x and j_y is independent of the scattering mechanism of Drude absorption and, as result, of temperature or wavelength.

9.3.2 Spin Photocurrent Due to the Relaxation Mechanism

The Zeeman effect also disturbs the balance between electron flows of opposite spins caused by the relaxation mechanism (Sec. 9.2.2) resulting in a net photocurrent. In the case of low spin polarization, the equilibrium electron spin per electron is determined by Eq. (9.3) and the net electric current caused by the relaxation mechanism is given by

$$\mathbf{j} = -4e \frac{\varepsilon_Z}{4\bar{\varepsilon}} \left(n_e \frac{\partial \mathbf{j}_s}{\partial n_e} \right)\,. \tag{9.10}$$

The pure spin current $\mathbf{j}_s$ is considered here as a function of the carrier density n_e. In particular, for the Boltzmann statistics, where $\mathbf{j}_s \propto n_e$ and, therefore, $n_e \partial \mathbf{j}_s / \partial n_e = \mathbf{j}_s$, Eq. (9.10) yields

$$\mathbf{j} = -e \frac{\varepsilon_Z}{k_B T} \mathbf{j}_s\,. \tag{9.11}$$

For the Fermi distribution, the derivative $\partial \mathbf{j}_s / \partial n_e$ vanishes if the spin current is caused solely by $\mathbf{k}$-linear terms in the matrix element of the electron–phonon interaction but it is nonzero if terms of higher order in $\mathbf{k}$ contribute to the spin current.

The microscopic theory of the magneto-induced photocurrent caused by the energy relaxation is developed in Ref. 30. The treatment is also based on the spin-density-matrix formalism and is presented here for acoustic phonon–mediated electron scattering. The energy relaxation of spin-polarized hot carriers in gyrotropic structures and in the presence of a magnetic field is accompanied by the generation of an electric current which is given by

$$\mathbf{j} = e \sum_{s\mathbf{k}\mathbf{k}'} \tau_p (\mathbf{v}_\mathbf{k} - \mathbf{v}_{\mathbf{k}'}) f_{s\mathbf{k}'} (1 - f_{s\mathbf{k}}) w_{s\mathbf{k} \leftarrow s\mathbf{k}'} , \tag{9.12}$$

where the index s designates the spin state, and $w_{s\mathbf{k}\leftarrow s\mathbf{k}'}$ the rate of phonon-induced electron scattering. The scattering rate has the form

$$w_{s\mathbf{k}\leftarrow s\mathbf{k}'} = \frac{2\pi}{\hbar} \sum_{\mathbf{q},\pm} \left| M^{\rm rel}_{s\mathbf{k},s\mathbf{k}'} \right|^2 \left(N_\mathbf{q} + \frac{1}{2} \pm \frac{1}{2} \right) \delta(\varepsilon_\mathbf{k} - \varepsilon_{\mathbf{k}'} \pm \hbar\Omega_\mathbf{q}) , \tag{9.13}$$

where $\Omega_\mathbf{q}$ and $\mathbf{q}$ are the frequency and wave vector of the phonon involved, $M^{\rm rel}_{s\mathbf{k},s\mathbf{k}'}$ is the matrix element of electron–phonon interaction, $N_\mathbf{q}$ the phonon occupation number, and the signs "$\pm$" correspond to the phonon emission and absorption.

Taking into account $\mathbf{k}$-linear contributions to the electron–phonon interaction, the matrix elements of electron scattering by phonons in (001)-grown QWs can be modeled by[30]

$$M^{\rm rel}_{\mathbf{k},\mathbf{k}'} = \mathcal{A}(q_z) + \mathcal{B}(q_z)[\sigma_x(k_y + k'_y) - \sigma_y(k_x + k'_x)] , \tag{9.14}$$

where $\mathcal{A}(q_z)$ and $\mathcal{B}(q_z)$ are material parameters determined by the QW structure, and q_z is the phonon wave vector component along the growth direction. The photocurrent is obtained assuming low electron gas heating by the radiation so that the electron temperature only slightly exceeds the lattice temperature. In the non-degenerated case and in the quasi-elastic approximation the photocurrent is given by

$$j_x = 4e\tau_p S_y \xi I\eta/\hbar , \quad j_y = -4e\tau_p S_x \xi I\eta/\hbar , \tag{9.15}$$

where $\xi = \sum_{q_z} A(q_z)B(q_z)|q_z| / \sum_{q_z} A^2(q_z)|q_z|$ is a parameter which is determined by the ratio of the spin-dependent and spin-independent parts of the electron–phonon interaction given by Eq. (9.14). In contrast to the current caused by the photoexcitation mechanism [see Eq. (9.9)] the photocurrent due to relaxation is independent of the radiation polarization state and can be generated even by unpolarized light.

9.3.3 Spin Photocurrents in Diluted Magnetic Semiconductors

Diluted magnetic semiconductors (DMS) are traditionally defined as diamagnetic semiconductors doped with a few to several atomic percent of some transition metal with unpaired d electrons. The exchange interaction between the localized electrons of d shells of the magnetic ions and delocalized band carrier states induce many spectacular features of DMS, such as giant Zeeman splitting of the electronic bands and the giant Faraday rotation (for reviews see Refs. 11, 37–39). Naturally, the conversion of the zero-bias spin separation into the net electric current is also strongly affected by the magnetic properties of DMS structures. Spin polarization of magnetic ions not only enhances the spin photocurrent due to the giant Zeeman effect but also disturbs the balance between electron flows with opposite spins due to spin-dependent scattering by localized magnetic ions. As for other spin-dependent phenomena in DMS the strength can be widely tuned by temperature, magnetic field and concentration of the magnetic ions. Moreover, the interplay of intrinsic and exchange effects may result in the sign inversion of the spin photocurrent.

First, we discuss the manifestation of the giant Zeeman effect. To be specific we consider (Cd, Mn)Te/(Cd, Mg)Te DMS quantum wells with Mn^{2+} magnetic ions. As we discussed in Secs. 9.3.1 and 9.3.2, the photocurrent generated due to both excitation and relaxation mechanisms is proportional to the energy separation of the spin-up and spin-down subbands [see Eqs. (9.2) and (9.10)], which in $Cd_{1-x}Mn_xTe$ QW reads[37]

$$\varepsilon_Z(B) = g_e \mu_B B + x S_0 N_0 \alpha \mathrm{B}_{5/2}\left(\frac{5\mu_B g_{Mn} B}{2k_B(T_{Mn}+T_0)}\right). \tag{9.16}$$

Here, the first term defines the intrinsic spin splitting with the electron g-factor $g_e = -1.64$.[40] The second term is caused by exchange and contributed by the Mn g-factor, $g_{Mn} = 2$, and the temperature of the Mn-spin system, T_{Mn}. Here, $\mathrm{B}_{5/2}(\gamma)$ is the modified Brillouin function, $N_0\alpha = 220\,\mathrm{meV}$ the exchange integral and N_0 the number of cations per unit volume. Phenomenological parameters S_0 and T_0 allow one to account for the Mn–Mn antiferromagnetic interactions within the magnetic ion system. The strong temperature dependence of the exchange contribution to the Zeeman splitting, given by the Brillouin function $\mathrm{B}_{5/2}(\gamma)$, determines the temperature behavior of the spin photocurrent. In particular, it results in the reversing of the photocurrent direction by temperature variation due to the fact that the temperature independent intrinsic and T-dependent exchange contributions to the Zeeman splitting have opposite signs.

While the giant Zeeman effect only quantitatively changes the photocurrent, the mechanism we consider now is specific for DMS and is caused by the well-known spin-dependent electron scattering by polarized magnetic ions.[37] This electron scattering mechanism in DMS is effective because Mn atoms are introduced into the QW region. In external magnetic fields, when the Mn^{2+} ions are spin polarized, the scattering rate of electrons with the spins aligned parallel and antiparallel to the Mn spins becomes different. This results in two unequal momentum relaxation times ($\tau_{p,+1/2}$ and $\tau_{p,-1/2}$) in the spin subbands. Since the electron fluxes $\mathbf{i}_{\pm 1/2}$ are proportional to $\tau_{p,\pm 1/2}$, the polarization of Mn spins leads to a net electric current. In Ref. 28 the photocurrent has been obtained assuming that the momentum relaxation of electrons is governed by their interaction with the Mn^{2+}

ion localized in a QW. The corresponding Hamiltonian is given by[37]

$$H_{e-\mathrm{Mn}} = \sum_i [u - \alpha(\mathbf{S}_i \cdot \sigma)]\delta(\mathbf{r} - \mathbf{R}_i)\,, \tag{9.17}$$

where i is the Mn^{2+} ion index, $\mathbf{S}_i$ the vector composed of the matrices of the angular momentum $5/2$, u the scattering potential without exchange interaction, $\mathbf{r}$ the electron coordinate, and $\mathbf{R}_i$ the Mn^{2+} ion position. The electron scattering by Mn potential, determined by u, is usually stronger than the exchange scattering determined by α. Note that the parameter α in Eq. (9.17) is also responsible for the giant Zeeman splitting in Eq. (9.16). Then, in the case of $|\alpha| \ll |u|$, we derive

$$\mathbf{j} = 4e\frac{\alpha}{u}\mathbf{j}_s S_{\mathrm{Mn}}\,, \tag{9.18}$$

where S_{Mn} is the average Mn spin along the magnetic field direction. The photocurrent due to the spin-dependent scattering has the same direction as that due to the giant Zeeman splitting because the average electron spin caused by the giant Zeeman effect is parallel to $\mathbf{S}_{\mathrm{Mn}}$.

The total electric current caused by an imbalance of the oppositely directed spin-polarized flows of electrons in the magnetic field is given by the sum of both contributions. This holds until the electron gas becomes completely spin polarized, which can be achieved in diluted magnetic semiconductors in reasonable magnetic fields. For the limiting case of full polarization all electrons occupy only one spin subband and one of the spin flows vanishes. In this case the photocurrent is due to the giant Zeeman effect only. It becomes independent of the magnetic field strength and its strengths are given by $\mathbf{j} = \mp 2e\mathbf{j}_s$, where $\mp$ corresponds to $\pm$ sign of the Zeeman splitting.

9.3.4 Phenomenological Theory of the Magneto-gyrotropic Photogalvanic Effect

In Secs. 9.3.1–9.3.3 we considered various microscopic mechanisms of the spin photocurrent formation. However, many macroscopic features of the magneto-gyrotropic photogalvanic effect, its behavior upon variation of radiation polarization, crystallographic orientation, experimental geometry, etc. can be described in the frame of a phenomenological theory.[35] It operates with conventional (*polar*) vectors and pseudo (*axial*) vectors, and does not depend on details of microscopic mechanisms.

Phenomenologically, the magnetic-field induced photocurrent in a linear approximation in the magnetic field $\mathbf{B}$ is given by

$$j_\lambda = I\Phi_{\lambda\mu\nu\eta}B_\mu e_\nu e^*_\eta\,, \tag{9.19}$$

where $\mathbf{\Phi}$ is a fourth-order tensor relating the electric current with the magnetic field. It has nonzero components in all systems lacking a center of inversion symmetry.

We analyze magneto-gyrotropic photogalvanic effects induced in asymmetric (001)-grown zinc-blende QW structures of the C_{2v} symmetry. For normally-incident linearly polarized radiation and an in-plane magnetic field, a configuration relevant to most experiments considered below, the phenomenological Eq. (9.19) reduces to[35]

$$j_x = S_1 B_y I + S_2 B_y (e_x^2 - e_y^2) I + 2S_3 B_x e_x e_y I \,, \tag{9.20}$$
$$j_y = S_1' B_x I + S_2' B_x (e_x^2 - e_y^2) I + 2S_3' B_y e_x e_y I \,.$$

Here, the parameters S_1 to S_3 and S_1' to S_3' are linearly independent invariant components of the tensor $\mathbf{\Phi}$ and the coordinate system used is given by Eq. (9.6).[c] For $\mathbf{B} \| y$ we have

$$j_x = j_1 \cos 2\alpha + j_2 \,, \quad j_y = j_3 \sin 2\alpha \,, \tag{9.21}$$

where $j_1 = S_2 B_y I$, $j_2 = S_1 B_y I$, $j_3 = 2S_3' B_y I$ and α is the angle between the polarization plane of the light and the x-axis. The azimuthal angle dependences of the transverse (j_x) and longitudinal (j_y) components of the photocurrent given by Eq. (9.21) coincide with that obtained microscopically for the photoexcitation mechanism [see Eqs. (9.9)]. For the photocurrent detected in the direction perpendicular to the magnetic field a polarization independent offset is allowed. Therefore, the effect can appear even with excitation by unpolarized radiation. This contribution is driven by the relaxation mechanism [see Eq. (9.15)]. Finally, we would like to note that the magneto-gyrotropic photogalvanic effect, in addition to the zero-bias spin separation considered here, may also arise from spin-independent orbital mechanisms. The latter comprise magnetic field induced photocurrents caused by a diamagnetic shift of energy bands[41] and diamagnetic corrections to the electron–phonon interaction,[42] which, in particular, can be dominating in systems with a weak spin-orbit coupling (see also Ref. 43 and references therein). Independently of the mechanism the magneto-gyrotropic photogalvanic effect is described by the same phenomenological equation (9.19). Thus, additional experiments aimed at exploring the microscopic origin are required.

9.4 EXPERIMENTS ON THE ZERO-BIAS SPIN SEPARATION

Generation of the spin photocurrent caused by the zero-bias spin separation in the presence of an external magnetic field (the magneto-gyrotropic photogalvanic effect) has been investigated in various low-dimensional materials like GaAs-, InAs- and DMS (Cd, Mn)Te-based quantum well structures.[26–28] In typical experiments terahertz radiation is directed onto square shaped doped quantum well structures subjected to an in-plane magnetic field. A voltage signal across a load resistor is picked up on a pair of contacts alloyed into the sample (see inset in Fig. 9.3). The

[c] All linearly independent invariant components of the tensor $\mathbf{\Phi}$ can be found in Ref. 35.

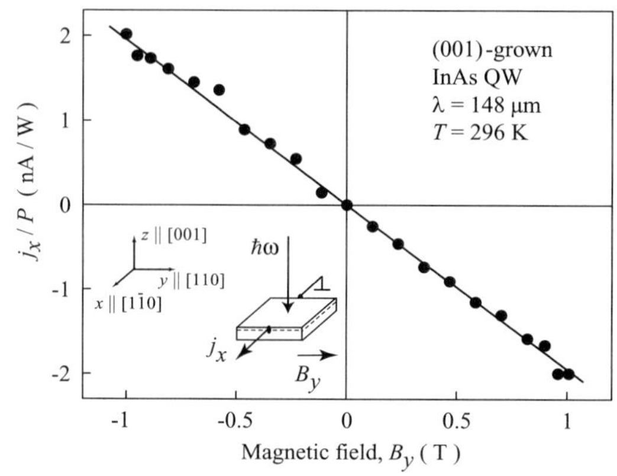

Figure 9.3. Magnetic field dependence of the photocurrent measured in an InAs QW at room temperature with the magnetic field **B** parallel to the y-axis. Data are given for normally incident unpolarized radiation of $P \approx 25\,$kW at the wavelength $\lambda = 148\,\mu$m. The current is measured in the direction *perpendicular* to **B**.

photoresponse can be detected in the external circuit with an oscilloscope or using the lock-in technique. The experiments are carried out applying radiation at normal incidence.[d]

In Fig. 9.3 a magneto-induced photocurrent is shown for (001)-grown InAs QWs at room temperature. The linear coupling between the photocurrent **j** and the magnetic field **B** is in agreement with the phenomenological Eq. (9.20). This behavior is typically observed for all non-magnetic QWs and for DMS structures at moderate fields and temperatures.

The most spectacular evidence for the spin dependent origin of the magnetic field induced photocurrent has been obtained in experiments on (Cd, Mn)Te/(Cd, Mg)Te DMS structures.[28] The photocurrent was detected at low power terahertz radiation of a continuous-wave laser operating at $\lambda = 118\,\mu$m and power P of about 1 mW. As an important result it was observed that cooling of the sample changes the signal polarity and increases its absolute values by more than two orders of magnitude. While at moderate temperatures the signal depends linearly on B, at low T the saturation of the photocurrent with rising magnetic field strength was observed (see the data for $T = 1.9\,$K in Fig. 9.4). Both the above temperature and magnetic field dependences are typical for DMS due to the exchange interaction and polarization of the Mn^{2+} ions localized in the quantum well (see Sec. 9.3.3). The sign inversion of the photocurrent with temperature decrease follows from the fact that both photocurrent contributions caused by the exchange interaction [see Eqs. (9.10) and (9.18)] have opposite signs to the photocurrent due to the intrinsic Zeeman effect and rapidly increase with decreasing temperature.

[d] In (001)-oriented structures this geometry allows one to exclude other effects known to cause photocurrents in QWs.[26] In structures of lower symmetry the possible background can be eliminated by treating the data in the form $j = [j(B) - j(-B)]/2$.

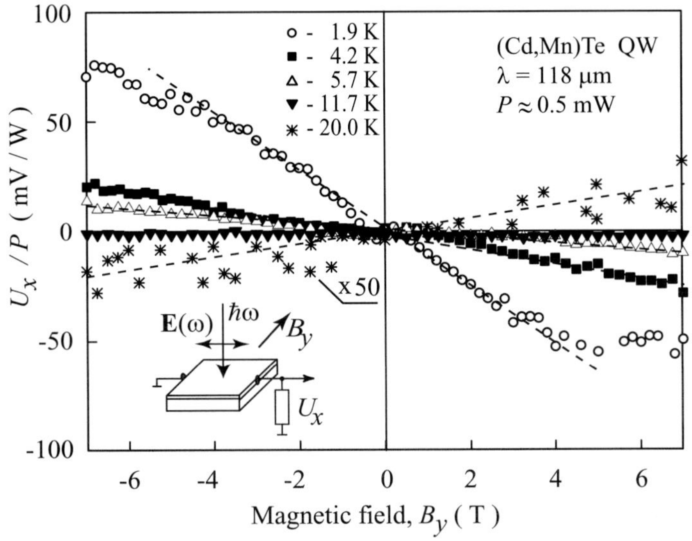

Figure 9.4. Magnetic field dependence of the voltage signal U_x normalized by the radiation power P measured in a (Cd, Mn)Te DMS QW in response to a low power excitation (after Ref. 28).

A substantial input of the photocurrent caused by the spin-dependent electron scattering due to polarized magnetic ions [see Eq. (9.18)] results in a much stronger temperature dependence than that of the giant Zeeman effect. The photocurrent saturation with rising magnetic field strength observed at low temperature is a consequence of the full polarization of Mn^{2+} ions. We note that for investigations of the spin photocurrents in DMS structures only low radiation power with P on the order of a milliwatt or even lower should be used.[28] Application of high power radiation heats the Mn^{2+} system and destroys magnetic ion polarization.[44]

The dependence of the photocurrent direction on the sign of electron g-factor was clearly demonstrated in measurements on non-magnetic GaAs/AlGaAs quantum wells of different width. It is well known that in such systems the g-factor is negative in wide enough QWs and positive if the quantum well width is smaller than $\approx$ 5 nm.[13] Experiments show that both polarization independent and polarization dependent contributions to the photocurrent in GaAs/AlGaAs QWs of 4 nm width have opposite sign with respect to that in wider QWs with a similar doping. This result agrees with Eqs. (9.2) and (9.3), demonstrating that the current direction is governed by the g-factor sign. The connection of the photocurrent to the zero-bias spin separation is also manifested by the temperature dependence of the photosignal in non-magnetic QWs shown in Fig. 9.5. In Ref. 26 it was shown that at high temperatures the current decreases as n_e/T with increasing temperature. At low temperature, however, the photocurrent becomes independent of temperature and, moreover, of the electron mobility and density. Such a behavior follows from the analysis of Eqs. (9.7) and (9.8) corresponding to the excitation

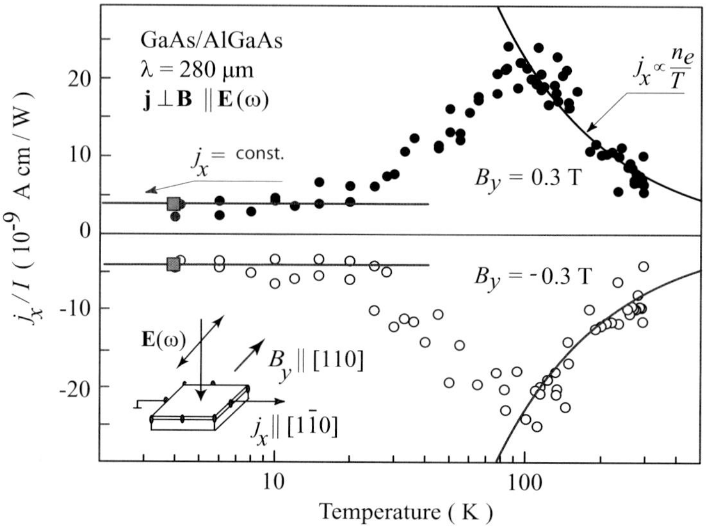

Figure 9.5. Temperature dependence of the photocurrent j_x normalized by the intensity I for a (001)-grown GaAs/AlGaAs heterojunction. The photocurrent $\mathbf{j} \perp \mathbf{B} \| y$ is excited by linearly polarized radiation of $\lambda = 280\,\mu$m under normal incidence. Full lines are fits to $j_x = \pm$const. at low T and $j_x \propto n_e/T$ at high T. The squares show the data obtained for samples additionally illuminated by visible and near infrared light. The data show that though both n_e and τ_p increase by a factor of two, the photocurrent remains unchanged. The inset shows the geometry of the experiment (after Ref. 26).

mechanism discussed in Sec. 9.3.1. These equations show that the temperature dependence of the photocurrent j is determined by the product of the average spin S and momentum relaxation time τ_p. Thus, for a fixed scattering mechanism, e.g., phonon or impurity scattering, the current is given by $j \propto \eta(\omega)\tau_p S$. At $\omega\tau_p \gg 1$, which is valid for most THz experiments on high mobility samples, absorbance is proportional to n_e/τ_p (see Ref. 45), and therefore $\eta(\omega)\tau_p$ is independent of the electron mobility. The only temperature-dependent parameter remaining is the equilibrium spin polarization given by Eq. (9.3). At low temperatures $S \propto 1/\varepsilon_F$ and does not depend on temperature. Furthermore, as the Fermi energy $\varepsilon_F \propto n_e$ the photocurrent becomes independent even of the electron density (see Fig. 9.5). At high temperatures electron energy distribution is sufficiently well described by the Bolzmann statistics and $S \propto 1/T$ [see Eq. (9.3)]. Therefore, the current j is proportional to n_e/T, in agreement with experiment. Such temperature dependence has also been detected for CdTe/(Cd, Mg)Te non-magnetic QWs as well for (Cd, Mn)Te/(Cd, Mg)Te DMS structures excited by high power radiation of a pulsed THz laser. In the latter case the photocurrent is due to intrinsic Zeeman splitting only and the temperature dependence of the DMS matches that of non-magnetic structures.

Both microscopical and phenomenological treatment of the magneto-gyrotropic photogalvanic effect show pronounced dependence of the photocurrent on the polarization state of the radiation. Depending on the microscopic

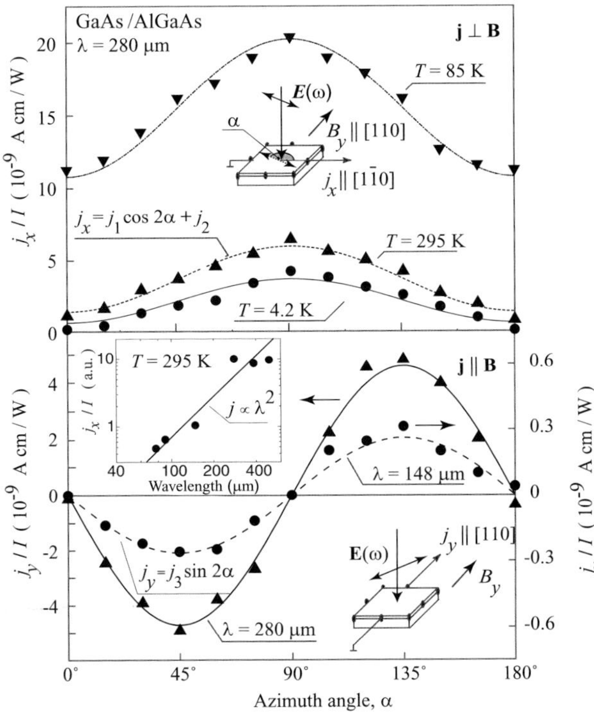

Figure 9.6. Photocurrents for a GaAs/AlGaAs heterojunction as a function of the azimuth angle α. Upper panel: photocurrent $\mathbf{j} \perp \mathbf{B} \| y$ at $\lambda = 280\,\mu$m and $T = 4.2$ K, 85 K and 295 K. Lines are fits according to $j_x = j_1 \cos 2\alpha + j_2$. Lower panel: photocurrent $\mathbf{j} \| \mathbf{B} \| y$ measured at room temperature for $\lambda = 148$ and 280 μm. Lines are fitted to $j_y = j_3 \sin 2\alpha$. Insets show the experimental geometries. An additional inset in the lower panel displays the wavelength dependence of the signal for transverse geometry; the full line shows $j_x \propto \lambda^2$ (after Ref. 26).

mechanism of the zero-bias separation, the photocurrent can be obtained applying unpolarized radiation (relaxation mechanism, Sec. 9.3.2) or polarized radiation (excitation mechanism, Sec. 9.3.1). Experimental data obtained upon changes of the azimuth angle α for the in-plane magnetic field aligned along the y-axis (see Fig. 9.6), clearly demonstrate agreement with phenomenological equations (9.21) and microscopic equations (9.7) and (9.8).[26, 35] In (001)-grown QWs an in-plane magnetic field applied along one of $\langle 110 \rangle$-directions results both in contributions of the photocurrent normal and parallel to **B**. The longitudinal current [$j_3 \sin 2\alpha$ in Eq. (9.21)] is caused by the excitation mechanism and depends on the radiation polarization state. The transverse photocurrent consists of two contributions: the polarization independent photocurrent [j_2 in Eq. (9.21)] due to the relaxation mechanism and the polarization dependent photocurrent [$j_1 \cos 2\alpha$ in Eq. (9.21)] caused by the excitation mechanism. The contributions j_1 and j_2 can easily be separated by subtracting and adding the photocurrents in response to the radiation field oriented along ($\alpha = 0°$) and normal ($\alpha = 90°$) to magnetic field **B**, respectively. For experiments with unpolarized radiation like in Fig. 9.3 the photocurrent

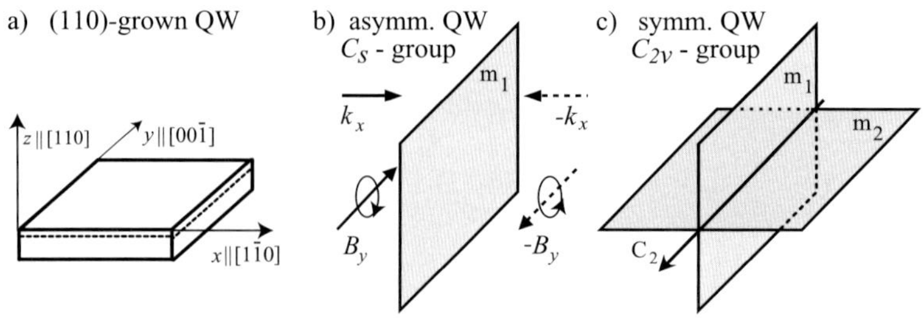

Figure 9.7. (a) Coordinate system and symmetry elements of a (b) asymmetrical and (c) symmetrical QW grown in the $z\|$ [110] direction. Arrows in (b) show reflection of the components of polar vector **j** and axial vector **B** by the mirror reflection plane m_1, demonstrating that the linear coupling of j_x and B_y is allowed for structures of this symmetry. The coupling $j_y \propto B_x$ can be obtained in a similar way. For (c) linear coupling of a current and an in-plane magnetic field is forbidden because a reflection by m_2 does not modify the components of an in-plane polar vector **j** but changes the polarity of an in-plane axial vector **B**.

can be detected in the transverse geometry only and is solely determined by the relaxation mechanism.

9.5 APPLICATION OF SPIN PHOTOCURRENTS

Spin photocurrents provide an experimental access to symmetry of low-dimensional structures. Indeed, the theory of the zero-bias spin separation (Sec. 9.2) reveals that the photocurrent is proportional to the degree of inversion asymmetry given by the values of $V_{\alpha\beta}$ in Eq. (9.1). The microscopic origin of these terms linear in electron wave vector is the structure inversion asymmetry and the bulk inversion asymmetry. BIA and SIA terms can interfere resulting in an anisotropy of the spin-dependent scattering and, consequently, of the photocurrent. Hence, the analysis of the photocurrent can be used to characterize the quantum well symmetry and the band spin splitting.

Applying this technique SIA/BIA has been investigated in great detail in (110)-oriented GaAs/AlGaAs heterostructures.[46] The particular feature of zinc-blende structure based (110)-grown QWs is that they are characterized by extraordinarily slow spin dephasing.[47–50] The reason for the long spin lifetime of several nanoseconds is the (110)-crystal orientation: the effective magnetic field due to spin-orbit coupling points into the growth direction[51] and spins oriented along this direction do not precess. Hence, the Dyakonov–Perel' spin relaxation mechanism,[52] which is based on the spin precession in the effective magnetic field and usually limits the spin lifetime of conduction electrons, is suppressed. If, however, (110)-grown QWs are asymmetric, the structural inversion symmetry is broken and Rashba spin-orbit coupling[53] causes an in-plane effective magnetic field, thus speeding-up spin dephasing.

The magneto-gyrotropic photogalvanic effect is an ideal tool to probe the symmetry of (110)-grown QWs. The photocurrent can only be observed for asymmetric

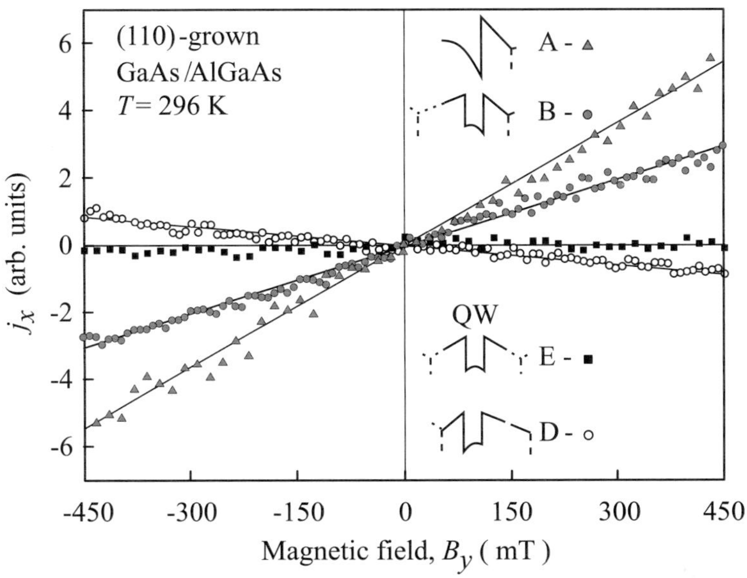

Figure 9.8. Magnetic field dependences of j_x measured in (110)-grown GaAs/AlGaAs heterostructures for the radiation polarized along the x-axis and an in-plane magnetic field, $\mathbf{B} \parallel y$. Insets sketch the band profile of QWs and doping position (vertical dashed lines). After Ref. 110.

structures but vanishes if QWs are symmetric. This follows from the symmetry arguments. For unpolarized radiation or for radiation polarized along the x-axis the photocurrent is phenomenologically determined solely by the coupling of the photocurrent (polar vector) with the magnetic field (axial vector).[e] In the asymmetric structures (C_s-point group) the only nonidentity symmetry element is the reflection plane m_1 [see Fig. 9.7(b)]. Reflection by m_1 transforms current component j_x and magnetic field component B_y ($j_x \to -j_x$, $B_y \to -B_y$) equally. Therefore, coupling $j_x \propto B_y$ is allowed. The same arguments are valid for coupling $j_y \propto B_x$. As a result, the generation of a magnetic field induced photocurrent is possible. Symmetric (110)-grown QWs which belong to the higher point group symmetry C_{2v} contain an additional mirror plane, m_2, being parallel to the QW plane [see Fig. 9.7(c)]. A reflection by m_2 does not modify the components of an in-plane polar vector $\mathbf{j}$ but changes the polarity of an in-plane axial vector $\mathbf{B}$. Therefore, in such systems linear coupling of a current and an in-plane magnetic field is forbidden.

In experiments the structural inversion asymmetry has been varied by the δ-doping position with respect to the QW. This asymmetric doping yields an asymmetric potential profile inside the QWs. The degree of SIA is reflected in the magnetic field dependence of the photocurrent displayed in Fig. 9.8. The currents shown in Fig. 9.8 are proportional to the applied field but the slope of $j_x(B_y)$ is sample dependent. The largest slope is obtained for the sample A with the

[e]Here, $x \parallel [1\bar{1}0]$, $y \parallel [00\bar{1}]$ and $z \parallel [110]$ coordinate system is used [see Fig. 9.7(a)].

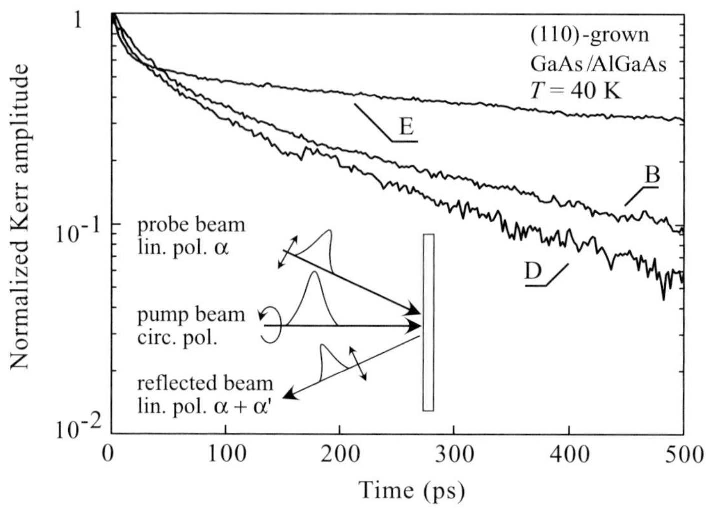

Figure 9.9. Time-resolved Kerr rotation measured in asymmetrically grown QW samples B, D and symmetrical structure E. We note that the fast decaying components in the signals at short times are due to spin relaxation of photoexcited holes. The inset shows the experimental geometry (after Ref. 46).

strongest asymmetry while the photocurrent vanishes for the symmetrically doped sample E. Furthermore, if the doping profile is reversed (from sample B to D), the slope of the photocurrent becomes reversed, too. In the case of sample E, the absence of a magnetic field induced photocurrent in an in-plane **B** indicates that the QW is highly symmetric and lacks the structure asymmetry. Hence, measurement of the photocurrents gives us an experimental handle to analyze the degree of structure inversion asymmetry and demonstrates that the position of the doping layer can be effectively used for tuning the SIA strength.

The structural inversion asymmetry determines the Rashba spin splitting and therefore controls the Dyakonov–Perel′ spin relaxation[52] for spins aligned along the z-direction. Any variation of SIA, e.g., due to asymmetric doping, should result in a variation of the spin relaxation time. The connection of the spin lifetimes τ_s with the structure asymmetry has been demonstrated applying samples characterized by the photocurrent measurements.[46] The spin lifetimes extracted from time-resolved Kerr rotation measurements are shown in Fig. 9.9 for symmetrically and asymmetrically doped QW. In agreement with the photocurrent measurements, indicating a larger degree of asymmetry of the samples B and D compared to E, τ_s in sample E is found to be more than three times larger than that in sample B and about two times larger than that in sample D.

In (110)-grown QWs zero SIA is obtained from the almost symmetrically doped QWs which set an upper limit of spin dephasing in GaAs QWs. This is in contrast to (001)-grown structures, where such a profile results in a substantial SIA.[54, 55] This essential difference stems from the lower growth temperature of (110)-oriented QWs which suppress the segregation process. The high growth

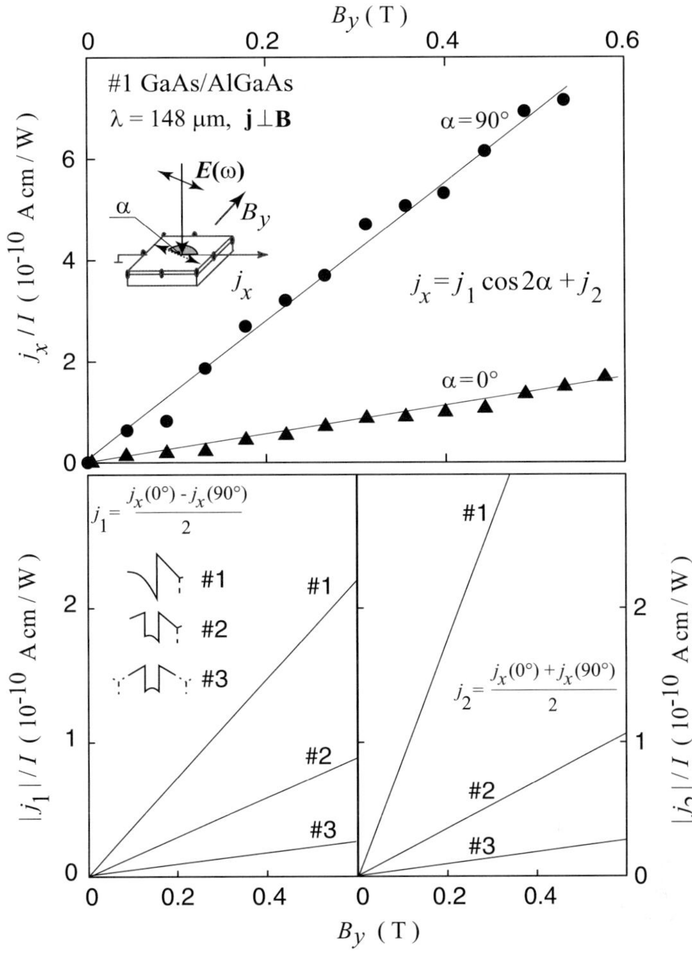

Figure 9.10. Magnetic field dependences of the transverse photocurrent for (001)-grown GaAs/AlGaAs structures. Upper panel: $j_x(B)$ normalized by radiation intensity I, measured for a heterojunction (sample 1) at room temperature for $\alpha = 90°$ and $\alpha = 0°$. Lower panels: $|j_1|/I$ (left panel) and $|j_2|/I$ (right panel) for samples 1–3. Insets sketch the band profile of the structures and doping position (vertical dashed lines) (after Ref. 26).

temperature of (001)-oriented heterostructures leads to substantial dopant migration in the growth direction[56] and results in structure asymmetry of symmetrically doped QWs. Therefore, the structure asymmetry in (001)-grown samples cannot be judged simply by relying on the growth process. The study of spin photocurrents can also provide direct access to the interference of SIA and BIA in such structures. This is demonstrated by Fig. 9.10, which shows j_1 and j_2 attributed to the photoexcitation and to the relaxation mechanisms, respectively. The data corresponds to three GaAs/AlGaAs low-dimensional structures which differ in their structural inversion asymmetry. Sample 1 is a heterojunction which has the strongest SIA contribution. Samples 2 and 3 are quantum wells of a 30 nm width, asymmetrically and symmetrically modulation doped, with larger and smaller strength of SIA, respectively. The fact that with decreasing degree of the structure inversion asymmetry the currents become lower is in excellent agreement with the picture of asymmetric scattering driven currents.

9.6 CONCLUDING REMARKS

In gyrotropic semiconductor structures spin-dependent asymmetry of electron scattering induces a pure spin current which results in the zero-bias spin separation. Two microscopic mechanisms of this effect are known so far. The excitation mechanism is based on radiation absorption and the resulting spin current can be detected only in optical experiments applying terahertz or microwave radiation. The relaxation mechanism, however, is more general and can be responsible for the spin separation in optical as well as in transport experiments. It is based on electron gas heating which can be achieved by absorption of radiation or by applying an electric field. In the latter case the zero-bias spin separation may overlap with the spin Hall effect. These two effects, however, can easily be distinguished in experiment by reversing the *d.c.* electric current polarity. While the spin Hall effect should invert its sign, the zero-bias spin separation is insensitive to the current direction. As for the future study of zero-bias spin separation and its conversion into a net electric current, experiments on structures with strong spin-orbit coupling, large g-factor or g-factor engineering would be desirable and informative. Such experiments would reveal a great deal about the momentum, energy and spin relaxation of nonequilibrium photoexcited carriers, band spin splitting, and so on.

ACKNOWLEDGMENTS

The authors thank E. L. Ivchenko, L. E. Golub, S. A. Tarasenko, P. Olbrich and W. Prettl for helpful discussions. This work is supported by the DFG (SFB689).

References

[1] F. Meier and B. P. Zakharchenya, Optical orientation, in *Modern Problems in Condensed Matter Sciences*, Vol. 8, eds. V. M. Agranovich and A. A. Maradudin (Elsevier Science Publ., Amsterdam, 1984).

[2] N. S. Averkiev, L. E. Golub and M. Willander, *J. Phys. Cond. Matt.* **14**, R271 (2002).

[3] D. D. Awschalom, D. Loss and N. Samarth, Semiconductor spintronics and quantum computation, in *Nanoscience and Technology*, eds. K. von Klitzing, H. Sakaki and R. Wiesendanger (Springer, Berlin, 2002).

[4] S. D. Ganichev and W. Prettl, *J. Phys. Cond. Matt.* **15**, R935 (2003).

[5] R. Winkler, *Spin-Orbit Coupling Effects in Two-Dimensional Electron and Hole Systems*, Springer Tracts in Modern Physics, Vol. 191 (Springer, Berlin, 2003).

[6] I. Zutic, J. Fabian and S. Das Sarma, *Rev. Mod. Phys.* **76**, 323 (2004).

[7] W. Zawadzki and P. Pfeffer, *Semicond. Sci. Technol.* **19**, R1 (2004).

[8] S. D. Ganichev and W. Prettl, *Intense Terahertz Excitation of Semiconductors* (Oxford University Press, Oxford, 2006).
[9] R. Winkler, Spin-dependent transport of carriers in semiconductors, in *Handbook of Magnetism and Advanced Magnetic Materials* (John Wiley & Sons, New York, 2007).
[10] J. Fabian *et al.*, *Acta Phys. Slov.* **57**, 565 (2007), [cond-mat/0711.1461].
[11] M. I. Dyakonov (ed.), *Spin Physics in Semiconductors* (Springer, Berlin, 2008).
[12] S. D. Ganichev *et al.*, *Nature* **417**, 153 (2002).
[13] E. L. Ivchenko, *Optical Spectroscopy of Semiconductor Nanostructures* (Alpha Science International, Harrow, 2005).
[14] S. D. Ganichev, *Phys. Rev. Lett.* **102**, 156602 (2009).
[15] E. L. Ivchenko and G. E. Pikus, *J. Exp. Theor. Phys. Lett.* **27**, 604 (1978).
[16] A. G. Aronov and Yu. B. Lyanda-Geller, *J. Exp. Theor. Phys. Lett.* **50**, 431 (1989).
[17] V. M. Edelstein, *Solid State Commun.* **73**, 233 (1990).
[18] A. G. Aronov, Yu. B. Lyanda-Geller and G. E. Pikus, *Sov. Phys. J. Exp. Theor. Phys.* **73**, 537 (1991).
[19] L. E. Vorob'ev *et al.*, *J. Exp. Theor. Phys. Lett.* **29**, 441 (1979).
[20] S. D. Ganichev *et al.*, cond-mat/0403641 (2004); see also *J. Magn. Magn. Mat.* **300**, 127 (2006).
[21] A. Yu. Silov *et al.*, *Appl. Phys. Lett.* **85**, 5929 (2004).
[22] Y. K. Kato *et al.*, *Phys. Rev. Lett.* **93**, 176601 (2004).
[23] R. D. R. Bhat *et al.*, *Phys. Rev. Lett.* **94**, 096603 (2005).
[24] S. A. Tarasenko and E. L. Ivchenko, *J. Exp. Theor. Phys. Lett.* **81**, 231 (2005).
[25] H. Zhao *et al.*, *Phys. Rev. B* **72**, 201302 (2005).
[26] S. D. Ganichev *et al.*, *Nature Phys.* **2**, 609 (2006).
[27] V. V. Bel'kov and S. D. Ganichev, *Semicond. Sci. Technol.* **23**, 114003 (2008).
[28] S. D. Ganichev *et al.*, *Phys. Rev. Lett.* **102**, 156602 (2009).
[29] E. L. Ivchenko and G. E. Pikus, *Bull. Acad. Sci. USSR Phys. Ser.* **47**, 81 (1983).
[30] S. D. Ganichev *et al.*, *Phys. Rev. B* **75**, 155317 (2007).
[31] Y. Kato *et al.*, *Science* **306**, 1910 (2004).
[32] J. Wunderlich *et al.*, *Phys. Rev. Lett.* **94**, 047204 (2005).
[33] M. Johnson and R. H. Silsbee, *Phys. Rev. Lett.* **55**, 1790 (1985).
[34] S. O. Valenzuela *et al.*, *Nature* **442**, 176 (2006).
[35] V. V. Bel'kov *et al.*, *J. Phys. Cond. Matt.* **17**, 3405 (2005).
[36] S. A. Tarasenko, *Phys. Rev. B* **73**, 115317 (2006).
[37] J. K. Furdyna, *J. Appl. Phys.* **64**, R29 (1988).
[38] T. Dietl, in *Handbook on Semiconductors*, Vol. 3b, ed. T. S. Moss (North-Holland, Amsterdam, 1994).
[39] W. J. Ossau and B. Kuhn-Heinrich, *Physica B* **184**, 442 (1993).
[40] A. A. Sirenko *et al.*, *Phys. Rev. B* **56**, 2114 (1997).
[41] A. A. Gorbatsevich *et al.*, *J. Exp. Theor. Phys. Lett.* **57**, 580 (1993).
[42] O. V. Kibis, *Phys. Lett. A* **244**, 432 (1998).
[43] H. Diehl *et al.*, *J. Phys. Cond. Matt.* **19**, 36232 (2007).
[44] D. Keller *et al.*, *Phys. Rev. B* **65**, 035313 (2002).
[45] K. Seeger, *Semiconductor Physics: An Introduction* (Springer, Berlin, 2004).
[46] V. V. Bel'kov *et al.*, *Phys. Rev. Lett.* **100**, 176806 (2008).
[47] Y. Ohno *et al.*, *Phys. Rev. Lett.* **83**, 4196 (1999).
[48] O. Z. Karimov *et al.*, *Phys. Rev. Lett.* **91**, 246601 (2003).
[49] S. Döhrmann *et al.*, *Phys. Rev. Lett.* **93**, 147405 (2004).

[50] K. C. Hall *et al.*, *Appl. Phys. Lett.* **86**, 202114 (2005).
[51] M. I. Dyakonov and V. Yu. Kachorovskii, *Sov. Phys. Semicond.* **20**, 110 (1986).
[52] M. I. Dyakonov and V. I. Perel', *Sov. Phys. Solid State* **13**, 3023 (1971).
[53] Yu. A. Bychkov and E. I. Rashba, *Sov. J. Exp. Theor. Phys. Lett.* **39**, 78 (1984).
[54] S. Giglberger *et al.*, *Phys. Rev. B* **75**, 035327 (2007).
[55] V. Lechner *et al.*, *Appl. Phys. Lett.* to be published (arXiv: 0903.1232).
[56] A. P. Mills *et al.*, *J. Appl. Phys.* **88**, 4056 (2000).

Chapter Ten

Electrical Spin Injection in Hybrid Ferromagnetic Metal/Semiconductor Structures

Pierre Renucci*, Henri Jaffrès†, Jean-Marie George†, Thierry Amand* and Xavier Marie*,‡

*Université de Toulouse, INSA-CNRS-UPS, LPCNO, 135 Avenue de Rangueil, 31077 Toulouse, France
†Unité Mixte de Physique CNRS/Thales, Route Départementale 128, 91767 Palaiseau Cedex, France and Université Paris-Sud 11, 91405 Orsay, France
‡marie@insa-toulouse.fr

An overview of the electrical spin injection from a ferromagnetic metal into a semiconductor is given. We present the fundamental theoretical obstacle known as the "impedance mismatch" problem and demonstrate from calculations based either on a diffuse or a "ballistic" approach that this obstacle can be overcome by the insertion of a tunnel barrier between the semiconductor and the ferromagnetic metal. We then review experimental results on the measurement of the electrons' spin polarization injected electrically in spin-light emitting diodes with tunnel barriers based on a Schottky barrier or a thin insulator layer such as alumina oxide or magnesium oxide.

10.1 INTRODUCTION

Efficient electrical spin injection from a ferromagnetic metal or a dilute magnetic semiconductor is the fundamental requirement of semiconductor-based spintronic devices. Spectacular results have been obtained using spin injectors based on dilute magnetic semiconductors, such as ZnMnSe.[1] Unfortunately, room temperature (RT) operation has been impossible so far because of the very low Curie temperature of these compounds.[2,3] High spin injection efficiencies have also been reported by using GaMnAs as the spin injector,[4] but still at relatively low temperature compared to 300 K. In contrast, electrical spin injection using FM metals as spin injectors has already been demonstrated at RT.[5]

Handbook of Spintronic Semiconductors by W M Chen & I A Buyanova

www.panstanford.com
978-981-4267-36-6

This chapter is organized as follows. In Sec. 10.2, we present the theory of electrical spin injection through a hybrid metal–semiconductor interface. The problem of "impedance mismatch" is reviewed and the solution brought about by a tunnel barrier between the semiconductor and the metal is explained. In Sec. 10.3, we present the experimental results on electrical spin injection in spin-light emitting diodes (spin-LED) with different tunnel barriers (Schottky, Al_2O_3 or MgO). Some perspectives on the electrical spin injection in semiconductors are given in Sec. 10.4.

10.2 THEORY OF SPIN INJECTION THROUGH A HYBRID METAL–SEMICONDUCTOR INTERFACE

This section is devoted to the mechanism of spin injection at the interface between a ferromagnetic conductor (FM) (3*d* magnetic metal for instance), and a semiconductor (SC). Firstly, we will consider the specific problem of spin injection in a diffusive picture beyond a ballistic range. This approach is adapted from the Valet–Fert theory[6] of spin injection in metallic magnetic multilayers related to giant magneto-resistance (GMR) effects and extended to the case of a single FM/SC interface.[7–10] This is of particular interest in aiming to understand the depolarization processes occurring at the FM side that prevent a massive injection of spins in the semiconductor part. This is known as the "impedance mismatch" problem.[11] Nonetheless, we will show how the inclusion of a spin-dependent interfacial resistance, e.g., a spin-conserving tunnel barrier (I) at the FM/SC interface to form a FM/I/SC system, can solve such impedance mismatch to restore a significant current spin-polarization in the semiconductor part. This results in a spin-splitting between spin-up and spin-down electro-chemical potentials as the interfacial resistance played by the tunnel junction increases up to a certain threshold value. This is the main issue of this section.

Secondly, we will proceed to a description of spin injection in a more intuitive single "ballistic" electron picture. We will show how the impedance mismatch can be viewed as a mechanism of "re-absorption" which can be solved by strongly reducing the interface transparency down to a certain limit (transmission of a tunnel barrier). We will ultimately derive the condition of spin injection between a ferromagnetic metal and a semiconductor as a function of some characteristic lengths (mean free path, semiconductor thickness, spin diffusion length) and the interface transparency. We will give a general expression of the current spin-polarization expected to be injected at the interface versus the latter's physical parameters.

10.2.1 *The Diffusive Regime of Spin Injection and the Problem of Impedance Mismatch*

Schmidt *et al.* were the first to put forward the difficulty of injecting, by electrical means, spins into a semiconductor from a ferromagnetic reservoir by taking advantage of the "natural" spin-polarization of the current in a bulk FM material.[11]

The physical issues involved in the problem of spin injection are presented in Fig. 10.1. As illustrated by Fig. 10.1(a), the flow of electrons is spin polarized on the far left in the ferromagnetic reservoir FM with a characteristic value β equal to a fraction of unity, and non-polarized on the far right in the non-magnetic conductor N.[6] Near the interface over a region of a certain extension, there must be a transfer of current between one of the spin channels (spin + channel in the figure) to the other. This transfer is possible because there is spin accumulation in the region of the interface, that is, a splitting $\Delta\mu$ between the electro-chemical potentials of the spin + and spin − carriers, as shown in Fig. 10.1(b).

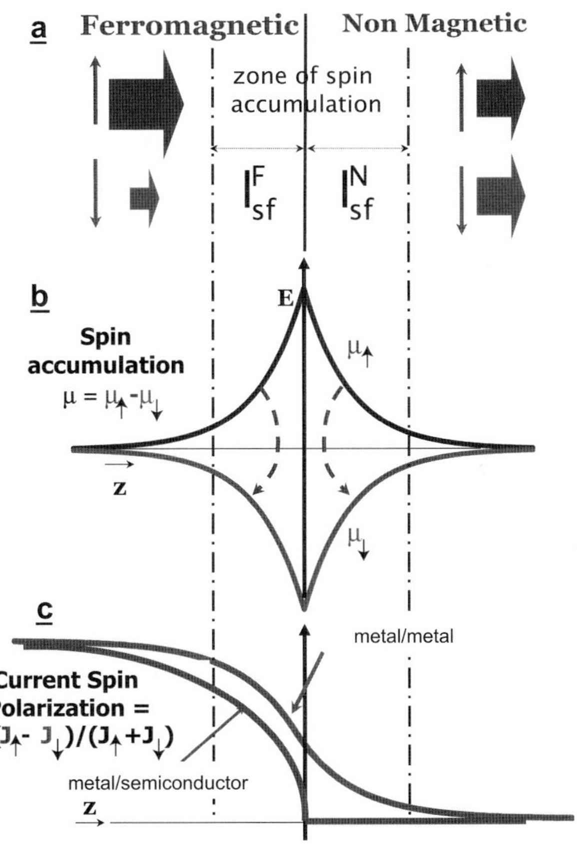

Figure 10.1. (a) Spin-up and spin-down current far from an interface between ferromagnetic and non-magnetic materials (outside the spin accumulation zone). (b) Splitting of the chemical potentials $\mu_\uparrow$ and $\mu_\downarrow$ near the FM/N interface. The arrows symbolize the spin flips induced by the out-of-equilibrium spin-split distribution and governing the depolarization of the electron current between the left and the right. With an opposite direction of the current, there is an inversion of the spin accumulation and opposite spin flips which polarizes the current across the spin accumulation zone. (c) Variation of the current spin polarization when there is an approximate balance between the spin flips on both sides (metal/metal) and when the spin flips on the left-hand side are predominant (metal/semiconductor for example) (from Ref. 10).

Such a spin accumulation profile near the interface is characterized by a positive gradient in the FM region (left-hand side) and a negative gradient in the non-magnetic region (right-hand side), giving rise to a respective negative diffusive spin-polarized current on the left and a positive spin-polarized current on the right. This results in a gradual drop of the spin-polarized current in the FM material when the electron flow approaches the interface and, inversely, an increase of the spin-polarization when the electron flow approaches the interface at the N side (from its zero value at infinity). Note that from a diffusive point of view these arguments hold when the electron flow is reversed. Such diffusive spin-currents are written simply as the product of the spin-average diffusive constant D^* and the gradient of the out-of equilibrium spin density (Δm) according to

$$\delta(j_+ - j_-) = eD^* \mathrm{div}(\Delta m). \tag{10.1}$$

In the presence of two-spin population characterized each by their own diffusion constant D_+ and D_- as well as their own spin-flip time, respectively $\tau_{+\to-}$ and $\tau_{-\to+}$, one can show without any further proofs that D^* is written

$$D^* = \frac{D_+(\tau_{+\to-} + \tau_{-\to+})D_-}{D_+\tau_{+\to-} + D_-\tau_{-\to+}}.$$

The steady state is reached when the number of spin flips generated by this out-of-equilibrium distribution is just what is needed to balance the incoming and outgoing spin fluxes as seen from the general law of spin conservation which yields

$$\mathrm{div}[\delta(j_+ - j_-)] = e\frac{\Delta m}{\tau_{sf}}, \tag{10.2}$$

where j_+ and j_- are the respective $+$ and $-$ spin current, Δm is the out-of-equilibrium spin density profile, e the electron charge, and τ_{sf} the spin-lattice relaxation time. In that sense, due to spin diffusion, the spin accumulation is not localized just at the interface but extends over a distance on the order of the spin diffusion length, respectively $l_{\mathrm{sf}}^{\mathrm{F}}$ on the left and $l_{\mathrm{sf}}^{\mathrm{N}}$ on the right. The spin diffusion length $l_{\mathrm{sf}}^{\mathrm{N}}$ in the non-magnetic material can be expressed as a function of $\tau_{\mathrm{sf}} = \frac{1}{2}(\tau_{+\to-}^{-1} + \tau_{-\to+}^{-1})^{-1}$ and the diffusion constant D^* or equivalently the density of states $2N(E_{\mathrm{F}})$ and the resistivity ρ_{N} by[8]

$$l_{\mathrm{sf}}^{\mathrm{N}} = \sqrt{\frac{D^*\tau_{\mathrm{sf}}}{3}} = \sqrt{\frac{\lambda\lambda_{\mathrm{sf}}}{6}} = \sqrt{\frac{\tau_{\mathrm{sf}}}{4e^2N(E_{\mathrm{F}})\rho_{\mathrm{N}}}} \tag{10.3}$$

in a metal or a degenerate Fermi gas semiconductor, and by

$$l_{\mathrm{sf}}^{\mathrm{N}} = \sqrt{\frac{k_{\mathrm{B}}T\tau_{\mathrm{sf}}}{2ne^2\rho_{\mathrm{N}}}} \tag{10.4}$$

in the non-degenerate regime of a semiconductor. There are similar but a little more complex expressions for the spin diffusion length $l_{\mathrm{sf}}^{\mathrm{F}}$ in ferromagnets involving a mixed diffusion constant between up and down spin channels.[12]

The solution of standard equations for diffusive spin dependent transport found by merging Eqs. (1) and (2) leads to an exponential decrease of the spin accumulation splitting $\Delta\mu$ as well as out-of-equilibrium spin density on both sides of the interface, respectively as $\exp(z/l_{\mathrm{sf}}^{\mathrm{F}})$ and $\exp(-z/l_{\mathrm{sf}}^{\mathrm{N}})$ as shown on in Fig. 10.1(b), with continuity at the interface of $\Delta\mu$ (but without any continuity of Δm because the density of states on each side of the interface is discontinuous!); here, there is no spin-dependent interface resistance.

An alternative view is that the progressive depolarization of the current at the FM side is due to the spin flips related to this spin accumulation. On the other hand, and quite astonishingly, the progressive polarization of the current in the N side from its zero value at infinity is also related to spin-flip processes. In other words, spin-flip processes in a non-magnetic material lead to a progressive polarization of the current! The intermediate level of polarization at the interface is then simply related to the respective proportion with of spin flips on the FM and N sides. When FM and N are two metals with about equivalent density of states and spin diffusion length (Co and Cu for example in current-perpendicular-to-plane CPP-GMR experiments), this number of spin-flips is about the same on each side and the spin-polarization at the interface is about half its maximum value far in the bulk FM, that is, about $\beta/2$.[6] On the other hand, when FM is a metal and N a semiconductor, the density of states (DOS) is much higher in FM and similar spin accumulation splittings on both sides correspond to a much higher spin accumulation density (number of accumulated spins) in FM. For similar spin relaxation times in FM and N, this leads to a much higher number of spin flips in FM, so that the depolarization of the electron current occurs in FM before the interface [see curve in Fig. 10.1(c)]. The same depolarization also occurs if the DOS are similar but the spin lifetime is much shorter in the ferromagnet. Quantitatively, it can be shown that the polarization of the current just at the interface is[7–9]

$$(\mathrm{SP})_1 = \frac{j_+ - j_-}{j_+ + j_-} = \frac{\beta}{1 + r_{\mathrm{N}}/r_{\mathrm{F}}}, \tag{10.5}$$

where, in the notation of Ref. 8, the specific resistances r_i are the product of the corresponding resistivity $\rho_i \propto N(E_{\mathrm{F}})^{-1}$ and spin diffusion length l_{sf}^i according to $r_{\mathrm{N}} = \rho_{\mathrm{N}} l_{\mathrm{sf}}^{\mathrm{N}}$ and $r_{\mathrm{F}} = \rho_{\mathrm{F}}^* l_{\mathrm{sf}}^{\mathrm{F}}$. The resistivities of the majority spin $(+)$ and minority spin $(-)$ channels in FM are written as $\rho_\pm^{\mathrm{N}} = 2\rho_{\mathrm{N}}$ and $\rho_\pm^{\mathrm{F}} = 2[1 \mp \beta]\rho_{\mathrm{F}}^*$. From Eq. (10.5) above, the current is almost completely depolarized when it enters the semiconductor as $r_{\mathrm{N}} \gg r_{\mathrm{F}}$, that is, for example, if the resistivity of metal is much smaller — hence the name "conductivity mismatch." But the same depolarization occurs also when the spin diffusion length is much longer in the semiconductor. This is illustrated by the classical picture of resistances $(\rho_+^{\mathrm{F}} l_{\mathrm{sf}}^{\mathrm{F}} + \rho_+^{\mathrm{N}} l_{\mathrm{sf}}^{\mathrm{N}})$ and $(\rho_-^{\mathrm{F}} l_{\mathrm{sf}}^{\mathrm{F}} + \rho_-^{\mathrm{N}} l_{\mathrm{sf}}^{\mathrm{N}})$ in parallel.

To restore the spin polarization of the current inside the semiconductor, it is then necessary to increase the proportion of spin flips on the N side by increasing the spin accumulation on the N side with respect to its value on the FM side. Such a discontinuity of the spin splitting can be brought by a spin-dependent interface resistance of the form $r_{+(-)} = 2r_{\mathrm{b}}^{*}[1-(+)\gamma]$ (typically a tunnel junction resistance; note that γ is the usual notation for the interface spin asymmetry in CPP-GMR, the notation otherwise being P in tunneling). This interface resistance introduces, in general, a sufficiently large spin-dependent drop (discontinuity) of the electrochemical potentials at the interface according to the following discontinuity equations:

$$\mu_{+(-)}(z=0^{+}) - \mu_{+(-)}(z=0^{-}) = r_{+(-)}j_{+(-)}(z=0), \tag{10.6}$$

and an enhancement of the spin accumulation in N increases the proportion of spin flips on the N side and restores the current polarization in N. With such a spin-dependent interface resistance, the spin polarization of the current at the interface (in the limit of small currents) equals[7–9]

$$(\mathrm{SP})_I = \left(\frac{j_+ - j_-}{j_+ + j_-}\right)_I = \frac{\beta r_{\mathrm{F}} + \gamma r_{\mathrm{b}}^{*}}{r_{\mathrm{F}} + r_{\mathrm{N}} + r_{\mathrm{b}}^{*}}, \tag{10.7}$$

and then decreases exponentially as $\exp(-z/l_{\mathrm{sf}}^{\mathrm{N}})$ on the non-magnetic side. Equation (10.7) expresses the gradual rise of the current spin-polarization as the interfacial resistance r_{b}^{*} increases as well as for electrons entering N and outgoing from N as already emphasized above. We see from Eq. (10.7) that an intermediate value of the current spin polarization is partly restored for $r_{\mathrm{b}}^{*} \cong r_{\mathrm{N}}$ as illustrated by the example of Fig. 10.2 and that the polarization reaches the spin asymmetry coefficient γ of the interface resistance for $r_{\mathrm{b}}^{*} \gg r_{\mathrm{N}} + r_{\mathrm{F}}$. It can be noted that Eq. (10.7) holds for degenerate and non-degenerate carriers.

In the same manner, the spin accumulation in N at the interface is given by

$$(\Delta\mu)_1 = e\frac{r_{\mathrm{N}}(\beta r_{\mathrm{F}} + \gamma r_{\mathrm{b}}^{*})}{r_{\mathrm{F}} + r_{\mathrm{N}} + r_{\mathrm{b}}^{*}} j, \tag{10.8}$$

where j is the electrical current density, and then decreases exponentially, as illustrated in Fig. 10.2.

10.2.2 Microscopic Picture

We shall now derive a microscopic picture of spin injection from a ferromagnetic reservoir (FM) into a semiconductor (SC) through an interface (I) of transparency T. In a single dimension problem (z), the boundary conditions at time $t=0$ is the injection of a spin, e.g., oriented in the up direction with a probability $(1+\gamma)/2$ at point $z=0$ of the interface.

We will consider for simplicity the case of a very long spin lifetime ($\tau_{\mathrm{sf}} = \infty$) in the semiconductor, which corresponds to a detector (a spin light emitting diode for example) placed at a distance from the interface well smaller than the spin lifetime within the semiconductor. Nonetheless, even in the limit of an infinite spin lifetime, one cannot avoid the general problem of retrodiffusion. A particle injected at the interface at a given time $t=0$ has a certain probability, due to backscattering processes, of being reabsorbed into the FM reservoir after a certain time t where its spin will be lost before its reinjection into the SC region. Such a reabsorption process is at the origin of the "impedance mismatch" which more correctly should be called "density of states mismatch" problem. Indeed, an equivalent density of states and equivalent spin lifetime at each side of the interface (metal/metal system) would result statistically, after diffusion processes, in an equal probability of spin-flip at each side and consequently in a reduction of the spin current by a factor of two.

In a general problem of diffusion *in one dimension* this out-of-equilibrium spin will be diffused in space (z) with a certain probability $p(z,t)dz$ of finding such a spin in a region between z and $z+dz$. The probability $p(z,t)$ obeys the general diffusion equation, which can be found by merging Eqs. (10.1) and (10.2), leading to

$$D^* \frac{\partial^2 p(z,t)}{\partial z^2} = \frac{\partial p(z,t)}{\partial t} \tag{10.9}$$

with the limiting condition $p(z,t=0) = \delta(z)$, which has the following solution:

$$p(z,t) = \frac{1}{\sqrt{4\pi D^* t}} \exp\left(-\frac{z^2}{4D^* t}\right). \tag{10.10}$$

If one considers multiple reflection processes upon the interface, the average exploration radius is written simply as $\sqrt{\overline{z^2}} = \sqrt{\int z^2 p(z,t)dz} = \sqrt{Dt}$, which is of course consistent with the standard diffusion problem. It follows that the time t_d needed to reach the detector placed at a distance d from the origin equals d^2/D.

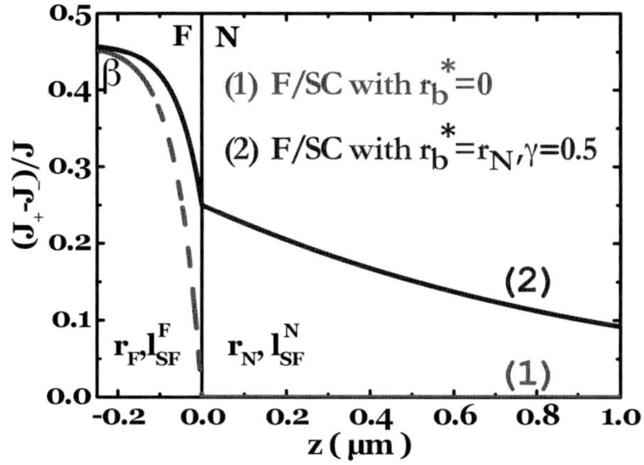

Figure 10.2. Current spin polarization at the interface between a ferromagnetic metal F and a semiconductor N. The calculation has been performed for F = Co with $r_F = \rho_F^* l_{sf}^F = 4.5 \times 10^{-15} \Omega m^2$, $\beta = 0.46$, $l_{sf}^F = 60$ nm from CPP-GMR data on Co, and for N = GaAs with $r_N = 4.5 \times 10^{-9} \Omega m^2$, $l_{sf}^N = 2\mu$ m derived from room temperature data on an n-type GaAs ($n = 10^{16}$ cm^{-3}). The blue solid line is calculated with spin dependent interface resistance ($r_b^* = r_N = 4 \times 10^{-9} \Omega m^2$, $\gamma = 0.5$) and the red dashed line without interface resistance (from Ref. 8).

What is maybe less obvious is counting the number of reflections that the injected spin undergoes after a certain time t. This number $N(t)$ equals $N(t) = \int \lambda p(0,t)\frac{dt}{\tau_{\mathrm{p}}}$, where λ is the electron mean free path, and τ_p the momentum relaxation time, which are linked to the diffusion constant D through the relation $D = \frac{\lambda^2}{\tau_{\mathrm{p}}}$. This number of reflections $N(t)$ then equals $N(t) \cong \sqrt{(4/\pi)(\tau/\tau_{\mathrm{p}})}$. Consequently, during its transit time t_d from the interface to the detector, the spin has undergone an average of $N(t_d) \cong \frac{d}{\lambda}$ reflections upon the interface in the limit of a very small interface transparency T. The immediate consequence is that the upper threshold limit for the interface transparency T to avoid spin-flip in the FM reservoir through retrodiffusion processes is given by $T < \lambda/d$. This condition is equivalent to $r_{\mathrm{b}}^* > \rho \times d$, without any further proofs, by the use of Landauer equation for interfacial conductance. This condition is somewhat different from the one derived in the previous paragraph, $r_{\mathrm{b}}^* > \rho \times l_{\mathrm{sf}}^{\mathrm{N}}$, which shows that the condition for spin injection is partially relaxed when the detector is placed at a distance d significantly smaller than $l_{\mathrm{sf}}^{\mathrm{N}}$. It also demonstrates that an injection of spins over a distance d less than the mean free path λ (ballistic injection regime) does not *a priori* require any spin-dependent interfacial resistance or tunnel junction.

To conclude, from the above arguments, one can foresee that the spin-polarization P_{s} of the electron flow at the position of a detector placed at a distance d from the interface should be close to

$$P_{\mathrm{s}} \approx \frac{\gamma}{1 + T\frac{d}{\lambda}} \approx \frac{\gamma}{1 + \frac{r_{\mathrm{N}}}{r_{\mathrm{b}}^*}\frac{d}{l_{\mathrm{sf}}^{\mathrm{N}}}}. \tag{10.11}$$

This could be experimentally demonstrated in future spin injection experiments with spin-LEDs by varying the distance between the light emitting region (quantum wells or quantum dots) and the interface. Nevertheless, one must realize that other physical parameters like electric field, band bending or hot-electron relaxation mechanisms could enhance or decrease the efficiency of spin injection. This should be taken into account in more accurate calculations.

10.3 MEASUREMENT OF THE ELECTRICAL SPIN INJECTION IN HYBRID DEVICES

10.3.1 General Considerations

As discussed in Sec. 10.2, an efficient electrical spin injection from an FM metal into a semiconductor can only be achieved in the presence of a tunnel barrier between them.[7,8] Two kinds of tunnel barriers have been successfully used. The first consists of a Schottky barrier obtained through the engineering of the doping concentration of the semiconductor at the interface with the metal.[5,13,14] The second type of tunnel barrier is composed of a thin insulator layer (Al_2O_3 or MgO) embedded between the FM layer and the semiconductor. In the following we review the experimental results of electrical spin injection in devices using these different types of tunnel barriers. The efficiency of the electrical spin injection has been measured in hybrid FM/tunnel barrier/Sc devices, called spin-LEDs, where the circular polarization of the electroluminescence probes the electron spin polarization degree in the semiconductor. This concept of combining electrical spin injection and optical detection was originally proposed by Aronov and Pikus.[15]

10.3.2 Spin-LEDs with Schottky Barriers

A typical spin-LED device structure with a Schottky tunnel barrier is shown in Fig. 10.3(a). It is grown by molecular beam epitaxy (MBE) on a p-doped GaAs (001) substrate with a 500 nm thick p-doped $Al_{0.08}Ga_{0.92}As$ buffer layer. The active layer consists of a 10 nm thick GaAs quantum well (QW) separated by 50 nm thick (bottom) and 40 nm thick (top) $Al_{0.08}Ga_{0.92}As$ barriers. On top of this intrinsic region, a 40 nm thick n-doped $Al_{0.08}Ga_{0.92}As$ layer ($N \sim 2.10^{18}\,cm^{-3}$) was grown. Then, 15 nm thick Co film was deposited by sputtering on the semiconductor structure before capping by a 2 nm thick Pt protective layer.[14] A Schottky barrier is thus formed at the interface between the FM and the SC.

The operation principle of a spin-LED is that the spin polarized electron in the FM metal will be injected electrically into the SC by applying a forward bias (with respect to the p–i–n junction) to the device. These electrons have a spin polarization along the growth axis of the structure (Oz) thanks to the application of an external magnetic field B along this axis. Once these electrons are captured by the QWs, they recombine radiatively with the unpolarized holes coming from the

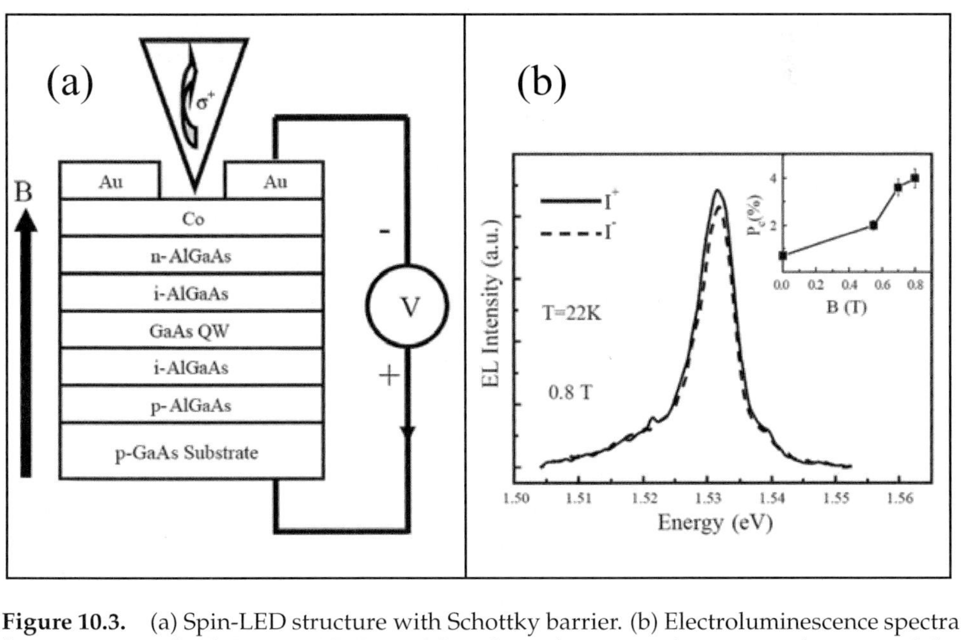

Figure 10.3. (a) Spin-LED structure with Schottky barrier. (b) Electroluminescence spectra for a magnetic field $B_z = 0.8\,T$. Inset: Circular polarization degree as a function of the external magnetic field (from Ref. 14).

substrate. The corresponding electroluminescence (EL) will be circularly polarized if the electrons kept their spin orientation after the injection, capture and recombination processes.

If spin polarized electrons aligned in the FM along Oz are injected into the semiconductor, the polarization of the electroluminescence of the spin-LED detected along Oz will easily be related to the spin polarization degree of the electrons in the QW thanks to the optical selection rules.[16] Note that if the luminescence is detected perpendicular to the growth axis, the electroluminescence corresponding to the recombination in the QW of electron and heavy-hole should not be circularly polarized due to the symmetry of the heavy-hole wave function[17]; this makes the interpretation of the experiments performed with this geometry very difficult.[18,19]

A typical EL spectrum of the spin-LED at 22 K is shown in Fig. 10.3(b). The peak at 1.53 eV corresponds to the radiative recombination of electrons and heavy-holes. The circular polarization is defined as $P_C = (I^+ - I^-)/(I^+ + I^-)$, where I^+ (I^-) is the right (left) circularly-polarized EL component peak intensity. When a magnetic field B is applied, P_C^{Exp} increases with increasing magnetic field as shown in the inset of Fig. 10.3(b). For $B = 0.8$ T, we measured $P_C^{Exp} = 4.0 \pm 0.5\%$ (depending on the investigated device,[20] this value varies between 3% and 5%). We have checked that the circular polarization due to magnetic circular dichroism (MCD)[21] through the Co layer is less than 1%. The spurious effect of electron Zeeman splitting in the QW can also be estimated to be less than 1% for this range of magnetic field and temperature.[5] Thus, we can conclude that the measured EL circular polarization is mainly due to the electrical spin injection through the FM layer.

In a first approach, one can consider that the circular polarization degree of the luminescence is identical to the spin polarization of the electrically injected carriers in the semiconductor QW structure according to quantum selection rules. In fact, this description is oversimplified because the measured continous wave (CW) electro-luminescence polarization depends on the electrical spin injection efficiency at the Co/I/AlGaAs interface *and* all the spin relaxation mechanisms which occur in the AlGaAs barrier and in the GaAs/AlGaAs QW.

In order to quantify the correction due to the spin relaxation into the SC part of the device, we have measured independently the circular polarization dynamics by time-resolved photoluminescence (TRPL) experiments in which the electron spins are optically injected in the same structure (with no FM layer). For these TRPL measurements, a mode-locked Ti:sapphire laser (1.5 ps pulse width) was used for the non-resonant circularly-polarized excitation at ~1.642 eV (i.e., in the barriers). The PL signal was detected by a 2D synchroscan streak camera, which provides an overall temporal resolution of less than 8 ps.[22] Figure 10.4(a) presents the time evolution of the photoluminescence components co-polarized (I^+) or counter-polarized (I^-) with the ($\sigma+$) polarized picosecond excitation laser. The corresponding PL circular polarization degree is also displayed; it shows a mono-exponential decay with a characteristic time $\tau_S \sim 400$ ps. We have

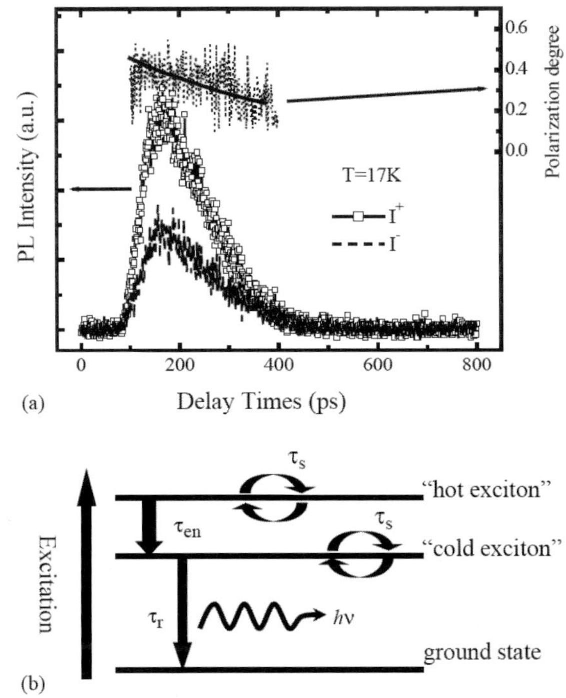

Figure 10.4. (a) Time evolution of the luminescence intensity co-polarized (I^+) and counter-polarized (I^-) with the picosecond $(\sigma+)$ polarized laser excitation pulse; the circular polarization degree dynamics is also displayed. (b) Schematic representation of the three-level model considered (see text) (from Ref. 14).

also fitted the luminescence intensity dynamics with a simple three-level model depicted in Fig. 10.4(b). We found a radiative recombination time $\tau{\sim}100$ ps and a capture/energy relaxation time $\tau_{en} \sim 50$ ps.

On the basis of this three-level model, it can be shown that the circular polarization measured in Continous Wave (CW) or time-integrated experiments is written simply as[14]

$$P_C = FP_{inj} \quad \text{with} \quad F = \frac{1}{\left(1+\frac{\tau}{\tau_s}\right)\left(1+\frac{\tau_{en}}{\tau_s}\right)}. \tag{10.12}$$

First, we tested this phenomenological model by comparing theoretical and experimental values of the polarization of the time-integrated photoluminescence under non-resonant excitation. According to the optical selection rules,[16] the injected electron spin polarization is $P_{Inj} = 50\%$ when exciting in the barriers with circularly-polarized light. The spin polarization of the time-integrated photoluminescence can then be calculated using Eq. (10.12) and the measured parameters τ, τ_{en} and τ_s; we find $P_C^{Cacl} = 35.8\%$, which is in very good agreement with the experimental result of ~36% (the spectrum is not shown here). This validates this simple model which takes into account the energy and spin relaxation processes in the structure. This correction will be increasingly important if the ratio τ/τ_s is large. Now, we apply this model to estimate the electrical spin injection efficiency at the Co/I/AlGaAs interface. Equation (10.12) yields an injected spin polarization $P_{Inj} = 5.6\%$ using the measured EL circular polarization $P_C^{Exp} = 4\%$ and the parameters τ, τ_{en} and τ_s determined previously with the TRPL experiments. The Co magnetic moment is saturated out-of-plane for an applied magnetic field of $B = 1.8\,T$. The electron spin polarization near the Fermi energy is then about $P_{Co} = 42\%$.[23] Below this value, the magnetization of Co increases linearly with external magnetic field. We thus deduce that the spin polarization of the Co contact is about $P_{Co} \sim 19\%$ for $B_z = 0.8\,T$. The "spin injection efficiency" η can then be calculated as follows: $\eta = P_{Inj}/P_{Co}$. We obtain $\eta \approx 30\%$. This shows that the large conductivity mismatch between the FM metal and the semiconductor which constituted the fundamental obstacle for spin injection[8] can clearly be overcome by the introduction of an interface resistance as explained in Sec. 10.2.[7,8] Note that the simple η parameter introduced here is just a figure of merit designed to compare the electrical spin injection in different devices; it does not take into account any spin filtering occurring at the interface between the FM metal and the semiconductor.[24]

Thanks to the optimization of the Schottky barrier, Jonker's group at the Naval Research Laboratory managed to measure a record EL circular polarization degree of $P_C^{Exp} = 32\%$ for $B = 4\,T$ at a temperature of 4 K with an Fe ferromagnetic spin injector.[25] The same group demonstrated electrical spin injection up to a temperature of 240 K[13] with an injection efficiency of about 30%.[a] Though most of the experimental measurements were performed with spin-LEDs based on QW spin detectors, the electrical spin injection in Schottky barrier spin-LEDs with a quantum–dot active region was also demonstrated.[26,27]

10.3.3 Spin-LEDs with Alumine Barriers

An alternative way to achieve an efficient electrical spin injection is to insert between the ferromagnetic metal and the semiconductor an insulating layer such as aluminium oxide[21,28,29] which acts as a tunnel barrier. This technique was first demonstrated by Motsnyi *et al.*[21] The results in terms of spin injection efficiency are comparable with the best ones obtained with a Schottky barrier, even if the charge injection efficiency is lower.[29] In this section, we address the ability to inject spin polarized electrons through a Co/Al_2O_3/GaAs tunnel barrier into *p*-doped InAs/GaAs quantum dots (QDs) embedded in a *p–i–n* GaAs LED.[30] The attempts to inject spin polarized carriers in zero-dimensional confined nanostructures are guided by the fact that achieving spintronic devices necessitates sufficiently long spin relaxation times to manipulate or store the spin orientation of the injected carriers. This is indeed the case for semiconductor QDs.[31] Electrical spin injection into InAs/GaAs intrinsic QDs has been first reported with spin aligner systems based on a magnetic semiconductor[32–34] or ferromagnetic metal/semiconductor

[a] Assuming that the electron spin polarization for Fe is $P_{Fe} = 44\%$.

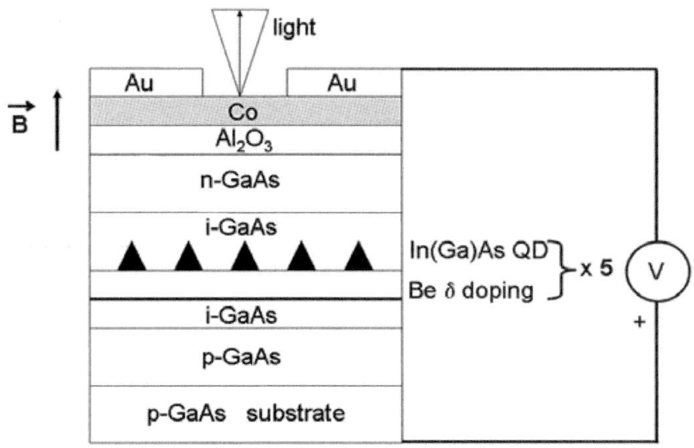

Figure 10.5. Spin-LED device with *p*-doped In(Ga)As quantum dots (from Ref. 30).

Schottky barrier.[26,27] Compared to intrinsic QDs, *p*-doped QDs are very promising for their very long electron spin relaxation times[35] under weak external magnetic fields (about 100 mT), and their spin properties at high temperature.[36,37]

A typical spin-LED structure with an aluminium oxide tunnel layer is displayed in Fig. 10.5. The intrinsic GaAs zone contains five QD planes, separated by 30 nm in order to keep them electronically and mechanically uncoupled. QDs are doped with one hole per dot on average thanks to Be δ-doped planes located 15 nm from the QD planes.

The Al_2O_3 tunnel barrier and cobalt ferromagnetic thin film (8 nm) were grown at room temperature in a sputtering chamber. The tunnel barrier was formed by oxidation of 1.5 nm Al under an O_2 and Ar plasma. A gold cap layer (2 nm) was deposited last to prevent the cobalt from oxidation. Measurements were performed on two samples S1 and S2. S1 is the structure described above, whereas S2 is a test structure without Al_20_3 and Co for time and polarization-resolved photoluminescence (TRPL) experiments.

The QDs embedded in the *p–i–n* junction were first characterized by TRPL. The photoluminescence circular polarization rate $P_C = (I^+ - I^-)/(I^+ + I^-)$ was measured as a function of time, after a picosecond σ^+ laser pulsed excitation. We checked that, after an excitation in the wetting layer ($\sim$1.441 eV), P_C exhibits the same behavior as in Ref. 35, where a polarization around 15% is observed at long delay times without any magnetic field, which is characteristic for *p*-doped QDs with approximately one hole per dot on average. In this kind of structures a very long electron spin relaxation time τ_s larger than 4 ns under a small longitudinal magnetic field B of more than 100 mT was measured.[35] This slow relaxation is due to quenching by the applied magnetic field of the effect of the hyperfine interaction between electron and nuclear spins, responsible for the electron spin relaxation in these systems. Figure 10.6 displays the kinetics of the two polarized components of luminescence and the corresponding circular polarization rate after a non-resonant

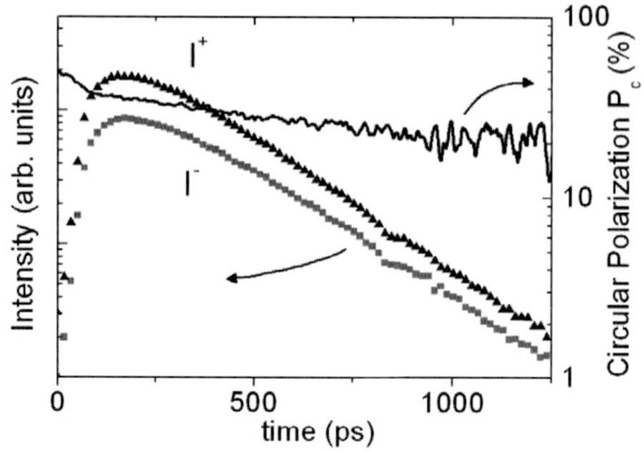

Figure 10.6. Left axis: I^+ (black) and I^- (gray) intensity components of PL (co- and counter-polarized with the laser) versus time at 10 K for sample S2. Right axis: Corresponding time-resolved circular polarization rate; logarithmic scale (from Ref. 30).

laser excitation in the GaAs barrier (at $\sim$1.549 eV) at $B = 0$. These optical excitation conditions are close to that encountered for an electrical injection of carriers, where the electrons are also non-resonantly injected in the QDs from the barrier. A slow decay of the polarization rate is measured, with a characteristic decay time of $\sim$2 ns. Note that, as mentioned previously, this decay should be even slower when a small longitudinal magnetic field of more than to 100 mT is applied, which is the case in the standard operating conditions of the device. As a consequence, the polarization decay is much slower than the carrier lifetime, and one can thus consider [see Eq. (10.12) with τ and $\tau_{en} \ll \tau_s$] that the electron spin relaxation is negligible once the electron is trapped in the fundamental level (S^e) in the dot. This makes these dots very suitable for probing the electron spin polarization immediately after this trapping.

The electroluminescence circular polarization P_C at the QD's peak emission is displayed in Figs. 10.7(a) and 7(c) and exhibits a value of 15% for $B = 2$ T. Note that it has been checked that magnetic circular dichroism is negligible (less than 2%) in a photoluminescence experiment under linearly polarized non-resonant excitation in the GaAs barriers. The spurious effect of Zeeman splitting in the dots can also be estimated to be less than 1% for this range of magnetic field and temperature.[34] In addition to QD emission, a spectral feature is observed at 1.49 eV in Fig. 10.7(a), attributed to a conduction band-neutral acceptor transition $(e - A_0)$ probably due to beryllium impurities[38] in GaAs. This optical transition constitutes an interesting additional probe of the electron spin polarization directly in the GaAs after injection through the tunnel barrier. Due to degeneracy,[39] one cannot distinguish between the heavy- and light-hole states for the impurity, and the electron spin polarization in GaAs should be twice[16] the circular polarization rate of about 13% measured for the luminescence of the transition $(e - A_0)$ for $B = 2$ T.

For the same magnetic field, the peak energy corresponding to QD emission (S^e–S^{hh} transition) exhibits an EL circular polarization of about 15%. For QDs, the polarization rate of luminescence should directly reflect the electron spin polarization in the dots, thanks to the splitting between heavy and light holes in these systems. The electrically injected electron spin polarization seems thus lower in the dots (about 15%) than in GaAs (about 26%). A spin relaxation during the trapping in the dots may explain this discrepancy, but another argument

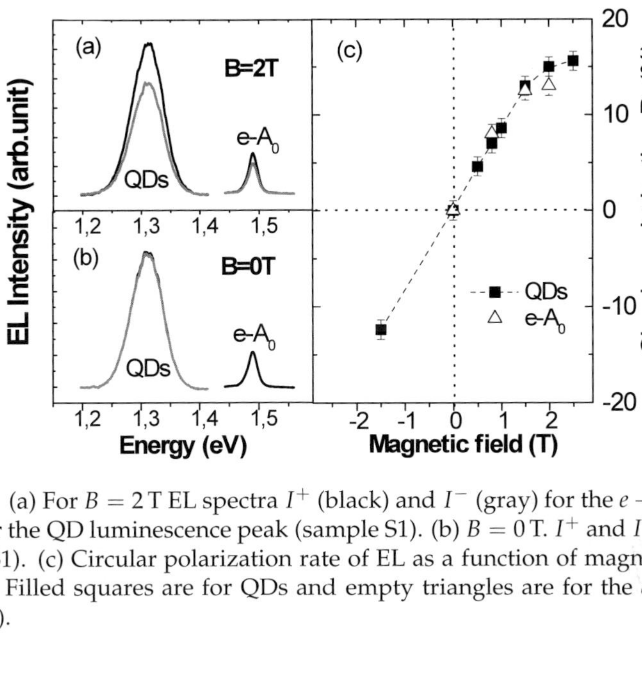

Figure 10.7. (a) For $B = 2\,\text{T}$ EL spectra I^+ (black) and I^- (gray) for the $e - A_0$ optical transition and for the QD luminescence peak (sample S1). (b) $B = 0\,\text{T}$. I^+ and I^- components of EL (sample S1). (c) Circular polarization rate of EL as a function of magnetic field at 1.7 K (sample S1). Filled squares are for QDs and empty triangles are for the $e - A_0$ transition (from Ref. 30).

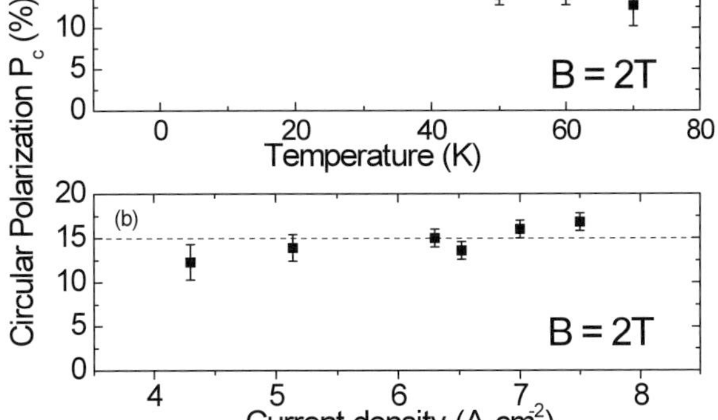

Figure 10.8. (a) Circular polarization rate of EL as a function of temperature at $B = 2\,\text{T}$ (sample S1). (b) Circular polarization rate of EL as a function of current density ($T = 1.7\,\text{K}$, $B = 2\,\text{T}$) (from Ref. 30).

should also be taken into account: under CW electrical excitation in the range of current density employed here (4.3 to 7.5 A cm^{-2}), part of the electroluminescence may arise not only from positively charged, but also from multi-charged, dots.[40,41] These additional contributions lead mainly to a reduction of the averaged EL circular polarization rate measured on an ensemble of QDs. Figure 10.7(c) displays the magnetic field dependence of the EL polarization rate for QDs and $(e - A_0)$ transitions. It is proportional for both transitions to the Co layer magnetization along the z (hard) axis. For $B = 2\,\text{T}$, the electron spin polarization P_{Co} in the Co layer is about 42%.[23] The factor of merit $\eta = P_C^{EI}/P_{Co}$ is at least 35% inside the dots. Finally, Fig. 10.8 displays the EL circular polarization stability with temperature up to 70 K [Fig. 10.8(a)] and electric current density from 4.3 to 7.5 A cm^{-2} [Fig. 10.8(b)]. All the works cited previously,[26,27,32–34] as well as this demonstration of an efficient electrical spin injection through the $Co/Al_2O_3/GaAs$ interface into p-doped InAs/GaAs QDs, are performed on ensembles of QDs. Very recently, optical detection in a single QD[42,43] of electrically injected spin polarized electrons has been successfully demonstrated (at very low temperature), paving the way for the collective initialization and individual readout of electron spin states in quantum registers.

10.3.4 Spin-LEDs with MgO Barriers

In 2005, Jiang *et al.*[44] demonstrated that MgO tunnel barriers, already widely used in tunnel magnetic resonance[45] based devices, yield very efficient spin injection into GaAs[46–48] owing to the symmetry selection rules during the tunneling transfer of carriers.[49] It has been shown that the electron spin polarization in semiconductors can be even larger than that at the Fermi level of the FM metal. In the following we discuss the role of different key parameters, such as the temperature growth of the oxide layer, the thickness of the layer, and the temperature of the device, in electron spin injection efficiency for a CoFeB/MgO spin injector.[50]

In order to study the influence of these parameters, a standard spin-LED, with a 10 nm AlGaAs/GaAs quantum well embedded in the intrinsic region of the p–i–n structure, is investigated.

The spin injector consists of an MgO tunnel barrier with different thicknesses (sample A: 1.4 nm, B2: 2.6 nm, C: 4.3 nm) grown with a magnetron sputtering system[51] at 300°C, followed immediately by a 3 nm $Co_{40}Fe_{40}B_{20}$ ferromagnetic contact capped with 5 nm Ta to prevent oxidation. A sample B1 (3 nm MgO) is also synthesized at RT in order to address the effect of the MgO growth temperature on the spin injection yield.

For sample B1, after 3 nm MgO grown at RT, a hollow diffused RHEED pattern indicates that the MgO is completely amorphous, as shown in the top inset of Fig. 10.9(a). For sample B2 (MgO grown at 300°C), the RHEED pictures are strongly modified highlighting a mono-crystalline spotty diffraction pattern, as shown in the top inset of Fig. 10.9(b). The interfacial structure of the whole spin-LED was further investigated by high-resolution transmission electron microscopy

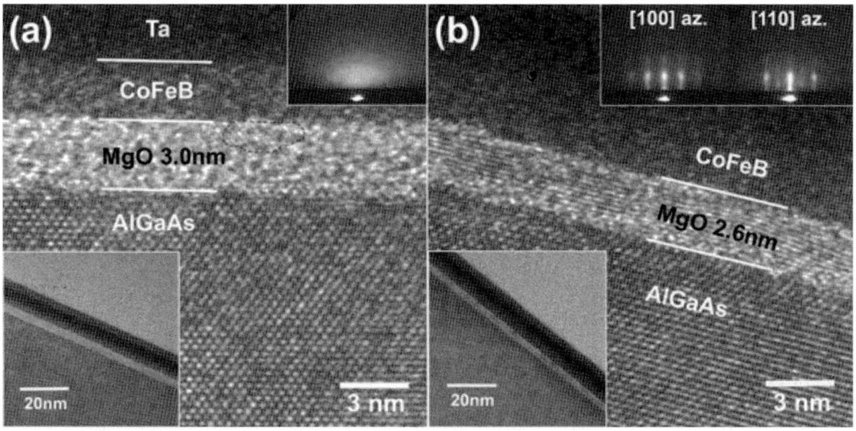

Figure 10.9. High magnification HRTEM cross-section images for (a) sample B1 and (b) B2. Low magnification images are shown in the bottom inset. RHEED patterns for samples after MgO deposition are shown in the top inset. For sample B2, the two RHEED images were obtained from (Al)GaAs [100] and [110] azimuths, respectively (from Ref. 50).

(HRTEM). The good homogeneity of the structures can be seen on the low magnification images in the bottom inset of Figs. 10.9(a) and 9(b). No clear difference in terms of the MgO roughness between RT and 300°C growth can be evidenced. Figure 10.9 displays the HRTEM high magnification images for sample B1 [Fig. 10.9(a)] and B2 [Fig. 10.9(b)]. For sample B1, the MgO is amorphous at the initial stage and presents some crystalline grains after 1.6 nm thickness. However, for sample B2, a MgO (001) texture is clearly observed over a widespread region. On the other hand, the 3 nm CoFeB layers on MgO are amorphous in both cases (a temperature of around 360°C is necessary for crystallization of a CoFeB layer on MgO[52]).

A typical EL spectra acquired for sample B2 at 15 K under a bias of $V_T = 3.1$ V is shown in the inset of Fig. 10.10(a) for $B = 0$ and 0.8 T. Whereas the heavy-hole exciton (XH) EL peak observed at 814 nm does not show any circular polarization at zero magnetic field, the EL polarization reaches a value as large as 23% under 0.8 T. The magnetic field dependence of P_C for samples B1 and B2 is shown in Fig. 10.10(a). One can observe in both cases a linear increase of P_C reflecting the progressive rotation of the CoFeB magnetization in the out-of-plane direction. Whereas sample B1 shows a maximum P_C of 16%, sample B2 exhibits a larger P_C of 23% at 0.8 T, e.g., beneath the magnetization saturation occurring at about 1.3 T from superconducting quantum interference device (SQUID) magnetometer measurements. An extrapolation procedure by matching the EL polarization results with the SQUID data in the linear regime allows a rough estimate of P_C at saturation of values of about 20% and 30% for samples B1 and B2, respectively. The result corresponding to the effect of thermal annealing performed at 200°C for 30 min on sample B2 is also added (filled circle), which gives a rise of P_C to 29.5%, which can be extrapolated to 38% after saturation. Such an enhancement of the EL polarization together with the result of the annealing effect is a signature of the relevant

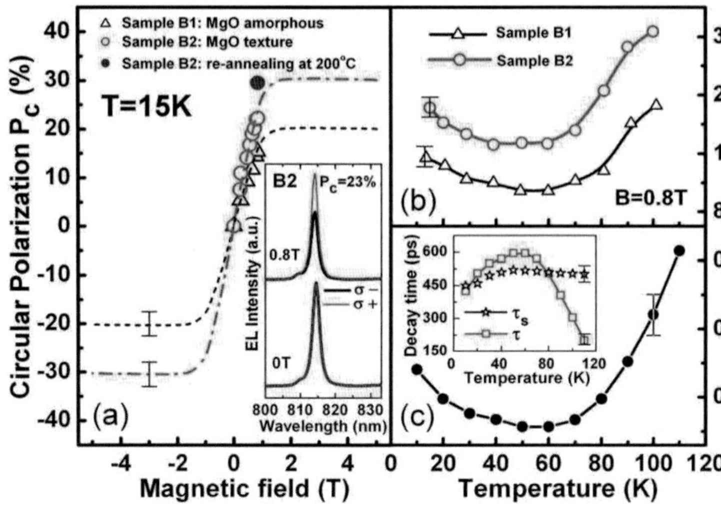

Figure 10.10. (a) Magnetic field dependence of P_C for sample B1 (open triangles) and B2 (open circles) at 15 K. The value of P_C for B2 after annealing at 200°C is indicated by a filled circle. The short dashed lines (B1) and dash dotted lines (B2) show the field dependence of CoFeB moments measured by SQUID at 10 K, which were scaled for comparison with P_C. Inset: EL spectra for sample B2 at 0 T (bottom) and 0.8 T (top). The intensities I^+ and I^- are shown by thin and thick lines, respectively. (b) Temperature dependence of P_C for sample B1 (open triangles) and B2 (open circles) under a magnetic field of 0.8 T. (c) Temperature dependence of the F factor deduced by TRPL measurements. Inset: Evolution of τ_S (open stars) and τ (open squares) with temperature (from Ref. 50).

character of the crystalline MgO on the spin-injection efficiency. Moreover, this high value of the EL circular polarization (comparable to that of Refs. 44 and 46 with CoFe) demonstrates the possibility of using CoFeB as a very efficient spin aligner [with a saturation field smaller than in CoFe (2.2T)].

In Fig. 10.10(b), the temperature dependence of the EL polarization is displayed for samples B1 and B2 from 15 K up to 100 K, above which the EL signal is too weak to be measured. A remarkable feature is the very similar non-monotonic shape for both samples. The thermal evolution shows first a decrease of P_C reaching a minimum at 60 K before increasing up to 100 K.[44,53] This behavior seems to reflect physical effects inherent in semiconductor LED structures rather than in the MgO layer. In order to check this assumption, TRPL measurements are performed on a *p–i–n* LED identical to that in samples B1 and B2, for non-resonant circularly polarized excitation in the AlGaAs barrier. It has then been possible to extract the spin relaxation time τ_s, as well as the carrier lifetime τ, which are presented in the inset of Fig. 10.10(c). The main trends, corroborating the aforementioned paper,[53] are the following: (i) a relative thermal independence of the spin lifetime τ_s on the order of 500 ps, and (ii) a strong thermal variation of τ that increases with temperature up to 60 K, and then decreases from 60 to 100 K. We plot in Fig. 10.10(c) the factor $F = 1/[1 + \tau(T)/\tau_s(T)]$ [we have simplified Eq. (10.12) by assuming $\tau_{en} \ll \tau_s$] that links the optical polarization P_C to the true electrically

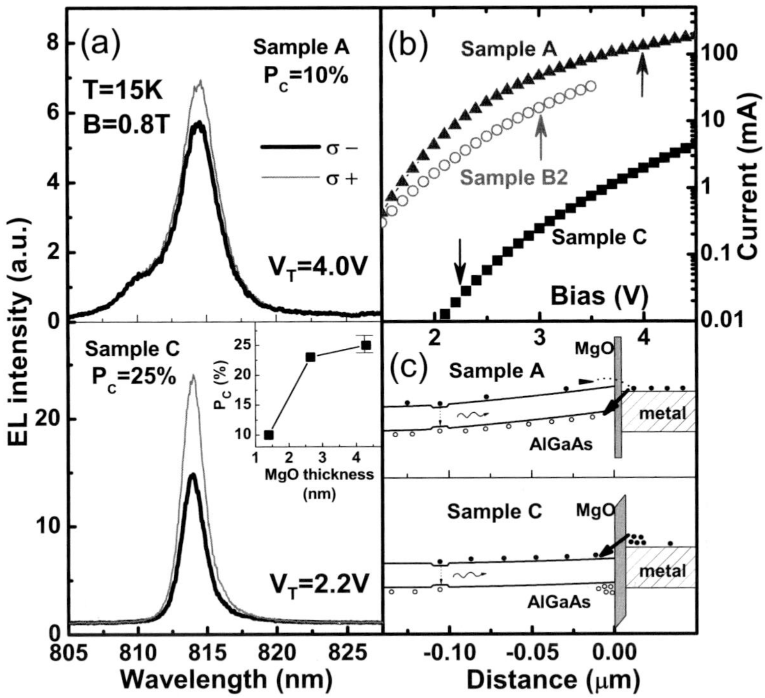

Figure 10.11. (a) EL spectra under a 0.8 T magnetic field at 15 K for samples A (top) and C (bottom). The intensities I^+ and I^- are shown by thin and thick lines, respectively. Inset: Dependence of P_C on the thickness of the MgO barrier. These values are measured at the optimum bias V_T for each sample. (b) *I–V* curves for sample A (filled triangles), B2 (open circles) and C (filled squares) on a vertical logarithmic scale. The optimum bias V_T is marked by an arrow for each sample. (c) Schematic band diagrams of spin-LED under V_T for samples A and C (from Ref. 50).

injected electron polarization, P_{inj}, according to $P_C(T) = F(T) \times P_{inj}$. The F factor explains by itself the characteristic thermal evolution of the measured EL polarization $P_C(T)$. The electrical spin injection process thus does not depend on temperature in the range of 15–100 K for both amorphous and textured MgO tunnel barriers.

Finally, we discuss the MgO thickness dependence of the electrical spin injection yield. To give essential trends, we only develop results corresponding to the two extreme MgO thicknesses (samples A and C), while the medium thickness (B2) shows an intermediate behavior. Figure 10.11(a) displays the polarization-resolved EL spectra for samples A and C acquired at an optimum bias of 4.0 V and 2.3 V, respectively. Raw data acquired at 15 K under a 0.8 T magnetic field give respective P_C of 10% for sample A and 25% for sample C. The whole results gathered in the inset of Fig. 10.11(a) show a pronounced increase of the spin polarization efficiency as the MgO layer is made thicker. To explain the MgO thickness-dependent spin injection efficiency, different mechanisms, or a combination of them, can be invoked. One of them could be related to band-bending effects in the semiconductor part of the device. The I–V curves for the samples with different MgO thickness are shown in Fig. 10.11(b). A further analysis of Figs. 10.11(a) and 11(b) reveals that a larger EL intensity by more than a factor of three for the thickest MgO sample (C) compared to the thinnest barrier (A) conjugated with a smaller current for sample C (30 μA) compared to sample A (126 mA) at the optimum bias. This exhibits a higher (non spin-resolved) electron injection yield for sample C, leading to a stronger EL signal with smaller current. Such features in Fig. 10.11(c) could be due to a distinct band-bending scheme within the semiconductor part of the spin-LED in the two samples, which is closely related to the difference of the hole currents established inside the semiconductor. For sample A, holes flowing from the p-doped GaAs substrate towards the emitter can tunnel through the thin MgO barrier to produce a surface recombination[29,54] thus limiting the electron injection into the conduction band of the semiconductor. Considering sample C, the thick MgO barrier prevents holes from tunneling towards the injector; thus, the holes accumulate at the interface with the barrier. This leads to a strong potential drop in the MgO barrier and a "flat-band" landscape in the semiconductor region, which is responsible for the high electron injection yield for sample C. The MgO thickness-dependent band profile in the semiconductor part could result, for the thinner one, in both the quantum-confined Stark effect,[55] which increases the electron lifetime in the QW (and thus decreases F and P_C), and the Bir–Aronov–Pikus[56] electron spin depolarization process induced by a parasitic hole current during the transit time from the injector to the QW. Besides these band-bending effects, the CoFeB/MgO injector spin filtering mechanism, based on the symmetries of wave functions involved in the coherent tunneling process through the thick crystalline MgO barrier, could also be invoked.[49] Finally, one cannot totally exclude, at this high level of development, possible effects due to interface defects whose density could vary with thickness.

10.3.5 Pulsed Electrical Injection and Time-Resolved Electroluminescence

All the electroluminescence experiments cited up to now have been realized under a quasi-stationary electrical injection operating mode. For future applications, it is also important to demonstrate very high speed pulsed operation. This could be a cornerstone in the generation of high speed spin currents as well as the fast initialization of quantum memories based on spin. In the following, we explore by time-resolved electroluminescence the ability of a spin-LED device similar to that discussed in Sec. 3.4 to be operated under pulsed electrical excitation. We analyze the ultra-fast electrical injection of spin-polarized electrons in a CoFeB/MgO/AlGaAs structure. For time-resolved electroluminescence, an electrical impulse forming circuit using a step recovery diode (SRD) provides a square pulse of 4 ns with rise and fall times of 100 ps. A streak scope with an S20 photocathode is synchronized with the pulsed generator to measure the

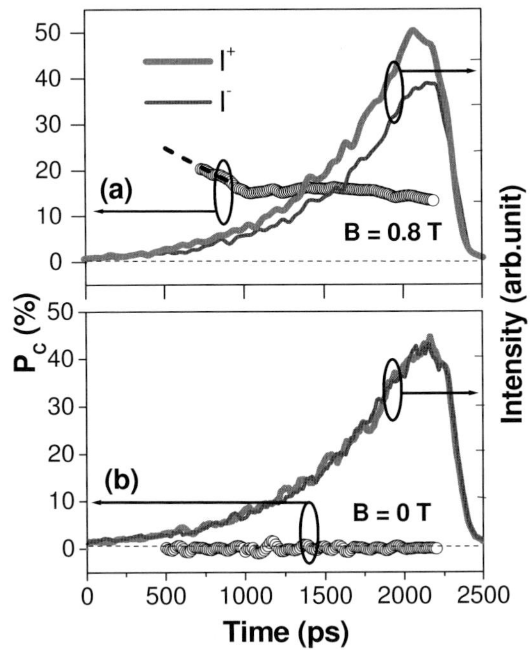

Figure 10.12. Right axis: Time-resolved EL intensity at 15 K with $\sigma+$ polarization (thin line) and σ^- polarization (thick line) with an applied bias $V = 4$ V for $B = 0.8$ T. Left axis: Time evolution of the EL circular polarization P_C (open circles) (a) for $B = 0.8$ T and (b) for $B = 0$. The dotted line is a guide for the eye.

time-resolved-electroluminescence (TREL) with a temporal resolution of 125 ps at the heavy-hole exciton (XH) peak.

Figure 10.12 (right axis) shows the I^+ and I^- intensities time-resolved EL spectra acquired at 15 K for $B = 0.8$ T (the applied bias is $V = 4$ V). The origin $t = 0$ is chosen when the EL signal appears, that is to say 1,700 ps after the rising edge of the electrical pulse. We observe first an increase of the EL signal with a rise time of about $1{,}250 \pm 125$ ps (measured between 10% and 90% of maximum signal), followed by a fast drop with a characteristic time at $(1/e)$ of 200 ± 125 ps close to the temporal resolution of our set-up. This slow EL rise time compared to the fast rise time (100 ps) of the applied voltage is due to parasitic resistances, inductances and capacitances of the device. The temporal window for EL (around 2.4 ns at the foot of the EL pulse) is shorter than the electrical pulse duration (around 4 ns). This could be due to the fact that the threshold voltage for EL emission is only reached during this window. A detailed analysis of the dynamical behavior of EL intensity is beyond the scope of this chapter. The time evolution of the EL circular polarization P_C can be deduced from $I^{\sigma+}$ and $I^{\sigma-}$ intensities and is displayed on the left axis of Fig. 10.12. When a longitudinal magnetic field of 0.8 T is applied, an electron circular polarization is evidenced [Fig. 10.12(a)], whereas no circular polarization can be detected for $B = 0$ [Fig. 10.12(b)]. This demonstrates a rapid establishment of the current-spin polarization below the nanosecond scale. Due to poor signal-to-noise ratio, it is however not possible to determine the circular polarization degree during the first 700 ps of the dynamics. From 700 ps to 1,000 ps, P_C undergoes a relaxation trend and then remains quasi-constant ($\approx 15 \pm 3\%$) during the whole dynamics.[57] This result is consistent with an electron spin relaxation time of 450 ps measured independently by TRPL,[50] which could be responsible for the circular polarization decrease during the rising time of EL, before the system reaches a polarization quasi-stationary regime after around 1 ns. The presence of a large circular polarization as soon as any EL signal is detectable proves unambiguously that the building time of the electronic spin polarization in the quantum well embedded in the intrinsic region of the diode is much faster than the building time of electroluminescence.

10.4 CONCLUSIONS AND OUTLOOK

The electrical spin injection from an FM metal into a semiconductor has now been demonstrated by different experimental teams around the world. In contrast to spin injectors based on diluted magnetic semiconductors, spin injection is possible at room temperature. However, the devices still require the application of a strong external magnetic field. The future for electrical spin injection necessitates developing magnetic layers with vertical remanent magnetization in order to inject electron spin into the semiconductor even in the absence of external magnetic field.[59] This would yield the operation of a vertical surface emitting laser with reduced threshold.[59,60] Finally, the recent successful injection of spin-polarized electrons from an Fe film into silicon is very promising.[61]

ACKNOWLEDGMENTS

The authors thank all the researchers who contributed to this work: B. Liu, L. Lombez, V. G. Truong, Y. Lu, P. Gallo, P. H. Binh, P. F. Braun, H. Carrere, M. Tran, C. Fontaine, A. Arnoult, A. Lemaître, R. Mattana, C. Deranlot, B. Urbaszek, Y. Zheng, D. Demaille and A. Fert.

References

[1] R. Fiederling, M. Keim, G. Reuscher, W. Ossau, G. Schmidt, A. Waag and L. W. Molenkamp, *Nature* **402**, 787 (1999).

[2] Y. Uspenskii, E. Kulatov, H. Mariette, H. Nakayama and H. Ohta, *J. Magn. Magn. Mater.* **258**, 248 (2003).

[3] G. Bouzerar, G. Ziman and J. Kudrnovsky, *Europhys. Lett.* **69**, 812 (2005).

[4] P. Van Dorpe, Z. Liu, W. Van Roy, V. F. Motsny, M. Sawicki, G. Borghs and J. DeBoeck, *Appl. Phys. Lett.* **84**, 3495 (2004).

[5] H. Zhu, M. Ramsteiner, H. Kostial, M. Wassermeier, H.-P. Schönherr and K. Ploog, *Phys. Rev. Lett.* **87**, 16601 (2001).

[6] T. Valet and A. Fert, *Phys. Rev. B* **48**, 7099 (1993).
[7] E. Rashba, *Phys. Rev. B* **62**, R46267 (2000).
[8] A. Fert and H. Jaffrès, *Phys. Rev. B* **64**, 184420 (2001).
[9] D. L. Smith and R. N. Silver, *Phys. Rev. B* **64**, 045323 (2001).
[10] A. Fert, J.-M. George, H. Jaffrès and R. Mattana, *IEEE Trans. Electr. Dev.* **54**(5), 921 (2007).
[11] G. Schmidt *et al.*, *Phys. Rev. B* **62**, 4790 (2000).
[12] A. Fert and S. F. Lee, *Phys. Rev. B* **53**, 6554 (1996).
[13] A. T. Hanbicki, B. T. Jonker, G. Itskos, G. Kieseoglou and A. Petrou, *Appl. Phys. Lett.* **80**, 1240 (2002).
[14] B. Liu, M. Senès, S. Couderc, J.-F. Bobo, X. Marie, T. Amand, C. Fontaine and A. Arnoult, *Physica E* **17**, 358 (2003).
[15] G. Aronov and G. E. Pikus, *Sov. Phys. Semicond.* **10**, 698 (1976).
[16] F. Meier and B. Zakharchenya, *Optical orientation*, in *Modern Problems in Condensed Matter Sciences*, Vol. 8 (North-Holland, Amsterdam, 1984).
[17] G. Bastard, *Wave Mechanics Applied to Semiconductor Heterostructures*, Monographie de Physique, Les Éditions de Physique (1988).
[18] Y. Ohno, D. K. Young, B. Beschoten, F. Matsukura, H. Ohno and D. D. Awschalom, *Nature* **402**, 790 (1999).
[19] R. Fiederling, P. Grabs, W. Ossau, G. Schmidt and L. W. Molenkamp, *Appl. Phys. Lett.* **82**, 2160 (2003).
[20] B. L. Liu, P. Renucci, H. Carrère, M. Sénès, X. Marie, T. Amand, J. F. Bobo, C. Fontaine, A. Arnoult and P. H. Binh, *Physica Status Solidi C*, **1**(3), 475 (2004).
[21] F. Motsnyi, J. De Boeck, J. Das, W. Van Roy, G. Borghs, E. Goovaerts and V. I. Safarov, *Appl. Phys. Lett.* **81**(2), 265 (2002).
[22] P. F. Braun, X. Marie, L. Lombez, B. Urbaszek, T. Amand, P. Renucci, V. K. Kalevich, K. V. Kavokin, O. Krebs, P. Voisin and Y. Masumoto, *Phys. Rev. Lett.* **94**, 116601 (2005).
[23] R. J. Soulen, Jr., J. M. Byers, M. S. Osofsky, B. Nadgorny, T. Ambrose, S. F. Cheng, P. R. Broussard, C. T. Tanaka, J. Nowak, J. S. Moodera, A. Barry and J. M. D. Coey, *Science* **282**, 85 (1998).
[24] R. Jansen, *Appl. Phys. Lett.* **81**, 2130 (2002).
[25] A. T. Hanbicki, O. M. J. van't Erve, R. Magno, G. Kieseoglou, C. H. Li, B. T. Jonker, G. Itskos, R. Mallory, M. Yasar and A. Petrou, *Appl. Phys. Lett.* **82**, 4092 (2003).
[26] C. H. Li, G. Kioseoglou, O. M. van't Erve, M. E. Ware, D. Gammon, R. M. Stroud, B. T. Jonker, R. Mallory, M. Yasar and A. Petrou, *Appl. Phys. Lett.* **86**, 132503 (2005).
[27] G. Itskos, E. Harbord, S. K. Clowes, E. Clarke, L. F. Cohen, R. Murray, P. Van Dorpe and W. Van Roy, *Appl. Phys. Lett.* **88**, 022113 (2006).
[28] T. Manago and H. Akinaga, *Appl. Phys. Lett.* **81**(4), 696 (2002).
[29] O. M. J. van't Erve, G. Kioseoglou, A. T. Hanbicki, C. H. Li, B. T. Jonker, R. Mallory, M. Yasar and A. Petrou, *Appl. Phys. Lett.* **84**(21), 4334 (2004).
[30] L. Lombez, P. Renucci, P. F. Braun, H. Carrère, X. Marie, T. Amand, B. Urbaszek, J. L. Gauffier, P. Gallo, T. Camps, A. Arnoult, C. Fontaine, C. Deranlot, R. Mattana, H. Jaffrès, J.-M. George and P. H. Binh **90**, 081111 (2007).
[31] M. Paillard, X. Marie, P. Renucci, T. Amand, A. Jbeli and J. M. Gerard, *Phys. Rev. Lett.* **86**, 1634 (2001).
[32] Y. Chye, M. E. White, E. Johnston-Halperin, B. D. Gerardot, D. D. Awschalom and P. M. Petroff, *Phys. Rev. B* **66**, R201301 (2002).
[33] S. Gosh and P. Battacharya, *Appl. Phys. Lett.* **80**, 4658 (2001).

[34] W. Löffler, D. Tröndle, J. Fallert, H. Kalt, D. Litvinov, D. Gerthsen, J. Lupaca-Schomber, T. Passow, B. Daniel, J. Kvietkova, M. Grün, C. Klingshirn and M. Hetterich, *Appl. Phys. Lett.* **88**, 062105 (2006).

[35] P. F. Braun, X. Marie, L. Lombez, B. Urbaszek, T. Amand, P. Renucci, V. K. Kalevich, K. V. Kavokin, O. Krebs, P. Voisin and Y. Masumoto, *Phys. Rev. Lett.* **94**, 116601 (2005).

[36] K. Gündogdu, K. C. Hall, T. F. Boggess, D. G. Deppe and O. B. Shchekin, *Appl. Phys. Lett.* **84**, 2793 (2004).

[37] S. Marcinkevicius, D. Siegert and Q. X. Zhao, *J. Appl. Phys.* **100**, 054310 (2006).

[38] R. A. Lewis, Y. J. Wang and M. Henini, *Physica B* **346**, 483 (2004).

[39] A. Baldereschi and N. O. Lipari, *Phys. Rev. B* **8**, 2697 (1973).

[40] A. V. Savelyev, A. I. Tartakovskii, M. S. Skolnick, D. J. Mowbray, M. V. Maximov, V. M. Ustinov and R. P. Seisyan, *Appl. Phys. Lett.* **88**, 111104 (2006).

[41] C. Monat, B. Alloing, C. Zinoni, L. H. Li and A. Fiore, *Nano Lett.* **6**, 1464 (2006).

[42] W. Löffler, M. Hetterich, C. Mauser, S. Li, T. Passow and H. Kalt, *Appl. Phys. Lett.* **90**, 232105 (2007).

[43] M. Ghali, R. Arians, T. Kümmell, G. Bacher, J. Wenish, S. Mahapatra and K. Brunner, *Appl. Phys. Lett.* **90**, 093110 (2007).

[44] X. Jiang, R. M. Shelby, R. M. Macfarlane, S. R. Bank, J. S. Harris and S. S. P. Parkin, *Phys. Rev. Lett.* **94**, 056601 (2005).

[45] J. Moodera, L. R. Kinder, R. M. Wong and R. Meservey, *Phys. Rev. Lett.* **74** 3273 (1995).

[46] R. Wang, X. Jiang, R. M. Shelby, R. M. Shelby, R. M. Macfarlane, S. S. P. Parkin, S. R. Bank and J. S. Harris, *Appl. Phys. Lett.* **86**, 052901 (2005).

[47] T. Manago, A. Sinsarp and H. Akinaga, *J. Appl. Phys.* **102**, 083914 (2007).

[48] A. Sinsarp, T. Manago, F. Takano and H. Akinaga, *Jpn. J. Appl. Phys.* **46**, L4 (2007).

[49] W. H. Butler, X.-G. Zhang, T. C. Schulthess and J. M. MacLaren, *Phys. Rev. B* **63**, 054416 (2001).

[50] Y. Lu, V. G. Truong, P. Renucci, M. Tran, H. Jaffrès, C. Deranlot, J.-M George, A. Lemaître, Y. Zheng, D. Demaille, P.-H. Binh, T. Amand and X. Marie, *Appl. Phys. Lett.* **63**, 054416 (2008).

[51] Y. Lu, C. Deranlot, A. Vaurès, F. Petroff, J.-M. George, Y. Zheng and D. Demaille, *Appl. Phys. Lett.* **91**, 222504 (2007).

[52] S. Yuasa, Y. Suzuki, T. Katayama and K. Ando, *Appl. Phys. Lett.* **87**, 242503 (2005).

[53] G. Salis, R. Wang, X. Jiang, R. M. Shelby, S. S. P. Parkin, S. R. Bank and J. S. Harris, *Appl. Phys. Lett.* **87**, 262503 (2005).

[54] W. Van Roy, P. Van Dorpe, J. De Boeck and G. Borghs, *Mat. Sci. Eng. B* **126**, 155 (2006).

[55] D. A. B. Miller, D. S. Chemla, T. C. Damen, A. C. Gossard, W. Wiegmann, T. H. Wood and C. A. Burrus, *Phys. Rev. Lett.* **53**, 2173 (1984).

[56] G. L. Bir, A. G. Aronov and G. E. Pikus, *Sov. Phys. J. Exp. Theor. Phys.* **42**, 705 (1976).

[57] V. G. Truong, P.-H. Binh, P. Renucci, M. Tran, Y. Lu, H. Jaffrès, J.-M. George, C. Deranlot, A. Lemaître, T. Amand and X. Marie, *Appl. Phys. Lett.* **94**, 141109 (2009).

[58] N. Gerhard, S. Hövel, C. Brenner, M. R. Hofman, F. Y. Lo, D. Reuter, A. D. Wieck, E. Schuster, W. Keune and K. Westerholt, *Appl. Phys. Lett.* **87**, 32502 (2005).

[59] J. Rudolph, D. Hagele, H. M. Gibbs, G. Khitrova, M. Ostreich, *Appl. Phys. Lett.* **82**, 4516 (2003).

[60] M. Holub, J. Shin, D. Saha and P. Bhattacharya, *Phys. Rev. Lett.* **98**, 146603 (2007).

[61] B. Jonker, G. Kioseoglou, A. Hanbicki, C. Li and P. Thompson, *Nature Phys.* **3**, 542 (2007).

Chapter Eleven

Magneto-optical Spectroscopy of Spin Injection and Spin Relaxation in Spin Light-Emitting Structures

Irina A. Buyanova* and Weimin M. Chen†

Department of Physics, Chemistry and Biology, Linköping University, S-581 83 Linköping, Sweden

**irb@ifm.liu.se; †wmc@ifm.liu.se*

In this chapter we review our recent results from in-depth investigations of physical mechanisms which govern the efficiency of several processes important for future spintronic devices, such as spin alignment within diluted magnetic semiconductors (DMS), spin injection from DMS to non-magnetic semiconductor spin detectors (SD) and also spin depolarization within SD. Spin-injection structures based on II-VIs [e.g., ZnMnSe/Zn(Cd)Se, Zn(Cd, Mg)O] and III-Vs [e.g., GaMnN/Ga(In)N] were studied as model cases.

11.1 INTRODUCTION

All-semiconductor spintronics has attracted increasing attention (for reviews see Refs. 1–3) due to its promise to combine new spin enabling functionality with well-established electronic and optical properties of semiconductors that have widely been explored in the present microelectronics and optoelectronics. Among many challenges, realizations of efficient spin polarization, spin injection and reliable spin detection are key conditions required for the success of this emerging technology. In this chapter we review our recent results from in-depth investigations of physical mechanisms which govern the efficiency of these processes in spin-injection structures. Several material systems will be discussed such as ZnMnSe/Zn(Cd)Se quantum well (QW) and quantum dot (QD) structures, GaMnN/Ga(In)N spin light emitting diode (LED) structures, and also ZnO-based structures.

The chapter is organized as follows. A brief description of the structures under study will be provided in Sec. 11.2. Section 11.3 will be devoted to key processes

Handbook of Spintronic Semiconductors by W M Chen & I A Buyanova

www.panstanford.com
978-981-4267-36-6

which determine spin injection efficiency in the spin LED from ZnMnSe/Zn(Cd)Se QW structures. These include dominant processes for optical spin injection from ZnMnSe spin injectors (SI) to a non-magnetic Zn(Cd)Se QW spin detector (SD), spin relaxation processes for excitons in the SI, and spin relaxation within the SD. In Sec. 11.4 spin injection processes from ZnMnSe DMS to a zero-dimensional SD based on CdSe QD will be analyzed. Spin depolarization and spin dynamics in GaMnN/Ga(In)N and Zn(Cd, Mg)O spin light emitting structures will be addressed in Secs. 11.5 and 11.6, respectively. The main conclusions of the present study will be summarized in Sec. 11.7.

11.2 SAMPLE STRUCTURES

11.2.1 ZnMnSe/Zn(Cd)Se QW Spin Injection Structures

A series of ZnMnSe/ZnCdSe structures was studied. They were all grown on GaAs substrates by molecular beam epitaxy (MBE). The active regions of the structures consist of a ZnMnSe DMS SI and a non-magnetic SD from a $Zn_{0.75}Cd_{0.25}Se$ QW (1.8–7 nm wide), with a non-magnetic ZnSe spacer (4–10 nm thick) inserted between the SI and SD. The DMS SI is a 10-period $Zn_{0.96}Mn_{0.04}Se$/CdSe (40 nm/0.8 monolayer) superlattice (SL) or a 10–15 nm thick $Zn_{0.95}Mn_{0.05}Se$ layer. The whole region was surrounded by wider band-gap $ZnS_{0.06}Se_{0.94}$ thick layers, preventing the leakage of carriers.

11.2.2 ZnMnSe/CdSe QD Spin Injection Structures

The basic design of these structures is similar to that described in the previous section. The $Zn_{0.93}Mn_{0.07}Se$ DMS SI with a thickness of 100 nm was grown on a ZnSe-buffer layer, followed by a spacer layer of ZnSe. The width of the spacer layer was intentionally varied between samples from 0.5 to 10 nm. Then, three monolayers of CdSe were deposited on the ZnSe spacer, forming self-assembled QDs. The dots were finally capped by a 10 nm thick ZnSe layer. The dot formation was confirmed from reflection high-energy electron diffraction (RHEED) and μ-photoluminescence (PL) measurements.

11.2.3 GaMnN/InGaN Spin Light Emitting Diode Structures

Two LED structures, namely spin-LED and ref-LED, were studied. They have an inverted geometry with n-type layers on the top and were grown on sapphire substrates starting with a 2 μm thick buffer layer of undoped semi-insulating GaN. The active region of the spin-LED consists of (i) a non-magnetic SD with five periods of $In_{0.4}Ga_{0.6}N$ (3 nm)/GaN:Si (10 nm) multiple QW and (ii) an SI with a 100 nm thick $Ga_{0.97}Mn_{0.03}N$ layer. The SI and SD are separated by a 20 nm thick non-magnetic GaN:Si spacer. In the ref-LED the SI was replaced by a 100 nm n-type Si-doped

GaN. The GaMnN SI exhibits ferromagnetic ordering up to room temperature, evident from the hysteresis behavior of in-plane magnetization.

11.2.4 ZnO-Based Structures

The following types of wurtzite ZnO-based materials were studied: (i) bulk n-type ZnO crystals fabricated by melt growth technique, from Cermet Inc; (ii) 0.2–0.8 μm thick epitaxial ZnO layers; (iii) epitaxial ZnCdO alloys with Cd compositions of 9.1% and 15.7% and thicknesses of 0.2–0.8 μm; and (iv) an n-$Mg_{0.1}Zn_{0.9}O$/ZnO/p-$Al_{0.16}Ga_{0.84}N$ LED structure. The thicknesses of the ZnMgO, ZnO and AlGaN epilayers were 150 nm, 50 nm and 40 nm, respectively. For the LED structure, spin relaxation within the ZnO and ZnMgO layers can be studied by selectively monitoring the corresponding PL emission. All epitaxial structures, i.e., (ii)–(iv), were grown by MBE on a c-plane sapphire substrate with a GaN/ZnO buffer layer.

11.3 ZnMnSe/ZnCdSe QW SPIN INJECTION STRUCTURES

Though the spectacular magnetic effects in the II-VI DMS structures such as (Zn, Cd, Mn)(Se, Te) are so far restricted to cryogenic temperatures, a detailed level of understanding of these DMS and quantum well structures, thanks to their rather mature modern epitaxial growth techniques, can provide a viable basis for in-depth investigations on the fundamental science underlying spin phenomena and for proof-of-concept spintronic devices.

Spin injection in these structures was reliably demonstrated in early studies,[4–8] from substantial changes of optical polarization of the SD measured under spin injection conditions as compared with intrinsic optical polarization of the SD (see Fig. 11.1). In applied magnetic fields, a weak, negative PL polarization defined by $100(\sigma^+ - \sigma^-)/(\sigma^+ + \sigma^-)$ was observed in the SD when the excitation photon energy was tuned to that of the resonant absorption in the QW SD. This is expected, as the σ^--active $|+1/2, -3/2\rangle$ state of the heavy hole (hh) exciton lies lower in energy with respect to the σ^+-active $|-1/2, +3/2\rangle$ hh exciton state. (Here the notation $|m_e, m_h\rangle$ is used to describe the excitonic state, and m_e and m_h denote the angular momentum projections of the electrons and holes of the excitons along the magnetic field direction, respectively.

On the other hand, a reversal of the PL polarization sign, observed under the above SI band-gap excitation conditions, was interpreted as a result of population inversion between the two exciton spin sublevels of the SD,[6–8] due to the injection of spin-polarized carriers from the SI.

The detected degree of spin polarization of the QW PL emission is determined by spin alignment within the SI, spin injection from the SI to the non-magnetic QW SD, and also spin relaxation within the SD. Therefore, each of the physical processes involved needs to be characterized and fully understood for full optimization of spin-enabling functionalities of the structures.

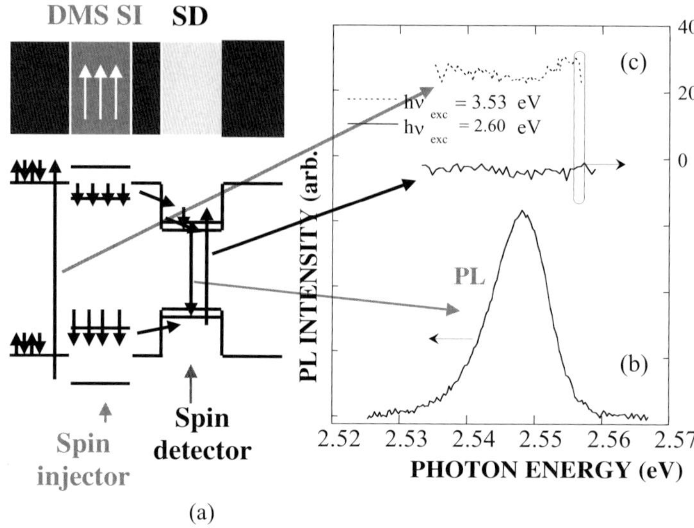

Figure 11.1. (a) Schematic illustration of the structural design of the investigated structures. (b) The ZnCdSe QW PL spectrum and (c) spectral dependence of the PL polarization, measured at an applied magnetic field of 3 T under optical excitation below (solid line) and above (dashed line) the DMS SI band-gap.

11.3.1 Spin Injection

Spin injection, i.e., transport of polarized spins from SI to SD, is one of the key processes in spin functional structures. The unambiguous experimental evidence for the efficient spin injection in II-VI QW structures was obtained by using tunable PL excitation spectroscopy.[6]

11.3.1.1 Spin injection: Experimental evidence

Representative PL excitation (PLE) spectra of the QW PL emission are shown in Fig. 11.2. Several regions of the PLE spectra, related to preferential light absorption within different spatial regions of the structures, can be distinguished. Photo-excitation with photon energies below 2.7 eV (at 0 T) corresponds to resonant absorption within the ZnCdSe QW SD. The excitonic absorption within the non-magnetic ZnSe spacer and the ZnSSe outer barrier gives rise to the PLE peaks at 2.81 eV and 2.85 eV, respectively. In addition, a pronounced PLE peak at the DMS SI exciton energy can be seen. The considerably higher strength of this peak as compared with that of the resonant excitation of the QW SD indicates that the dominant portion of the carriers/excitons under such excitation conditions is supplied by the SI.

In an applied magnetic field B, the SI-related peak exhibits a giant Zeeman splitting into two components (see Fig. 11.2). As expected, the σ^+-active $|-1/2, +3/2\rangle$ state of the hh exciton has the lowest energy in the ZnMnSe, confirmed by the performed PLE measurements using circularly polarized light (not

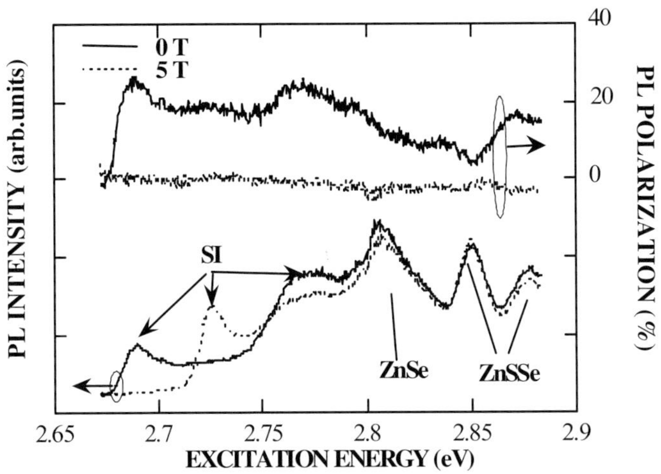

Figure 11.2. PL excitation spectra of the ZnCdSe QW PL emission (lower part) and PL polarization (upper part) taken at 0 T and 5 T. From Ref. 6.

shown in the figure). Simultaneously, the QW SD PL acquires circular polarization. Based on the degree of the QW SD PL polarization detected in an applied magnetic field under resonant photo-excitation at the lower $|-1/2, +3/2\rangle$ spin state of the DMS, the spin injection efficiency of excitons/free carries from the magnetic SI to the non-magnetic SD can be estimated as ~30–35%. This is because the selective excitation of the lowest σ^+-active $|-1/2, +3/2\rangle$ state assures full spin alignment in the SI. (A contribution within the same spectral region from the direct carrier absorption within the QW SD, leading to the opposite PL polarization, can be taken into account by subtracting the background signal.) The spin injection efficiency was found to be independent of the spacer thickness indicating that the spin polarization created by the SI remained preserved after injection into the QW SD through a barrier of ZnSe as thick as 10 nm.

11.3.1.2 Tuning functionality of the spin injection structures: Spin filtering versus spin switching

Depending on the desired application, spins of the carriers supplied to SD should be either fixed to a certain orientation or controlled at will. In principle, the desired device functionality can be realized by tuning the ratio between rates of spin alignment within the DMS and carrier escape from it. If spin alignment governed by spin relaxation precedes carrier escape, the DMS layer should act as an efficient spin filter. The spin switching function could, on the other hand, be achieved in the opposite case. Therefore, requirements for an efficient spin switch or spin filter can be met by optimizing structural design.[7]

We were able to attain the desired spin functionality by tuning the carrier escape rate from the DMS ZnMnSe to the ZnCdSe QW SD by varying the thickness of the ZnSe spacer and the energy barrier height between the DMS and the SD.

A preferential supply of carriers to a certain spin state of the DMS was achieved by optical means using a tunable excitation source. The resulting spin orientation of carriers was elucidated by measuring the PL polarization from the non-magnetic QW SD.

The following spin control functions were demonstrated as a function of the structural design. The spin switching was realized when using a $Zn_{0.95}Mn_{0.05}Se$ layer (10–15 nm) as a spin aligner, evident from a switch in sign of PL polarization of the non-magnetic QW SD when the photoexcitation changes between the two spin states of the DMS [see Fig. 11.3(a)]. The achieved spin switching was suggested to reflect fast carrier escape from the DMS due to the absence of the energy barrier between the upper spin state of the DMS layer and the SD. On the other hand, spin alignment was accomplished in the tunneling structure containing a DMS of a 10 period ZnMnSe/CdSe (40 nm/0.8 monolayer) SL and an 8 nm thick ZnSe spacer, ensured by the presence of the 0.2 eV energy barrier for the spin injection to the SD [Fig. 11.3(b)]. A decrease in the carrier escape time by reducing spacer thickness below 8 nm was shown to cause incomplete spin alignment in the DMS.

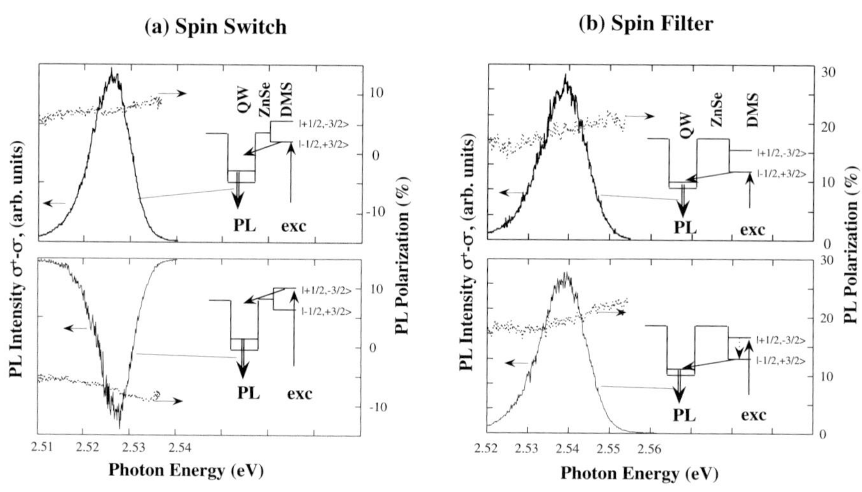

Figure 11.3. (a) Demonstration of spin switch functionality of the spin injector, when spin injection is designed to precede spin relaxation within the DMS SI (a ZnMnSe/CdSe SL). Depending on the excitation photon energies that were tuned to resonantly excite the lower σ^+-active $|-1/2,+3/2\rangle$ or the upper σ^--active $|+1/2,-3/2\rangle$ states of the hh exciton in the DMS, oppositely polarized spins were injected from the DMS to the ZnCdSe SD. In the insets, a schematic picture is given for the excitonic band-gaps of the structure, together with the optical excitation, the exciton spin injection (the solid arrows), and the QW PL transitions (the double arrows). (b) Demonstration of spin alignment functionality of the spin injector, when spin relaxation within the DMS SI (a ZnMnSe layer) is designed to precede spin injection. The differential polarization signal $\sigma^+ - \sigma^-$ (the solid curves) and polarization degree (the dotted curves) of the PL emission from the ZnCdSe QW SD were measured at 2 K in an applied magnetic field of 0.5 T. Independent of the excitation photon energies tuned to resonantly excite the lower or upper states of the hh exciton in the DMS, the spins injected from the DMS to the ZnCdSe SD have the same polarization. From Ref. 7.

11.3.1.3 Spin injection: Mechanisms

Though diffusion and tunneling of carriers/excitons are intuitively expected to be common physical processes underpinning spin injection, the exact mechanism has seldom been verified. In order to gain in-depth understanding of the exact physical mechanism responsible for optical spin injection from a ZnMnSe DMS SI to a non-magnetic ZnCdSe SD, we have carried out time-resolved magneto-optical studies of the spin injection process.[8]

Contributions of various mechanisms for optical spin injection were deliberately tuned and addressed utilizing a band engineering approach. For example, the role of tunneling in spin injection was evaluated by changing the spacer layer width between SI and SD. The importance of individual carrier injection versus injection of excitons as a whole was evaluated by comparing spin injection in the structures where one of these mechanisms was intentionally suppressed. A schematic picture of the relevant structures is shown in Fig. 11.4.

Both structures have a design as shown in Fig. 11.1(a) and only differ in their SD width, i.e., 7 nm and 1.85 nm for the type-I and type-II structures, respectively. In the type-I structures, the type-I band alignment between the SD and SI was realized for all applied magnetic fields, due to a wide SD width and thus weaker carrier confinement. This permits spin injection from the SI to the SD to occur via injection of individual carriers/excitons, denoted by open/filled horizontal arrows in Fig. 11.4(a). To evaluate the contributions of tunneling in spin injection, the spacer/barrier width (w_b) in these structures was varied between 4 and 8 nm. On the other hand, in the type-II structures, the type-II band alignment realized in an applied magnetic field prohibited tunneling of holes from the SI to the SD. Spin injection via excitons was, however, still allowed when the energy of the excitons

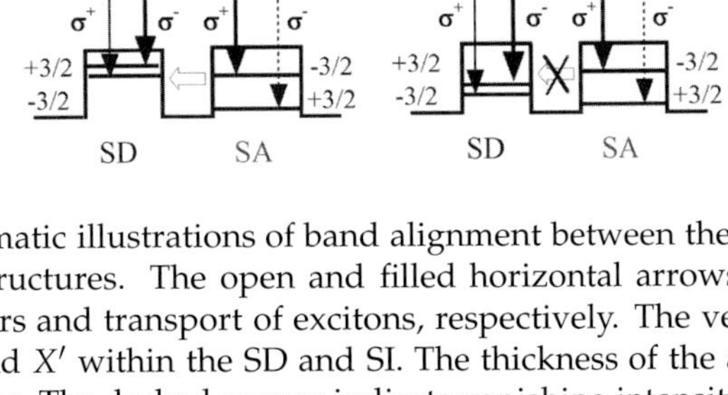

Figure 11.4. Schematic illustrations of band alignment between the SI and SD in (a) type-I and (b) type-II structures. The open and filled horizontal arrows symbolize tunneling of individual carriers and transport of excitons, respectively. The vertical arrows are optical transitions X and X' within the SD and SI. The thickness of the arrows symbolizes the related PL intensities. The dashed arrows indicate vanishing intensity of the corresponding transitions due to low population of the involved state. The spin states $\pm 1/2$ and $\pm 3/2$ refer to $|m_c\rangle$ and $|m_v\rangle$ of the conduction and valence band electrons, respectively. From Ref. 8.

in the SI (X') lies above that of the excitons in the SD (X), which is the case here even at 6 T [Fig. 11.4(b)].

To obtain information on the detailed dynamics of spin injection, in-depth studies by transient magneto-PL spectroscopy have been undertaken.[8] The corresponding results are summarized in Fig. 11.5, where the decay curves of PL and PL polarization from three representative structures are shown. Without injection from the SI (i.e., when the excitation photon energy $h\nu < E_g^{SI}$), the SD PL in applied magnetic fields is negatively polarized, determined by the intrinsic properties of the SD hh exciton, similar to the results of CW-measurements discussed in Sec. 11.3.1.1. The "intrinsic" decay of the SD excitons under this excitation condition is single exponential with a decay time of about 100 ps [shown by the dashed-dotted curve in Fig. 11.5(a)]. On the other hand, a substantial positive polarization of the SD PL developed when $B > 0$ T and $h\nu \geq E_g^{SI}$. Two temporal components of the SD PL polarization were distinguished under these conditions. The first component is responsible for the very fast rise (<10 ps, limited by the temporal resolution of the set-up) of the polarization degree to 20%. This component was only observed in the type-I structures with 4 nm spacer [Fig. 11.5(b)]. In addition, a slower rising of the PL polarization to 45% was observed. This reflects polarization properties of a slow feeding process from the SI to SD which gives rise to a slow component of the PL decay.

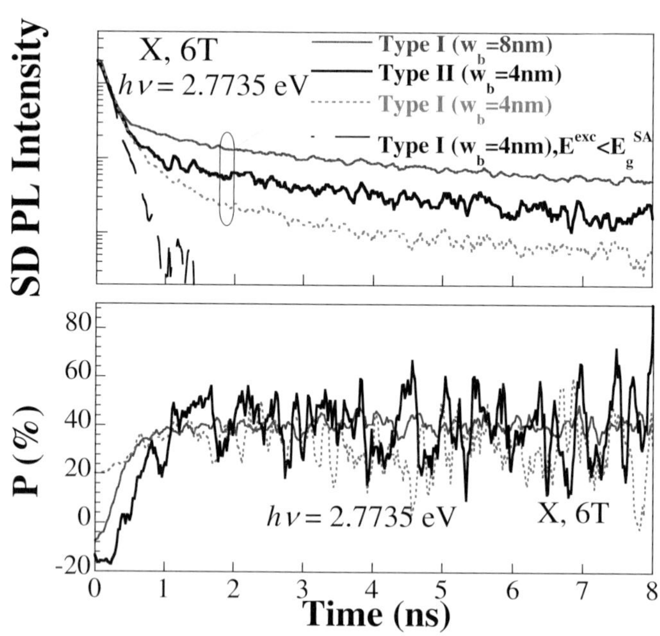

Figure 11.5. (a) Decay curves of the SD PL measured from three representative structures with the specified ZnSe barrier width (w_b), taken at 2 K and with the excitation photon energy at 2.7735 eV, i.e., with the excitation photon energy exceeding the SI band-gap. Also shown is "intrinsic" decay of the SD PL measured under resonant excitation of the SD with the photon energy below the SI excitonic band-gap at 2.725 eV. (b) PL polarizations of X from the same structures as shown in (a), taken at 6 T and at 2.7735 eV. From Ref. 8.

The existence of distinctly different components in the PL transient points towards several mechanisms of spin injection:

(a) *Fast component*. A strong dependence of the corresponding spin injection process on the width of the spacer layer inserted between the SI and SD indicated that it originates from tunneling. The observed suppression of the process in the type-II structures is evidence that the fast spin injection occurs due to tunneling of individual electrons and holes from the SI to SD.
(b) *Slow component*. The slow injection process is highly spin preserving and, therefore, dominates spin injection. It is also observed in all the studied structures, which suggests that a common physical mechanism is involved. The physical origin of this dominant injection process has been suggested[8] as being due to energy transfer. Indeed, tunneling of individual electrons and holes from the SI to SD can be ruled out, as it is improbable at 2 K in type-II structures where tunneling of holes is prohibited [Fig. 11.4(b)]. Tunneling of excitons as a whole can also be discarded as it should exponentially depend on the barrier width, which contradicts the experimental findings.

This leaves the other commonly occurring process — energy transfer — as perhaps the only plausible candidate. The most commonly discussed physical interaction responsible for the energy transfer is an electric dipole–dipole interaction.[9,10] The transfer rate between two adjacent QWs has been shown to vary as R^{-4},[11] i.e., to be strongly dependent on the separation between the two QWs. In contrast, the energy transfer rate was found to be insensitive to R in our experiments.

Recently, another physical interaction in terms of exciton–photon coupling was theoretically evaluated,[11] leading to the so-called photon-exchange energy transfer. It was shown that the rate of such energy transfer is a slowly varying function of R, consistent with our experimental observations.

The observed spin injection should then arise from some sort of localized electronic excitations (LEE) in the SI, such as localized excitons or localized electron–hole pairs. This is because the lifetime for the free excitons in SI is $\sim$50 ps, from PL transient measurements,[8] which is far too short to account for the very slow feeding process from the SI. The existence of these localized excitations was indeed confirmed by the time-resolved PL measurements and is not surprising in view of possible strain variation, interface roughness, and compositional fluctuation in the DMS.

Thus, we conclude that the photon-exchange energy transfer via the localized excitations is the most plausible physical process responsible for the observed optical spin injection. This finding renders attempts to improve spin injection efficiency, by optimizing conditions for tunneling, ineffective.

11.3.2 Spin Alignment

The success of future spintronic devices requires precise knowledge about ways to create and to control spin-polarized carriers. Spin orientation (or polarization) of carriers and excitons in semiconductors is governed by spin relaxation processes, which are caused by various physical processes admixing spin states.[12–16] The exact mechanisms, however, are rather complicated and sensitive to many electronic and structural parameters such as band structure, doping, strain, dimensionality, etc.[17–22] Recently, we have investigated[23,24] exciton spin relaxation in ZnMnSe based DMS and its possible physical mechanism, over a range of magnetic fields B where the energy of exciton spin splitting crosses the LO phonon energy — a regime where very little is known about spin-dependent processes in a semiconductor. Three types of magneto-optical experimental approaches, namely, tunable excitation spectroscopy, hot PL, and time-resolved PL measurements, were employed.

11.3.2.1 Spin alignment: Tunable excitation spectroscopy

This approach utilizes the idea that the degree of optical polarization of the non-magnetic SD under spin injection conditions reflects the combined efficiency of the spin alignment within the SI (determined by spin relaxation within the DMS layer) and the spin injection efficiency from the SI to the non-magnetic SD. For example, the degree of the SD PL polarization detected in an applied magnetic field under resonant photo-excitation in the lower $|-1/2, +3/2\rangle$ spin state in DMS is related to the spin injection efficiency of excitons/free carriers from the magnetic SI to the non-magnetic SD [process I in Fig. 11.6(a)]. Spin relaxation leading to the higher $|+1/2, -3/2\rangle$ state of the DMS is energetically unfavorable and can be neglected starting at a rather weak magnetic field, as a result of giant Zeeman splitting. This allowed a detailed analysis of the spin injection efficiency in different structures, as was discussed in Sec. 11.3.1.

On the other hand, the degree of the SD PL polarization detected under resonant excitation at the upper lying σ^--active $|+1/2, -3/2\rangle$ state of the SI is determined by two competing processes:

(i) direct injection of photo-created excitons/carriers from the upper spin state into the SD which contributes to the σ^- polarization of the SD PL [process II in Fig. 11.6(a)], and
(ii) spin relaxation down to the lowest σ^+-active spin state [process III shown by the dashed arrow in Fig. 11.6(a)] before injection into the SD that leads to the σ^+ polarization of the SD PL. Therefore, the SD PL polarization under the resonant excitation at the upper $|+1/2, -3/2\rangle$ spin state of the DMS exciton can be taken as a measure of spin relaxation within the two spin states of the DMS hh exciton.

The results of these experiments are demonstrated in Figs. 11.6(b) and 11.7. The SD PL polarization detected when the excitation photon energy is resonant with the σ^+-active $|-1/2, +3/2\rangle$ spin state of the DMS exciton is positive and almost independent of B or DMS exciton spin splitting. This is not surprising as only one spin orientation of the DMS excitons is excited and supplied to the SD due to the spin injection from the DMS (process I in Fig. 11.6). (The limited polarization degree of about 32% is believed to be due to spin relaxation within the SD as will be discussed in detail in Sec. 3.3.)

On the other hand, when the excitation photon energy is tuned to the $|+1/2, -3/2\rangle$ spin state of the DMS exciton, the change induced by spin injection in the SD PL polarization becomes more complicated. Even though a positive sign of the SD PL polarization is observed over the entire range (0–5 T) of B, the degree of polarization and, therefore, the rate of the spin relaxation within the DMS SI depend on the DMS exciton spin spitting (or B) (Fig. 11.7).

For large B (and thus a large spin splitting of the DMS exciton), spin relaxation is apparently much faster than spin injection, i.e., process III dominates over process II. This is obvious from similar values of the SD PL polarization detected upon optical excitations resonant with the upper and lower spin levels of the DMS exciton. On the other hand, for small spin splitting of the DMS exciton a reduction

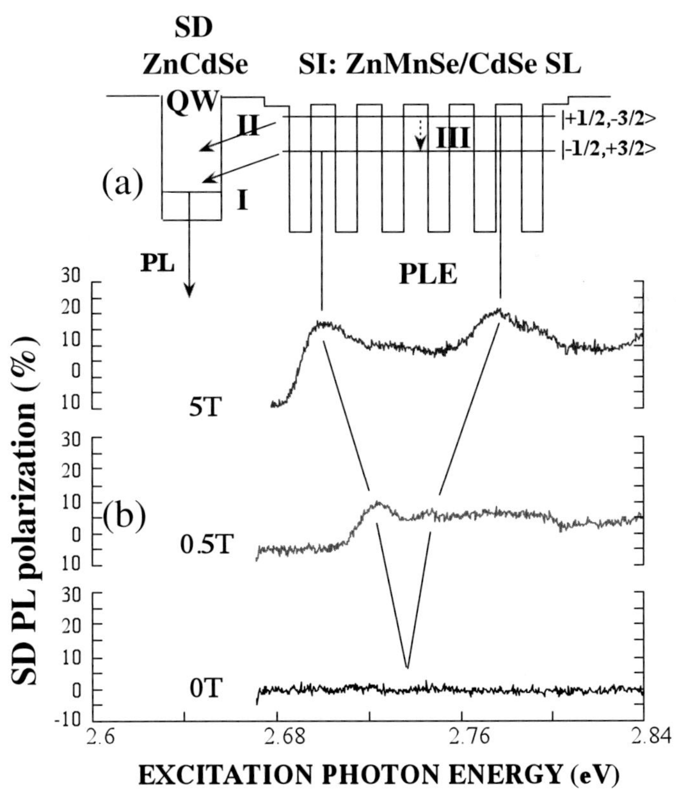

Figure 11.6. (a) A schematic energy diagram of the studied spin injection structures based on II-VI QWs and the investigated physical processes. Process I (II) denotes spin injection from the lower (upper) spin state of the hh exciton in DMS. Process III denotes spin relaxation between the spin states of the hh exciton. (b) Polarization of the SD PL measured at 5 T as a function of the excitation photon energy.

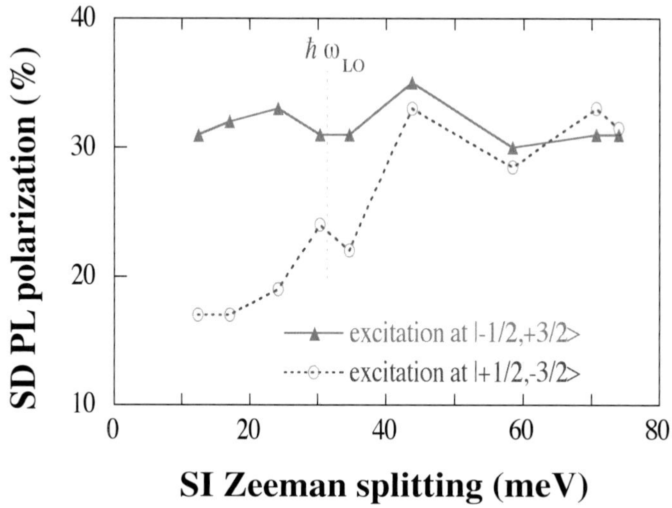

Figure 11.7. Corrected polarization of the SD PL measured under the resonant excitation at the upper (open circles) and lower (filled triangles) spin states of the DMS hh exciton, as a function of the Zeeman splitting of the two spin states. The values are obtained after subtracting the intrinsic PL polarization of the SD. The lines are guides for the eye. From Ref. 23.

of spin relaxation is apparent from a weaker SD PL polarization when optical excitation is tuned at the upper spin level of the DMS as compared to that when optical excitation is tuned at the lower spin level (see Figs. 11.6 and 11.7). This makes possible spin injection from the upper spin level (process II), which compensates the spin polarization provided by process I. The transition between the two regimes occurs when the spin splitting of the DMS excitons is close to the LO-phonon energy (31.6 meV), indicating a possible role of LO phonons in spin relaxation.

11.3.2.2 Spin alignment: Hot PL approach

In the second experimental approach, hot PL experiments were employed to examine the possible LO-assisted exciton spin relaxation (see Fig. 11.8 for an illustration of the principle). In these experiments, only the upper spin state of the DMS hh exciton is selectively excited close to $K = 0$.

Energy relaxation from this state can occur via an LO-assisted spin flip provided that (i) the spin splitting is larger than the LO phonon energy and (ii) this transition is the most efficient one among all available energy relaxation processes. This would generate a large number of hot excitons in the lower spin band exactly at one LO energy below the excitation photon energy (Fig. 11.8). The excitons can then undergo further spin-preserving intraband energy relaxation assisted by acoustic phonons, which leads to a thermalized distribution of the excitons at the bottom of the spin band. Accordingly, two PL features related to the lower spin band should be seen, i.e., a broad PL band corresponding to the thermalized excitons and a sharp hot PL line arising from the non-equilibrium distribution of the hot excitons exactly at one LO phonon energy below the excitation energy.

A typical hot PL spectrum, obtained under the resonant σ^- optical excitation at the upper spin level of the DMS exciton, is shown in Fig. 11.9, when the spin splitting of the two spin levels exceeds the LO phonon energy. It contains a broad PL band arising from the thermalized excitons of the lower spin state, and a sharp hot PL line exactly at one LO phonon energy below the pumping energy. The polarization of the hot PL line, obtained after subtracting the contribution from the broad PL band, has the same sign as the lower spin state but is opposite to that of the pumping light. This demonstrates an exciton spin flip via the emission of one LO phonon, providing direct evidence for the involvement of the LO phonon in exciton spin relaxation. The observed LO assisted spin relaxation is in excellent agreement with the results from the spin injection experiments described above. When the splitting of two spin levels of the DMS hh exciton is less than the LO phonon energy, on the other hand, the observed hot PL line as expected retains the polarization of the pumping light (not shown in the figure). In this case the hot PL line is due to the LO phonon replica of the PL arising from the upper spin state of the DMS hh exciton that is resonantly excited by the pumping light.

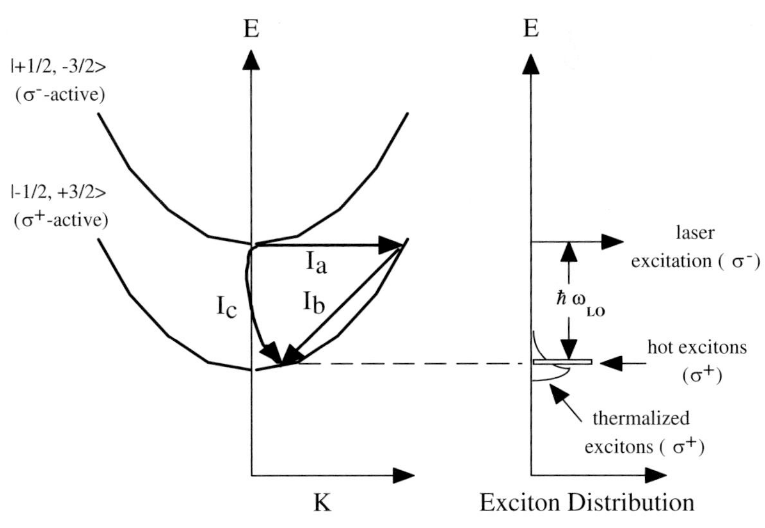

Figure 11.8. Left: A schematic diagram of the two optically active hh exciton bands of the DMS SI in k-space. Process I_a represents a nearly elastic spin scattering. Process I_b shows an LO-assisted, spin-preserving exciton kinetic energy relaxation. Process I_c depicts a single quantum process of the LO-assisted spin relaxation. Right: The exciton distribution in energy under optical pumping in the hot PL experiment. From Ref. 23.

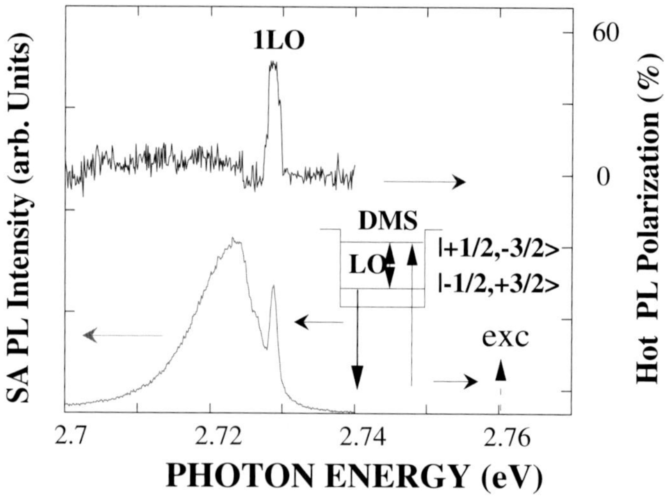

Figure 11.9. Lower part: The SI PL spectrum detected under resonant pumping of the upper spin state of the DMS hh exciton. The photon energy of the excitation light with the matching σ^- polarization is tuned exactly at the energy of the upper spin state (at $K = 0$) of the DMS hh exciton to ensure that only the spin orientation corresponding to that state is selectively generated. Upper part: The polarization of the 1LO hot PL line obtained after subtracting the polarization of the broad PL band. From Ref. 23.

11.3.2.3 Spin alignment: Time resolved PL

The exact rate of the LO-assisted exciton spin relaxation has been determined from time-resolved magneto-optical studies.[24] Representative PL decays measured at 1 T from both spin-split excitonic states of the DMS hh exciton are shown in Fig. 11.10 and yield effective lifetimes of 21 and 28 ps for the upper $|+1/2, -3/2\rangle$ and lower $|-1/2, +3/2\rangle$ excitons, respectively. Assuming that the observed shortening of the exciton lifetime for the $|+1/2, -3/2\rangle$ state as compared with that for the $|-1/2, +3/2\rangle$ state is solely attributed to the spin-flip process, i.e.,

$$\frac{1}{\tau^{u(l)}} = \frac{1}{\tau_{\mathrm{d}}} + \frac{1}{\tau_{\mathrm{SF}}\left(1 + \exp\left(\mp\frac{\Delta}{k_{\mathrm{B}}T}\right)\right)}, \tag{11.1}$$

a spin-flip time τ_{SF} of 84 ps (when the spin splitting $\Delta \approx 18$ meV) can be deduced. Here $\tau^{u(l)}$ is the measured exciton lifetime from the corresponding spin state, and τ_{d} is the characteristic time for all exciton decay processes (except for the spin-flip transition) deduced from the decay time of the lower spin state. The measured dependence of the τ_{SF} on the spin splitting of the DMS exciton states is shown in the inset of Fig. 11.10. Strong acceleration of the spin flip transitions was observed with increasing Δ, which implies that a new spin relaxation process of high efficiency becomes activated when the spin splitting of the exciton states exceeds the LO energy.

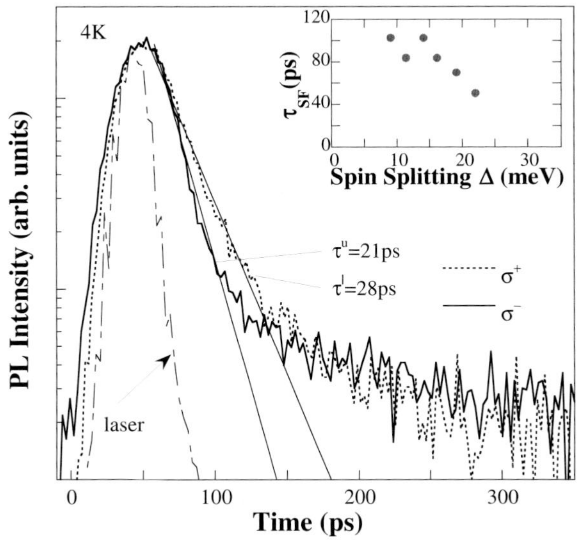

Figure 11.10. Decay curves of the integrated PL intensities of the two spin split states of the SI hh exciton measured at 1.0 T. The thin solid lines are fitting curves with $\tau_d = 28$ ps (or 21 ps) and $\tau_{SF} = 84$ ps. The inset shows spin relaxation times deduced from the measured decays according to Eq. (1) as a function of the spin splitting of the SI hh exciton states. The PL was excited by second harmonic femtosecond pulses of a Ti:sapphire laser tuned to 2.85 eV. The time resolution of the system is about 15 ps. From Ref. 24.

11.3.2.4 Spin alignment: Possible mechanisms for the LO-assisted spin relaxation

Let us now discuss possible physical mechanisms responsible for the fast LO phonon assisted spin relaxation from the $|+1/2, -3/2\rangle$ state to the $|-1/2, +3/2\rangle$ state of the DMS hh exciton.

It can be caused by a two-step process that includes (i) a nearly elastic spin scattering of an electron or hole induced by the strong exchange interaction with Mn magnetic impurities (process Ia in Fig. 11.8), which induces the $|+1/2, -3/2\rangle \rightarrow |-1/2, +3/2\rangle$ transition with only a small energy change, followed by (ii) an LO-assisted spin-preserving kinetic energy relaxation (process Ib). The first step, however, requires mixing between the hh and light-hole (lh) valence band states. This mixing is expected to be rather weak for the strained ZnMnSe SI studied here, where the energy separation between the lh and hh states is about 30 meV.

Another possibility is an inelastic process involving a direct LO-assisted spin flip (process Ic in Fig. 11.8). Though this model can explain the experimental findings from both spin injection and hot PL experiments, it is a higher-order process requiring some magnetic mediator between the LO phonons and the carrier spins.

An alternative explanation for the observed LO-assisted spin relaxation was most recently suggested in Ref. 25, where it was ascribed to a second-order process which involves two first-order processes, i.e., the exciton–phonon coupling and the coupling of spin-split exciton states via the electron–hole exchange interaction. This process can reasonably explain the suppression of spin relaxation for $K = 0$ excitons when the spin splitting is smaller than the LO phonon energy observed in our experiments. It also provides reasonable values of the LO-assisted spin flip time, which is around 20 ps for ZnSe-based structures.

11.3.3 Spin Detection

The results presented in Secs. 11.3.1 and 11.3.2 show that spin injection efficiency in the investigated structures remains limited (below 30% during CW measurements), even under conditions when complete spin alignment is achieved within the SI. By varying the thickness of the ZnSe spacer inserted between the SI and the SD, we have also shown that the spacer does not cause noticeable spin loss during spin injection.[6] It may be surmised that efficient spin depolarization within the SD could be a likely and important source of the spin loss leading to the observed limited efficiency of optical spin injection in these structures.

11.3.3.1 Spin detection: CW optical orientation spectroscopy

To obtain information on the degree of spin relaxation and the resulting spin depolarization within the ZnCdSe QW SD, we have employed CW and time-resolved optical spin orientation spectroscopy in combination with tunable laser excitation.[26] In these measurements, excitation photon energy $h\nu$ was restricted to the range $E_g^{SD} < h\nu < E_g^{SI}$, where $E_g^{SD} = 2.525\,\text{eV}$ is the excitonic band-gap of the SD and $E_g^{SI} = 2.8\,\text{eV}$. This allowed the generation of free carriers/excitons within the SD without spin injection from the DMS and, therefore, enabled us to address the role of spin depolarization within the SD in limiting apparent optical spin injection efficiency in ZnMnSe/ZnCdSe spin light emitting structures.

In optical spin orientation experiments, a preferred spin orientation of excitons and charge carriers is generated by a corresponding circularly polarized light without application of an external magnetic field. Spin relaxation of the SD can be monitored through the degree of the circular polarizations of the resulting PL.[27] Indeed, in the ZnCdSe semiconductor with a large spin-orbit (S-O) interaction, the top of the valence band (VB) consists of hh and lh states. Absorption of a particular circularly polarized (e.g., σ^+) light, which is employed in the optical spin orientation experiment, can create a certain degree of spin polarization of carriers. For example, 100% polarized optical excitation can generate 50% of electron spin polarization in the conduction band (CB) due to the difference in the oscillator strength of the hh- and lh-related optical transitions [see Fig. 11.11(a)]. The electron spin polarization is defined by $P_e = 100(n^+ - n^-)/(n^+ + n^-)$, where n^+ and n^- denote the populations of spin-up and spin-down electrons, respectively. If both photo-excited electrons and holes retain their spin orientations without spin relaxation, the reverse PL process should lead to 100% polarization in optical detection. If complete spin relaxation occurs between the hole states, e.g.,

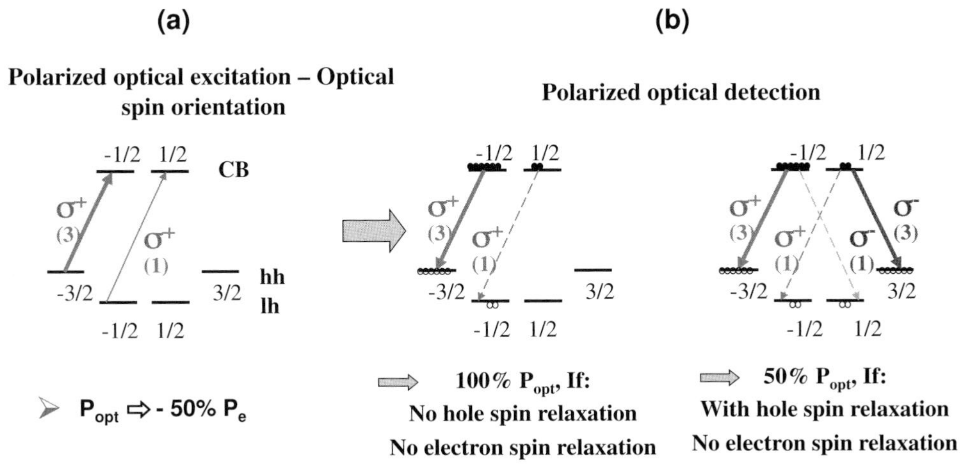

Figure 11.11. Schematic band diagrams as well as schemes of (a) polarized optical excitation and (b) polarized optical detection for the ZnCdSe QW SD with a large spin-orbit interaction. The relative oscillator strengths of the optical transitions are given by the values in parenthesis. The conduction band and valence band states are indicated by their angular momentum projections m_c and m_v. From Ref. 27.

due to strong hh–lh mixing, the polarization degree of optical detection should decrease down to 50% when only the hh states are populated [see Fig. 11.11(b)]. This is the case of the ZnCdSe QW SD studied here, where the energy separation between the lh and hh states is about 30 meV due to the strain, and only the hh state can be populated at low measurement temperatures.

PL polarization of the ZnCdSe SD measured as a function of the excitation photon energy of the σ^+-light is shown in Fig. 11.12. The maximum degree of the PL polarization is about 20% and decreases with increasing excitation energy. The observed values are substantially lower than the 50% polarization degree expected for the case when electron spins are preserved, whereas hole spins are completely depolarized. This indicates efficient relaxation of both electron and hole spins. Moreover, the observed reduction of PL polarization with increasing excitation photon energy implies that spin relaxation occurs during energy relaxation of the photo-excited carriers.

11.3.3.2 Spin detection: Time-resolved optical orientation spectroscopy

To gain insight into the physical processes responsible for the observed optical depolarization, time-resolved optical spin orientation studies were carried out.[26] Representative results are shown in Fig. 11.13, where (a) typical decay curves of the SD PL intensity as well as (b) the corresponding optical polarization are shown under the σ^+-excitation at two different energies.

The decay of the SD PL in all investigated structures is singly exponential with a characteristic time of about 100 ps, which represents the "intrinsic" decay time of the SD. Several spin depolarization processes accompanying this PL decay can be

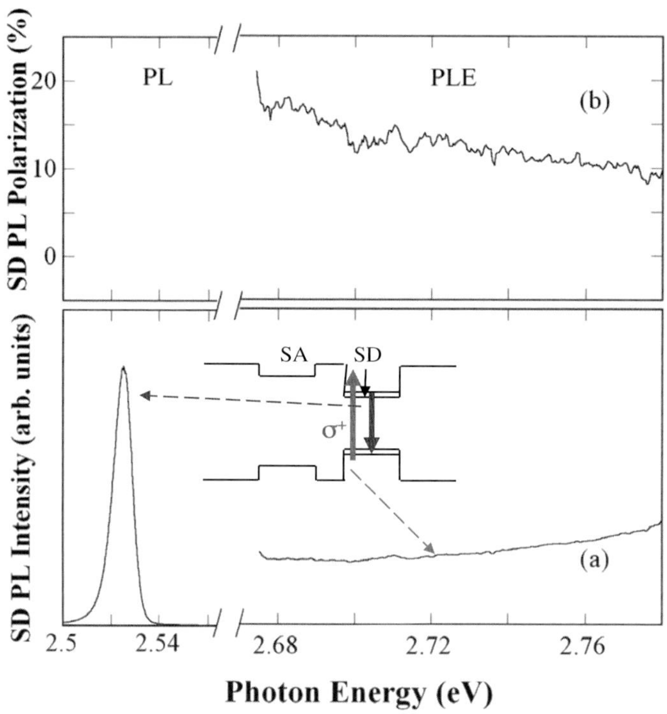

Figure 11.12. (a) PL and PLE excitation spectra of the ZnCdSe QW SD from the studied structures, taken at 2 K. The PLE spectrum was obtained by measuring the SD PL intensity. The inset in (a) shows a schematic diagram of the sample structure and also optical transitions monitored in the PL and PLE spectra. (b) Polarization degree of the SD PL as a function of the excitation photon energy. The excitation light was σ^+-polarized. From Ref. 26.

distinguished, as the decay of the PL polarization under the σ^+-excitation displays a multiply exponential character [Fig. 11.13(b)]. The first and predominant process is faster than the instrumental response of 10 ps and is responsible for an initial reduction of the polarization down to about 30%. This is followed by a moderately fast process with a time constant of 30 ps, and also a slow spin relaxation with the corresponding time of ~850 ps.

Also obvious from Fig. 11.13(b), the PL polarization decreases with increasing excitation photon energy of the σ^+-light, consistent with the results of the CW experiments (Fig. 11.12). This decrease predominantly occurs at a time scale of less than 10 ps, and therefore should be related to the relaxation of hot excitons and hot carriers that were generated by the optical excitation.

11.3.3.3 Spin detection: Possible mechanisms for spin depolarization

The results presented above indicate the existence of several spin relaxation mechanisms within the SD, which contribute to the loss of polarization of the SD PL. The most efficient PL depolarization processes ($<$ 30 ps) are sensitive to the exci-

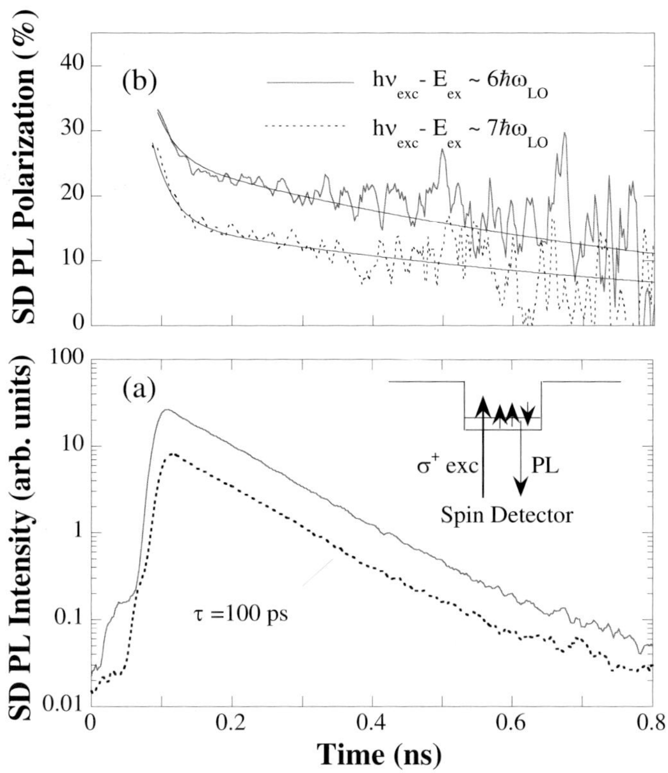

Figure 11.13. (a) Typical 4 K PL decay curves from the ZnCdSe QW SD under the σ^+-excitation. The excitation energies were at about six (solid line) and seven (dashed line) LO phonon energies above the bottom of the hh exciton band. (b) The corresponding decay curves of the PL polarization. $h\nu_{exc}$ denotes the excitation photon energy of the σ^+-light, and E_{ex} is the energy of the SD hh exciton at $K = 0$. The solid lines are the fitting curves obtained by including the contributions from the two depolarization components of 850 ps and 30 ps that can be resolved within the experimental resolution. From Ref. 26.

tation photon energy, which is in agreement with earlier studies of bulk CdS and CdSe[28,29] as well as quantum structures involving ZnMnSe DMS.[22,23,30] Therefore, the involved spin relaxation process is likely efficient for excitons/carriers with a high momentum K that accompanies the energy relaxation. The slow optical depolarization process (~850 ps) is, on the other hand, tentatively argued to be due to spin relaxation at $K = 0$. Several physical mechanisms could be responsible for the observed spin relaxation such as the D'yakonov–Perel' mechanism,[14] or the long-range exchange interaction.[16] Both predict a strong enhancement of spin relaxation for hot carriers/excitons with high K, consistent with experimental findings.

11.4 SPIN POLARIZATION IN ZnMnSe/CdSe QD SPIN INJECTION STRUCTURES

Semiconductor QDs are particularly promising for solid-state qubits and spin detection due to their slow spin relaxation and high efficiency of optical transitions. Unfortunately, both carrier- and spin-injection processes and physical mechanisms for spin loss in QD systems are still not fully understood. Most recently, we have performed comprehensive characterization studies of optical carrier/exciton and spin injection processes from a DMS layer of ZnMnSe into CdSe self-assembled QDs through a ZnSe barrier, by using tunable excitation and time-resolved magneto-PL measurements.[31,32] A schematic picture of the investigated structure is shown in Fig. 11.14.

Similar to the case of the QW SD, pronounced PLE peaks that originate from the lowest exciton states $|-1/2, +3/2\rangle$ and $|+1/2, -3/2\rangle$ of the DMS were seen in the PLE spectra of the QD SD, indicating efficient injection of carriers created in the DMS to the QD (see Fig. 11.15). Spin injection under such conditions was concluded from the observed reversal in the QD PL polarization sign, which indicated population inversion between the two Zeeman sublevels in the QD achieved via the injection of excitons/carriers from the DMS.

The spin polarization degree of excitons recombining in the QD and the extent of spin loss were estimated from a detailed analysis of the polarization using the following coupled rate equations under the condition of resonant excitation by unpolarized light at the lowest exciton spin state $|-1/2, +3/2\rangle$ of the DMS [Fig. 11.16(a)]:

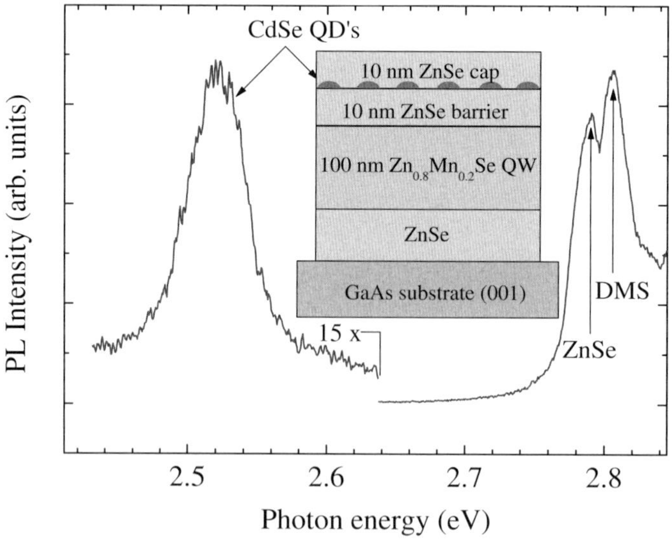

Figure 11.14. Exciton PL spectra of the studied structure, where the PL peaks from the CdSe QD, the ZnMnSe DMS and ZnSe are clearly resolved. The spectra are taken at 0 T and 2 K. PL was excited by linearly polarized light at 2.95 eV. The inset shows a schematic drawing of the structure. From Ref. 32.

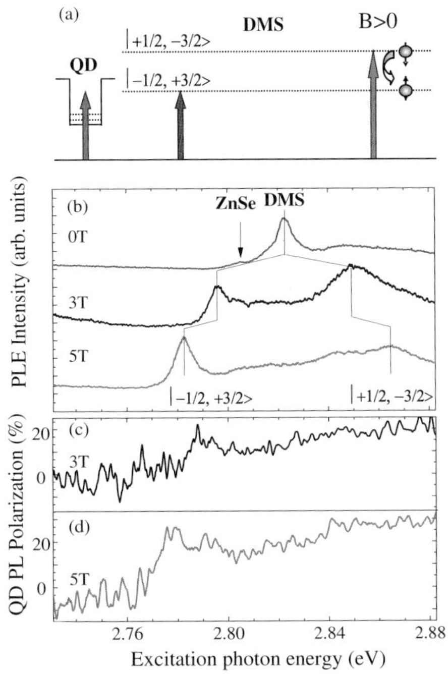

Figure 11.15. (a) A schematic picture of the optical excitation at 5 T. The excitation spectra of (b) the CdSe QD PL emission and (c, d) PL polarization at various applied magnetic fields are shown. The detection energy during all measurements was set at 2.54 eV. From Ref. 32.

$$\frac{dN_+^{\mathrm{QD}}}{dt} = G + KN_+^{\mathrm{DMS}} - \frac{N_+^{\mathrm{QD}}}{\tau_{\mathrm{r}}^{\mathrm{QD}}} - \frac{N_+^{\mathrm{QD}}}{\tau_{\mathrm{t}}^{\mathrm{QD}}} - \frac{N_+^{\mathrm{QD}}}{\tau_{\mathrm{s-f}}^{\mathrm{QD}}(1+e^{-\Delta E/kT})} + \frac{N_-^{\mathrm{QD}}}{\tau_{\mathrm{s-f}}^{\mathrm{QD}}(1+e^{\Delta E/kT})},$$

$$\frac{dN_-^{\mathrm{QD}}}{dt} = G + (1-K)N_+^{\mathrm{DMS}} - \frac{N_-^{\mathrm{QD}}}{\tau_{\mathrm{r}}^{\mathrm{QD}}} - \frac{N_-^{\mathrm{QD}}}{\tau_{\mathrm{t}}^{\mathrm{QD}}} + \frac{N_+^{\mathrm{QD}}}{\tau_{\mathrm{s-f}}^{\mathrm{QD}}(1+e^{-\Delta E/kT})} - \frac{N_-^{\mathrm{QD}}}{\tau_{\mathrm{s-f}}^{\mathrm{QD}}(1+e^{\Delta E/kT})}. \tag{11.2}$$

Such an excitation condition ensures that only σ^+-active exciton spins are generated, leading to complete spin alignment in the DMS, so that it can be ruled out as a possible source of spin loss. The first two terms on the right-hand side of Eq. (11.2) represent the direct generation of excitons within the QD by laser light and DMS feeding of the QD, respectively [Fig. 11.16(a)]. N_+^{DMS} is the total number of the excitons injected from the DMS to the QDs per second. $K \equiv N_+^{\mathrm{injected}}/N_+^{\mathrm{DMS}}$ is a dimensionless factor describing the spin conservation of the injected excitons,

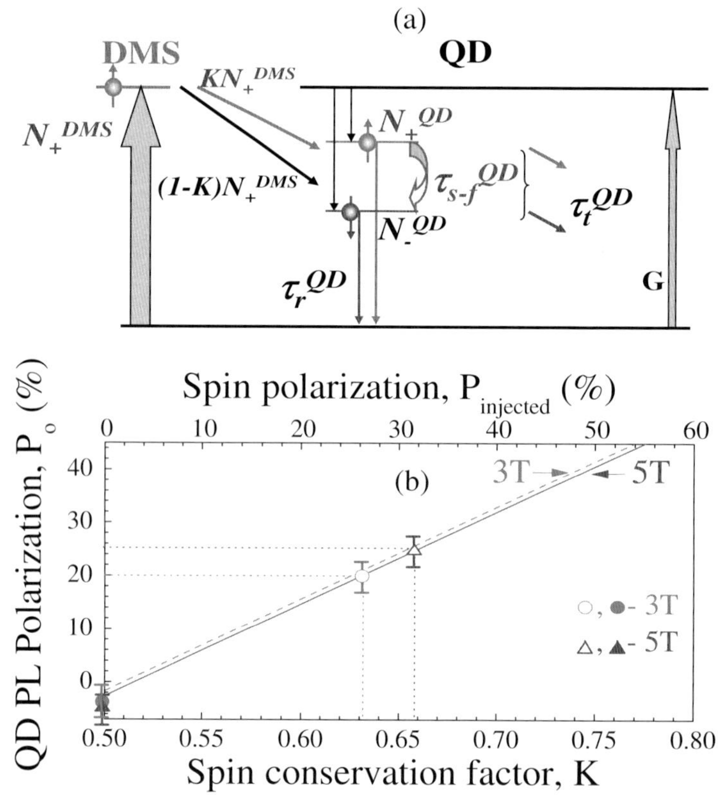

Figure 11.16. (a) A schematic picture of the spin injection, optical excitation and recombination processes. (b) PL polarization of the CdSe QD as a function of the spin conservation factor K and the polarization of the injected spins P_{injected}. The solid line indicates the simulated QD polarization based on Eq. (11.3). The open and filled circles are the experimentally determined values at 5 T, obtained when the excitation photon energy was tuned at and below the lowest exciton spin state of the DMS, respectively. From Ref. 32.

where N_+^{injected} denotes the number of the injected excitons that still retain the σ^+ polarization. $K = 1$ and $K = 0.5$ represent the two extreme cases of complete spin conservation and total spin loss during the spin injection, respectively. The next two terms in Eq. (11.2) account for exciton recombination and inter-dot exciton transfer from the dots with higher exciton energies to the ones with lower energies, when the polarization properties of the QD PL at the highest energy are analyzed here. The last two terms model the spin-flip process between the two spin sublevels of the QDs. $\tau_{\mathrm{t}}^{\mathrm{QD}}$, $\tau_{\mathrm{r}}^{\mathrm{QD}}$ and $\tau_{\mathrm{s-f}}^{\mathrm{QD}}$ represent inter-dot exciton transfer time, exciton lifetime, and exciton spin-relaxation time in the QDs. They have been independently determined in our time-resolved PL measurements[31]: $\tau_{\mathrm{t}}^{\mathrm{OD}} = 50\,\mathrm{ps}$; $\tau_{\mathrm{r}}^{\mathrm{OD}} = 200\,\mathrm{ps}$ and $\tau_{\mathrm{s-f}}^{\mathrm{OD}} = 800\,\mathrm{ps}$. The Zeeman splitting between the spin levels in the QDs at 5 T is estimated to be $\Delta E = 0.5\,\mathrm{meV}$ from the Zeeman splitting observed in μ-PL experiments. kT denotes the thermal energy of the excitons. From this simple model, we were able to deduce an expression for the polarization of the QDs:

$$P = \underbrace{\frac{\dfrac{1-e^{\Delta E/kT}}{1+e^{\Delta E/kT}}\dfrac{1}{\tau_{\text{s-f}}^{\text{QD}}}}{\dfrac{1}{\tau_{\text{s-f}}^{\text{QD}}}+\dfrac{1}{\tau_{\text{r}}^{\text{QD}}}+\dfrac{1}{\tau_{\text{t}}^{\text{QD}}}}}_{\text{INTRINSIC}} + \underbrace{\frac{\dfrac{2K-1}{1+G/N_{+}^{\text{DMS}}}\left(\dfrac{1}{\tau_{\text{r}}^{\text{QD}}}+\dfrac{1}{\tau_{\text{t}}^{\text{QD}}}\right)}{\dfrac{1}{\tau_{\text{s-f}}^{\text{QD}}}+\dfrac{1}{\tau_{\text{r}}^{\text{QD}}}+\dfrac{1}{\tau_{\text{t}}^{\text{QD}}}}}_{\text{DMS-INDUCED}} \tag{11.3}$$

We note that it contains two parts. The first part describes the intrinsic QD polarization whilst the second accounts for the spin injection from the DMS. The results of the calculations are depicted in Fig. 11.16(b), where the QD polarization degree P is plotted as a function of K. The calculated polarization degree ($P_{\text{o}} = -3\%$) at $K = 0.5$, equivalent to the case without spin injection from the DMS, is in good agreement with the observed intrinsic QD polarization degree ($P_{\text{o}} = -5\%$). Furthermore, we can see that the experimentally obtained QD polarization degree of $+25\%$ (during the resonant excitation of the DMS) yields $K \sim 0.66$. In other words, about 34% of the excitons lose their spin orientation during the spin injection. From this, the polarization of the injected excitons that reaches the QD can be estimated as $P_{\text{injected}} = (N_{+}^{\text{injected}} - N_{-}^{\text{injected}})/N_{+}^{\text{DMS}} = 2K - 1 = 32\%$, decreasing from 100% in the DMS before the injection. Possible reasons for spin scattering during the spin injection include the geometrical shape and size of the QD, defects, roughness and potential of the interfaces along the path of the spin injection, and unexpected fast spin relaxation accompanying energy relaxation within the QD.

11.5 GaMnN/InGaN SPIN INJECTION STRUCTURES

Recent predictions of room temperature ferromagnetism in GaMnN alloys[33–36] have triggered substantial efforts world-wide in growth of this material and other DMS materials based on wide band-gap GaN. A number of groups have successfully obtained single-phase GaMnN (see Ref. 37 and references therein), which is ferromagnetic at or above room temperature. This has offered new opportunities to add spin functionalities to recently developed GaN/InGaN LEDs and GaN-based field effect transistors. The feasibility of such devices is further encouraged by the recent observation of a long (20 ns) electron spin lifetime in GaN, due to a weak spin-orbit coupling in the material and, thus, decoupling of momentum- and spin-scattering processes.[38] Surprisingly, however, no spin injection in GaN-based spin injection structures has so far been demonstrated. This could be due to various reasons, such as poor spin alignment within the SI and fast spin loss within the SD. We have recently employed optical spin orientation and time-resolved magneto-optical spectroscopy to investigate the origin of spin depolarization in the InGaN QW SD, which has been used for detection of spin injection in GaMnN/InGaN spin-LEDs.[39,40]

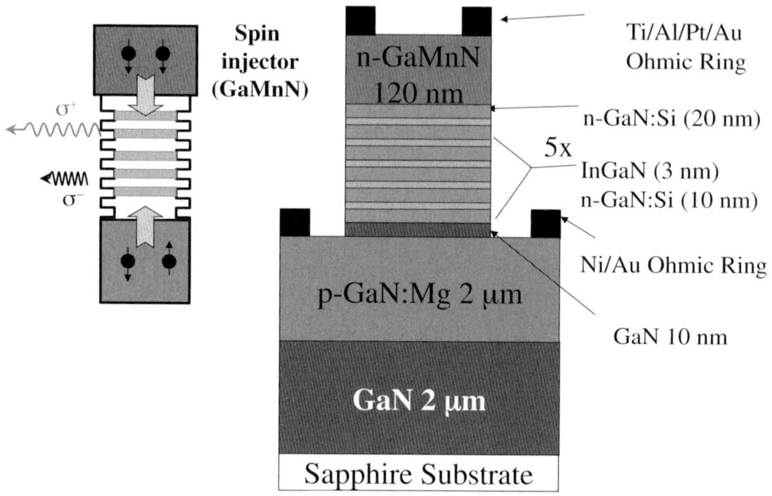

Figure 11.17. A schematic cross-sectional diagram of the GaMnN/InGaN spin-LED structures studied. (Courtesy of S. J. Pearton.)

11.5.1 Spin Injection

A schematic diagram of the studied spin LED structure is shown in Fig. 11.17. The efficiency of spin injection in this structure during electrical and optical injection was evaluated from the polarization of light emission measured by electro-luminescence (EL) or PL. To ensure that during the PL measurements free carriers/excitons participating in the PL are predominantly supplied by the ferromagnetic SI, the excitation photon energy was set at 5.08 eV (244 nm), i.e., well above the band-gap of the DMS layer.

The EL and PL spectra of the spin-LED at RT are shown in Figs. 11.18(a) and 11.18(b), respectively, and contain the emission band at about 2.7 eV from the InGaN MQW. In spite of the ferromagnetic behavior of the GaMnN layer, no polarization of light emission at RT was observed in applied magnetic fields ranging from 0 to 5 T. This could be related to cancellation of the emission polarization due to a spectral overlap of the A and B exciton emissions with opposite polarization, when both valence band states are populated with a small energy separation.

To avoid this complication, PL experiments were also repeated at 2 K but again did not provide convincing experimental evidence for spin injection in the investigated structures.[39] This could reflect (i) poor performance of the SI, (ii) spin loss during injection, and (iii) strong depolarization within the SD.

11.5.2 Spin Detection

In order to evaluate the possible loss of spin polarization within the non-magnetic SD, we have carried out optical spin orientation experiments[39] and also transient magneto-optical measurements with and without an external magnetic field.[40]

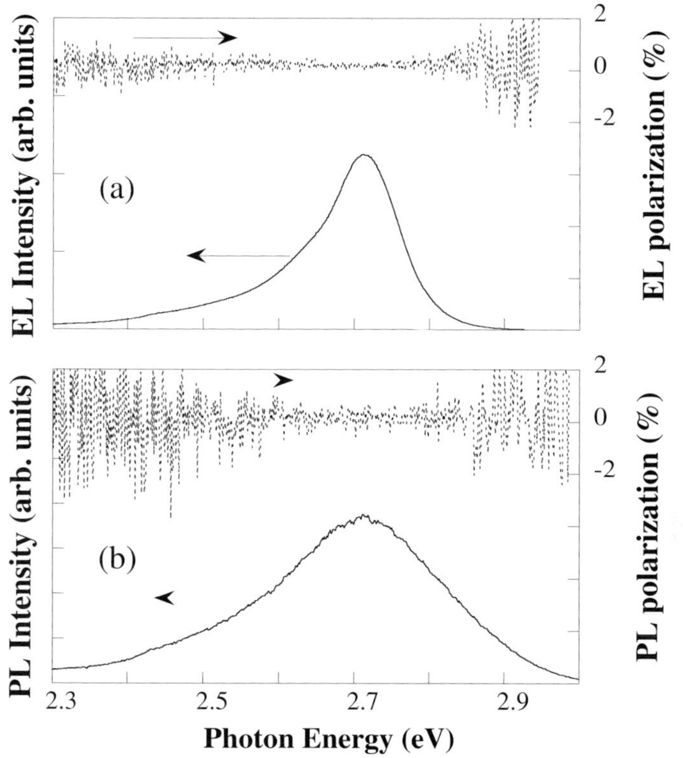

Figure 11.18. Room temperature (a) EL and (b) PL spectra measured from the same GaMnN/InGaN spin-LED structure. Also shown is their polarization at 5 T [the upper part of (a) and (b)]. PL was excited by a linearly polarized light as described in the text. From Ref. 39.

11.5.2.1 Spin detection: Optical orientation measurements

In wurtzite InGaN with a weak S-O interaction the top of the VB is known to consist of the so-called A, B and C states.[41] When $\mathbf{k}_{\text{photon}} \| \langle 0001 \rangle$ the band-to-band transitions are only allowed from the A and B VB states. The difference in the oscillator strengths between the CB $\leftrightarrow$ A and CB $\leftrightarrow$ B optical transitions is approximately 1:0.94[41] for the layers with compressive in-plane strain, as in the case of the InGaN QWs employed in these studies. This will limit carrier spin polarization created by a particular circularly polarized light in optical spin orientation experiments to about 3% if spin relaxation is neglected (see Fig. 11.19). Likewise, optical polarization is also limited to about 3% even if the electron spins are completely polarized assuming an equal population of the A and B VB states in, for example, electrical spin injection experiments at room temperature. At low temperatures, a higher degree of optical polarization is expected when the hole populations of the A and B states differ significantly. The failure to detect any optical polarization in the earlier CW[42] and time-resolved optical spin orientation experiments without magnetic fields[39] could therefore be interpreted as an indication

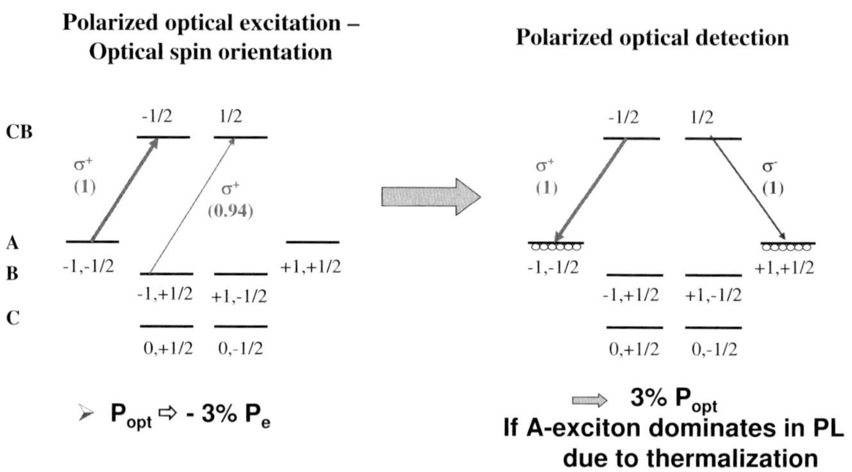

Figure 11.19. Schematic pictures of the band diagrams as well as schemes of (a) polarized optical excitation and (b) polarized optical detection for the InGaN QW SD with a weak spin-orbit interaction, relevant to optical spin orientation experiments. The values in parenthesis represent the relative oscillator strengths of the optical transitions. From Ref. 27.

that it is not merely because of complete cancellation of optical polarization by the A and B excitons. Fast spin relaxation within the InGaN QW spin detector is in fact an important source of spin loss. This is contrary to what is expected for a semiconductor with a weak S-O interaction.

11.5.2.2 Spin detection: Time-resolved magneto-optical spectroscopy

The application of magnetic fields causes splitting of the $|+1/2;-1,-1/2\rangle$ and $|-1/2;+1,+1/2\rangle$ optically-active states of the A exciton. This provides an additional channel for the exciton spin relaxation and allows one to access spin dynamics in the SD.

Representative results of the performed transient studies are summarized in Fig. 11.20. At 0 T, PL is unpolarized, as expected with vanishing Zeeman splittings. In an applied magnetic field, on the other hand, the σ^+ and σ^- PL components exhibit consistent differences in their decays, which leads to a detectable optical polarization, e.g., up to about 10% at 4 T [Fig. 11.20(b)].

For the A exciton, the intensities of the σ^+ and σ^- PL components should be proportional to the populations of the $|-1/2;+1,+1/2\rangle$ and $|+1/2;-1,-1/2\rangle$ sublevels, respectively. An intensity drop of the σ^- PL component, which is observed within the first 50 ps, in addition to that related to its decay time must arise from a population transfer from the $|+1/2;-1,-1/2\rangle$ state to other sublevels of the A exciton. Such transfer can only be mediated by spin relaxation which is completed within 50 ps after the excitation light pulse.

The exact origin of the observed spin relaxation as being due to the resulting spin flips of individual carriers forming the exciton was concluded from our simulations by solving coupled differential rate equations.[40]

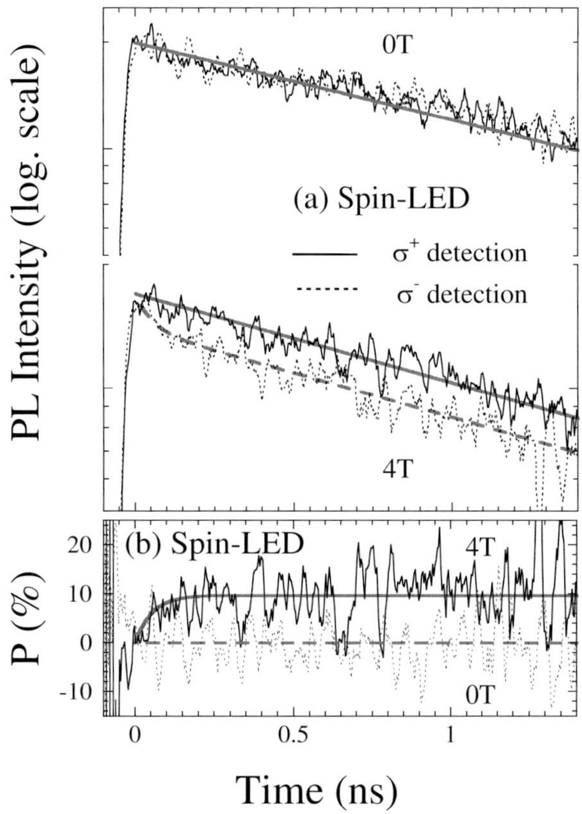

Figure 11.20. Typical decay curves of the σ^+ and σ^- PL components of the InGaN QW SD, taking the spin-LED as an example. The decays were measured at 2 K in magnetic fields at the strength indicated in the figure. Except for the σ^- PL component at 4 T, which is fitted by two exponents (50 ps and 2 ns), a single exponent of 2 ns can satisfactorily describe the PL decays. (b) PL polarizations from the spin-LED as a function of delay time at 0 T and 4 T. The thick solid and dashed lines are fitting curves at 0 T and 4 T, respectively, with identical fitting parameters as in (a). From Ref. 40.

11.5.2.3 Spin detection: Possible mechanisms for spin depolarization

The observed fast spin relaxation in the InGaN SD is rather unexpected if one simply takes into account the weak S-O interaction in InGaN. But it can be attributed[43] to a strong Rashba effect due to structural inversion asymmetry in the presence of a large piezo-electric field inherent to the InGaN QW (the DP mechanism).[14] The EY mechanism[12,13] may also play an important role due to strong compositional and structural non-uniformity and a high defect density known in InGaN. These processes can be greatly promoted during momentum and energy relaxation at high *K*-vectors.

11.6 ZnO-BASED SPIN FUNCTIONAL STRUCTURES

ZnO and related alloys are attracting increasing attention since they are excellent candidates for use in visible and ultraviolet (UV) light emitters, in transparent electronics, and also in chemical and bio sensing.[44–47] The demonstration of room temperature ferromagnetism in its DMS counterpart has also shown the potential of this material system for future applications in spintronics.[47–49] However, progress towards the realization of spin-dependent device structures from ZnO, e.g., ZnO-based spin-LEDs, has so far only been marginal, hindered not only by the difficulties in material preparation and control but also by a poor understanding of the spin-dependent phenomena in the materials. The reasons for the failure to obtain spin injection may be manifold. It could be due to an ineffective spin aligner, i.e., a DMS based on ZnO, which fails to create spin polarization of carriers. It could also be caused simply by our limited abilities in spin detection with respect to the materials studied.

Recently, we have addressed this important issue through spin-polarized, time-resolved optical orientation and magneto-optical spectroscopy.[50] Two of the most important factors that limit the efficiency of spin detection in this material system were identified as being related to the fundamental band structure and spin relaxation, respectively.

11.6.1 Spin Detection: Optical Orientation Measurements

The fundamental band structure of wurtzite ZnO is similar to that of the wurtzite GaN discussed in Sec. 11.5.2.1. It is characterized by a weak S-O interaction ($|\Delta_{\mathrm{so}}| = 4$–$15\,\mathrm{meV}$) and a stronger crystal field ($\Delta_{\mathrm{c}} = 30$–$40\,\mathrm{meV}$).[44–47,51,52] There has been a long-standing debate, though, on the exact nature of the A and B VB states in terms of the exact ordering of the Γ_7^v and Γ_9^v states that differ in the relative orientations of spin and orbital angular momentum (i.e., signs of m_l^v and m_s^v). In the model of a normal VB structure, the A and B states are represented by Γ_9^v and Γ_7^v,[53] respectively, as seen in many other similar semiconductors with a wurtzite structure like GaN. This ordering is reversed, on the other hand, in the model of an inverted VB structure.[54,55] In both cases, however, the implications of the fundamental band structure for spin detection remain the same. Indeed, as the A and B VB states of the same spin orientation have opposite orbital orientations, the corresponding spin-conserving optical transitions are oppositely circularly polarized (see Fig. 11.19). As a direct consequence, even complete spin polarization of electrons or holes will fail to lead to any sizable optical polarization at room temperature, as the optical polarizations involving the A and B VB states with a similar hole population effectively cancel each other. This means that ZnO with such a simple picture of the VB structure is not suitable for spin detection at room temperature. In practice, however, the oscillator strength of the optical transitions involving the A and B VB states can be altered to various degrees by a combined effect of the S-O interaction, the strength and anisotropy of the in-plane strain field, the electron–hole exchange interaction, etc. In particular, the electron–hole exchange interaction in ZnO is unusually strong[54] as compared to other semiconductors, e.g., about one and two orders of magnitude stronger than that in GaN and GaAs, respectively. As a result, the oscillator strengths of the optical transitions involving the A and B VB states can be different leading to incomplete cancellation of their optical polarization. However, the degree of optical polarization is likely too week to be detected, as was indeed observed in optical orientation experiments.

If the photon energy of the polarized excitation light (e.g., σ^+) is higher than the transition energies of both CB(Γ_7^c) $\leftrightarrow$ A and CB(Γ_7^c) $\leftrightarrow$ B, both spin orientations of CB electrons and VB holes can be generated. If the oscillator strengths of both transitions are similar, no net spin polarization of either electrons or holes can be created. This was confirmed by our optical orientation experiments performed on both CdZnO alloys and the *n*-MgZnO/ZnO/*p*-AlGaN LED, where no optical polarization can be detected at 0 T.

If the photon energy of the polarized excitation light (e.g., σ^+) is higher than the transition energy of CB(Γ_7^c) $\leftrightarrow$ A but lower than that of CB(Γ_7^c) $\leftrightarrow$ B, on the other hand, only one spin orientation of electrons and holes is created. This will lead to detectable optical polarization if spin relaxation does not completely depolarize the spins. However, even under these conditions no spin polarization of the free excitons was detected,[56,57] which was attributed to extremely fast spin relaxation (faster than their instrument response time of 30 ps). Donor bound excitons (DBE), on the other hand, were able to create spin polarization by optical orientation. The spin relaxation time of the donor bound excitons was estimated to be ~275 ps.

The present optical orientation studies together with the reported work by Lagarde *et al.*[56,57] have demonstrated that the band structure property of ZnO-based materials is not the only limiting factor for effective spin detection. Efficient spin relaxation is in fact another important factor that has to be taken seriously.

11.6.2 Spin Detection: Time-Resolved Magneto-optical Spectroscopy

As optical orientation has failed to generate any spin polarization in our study presented above, we have resorted to the application of an external magnet field to polarize electron and hole spins in order to investigate the extent of spin relaxation in ZnO-based materials.

Figure 11.21 illustrates the general principle of this approach, taking as an example a neutral donor bound exciton (DBE) denoted as D^0X. The DBE was found to dominate radiative recombination in the studied structures, due to residual *n*-type doping provided by impurities and defects. Since the total electron spin of the DBE state is zero (i.e., spins of the two electrons paired off), the measured spin relaxation time represents hole spin flips within the DBE. As can be seen from a comparison of Figs. 11.21(a) and 11.21(b), this is truly independent of the ordering of the Γ_7^v and Γ_9^v valence band states in ZnO. In the cases of the ionized donor

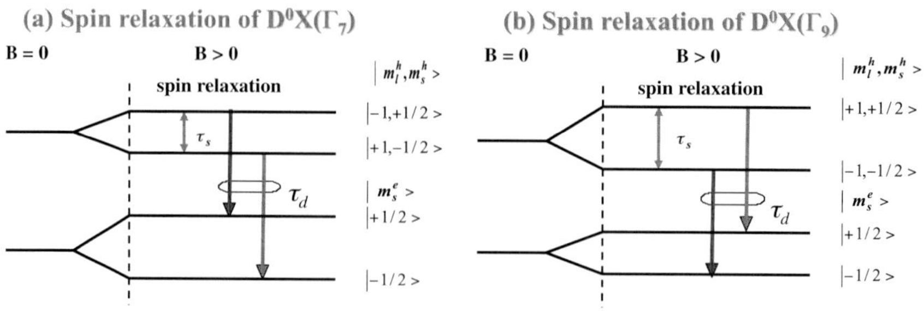

Figure 11.21. Schematic pictures of neutral donor bound exciton (D^0X) states and related optical transitions in wurtzite ZnO, shown for the models with (a) an inverted valence band structure and (b) a normal valence band structure. The blue and red arrows denote optical transitions with the σ^- and σ^+ circular polarizations, respectively, monitored in the configuration of $\mathbf{B}\|\mathbf{k}\|c$ employed here. The green arrow denotes spin relaxation of holes within the DBE. τ_d and τ_s are PL decay time and spin relaxation time, respectively. τ_s for the DBE is governed by hole spin flips.

bound exciton (D^+X) and FE, both electron and hole spin flips contribute to the exciton spin relaxation.

Figure 11.22 shows the results of the measurements taken for the bulk ZnO sample. The corresponding PL spectrum is dominated by the DBE transitions and also contains contributions from much weaker D^+X and FE emissions. All transitions strongly overlap in time-resolved PL measurements due to limited spectral resolution [Fig. 11.22(a)], but are clearly resolved in CW PL spectra measured with a higher spectral resolution [shown in the inset in Fig. 11.22(a)]. The observed higher intensity of the DBE transitions is typical for ZnO, due to a high concentration of residual donors. PL decay is nearly exponential with a decay time $\tau_d = 550$ ps, consistent with previously published results.[58,59] At 0 T, the PL is unpolarized as expected due to degeneracy of the spin sub-levels. On the other hand, it acquires substantial circular polarization under applied magnetic fields that reaches ~16% at $B = 5$ T [see Fig. 11.22(b)]. This polarization develops because of the accelerated initial decay of the σ^- polarized PL component due to a population transfer from the upper-lying to the lower-lying spin sub-level of the DBE, mediated by hole spin relaxation. By analyzing the temporal rise of the PL polarization, the spin relaxation time for the DBE can be estimated as $\tau_s = 140$ ps. The deduced time is rather short, which points to a high efficiency of the spin-flip process involved.

In order to evaluate the effects of strain on spin relaxation, similar measurements were performed for the thick ZnO epilayers and thin (40 nm) ZnO active layer within the LED structure. The corresponding results are summarized in Table 11.1. The exciton spin dynamics in the thick ZnO epilayers is found to be very similar to that observed in the bulk ZnO. It becomes slightly faster, i.e., $\tau_s = 80$–100 ps, for the ZnO layer within the LED structure. This is in spite of the fact that the PL decay in this sample is the longest among the investigated ZnO materials, which indicates good crystalline quality and a low defect density. We therefore

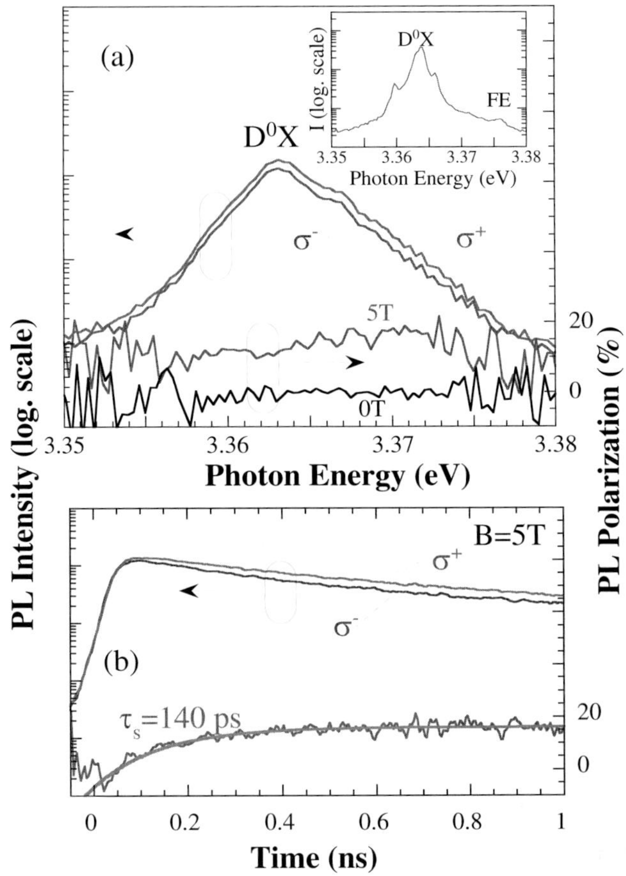

Figure 11.22. (a) Time-integrated PL spectra from the bulk ZnO monitored at $B = 5\,\mathrm{T}$ with circular polarizations, as specified in the figure. The green line represents the polarization degree. The inset shows a high-resolution PL spectrum from the same sample measured under CW excitation. (b) Spectrally integrated decay curves of the circularly polarized PL and PL polarization. The thick solid line in the lower part of (b) is a fitting curve of the PL polarization using the indicated spin relaxation time. The excitation photon energy was 4.65 eV.

tentatively attribute the observed slight acceleration of the spin relaxation process to strain effects within the multilayer structure.

More significant acceleration of the spin relaxation was found in ZnO-based alloys, such as ZnCdO and ZnMgO, where τ_s were found to be on the order of 45–60 ps (see Table 11.1).

11.6.3 Spin Detection: Possible Mechanisms for Spin Depolarization

Our results reveal that exciton spin relaxation in the ZnO-based SD is very fast, i.e., on the order of 100 ps in bulk strain-free ZnO at 4 K. It becomes further accelerated by strain and, even more severely, by alloy disorder. We can now briefly speculate about possible mechanisms responsible for these findings. Three spin relaxation mechanisms are known to be relevant for free carriers/excitons, namely, the Elliott–Yafet,[12,13] D'yakonov–Perel'[14] and Bir–Aronov–Pikus (BAP)[15] mechanisms. The first two mechanisms are both of spin-orbit interaction origin induced by host ions (or impurities) and a lack of inversion symmetry, respectively. The third mechanism is due to the exchange interaction between electrons and holes and is especially relevant for spin relaxation of excitons. Besides these relaxation channels the long-range exciton–exciton exchange interaction[16] can be important for FE, assisted by a large electron–hole exchange interaction in ZnO. As to the EY and DP processes, they are known to exhibit opposite dependences on momentum relaxation, i.e., whereas the former is largely promoted by scattering, the latter is suppressed by fast momentum relaxation. The observed acceleration of spin relaxation in Zn(Mg, Cd)O alloys could suggest that spin loss in the ZnO-based materials is largely mediated by the EY process. In spite of the large band-gap of these materials, this process may become promoted by high residual impurities and defect densities, as well as strong compositional non-uniformity. The observed slight strain dependence may indicate a weak DP effect induced by a piezoelectric field that is sensitive to the strain field in the materials. We should also note that a much longer spin decoherence time T_2^* (>1 ns) was observed for electrons in the absence of holes in n-type ZnO,[60,61] as compared with that for excitons in the presence of both electrons and holes. This suggests that the Bir–Aronov–Pikus mechanism could also play an important role in spin relaxation of carriers/excitons, which can be assisted by a large electron–holes exchange interaction in ZnO.

11.7 CONCLUSIONS AND OUTLOOK

We have briefly reviewed our recent results from in-depth investigations of mechanisms which govern the efficiency of several physical processes that are important for future spintronic devises, such as spin alignment within DMS, spin injection from DMS to non-magnetic SD and also spin depolarization within SD.

Table 11.1 Decay times and exciton spin relaxation times deduced from the time-resolved magneto-optical measurements in the studied structures.

			Spin-LED		$Zn_{1-x}Cd_xO$	
Structure	ZnO Bulk	Epi ZnO 0.2–0.8 μm thick	ZnO, strained, 50 nm thick	MgZnO	$x = 0.091$	$x = 0.157$
Decay time (ps)	550	100	740	550	333	568
Spin relaxation time (ps)	140	180	80–100	50	60	45

Spin-injection structures based on II-VIs [e.g., ZnMnSe/Zn(Cd)Se and Zn(Cd, Mg)O] and III-Vs [e.g., GaMnN/Ga(In)N] were employed as model cases. Exciton spin relaxation within the ZnMnSe DMS, important for spin alignment, was found to critically depend on Zeeman splitting of the exciton states and is largely facilitated by the involvement of LO phonons. Optical spin injection in ZnMnSe/Zn(Cd)Se was shown to be governed by (i) the commonly believed tunneling of individual carriers or excitons and (ii) *energy transfer* via localized excitons and spatially separated localized electron–hole pairs located within DMS. Unexpectedly, the latter mechanism is in fact found to dominate spin injection. We have also demonstrated that for both ZnCdSe and InGaN QW SD as well as for ZnO-based materials, efficient spin relaxation leads to severe spin depolarization. The dominant spin relaxation occurs during energy and momentum relaxation of hot excitons and hot carriers, driven by the physical mechanisms that are promoted by high momenta of excitonsand carriers such as the D'yakonov–Perel' mechanism and the long-range exciton exchange interaction, etc. Therefore, in parallel with the development of suitable spin injectors, sufficient research efforts should be devoted to identifying suitable SD and suppressing spin relaxation in the SD.

ACKNOWLEDGMENTS

The authors would like to thank D. Dagnelund, G. Yu. Rudko, K. Nishibayashi, K. Kayanuma, K. Seo, A. Murayama, Y. Oka, G. Thaler, R. Frazier, F. Ren, C.-C. Pan, G.-T. Chen, J.-I. Chyi, J. M. Zavada, S. J. Pearton, D. Norton, C. W. Tu, A. A. Toropov, S. V. Sorokin and S. V. Ivanov for their valuable contributions to the work presented in this review.

References

[1] S. A. Wolf, D. D. Awschalom, R. A. Buhrman, J. M. Daughton, S. von Molnar, M. L. Roukes, A. Y. Chtchelkanova and D. M. Treger, *Science* **294**, 1488 (2001).

[2] I. Zutic, J. Fabian and S. Das Sarma, *Rev. Mod. Phys.* **76**, 323 (2004).

[3] D. D. Awschalom and M. E. Flatté, *Nature Phys.* **3**, 153 (2007).

[4] R. Fiederling, M. Keim, G. Reuscher, W. Ossau, G. Schmidt, A. Waag and L. W. Molenkamp, *Nature* **402**, 787 (1999).

[5] M. Oestreich, J. Mübner, D. Hägele, P. J. Klar, W. Heimbrodt, W. W. Rühle, D. E. Ashenford and B. Lunn, *Appl. Phys. Lett.* **74**, 1251 (1999).

[6] I. A. Buyanova, W. M. Chen, G. Ivanov, B. Monemar, A. A. Toropov, Y. Terent'ev, S. V. Sorokin, A. V. Lebedev, S. V. Ivanov and P. S. Kop'ev, *Appl. Phys. Lett.* **81**, 2196 (2002).

[7] I. A. Buyanova, G. Yu. Rudko, W. M. Chen, A. A. Toropov, S. V. Sorokin, S. V. Ivanov and P. S. Kop'ev, *Appl. Phys. Lett.* **82**, 1700 (2003).

[8] W. M. Chen, I. A. Buyanova, K. Kayanuma, K. Nishibayashi, K. Seo, A. Murayama, Y. Oka, A. A. Toropov, A. V. Lebedev, S. V. Sorokin and S. V. Ivanov, *Phys. Rev. B* **72**, 073206 (2005).

[9] T. Förster, *Ann. Phys.* **2**, 55 (1948).
[10] D. L. Dexter, *J. Chem. Phys.* **21**, 836 (1953).
[11] S. K. Lyo, *Phys. Rev. B* **62**, 13641 (2000).
[12] R. J. Elliott, *Phys. Rev.* **96**, 266 (1954).
[13] Y. Yafet, in *Solid State Physics*, Vol. 14, eds. F. Seitz and D. Turnbull (Academic, New York, 1963).
[14] M. I. D'yakonov and V. I. Perel', *Sov. Phys. J. Exp. Theor. Phys.* **33**, 1053 (1971); *Sov. Phys. Solid State* **13**, 3023 (1972).
[15] G. L. Bir, A. G. Aronov and G. E. Pikus, *Sov. Phys. J. Exp. Theor. Phys.* **42**, 705 (1976).
[16] M. Z. Maialle, E. A. de Andrada e Silva and L. J. Sham, *Phys. Rev. B* **47**, 15776 (1993).
[17] J. Fabian and S. Das Sarma, *Phys. Rev. Lett.* **83**, 1211 (1999).
[18] Y. Ohno, R. Terauchi, T. Adachi, F. Matsukura and H. Ohno, *Phys. Rev. Lett.* **83**, 4196 (1999).
[19] M. Paillard, X. Marie, P. Renucci, T. Amand, A. Jbeli and J. M. Gérard, *Phys. Rev. Lett.* **86**, 1634 (2001).
[20] J. S. Sandhu, A. P. Heberle, J. J. Baumberg and J. R. A. Cleaver, *Phys. Rev. Lett.* **86**, 2150 (2001).
[21] R. Ferreira and G. Bastard, *Phys. Rev. B* **43**, 9687 (1991).
[22] C. D. Poweleit, A. R. Hodges, T.-B. Sun, L. M. Smith and B. T. Jonker, *Phys. Rev. B* **59**, 7610 (1999).
[23] W. M. Chen, I. A. Buyanova, G. Yu Rudko, A. G. Mal'shukov, K. A. Chao, A. A. Toropov, Ya. Terent'ev, S. V. Sorokin, A. V. Lebedev, S. V. Ivanov and P. S. Kop'ev, *Phys. Rev. B* **67**, 125313 (2003).
[24] I. A. Buyanova, G. Yu. Rudko, W. M. Chen, K. Kayanuma, A. Murayama, Y. Oka, A. A. Toropov, S. V. Sorokin and S. V. Ivanov, *Phys. Rev. B* **71**, 165203 (2005).
[25] E. Tsitsishvili, R. V. Baltz and H. Kalt, *Phys. Rev. B* **71**, 155320 (2005).
[26] W. M. Chen, I. A. Buyanova, K. Kayanuma, Z. H. Chen, A. Murayama, Y. Oka, A. A. Toropov, S. V. Sorokin, S. V. Ivanov and P. S. Kop'ev, *Appl. Phys. Lett.* **85**, 5260 (2004).
[27] I. A. Buyanova, W. M. Chen, Y. Oka, C. R. Abernathy and S. J. Pearton, *Phys. Status Solidi A* **204**, 159 (2007).
[28] S. A. Permogorov, Ya. V. Morozenko and B. A. Kazennov, *Sov. Phys. Solid State* **17**, 1974 (1975).
[29] E. L. Ivchenko, G. E. Pikus and L. V. Takunov, *Sov. Phys. Solid State* **20**, 1502 (1978).
[30] D. A. Tulchinsky, J. J. Baumberg, D. D. Awschalom, N. Samarth, H. Luo and J. K. Furdyna, *Phys. Rev. B* **50**, 10851 (1994).
[31] A. Murayama, T. Furuta, K. Hyomi, I. Souma, Y. Oka, D. Dagnelund, I. A. Buyanova and W. M. Chen, *Phys. Rev. B* **75**, 195308 (2007).
[32] D. Dagnelund, I. A. Buyanova, W. M. Chen, A. Murayama, T. Furuta, K. Hyomi, I. Souma and Y. Oka, *Phys. Rev. B* **77**, 035437 (2008).
[33] T. Dietl, H. Ohno, F. Matsukura, J. Cibert and D. Ferrand, *Science* **287**, 1019 (2000).
[34] K. Sato and H. Katayama-Yoshida, *Jpn. J. Appl. Phys.* (*Part 2*), **40**, L485 (2001).
[35] L. Kronik, M. Jain and J. R. Chelikowsky, *Phys. Rev. B* **66**, 041203(R) (2002).
[36] G. P. Das, B. K. Rao and P. Jena, *Phys. Rev. B* **68**, 035207 (2003).
[37] S. J. Pearton, C. R. Abernathy, G. T. Thaler, R. M. Frazier, D. P. Norton, F. Ren, Y. D. Park, J. M. Zavada, I. A. Buyanova, W. M. Chen and A. F. Hebard, *J. Phys. Cond. Matt.* **16**, R209 (2004).
[38] B. Beschoten, E. Johnston-Halperin, D. K. Yong, M. Poggio, J. E. Grimaldi, S. Keller, S. P. DenBaars, U. K. Mishra, E. L. Hu and D. D. Awshalom, *Phys. Rev. B* **63**, 121202(R) (2001).

[39] I. A. Buyanova, M. Izadifard, W. M. Chen, J. Kim, F. Ren, G. Thaler, C.R. Abernathy, S. J. Pearton, C.-C. Pan, G.-T. Chen, J.-I. Chyi and J. M. Zavada, *Appl. Phys. Lett.* **84**, 2599 (2004).

[40] W. M. Chen, I. A. Buyanova, K. Nishibayashi, K. Kayanuma, K. Seo, A. Murayama, Y. Oka, G. Thaler, R. Frazier, C. R. Abernathy, F. Ren, S. J. Pearton, C.-C. Pan, G.-T. Chen and J.-I. Chyi, *Appl. Phys. Lett.* **87**, 192107 (2005).

[41] B. J. Skromme, *Mater. Sci. Eng. B* **50**, 117 (1997).

[42] A. Tackeuchi, T. Kuroda, A. Shikanai, T. Sota, A. Kuramata and K. Domen, *Physica E* **7**, 1011 (2000).

[43] F. V. Kyrychenko, C. J. Stanton, C. R. Abernathy, S. J. Pearton, F. Ren, G. Thaler, R. Frazier, I. A. Buyanova and W. M. Chen, *AIP Conf. Proc.* **772**, 1319 (2005).

[44] S. J. Pearton, D. P. Norton, K. Ip, Y. W. Heo and T. Steiner, *Prog. Mater. Sci.* **59**, 293 (2005).

[45] Ü. Özgür, Y. I. Alivov, C. Liu, A. Teke, M. A. Reshchikov, S. Doğan, V. Avrutin, S. J. Cho and H. Morkoc, *J. Appl. Phys.* **98**, 041301 (2005).

[46] C. Klingshirn, *Phys. Status Solidi B* **244**, 3027 (2007).

[47] C. Jagadish and S. J. Pearton (eds.), *Zinc Oxide Bulk, Thin Films and Nanostructures: Processing, Properties, and Applications* (Elsevier, 2006).

[48] S. J. Pearton, D. P. Norton, M. P. Ivill, A. F. Hebard, J. M. Zavada, W. M. Chen and I. A. Buyanova, *IEEE Trans. Electr. Dev.* **54**, 1040 (2007).

[49] S. J. Pearton, D. P. Norton, M. P. Ivill, A. F. Hebard, J. M. Zavada, W. M. Chen and I. A. Buyanova, *J. Electronic Materials* **36**, 462 (2007).

[50] W. M. Chen, I. A. Buyanova, A. Murayama, T. Furuta, Y. Oka, D. P. Norton, S. J. Pearton, A. Osinsky and J. W. Dong, *Appl. Phys. Lett.* **92**, 092103 (2008).

[51] B. Gil, *Phys. Rev. B* **64**, 201310 (2001).

[52] W. R. L. Lambrecht, A. V. Rodina, S. Limpijumnong, B. Segall and B. K. Meyer, *Phys. Rev. B* **65**, 075207 (2002).

[53] D. C. Reynolds, C. W. Litton and T. C. Collins, *Phys. Rev. A* **140**, 1726 (1965).

[54] D. G. Thomas, *J. Phys. Chem.* **15**, 86 (1960).

[55] B. K. Meyer, H. Alves, D. M. Hofmann, W. Kriegseis, D. Forster, F. Bertram, J. Christen, A. Hoffmann, M. Straßburg, M. Dworzak, U. Haboeck and A. V. Rodina, *Phys. Status Solidi B* **241**, 231 (2004).

[56] D. Lagarde, L. Lombez, A. Balocchi, P. Renucci, H. Carrère, T. Amand, X. Marie, Z. X. Mei, X. L. Du and Q. K. Xue, *Phys. Status Solidi C* **4**, 472 (2007).

[57] D. Lagarde, A. Ballocchi, H. Carrere, F. Zhao, T. Amand, X. Marie, Z. X. Wei, X. L. De and Q. K. Xue, *Phys. Rev. B* **78**, 033203 (2008).

[58] D. C. Reynolds, D. C. Look, B. Jogai, J. E. Hoelscher, R. E. Sherriff, M. T. Harris, and M. J. Callahan, *J. Appl. Phys.* **88**, 2152 (2000).

[59] F. Bertram, J. Christe, A. Dagar and A. Krost, *Appl. Phys. Lett.* **90**, 041917 (2007).

[60] S. Ghosh, V. Sih, W. H. Lau, D. D. Awschalom, S.-Y. Bae, S. Wang, S. Vaidya and G. Chapline, *Appl. Phys. Lett.* **86**, 232507 (2005).

[61] W. K. Liu, K. M. Whitaker, A. L. Smith, K. R. Kittilsteved, B. H. Robinson, and D. R. Gamelin, *Phys. Rev. Lett.* **98**, 186804 (2007).

Color Plates

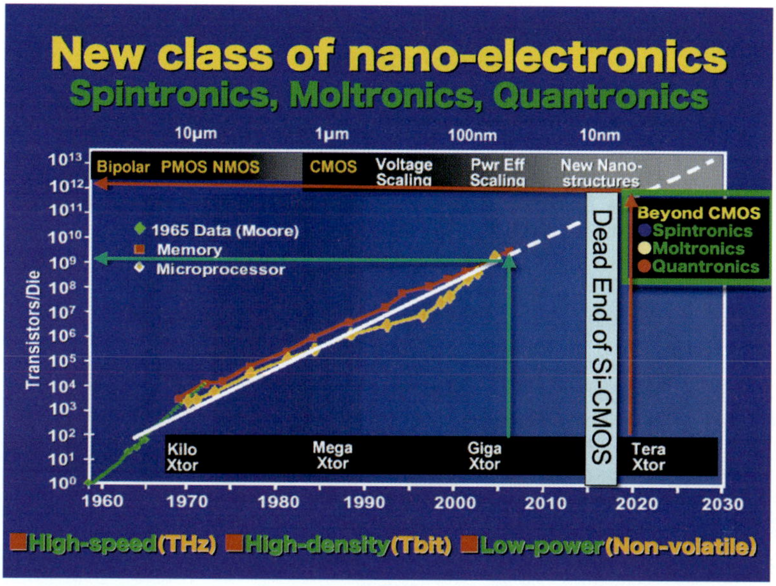

Figure 1.1

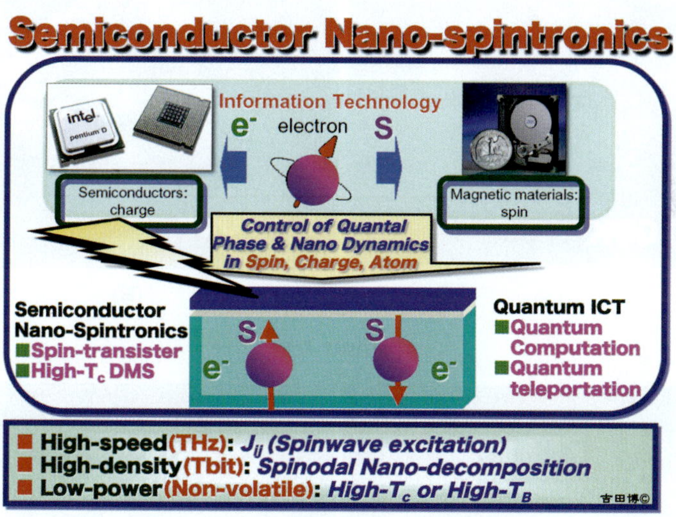

Figure 1.2

Figure 1.3

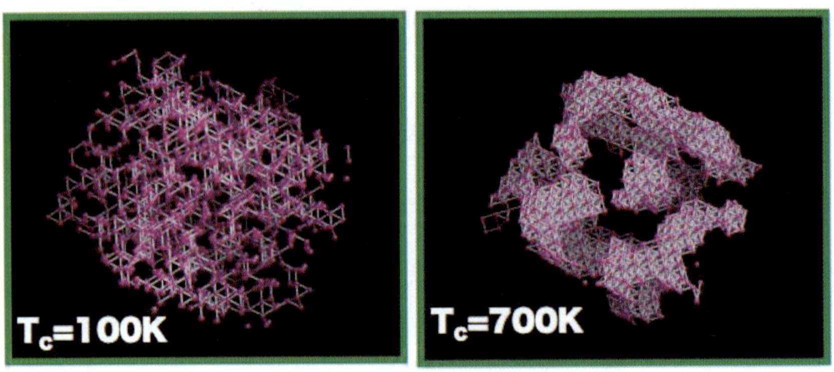

(a) (b)

Figure 1.4

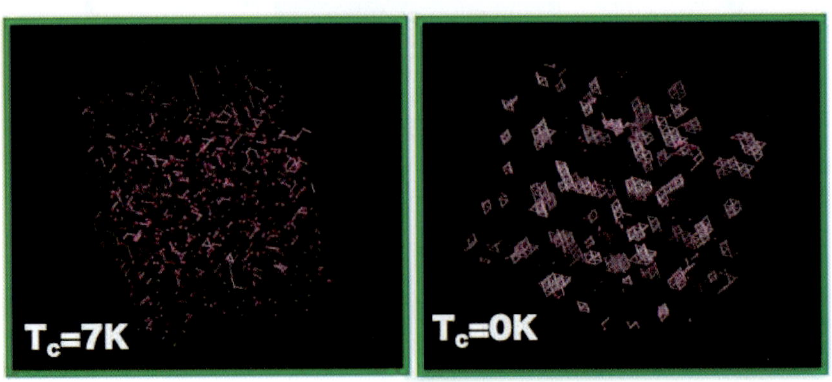

(a) (b)

Figure 1.5

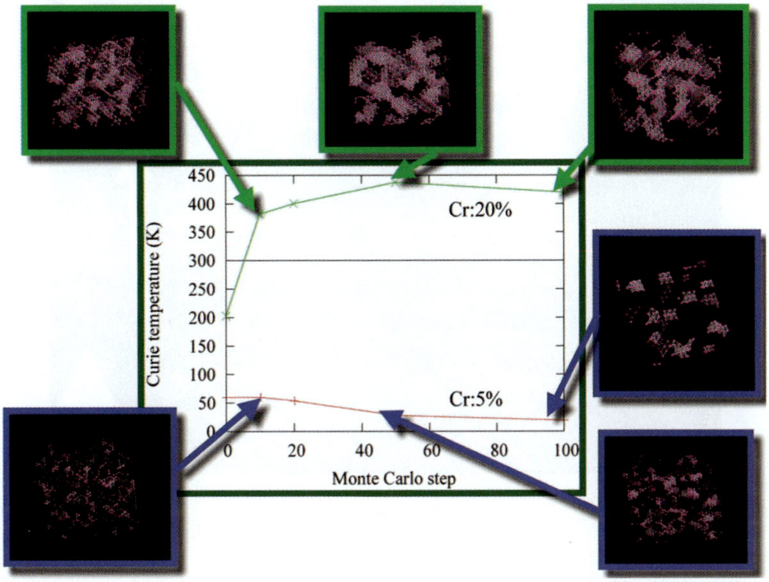

Figure 1.6

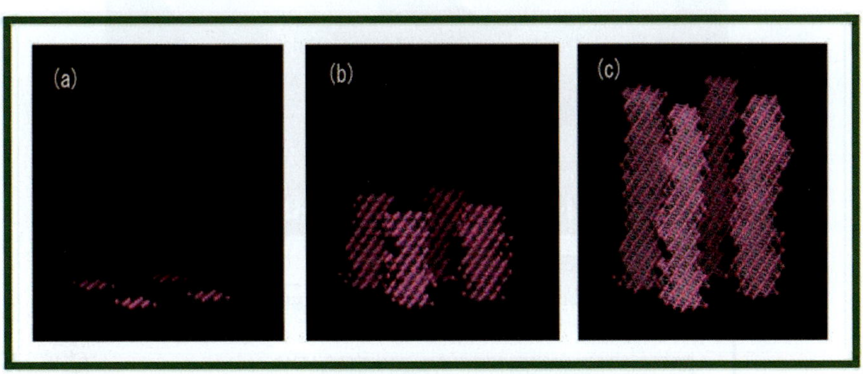

Figure 1.7

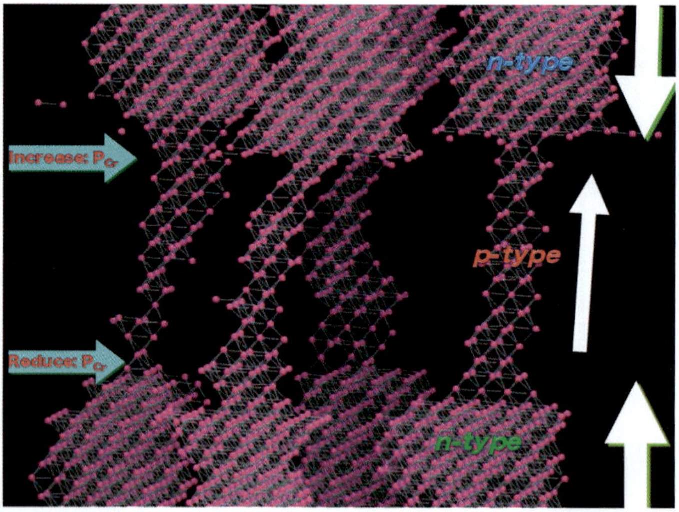

Figure 1.8

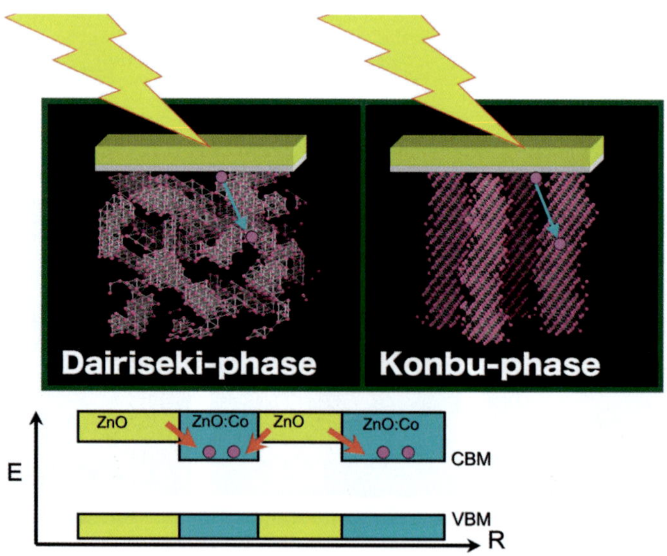

Figure 1.9

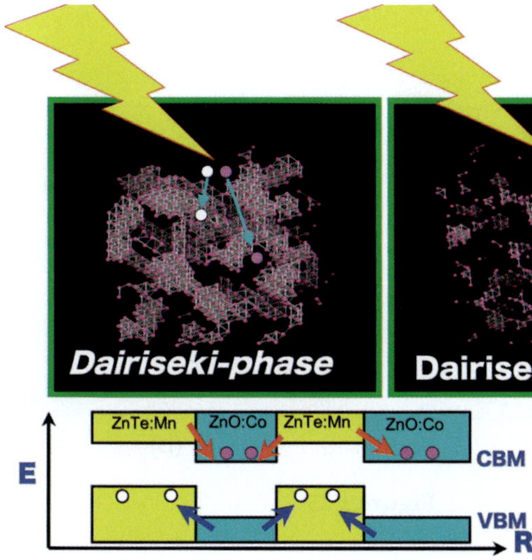

Figure 1.10

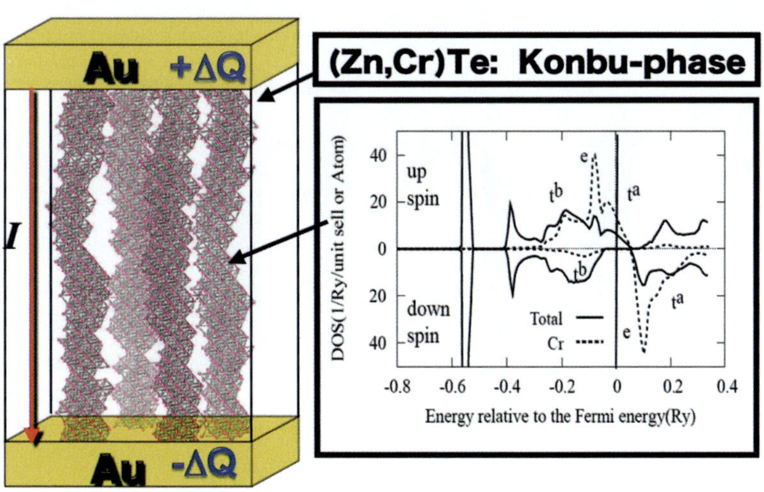

Figure 1.11

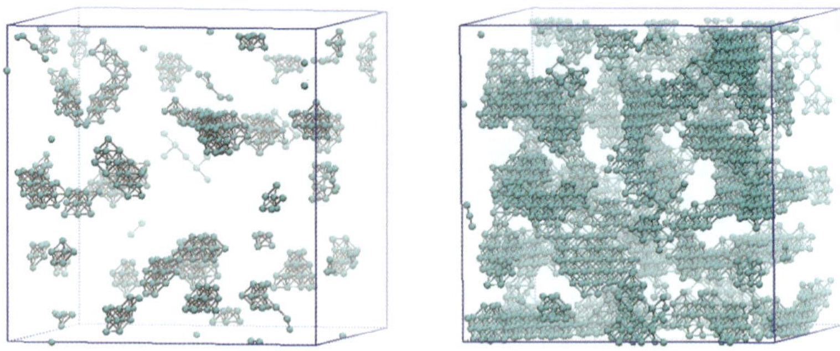

(a) (b)

Figure 1.25

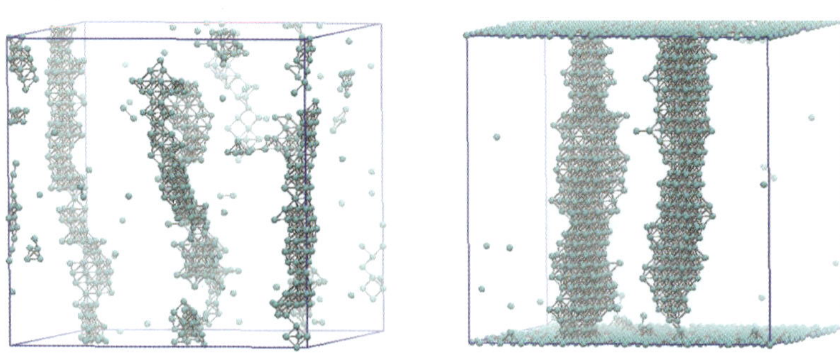

(a) (b)

Figure 1.27

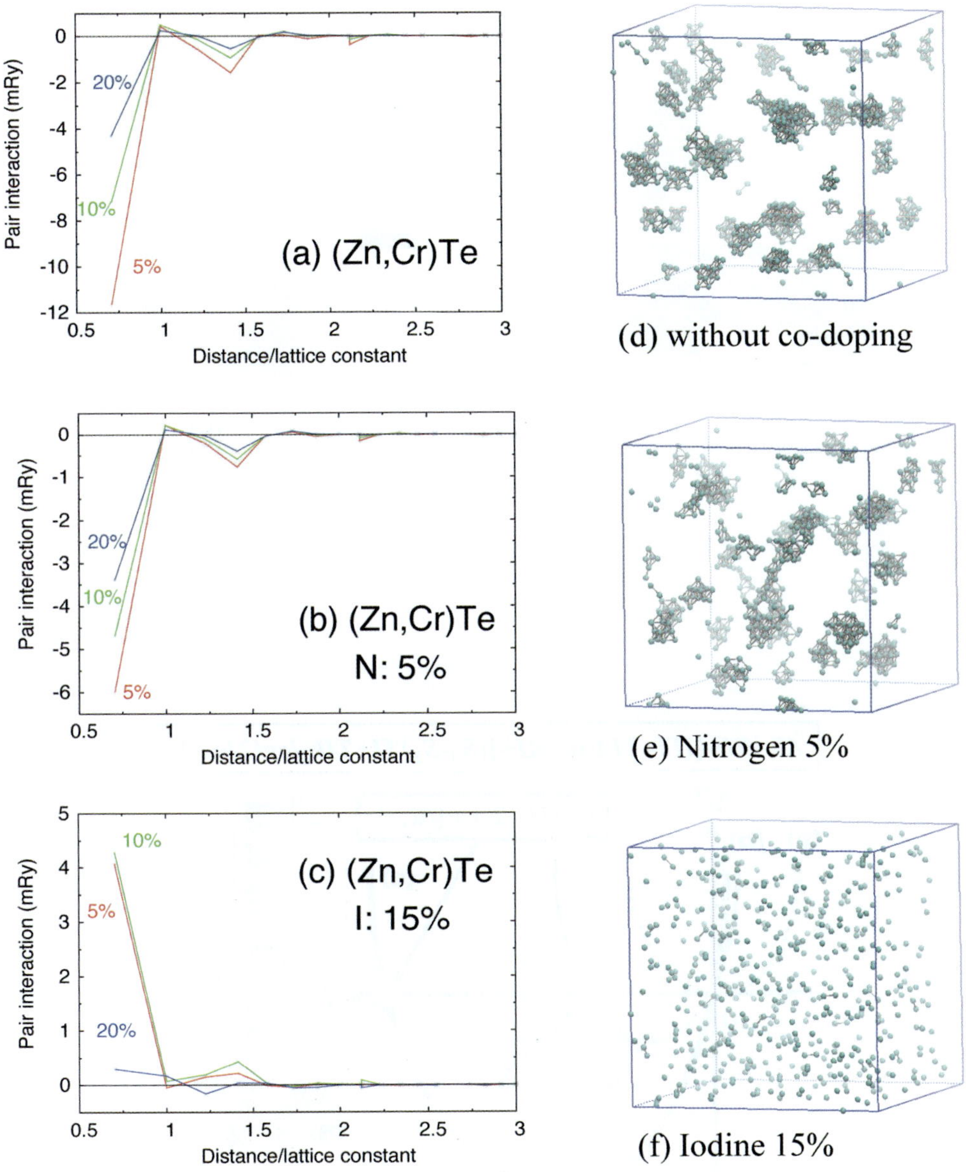

Pair interaction (mRy)
Distance/lattice constant
20%
10%
5%
(a) (Zn,Cr)Te
(b) (Zn,Cr)Te
N: 5%
(c) (Zn,Cr)Te
I: 15%
(d) without co-doping
(e) Nitrogen 5%
(f) Iodine 15%

Figure 1.29

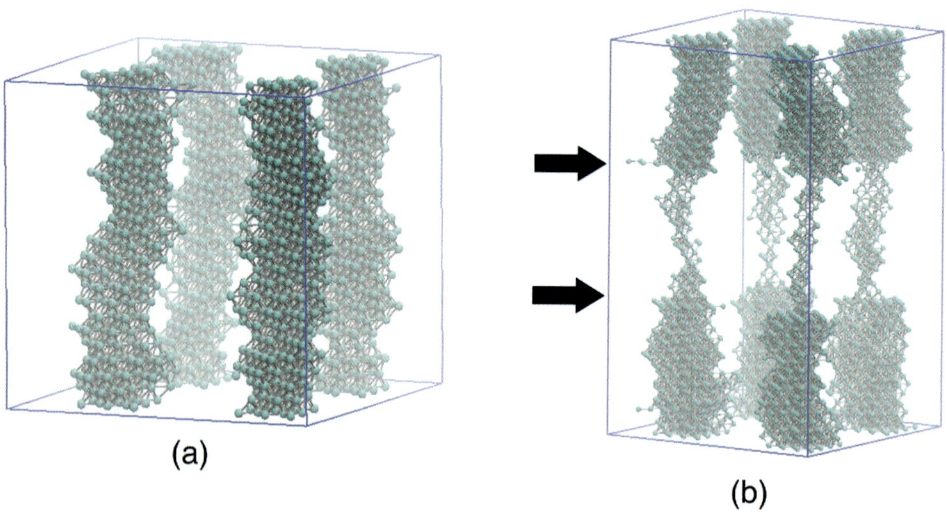

Figure 1.31

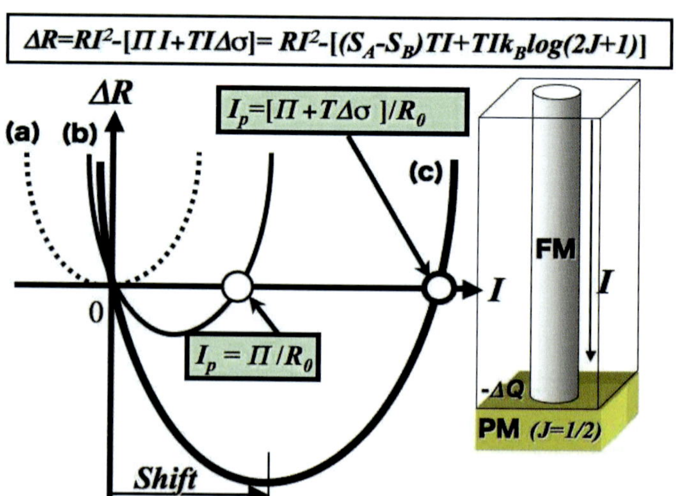

Figure 1.33

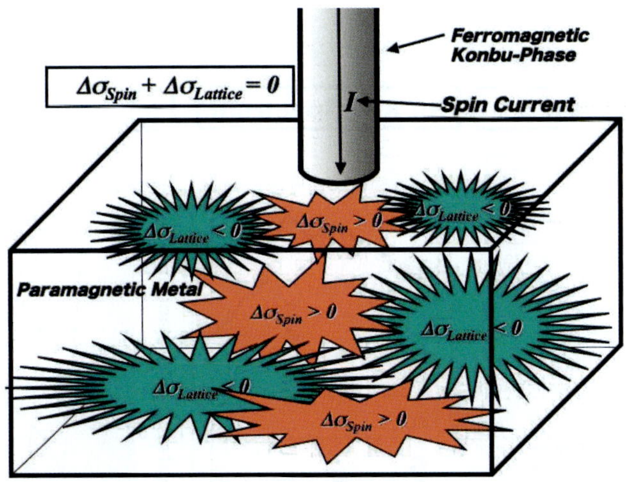

Figure 1.34

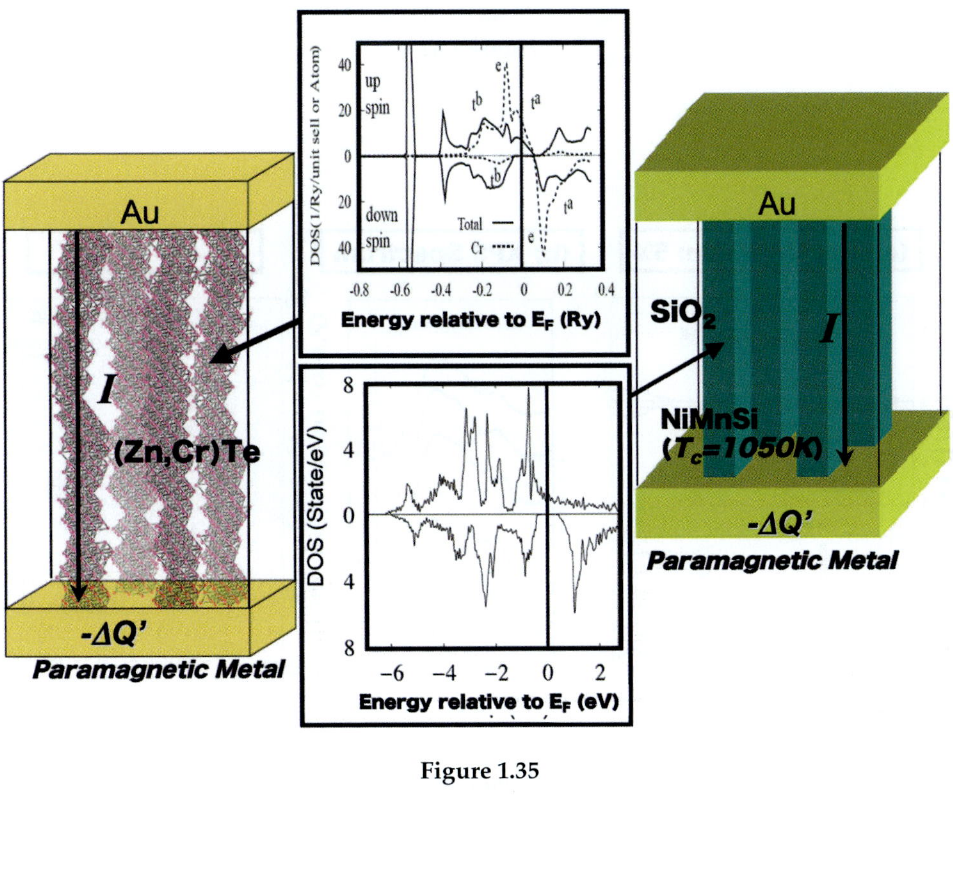

Figure 1.35

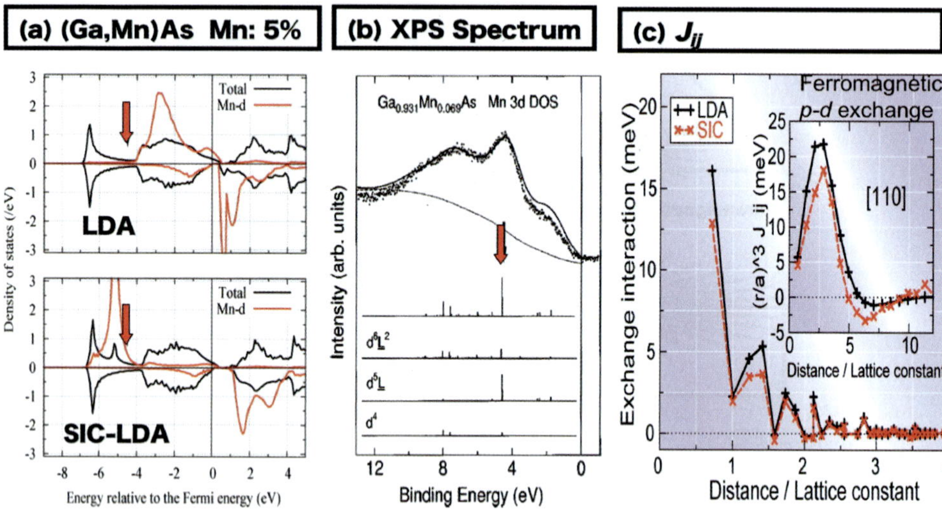

Figure 1.42

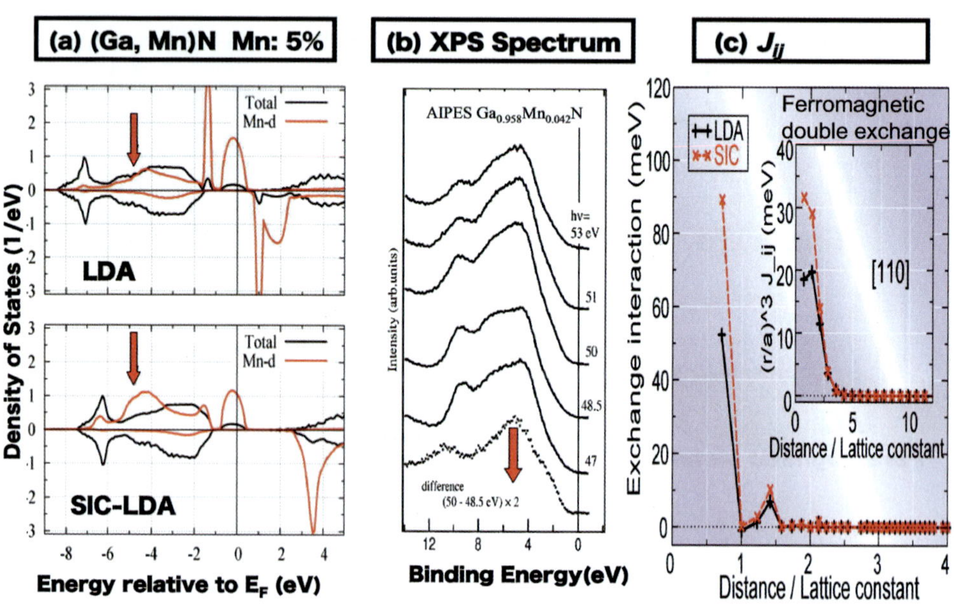

Figure 1.43

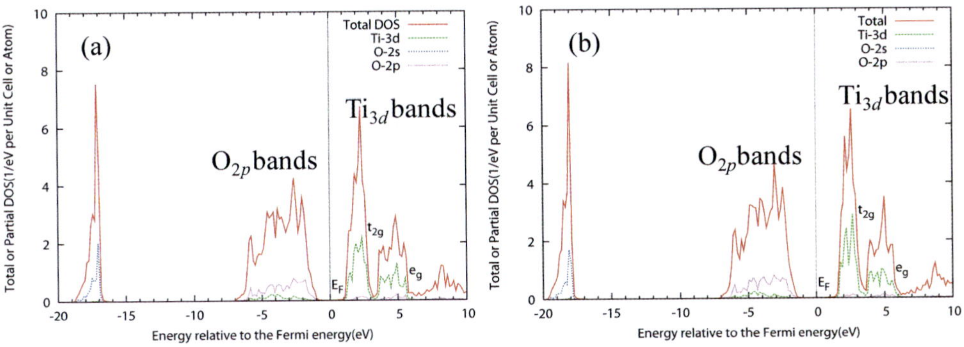

Figure 1.49

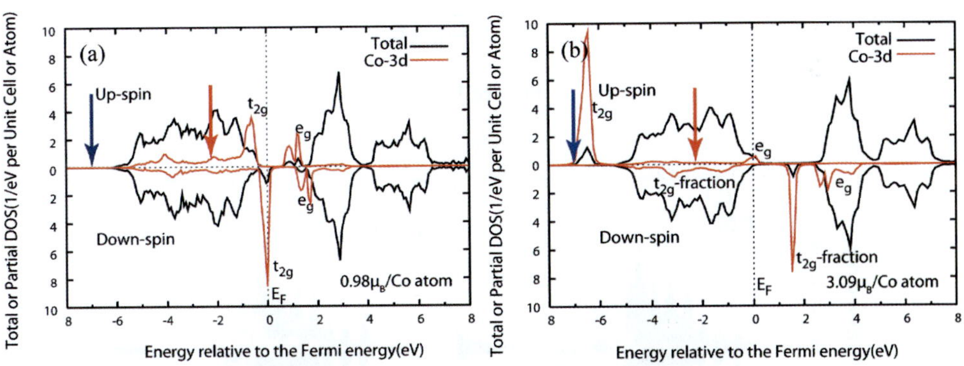

Figure 1.50

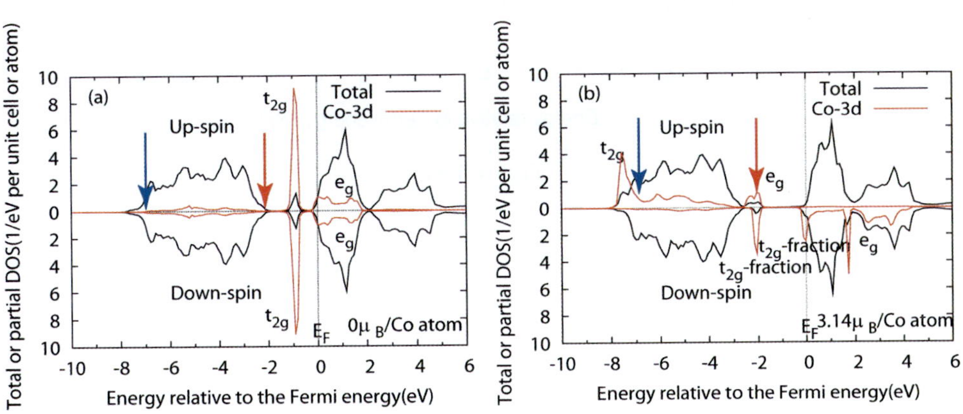

Figure 1.52

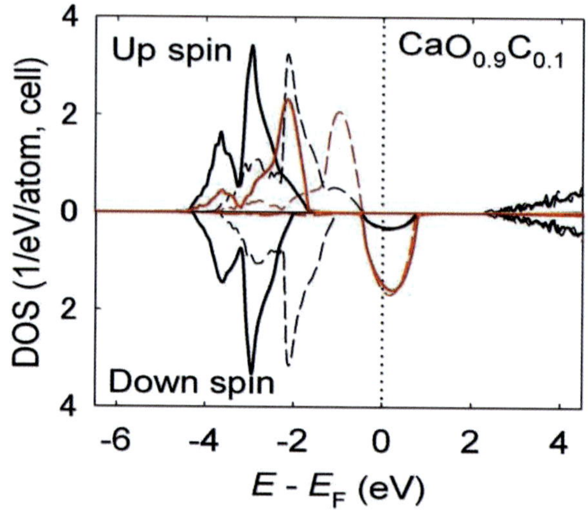

Figure 1.54

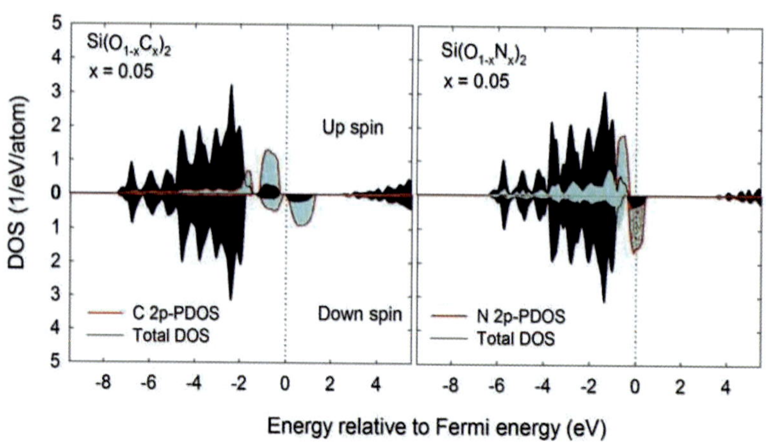

Figure 1.55

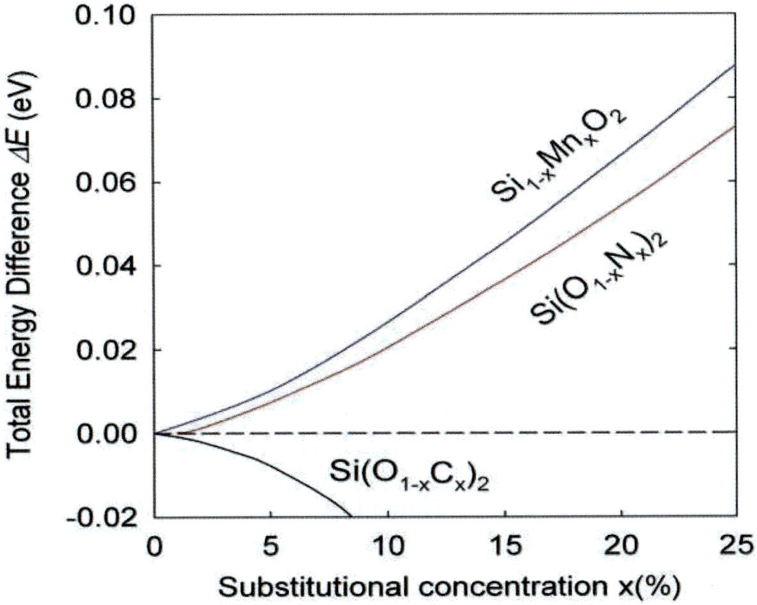

Figure 1.56

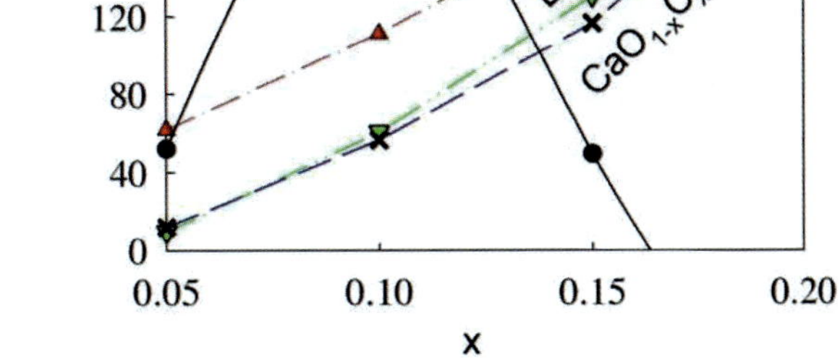

Figure 1.59

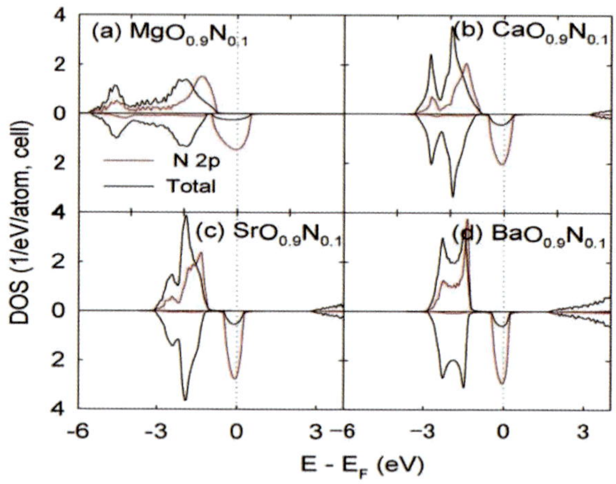

Figure 1.60

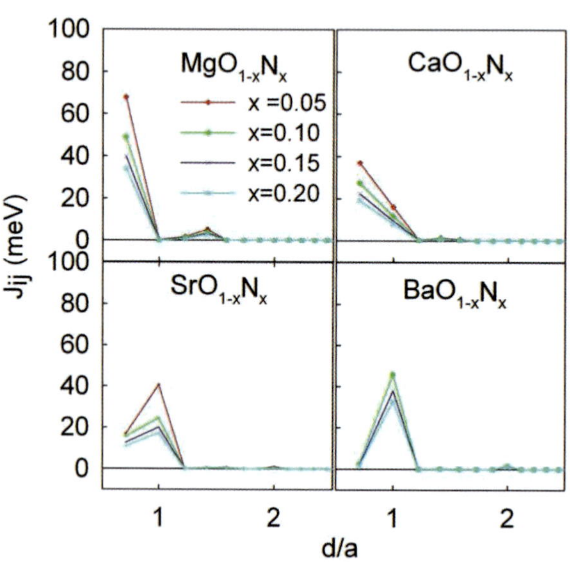

Figure 1.61

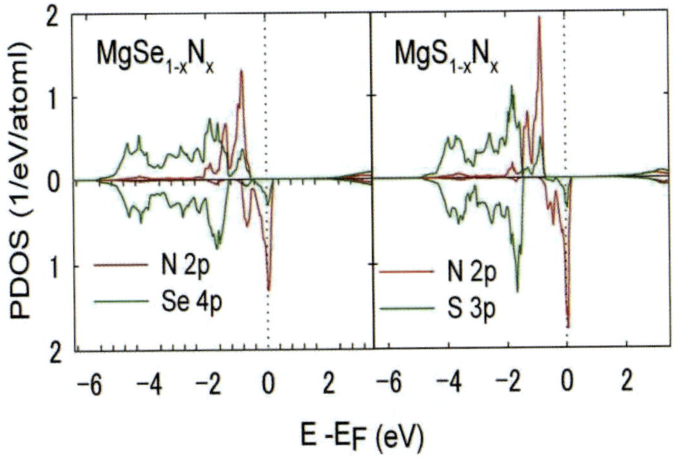

Figure 1.62

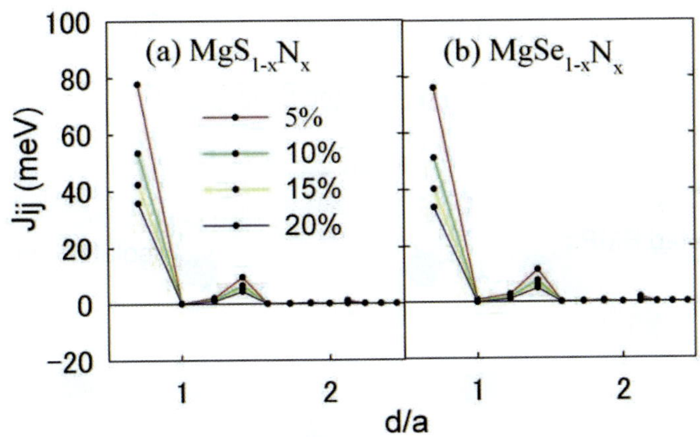

Figure 1.63

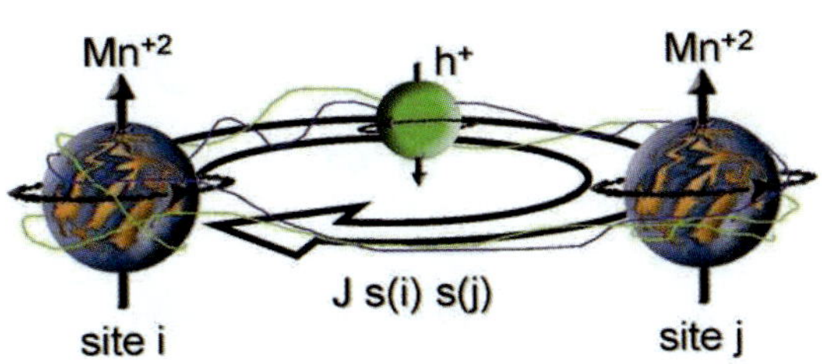

Figure 2.1

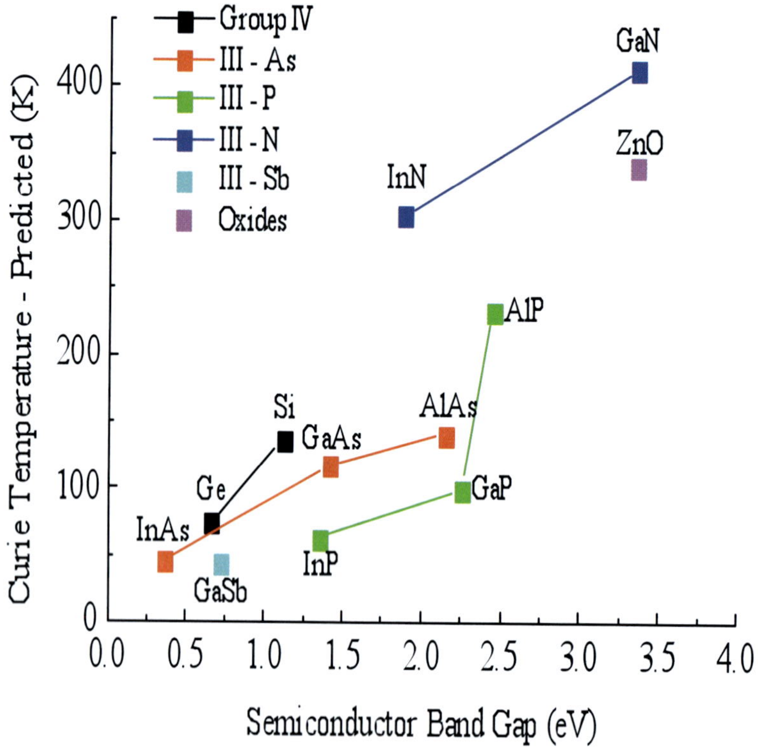

Figure 2.2

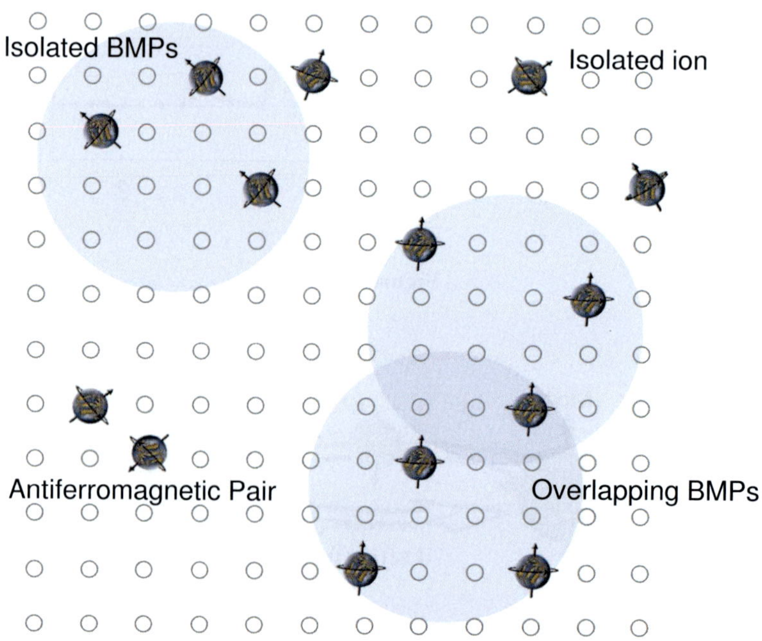

Figure 2.3

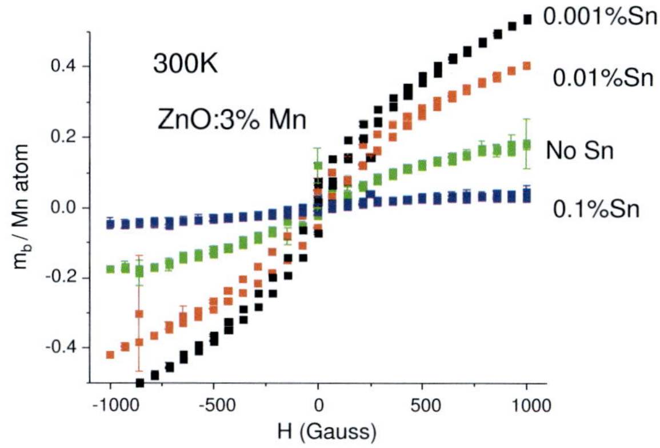

Figure 2.5

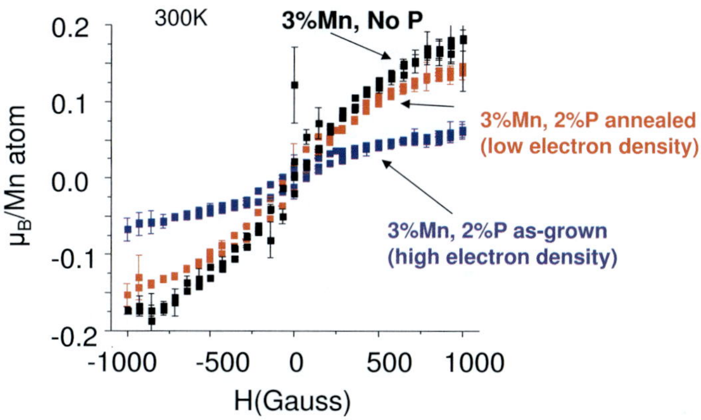

Figure 2.6

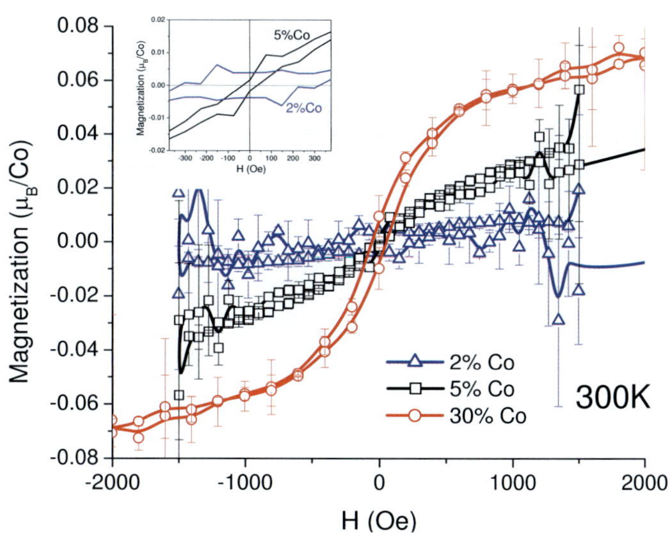

Figure 2.10

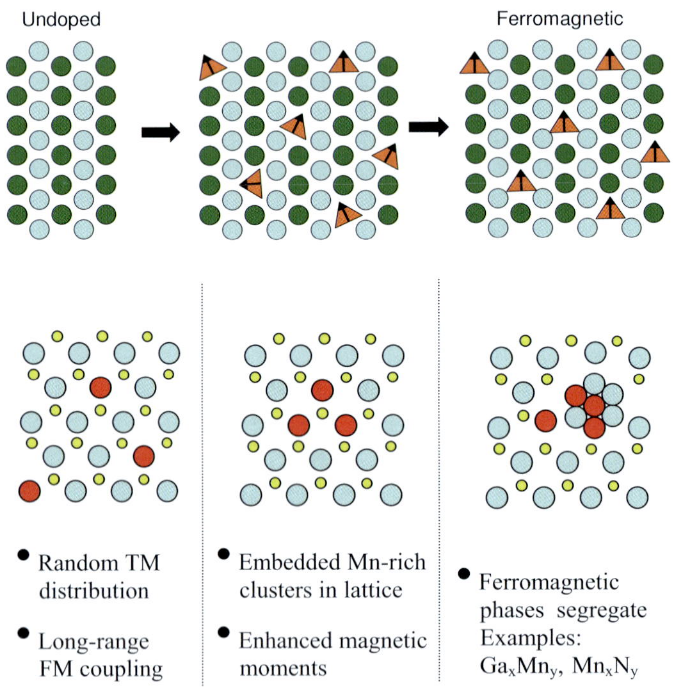

Figure 3.1

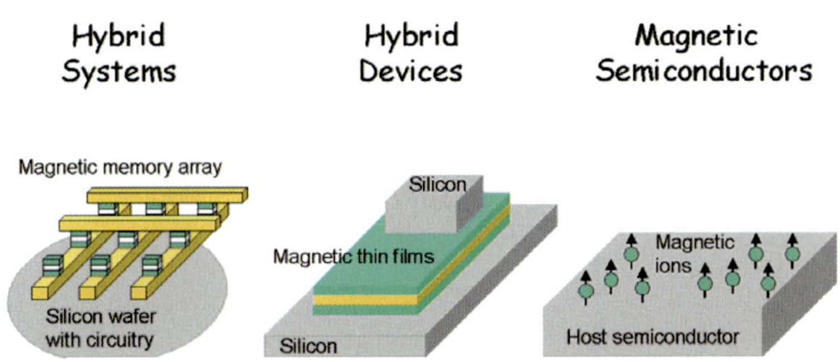

Figure 3.4

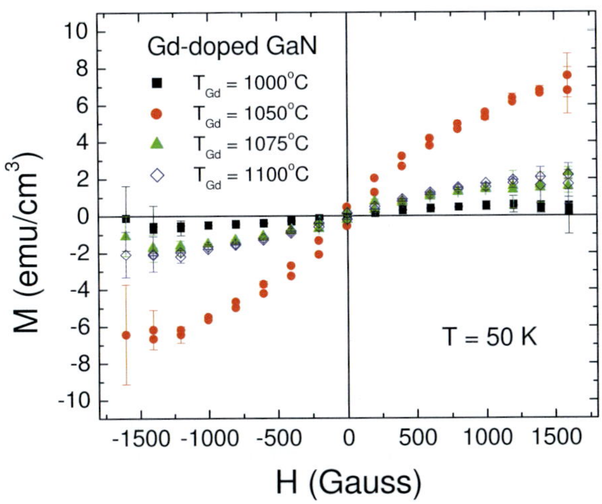

Figure 3.6

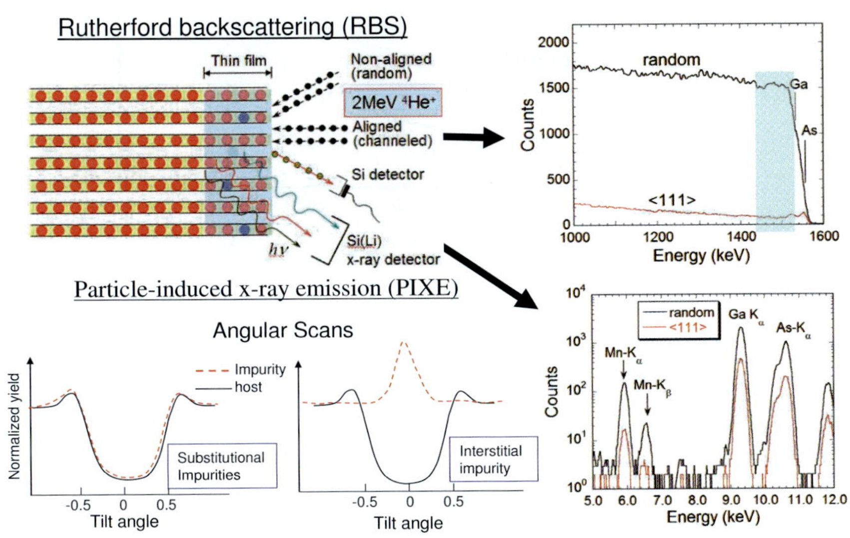

Figure 4.1

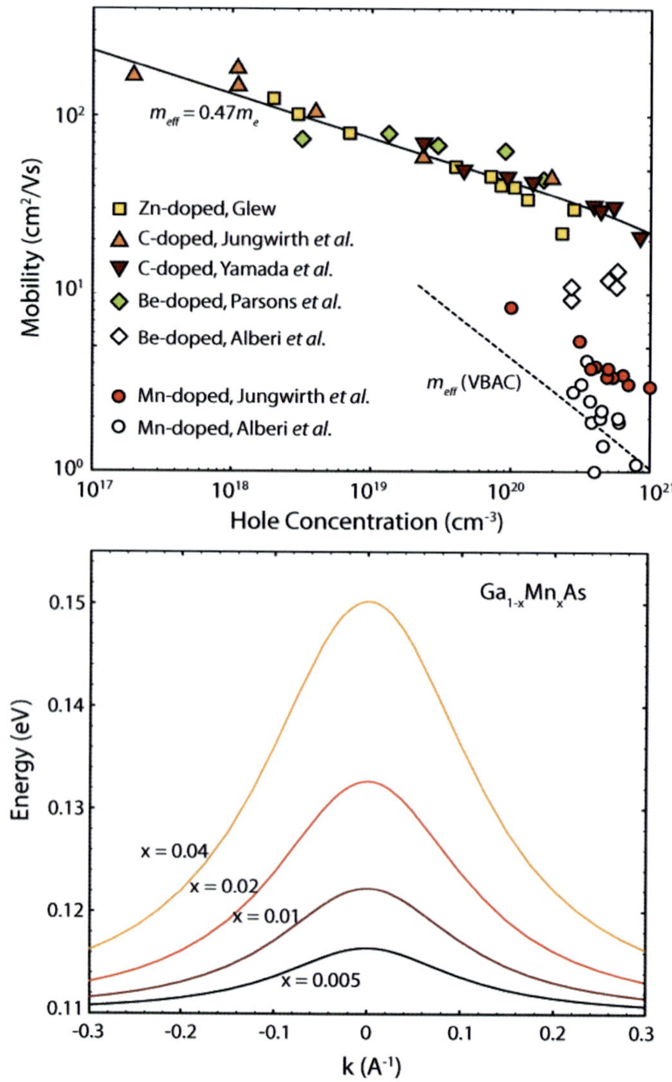

Mobility (cm²/Vs)
Hole Concentration (cm⁻³)
m_eff = 0.47m_e
m_eff (VBAC)
Zn-doped, Glew
C-doped, Jungwirth et al.
C-doped, Yamada et al.
Be-doped, Parsons et al.
Be-doped, Alberi et al.
Mn-doped, Jungwirth et al.
Mn-doped, Alberi et al.
Ga1-xMnxAs
Energy (eV)
k (A⁻¹)
x = 0.04
x = 0.02
x = 0.01
x = 0.005

Figure 4.17

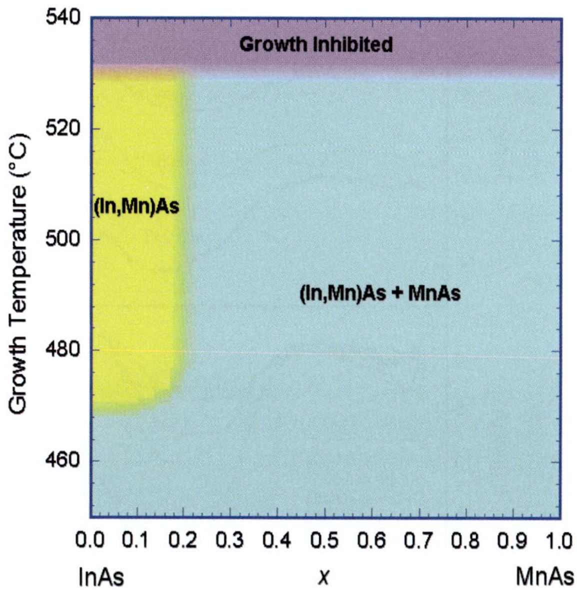

Figure 6.1

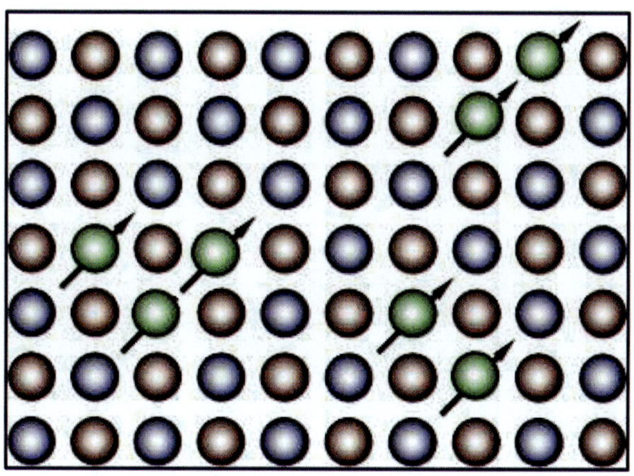

Figure 6.2

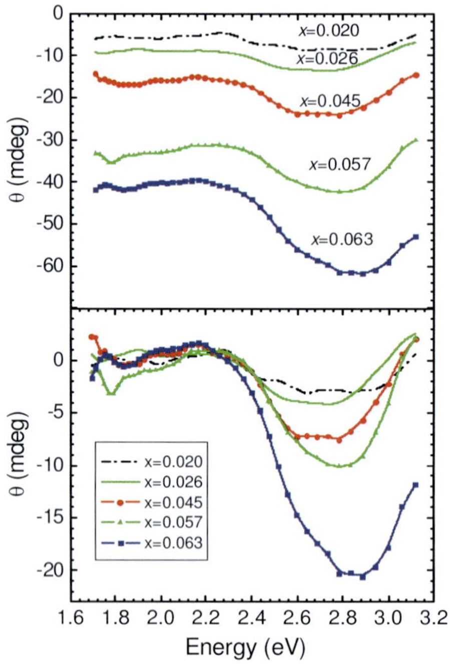

Figure 6.6

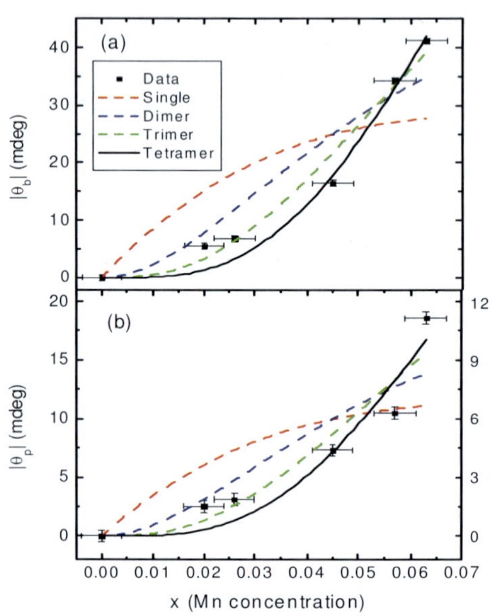

Figure 6.7

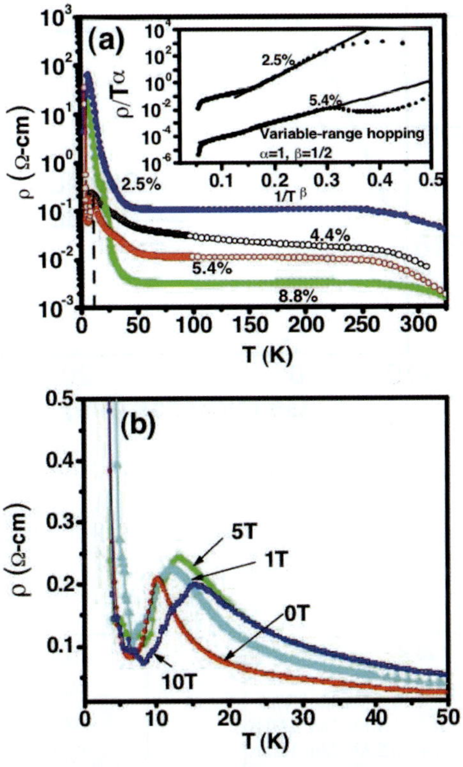

Figure 7.7

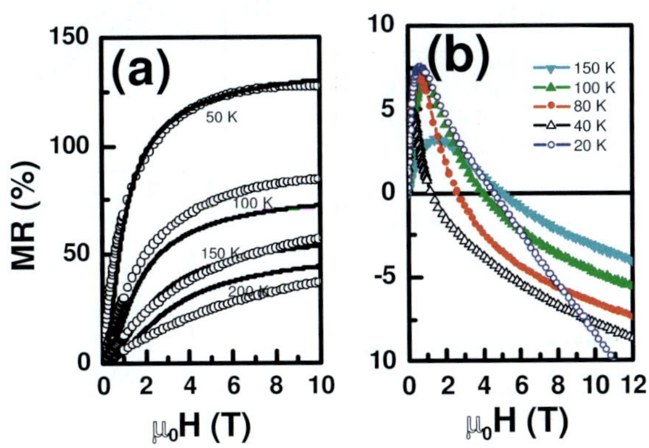

Figure 7.8

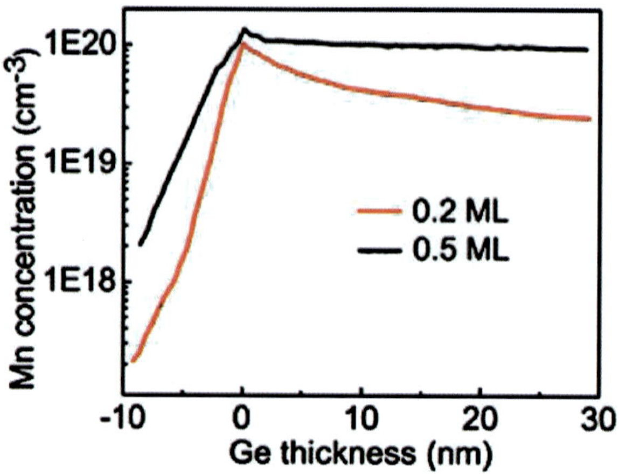

Figure 7.13

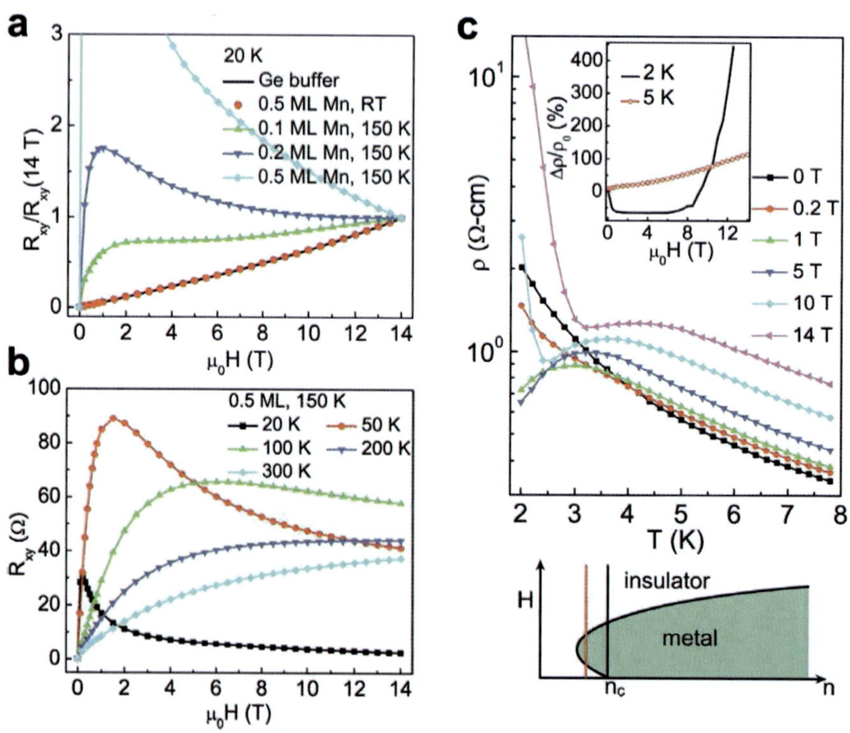

Figure 7.15

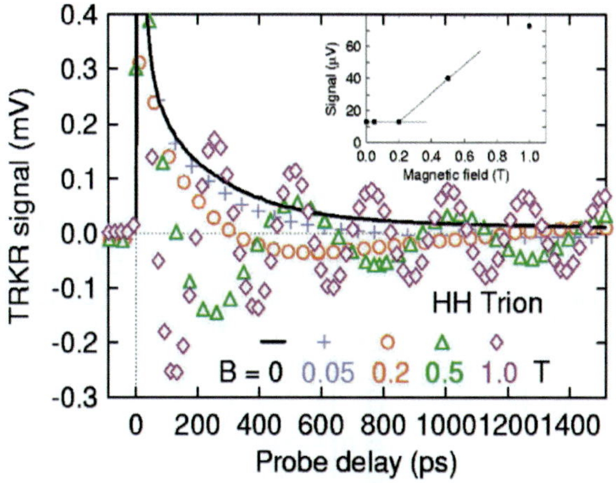

Figure 8.2

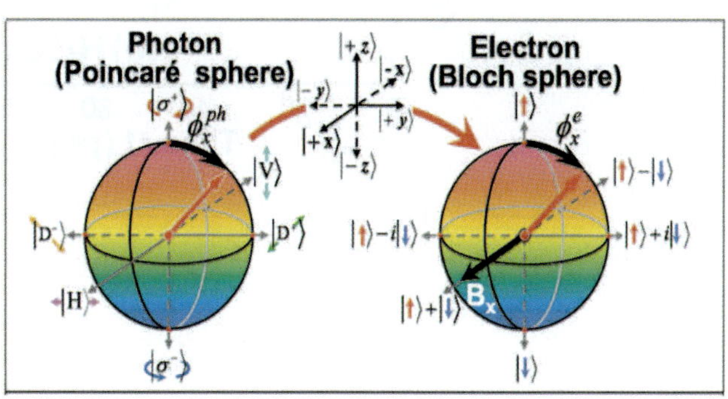

Figure 8.8

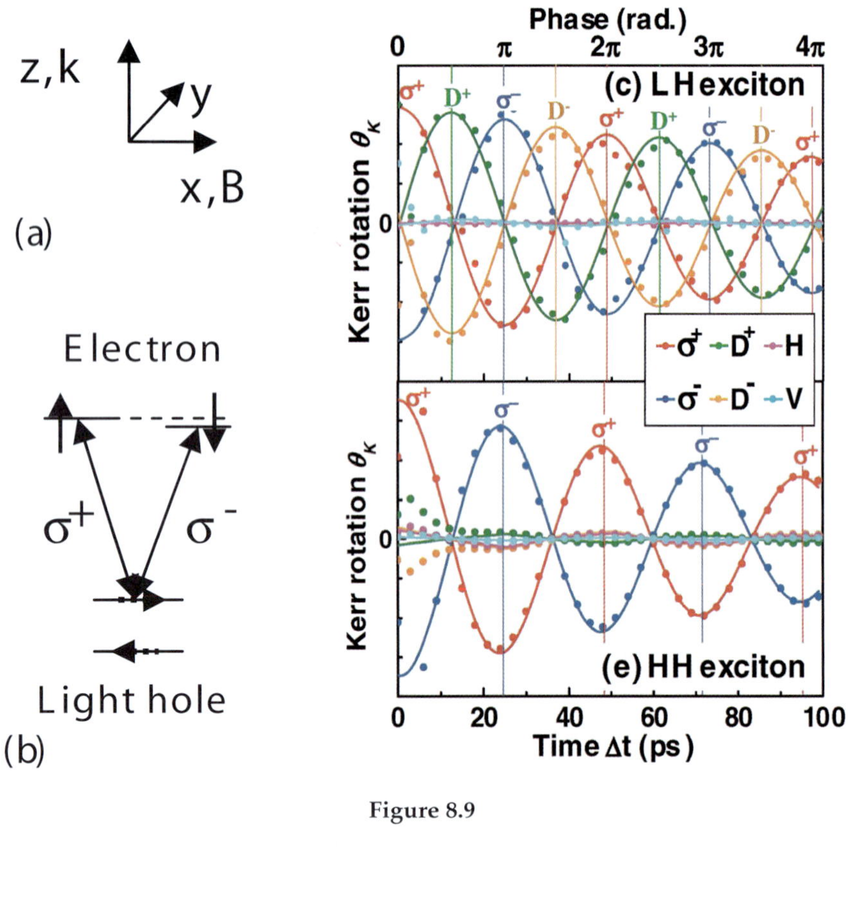
z, k
y
x, B
(a)
Electron
σ+
σ -
Light hole
(b)
Phase (rad.)
0
π
2π
3π
4π
(c) LH exciton
Kerr rotation θ_K
σ+ D+ H
σ- D- V
(e) HH exciton
Time Δt (ps)
0
20
40
60
80
100

Figure 8.9

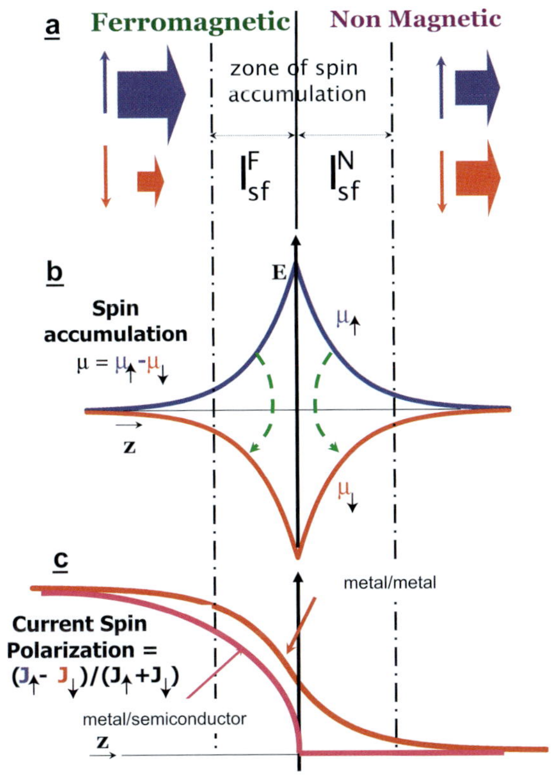

Figure 10.1

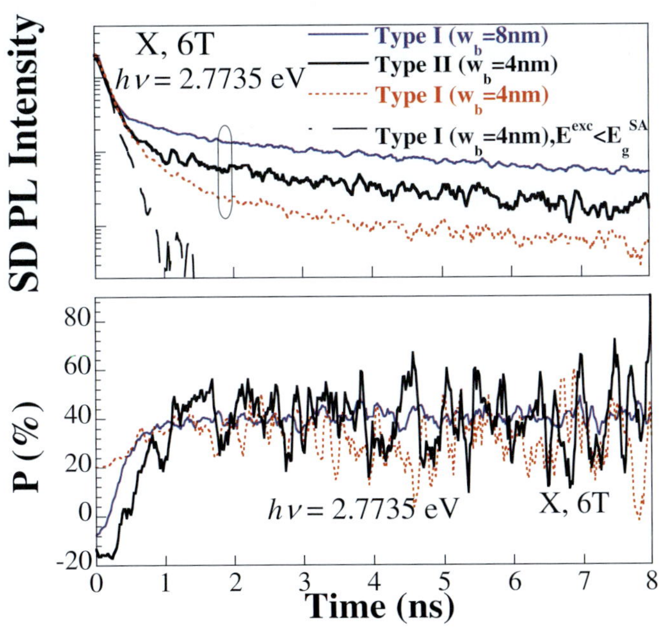

Figure 11.5

Index